Lecture Notes in Applied Mathematics and Mechanics

Volume 1

Series Editors

Alexander Mielke, Humboldt-Universität zu Berlin, Berlin, Germany
e-mail: mielke@wias-berlin.de

Bob Svendsen, RWTH Aachen University, Aachen, Germany
e-mail: bob.svendsen@rwth-aachen.de

For further volumes:
http://www.springer.com/series/11915

About this Series

The Lecture Notes in Applied Mathematics and Mechanics LAMM are intended for an interdisciplinary readership in the fields of applied mathematics and mechanics. This series is published under the auspices of the International Association of Applied Mathematics and Mechanics (IAAMM; German GAMM).

Topics of interest include for example focus areas of the IAAMM such as: foundations of mechanics, thermodynamics, material theory and modeling, multibody dynamics, structural mechanics, solid mechanics, biomechanics, damage, fracture, multiscale modeling and homogenization, fluid mechanics, gas dynamics, laminar flows and transition, turbulence and reactive flows, interface flows, acoustics, waves, applied analysis, mathematical modeling, calculus of variations, variational principles applied operator theory, evolutionary equations, applied stochastics, systems with uncertainty, dynamical systems, control theory, optimization, applied and numerical linear algebra, analysis and numerics of ordinary and partial differential equations.

Each contribution to the series is intended to be accessible to researchers in mathematics and mechanics and is written in English. The aim of the series is to provide introductory texts for modern developments in applied mathematics and mechanics contributing to cross-fertilization. The Lecture Notes are aimed at researchers as well as advanced masters and PhD students in both mechanics and mathematics. Contributions to the series are self-contained and focused on a few central themes. The goal of each contribution is the communication of modern ideas and principles rather than on completeness or detailed proofs. Like lecture notes from a course, a well-chosen example is preferable to an abstract framework that cannot be comprehended without deeper involvement. The typical length of each contribution is between 100 and 300 pages. If the lecture notes represent the proceedings of a summer school with several contributors, a unified, consistent presentation and style are required (e.g., common notation). In exceptional cases, doctoral theses may be accepted, if they fulfill the above-mentioned criteria.

Potential contributors should contact the appropriate editor with a title, table of contents, and a sample chapter. Full manuscripts accepted by the editors will then be peer-reviewed.

Erwin Stein
Editor

The History of Theoretical, Material and Computational Mechanics - Mathematics Meets Mechanics and Engineering

Editor
Erwin Stein
Institute of Mechanics and Computational
  Mechanics (IBNM)
Gottfried Wilhelm Leibniz Universität
  Hannover
Hannover
Germany

ISSN 2197-6724 ISSN 2197-6732 (electronic)
ISBN 978-3-642-39904-6 ISBN 978-3-642-39905-3 (eBook)
DOI 10.1007/978-3-642-39905-3
Springer Heidelberg New York Dordrecht London

Library of Congress Control Number: 2013949170

© Springer-Verlag Berlin Heidelberg 2014
This work is subject to copyright. All rights are reserved by the Publisher, whether the whole or part of the material is concerned, specifically the rights of translation, reprinting, reuse of illustrations, recitation, broadcasting, reproduction on microfilms or in any other physical way, and transmission or information storage and retrieval, electronic adaptation, computer software, or by similar or dissimilar methodology now known or hereafter developed. Exempted from this legal reservation are brief excerpts in connection with reviews or scholarly analysis or material supplied specifically for the purpose of being entered and executed on a computer system, for exclusive use by the purchaser of the work. Duplication of this publication or parts thereof is permitted only under the provisions of the Copyright Law of the Publisher's location, in its current version, and permission for use must always be obtained from Springer. Permissions for use may be obtained through RightsLink at the Copyright Clearance Center. Violations are liable to prosecution under the respective Copyright Law.
The use of general descriptive names, registered names, trademarks, service marks, etc. in this publication does not imply, even in the absence of a specific statement, that such names are exempt from the relevant protective laws and regulations and therefore free for general use.
While the advice and information in this book are believed to be true and accurate at the date of publication, neither the authors nor the editors nor the publisher can accept any legal responsibility for any errors or omissions that may be made. The publisher makes no warranty, express or implied, with respect to the material contained herein.

Printed on acid-free paper

Springer is part of Springer Science+Business Media (www.springer.com)

Foreword

In 2008 Professor Stein, the editor of this volume, applied for setting up a new section in the yearly GAMM conference related to the history of mechanics. This suggestion was approved by the Board of GAMM and the first session on history of mechanics started in 2010.

Lectures and contributions that were presented in these sessions are the backbone of this first volume of LAMM. There is no better way to start the series of GAMM lecture notes to reflect the history of the research field.

The contributions in this volume discuss different aspects of mechanics. They are related to solid and fluid mechanics in general and to specific problems in these areas including the development of numerical solution techniques. Thus this first addition of LAMM provides an overview on the field of mechanics and describes the wide area of applications within GAMM.

Finally I like to thank the editor, Professor Erwin Stein, for his continuous effort and his hard work to make this volume possible.

Hannover, May 2013

Peter Wriggers
Vice-President of GAMM

Preface

This collection of 23 articles is the output of lectures in special sessions on "The History of Theoretical, Material and Computational Mechanics" within the yearly conferences of the GAMM in the years 2010 in Karlsruhe, Germany, 2011 in Graz, Austria, and in 2012 in Darmstadt, Germany; GAMM is the "Association for Applied Mathematics and Mechanics", founded in 1922 by Ludwig Prandtl and Richard von Mises.

Guiding topics for the yearly sections were proposed and leading scientists invited as keynote-lecturers. This is reflected in the four parts of this book. In their sequence and in the total concept the published articles provide a certain completeness and logical consistency within the selected topics of theoretical, material, applied and computational mechanics.

I am indebted to the co-chairmen of the sections, Professor Oskar Mahrenholtz in 2010 and 2011, and Professor Lothar Gaul in 2013. It should be mentioned that each of the three sections had two sessions, each with about 150 attendees which shows the great interest of the conference participants.

The success of the new historical sections motivated the other authors and me to publish them in a book, also stimulated by Professor Peter Wriggers, President of GAMM in the period from 2008 to 2010. I also thank him for writing a foreword.

The rich history of theoretical, material, applied and computational mechanics of solids, structures and fluids should be of vivid interest for the community of mechanicians working in science and technology as well as of applied mathematicians. This is important for the self-conception of students and practitioners in order to know and realize on which shoulders we stand and how long it often took to arrive at simple-looking formulas for describing dominant effects in loading and deformation processes of engineering structures and in fluid flow processes, and moreover to derive rather general mathematical models – despite the ambitions and efforts of eminent scientists over decades and even centuries.

Following, the four parts of the book are briefly commented.

In Part I, the origins and developments of conservation principles in mechanics and related variational methods are treated together with challenging applications from the 17th to the 20th century.

Part II treats general as well as more specific aspects of material theories of deforming solid continua and porous soils, e.g. the foundation of classical theories of elastoplastic deformations, the development of theories and analysis for contact with friction and plastic deformations, as well as the formation and progress of fracture in brittle and ductile solid materials.

Part III presents important theoretical and engineering developments in fluid mechanics, beginning with remarkable inventions in the old Egypt, the dominating role of the Navier-Stokes PDEs for fluid flows and their complex solutions for a wide field of parameters as well as the invention of pumps and turbines in the 19th and 20th century.

And finally, Part IV gives a survey on the development of direct variational (numerical) methods – the Finite Element Method – in the 20th century with many extensions and generalizations, requiring a strong coupling of engineering, mathematical and computer science aspects. These three articles are restricted to static and dynamic elastic continua, according to page limitations of the book.

One may ask whether the well-written historical essays on a period of about $3\,1/2$ centuries of research in mechanics can highlight overriding insight to the motivation, the connections, the progress and the setbacks of so many eminent scientists in the past. Additionally, it has to be regarded that a master plan for the contents of the book could only be realized roughly, viewing the open calls for contributions to the related historical sections of GAMM conferences.

Nevertheless, the structure and the contents of the book are above all characterized by the invited lectures (chapters) of well-known scientists in their fields.

However, in order to know the real genesis of the scientific truth, we would have to ask all those splendid researchers behind the huge work about their motivations and goals, which – of course – is not possible.

Instead, we reflected essential individual achievements as parts and driving forces of the integral subject "Mechanics" with their important and distinct positions in the whole framework of this discipline. Thus, each chapter can be widely understood independently from the others.

It is my pleasant duty to deeply thank all authors for elaborating their articles on a high standard and publishing them in this book. The friendly collaboration over nearly a year provided the nice feeling of partnership.

We are thankful to Wiley Publishing Company for admitting republications of five over-worked and extended articles published in the "GAMM-Mitteilungen", Vol. 34 (2011) (Issue 2) and an article published in "ZAMM", Vol. 92 (2012), pp. 683–708. Further, I thank the editors of the Polish "Journal of Computer Assisted Methods in Engineering Science" for permitting publication of the abbreviated and revised article Vol. 19 (2012) No. 1, pp. 7–91.

The authors and the editor appreciate the publication of the book as Volume 1 of the new series "Lecture Notes in Applied Mathematics and Mechanics (LNAMM)". We thank Dr. Thomas Ditzinger, Springer-Verlag, for his advice and helpful collaboration.

Hannover, May 2013   Erwin Stein, Editor

Contents

Part I: Mechanical Conservation Principles, Variational Calculus and Engineering Applications from the 17th to the 20th Century

The Origins of Mechanical Conservation Principles and Variational Calculus in the 17th Century 3
Erwin Stein

Principles of Least Action and of Least Constraint 23
Ekkehard Ramm

Lagrange's "Récherches sur la libration de la lune"– From the Principle of Least Action to Lagrange's Principle 45
Hartmut Bremer

The Development of Analytical Mechanics by Euler, Lagrange and Hamilton – From a Student's Point of View ... 61
Maximilian Gerstner, Patrick R. Schmitt, Paul Steinmann

Heun and Hamel – Representatives of Mechanics around 1900 73
Hartmut Bremer

The Machine of Bohnenberger 81
Jörg F. Wagner, Andor Trierenberg

On the Historical Development of Human Walking Dynamics 101
Werner Schiehlen

Part II: Material Theories of Solid Continua and Solutions of Engineering Problems

On the History of Material Theory – A Critical Review 119
Albrecht Bertram

Some Remarks on the History of Plasticity – Heinrich Hencky, a Pioneer of the Early Years 133
Otto T. Bruhns

Prandtl-Tomlinson Model: A Simple Model Which Made History 153
Valentin L. Popov, J.A.T. Gray

A Historical View on Shakedown Theory 169
Dieter Weichert, Alan Ponter

Some Remarks on the History of Fracture Mechanics 195
Dietmar Gross

Porous Media in the Light of History 211
Wolfgang Ehlers

Parameter Identification in Continuum Mechanics: From Hand-Fitting to Stochastic Modelling 229
Rolf Mahnken

Historical Development of the Knowledge of Shock and Blast Waves 249
Torsten Döge, Norbert Gebbeken

The Historical Development of the Strength of Ships 267
Eike Lehmann

Part III: Theories, Engineering Solutions and Applications in Fluid Dynamics

The Development of Fluid Mechanics from Archimedes to Stokes and Reynolds 299
Oskar Mahrenholtz

The Millennium-Problem of Fluid Mechanics – The Solution of the Navier-Stokes Equations 317
Egon Krause

On Non-uniqueness Issues Associated with Fröhlich's Solution for Boussinesq's Concentrated Force Problem for an Isotropic Elastic Halfspace.................................. 343
A. Patrick S. Selvadurai

Essential Contributions of Austria to Fluid Dynamics Prior
to the End of World War II 355
Helmut Sockel

Part IV: Numerical Methods in Solid Mechanics from Engineering Intuition and Variational Calculus

From Newton's Principia via Lord Rayleigh's Theory
of Sound to Finite Elements 385
Lothar Gaul

History of the Finite Element Method – Mathematics
Meets Mechanics – Part I: Engineering Developments 399
Erwin Stein

History of the Finite Element Method – Mathematics
Meets Mechanics – Part II: Mathematical Foundation of
Primal FEM for Elastic Deformations, Error Analysis and
Adaptivity 443
Erwin Stein

Author Index 479

Subject Index 481

//
Part I
Mechanical Conservation Principles, Variational Calculus and Engineering Applications from the 17th to the 20th Century

The Origins of Mechanical Conservation Principles and Variational Calculus in the 17th Century

Erwin Stein

Abstract. The 17th century is considered as the cradle of modern natural sciences and technology as well as the begin of the age of enlightenment with the invention of analytical geometry by Descartes (1637), infinitesimal calculus by Newton (1668) and Leibniz (1674), and based on the rational mechanics by Newton (1687), initiated by Galilei (1638). In 1696, Johann Bernoulli posed the so-called brachistochrone problem in Acta Eruditorum, asking for solutions within a year's time. Seven solutions were submitted and published in 1697, the most famous one by his brother Jacob Bernoulli, anticipating Euler's idea of discrete equidistant support points and triangular test functions between three neighboured points, followed by the infinitesimal limit. Johann Bernoulli himself presented two intelligent solutions by joining geometrical and mechanical observations. Leibniz submitted a geometrical integration method for the differential equation of the cycloid and, what is important for this article, a short draft of a discrete or "direct variational" numerical approximation method, also using triangular test functions between neighboured support points with finite distances. This can be considered as a precursor of the finite element method. In connection with the brachistochrone, more general tautochrony problems were investigated, e.g. by Huygens and Newton. In conclusion many important developments of energy methods in mechanics using variational methods were already invented in the 17th century.

1   The 17th Century as the Cradle of Modern Natural Science and Mathematics

The late scholastic philosophy of the 16th century in central Europe, dominated by the catholic theology and based on the thinking of Aristoteles and Augustinus,

---

Erwin Stein
Institute of Mechanics and Computational Mechanics, Leibniz Universität Hannover,
Appelstr. 9A, 30167 Hannover, Germany
e-mail: stein@ibnm.uni-hannover.de

imposed severe restrictions on the progress of natural science and the human inventive genius for creating new useful technical tools. The Italian Renaissance of the 15th and 16th century already brought a fundamental change of human identity, orientation and self-assured thinking with the claim that man – not the Gods or God – had invented and still were inventing helpful artifacts.

Inspired by the ancient Greek culture, especially based on the New-Platonism, an autonomous thinking and creative abilities became attractive, and a new typus of gifted craftsmen and artistic engineers created revolutionary experiments of living and inanimate nature, inspired by this insight they made spectacular technical inventions, among them Brunelleschi, the architect and engineer of the Duomo of Florence, and Leonardo da Vinci, whose fascinating technical inventions were far ahead of his time. The first technical patents were conferred to inventors in Florence in the 16th century.

In central Europe Gutenberg, Paracelsus and especially Copernicus prepared the new age of natural science and technology. In 1620, Bacon, who has been called the father of empiricism, published his *Novum Organum*, [1], (addressing Aristoteles' Organum) in which he established inductive methodologies for scientific inquiring. He fought against prejudices and preconceived ideas.

And then, Descartes established the new mechanistic philosophy of rationalism and doubt with the dualism of the two different substances: matter (body) and mind. Later he asserted that these substances are not separated but build a single identity, Descartes (1637), [2]. Descartes marks the beginning of the philosophy of enlightenment, highlighted by his statement *"cogito ergo sum"*.

Spinoza, a lense grinder, was active in the Dutch Jewish Community and developed his so-called pantheistic philosophy from a deep critical study of the Christian Bible, Spinoza (1670), [3]. He provided an alternative to materialism, atheism and deism, claiming the identity of spirit and nature, so to say a religion of nature, Spinoza (1677), [4]. Spinoza was heavily attacked by the Catholic Church; all his publications were indexed, and being called a *spinocist* at that time was comparable to an *atheist* with the consequence of persecution.

Leibniz was a multi-ingenious scholar in all branches of science at his time and highly interested in new technical inventions, and applications for practical use in his holistic and universal thinking and the postulates *"theoria cum praxi"* and *"commune bonum"*, based on systematic collections of former scientific cognition and new findings in a universal frame, combined with new technical inventions and the improved production of goods in new manufactures. And he also contributed essentially to the new rational philosophy, guided by his postulates *"nihil sine ratione"*, *"nihil fit sine causa sufficiente"*, and *"the continuity principle"*. He was highly motivated to smooth down and to settle controversial political and religious convictions and ideas in order to achieve piece in the European states and to unify the Christian churches as a *"pacidius"* (a peacemaker), as he conceived himself. In his quasi-axiomatic *monadology*, Leibniz (1714), [5], with 90 short paragraphs, his theology different from Descartes is framed by the conviction that God as the highest monade created the universe as the best of all thinkable ones in conjunction with optimal natural laws. Thus, the creation and the development of the universe relies on rationality and mathematical logic. He was sure that reasoning of natural science inevitably leads to metaphysics: *nihil est in intellektu quod non fuerat in sensu, excipe: nisi ipse intellectu*, Leibniz (1686), [6]. Thus, the discrepancy of body and soul can be overcome in this metaphysical draft of the universe. He created a new paradigm from Christian salvation history to apprenticeship of wisdom, thus overcoming the Christian stigmas of the Original Sin and the Last Judgement.

There is no doubt that Isaac Newton outshines all physicists in the 17th and the following two centuries by the creation of new natural science in his famous principia (1687) [7]. In the introduction, he claims: the old (Greek) developed the *mechanica practica* but I created the *mechanica rationalis* (rational mechanics) which was the origin and the *bible* for the 18th and 19th century. C. Truesdell wrote a remarkable appraisal on the ingenious work of Newton in [8].

About four years after Newton, in 1674, Leibniz independently invented the infinitesimal calculus in Paris, published in 1684, [9], and gave a much deeper understanding of the infinitesimal limit for integration, using already the later Riemannian sums from the 19th century, Leibniz (1676), [10], unpublished until the 20th century.

Moreover he falsificated Descartes's findings for the *"true measure of the living force"*, who assumed erroneously that the product of mass and velocity of a moved body (which is not a scalar) ought to be a conservation quantity, and he discovered the kinetic energy $1/2m \cdot v^2$, Leibniz (1686), [11], first without the factor $1/2$, as the wanted conservation quantity of a straight on moved body with mass m and velocity v in quasi-static state.

With his important *continuity principle* he investigated short times before and after the impact of two bodies and thus found the error in Descartes' assumptions for his impact laws, see Szabo (1987), [12], also applying Galilei's finding of the velocity $v = \sqrt{2gh}$ of a falling or frictionless down gliding body due to gravity, according to the potential property of the kinetic energy of this mass.

Leibniz had the teleological vision that the physical laws of nature fulfil extremal principles for certain (scalar) conservation quantities, according to his postulate of ours as the best of all possible worlds.

The very first conservation principle in mechanics, the principle of minimum potential energy was established by Torricelli, secretary to Galilei, about 1630. He postulated that the gravity centre of an assembly of masses with arbitrary connections and boundary conditions finds its stable static equilibrium in a configuration for which the gravity centre takes the deepest possible position, published only in 1919, [13].

The birth of *variational calculus* can be dated with Jacob Bernoulli's ingenious solution of the brachistochrone problem in 1697, [14], first stated and approximately solved by Galilei in his *Discorsi* from 1638, [15], then again posed by Jacob's younger brother Johann Bernoulli in *Acta Eruditorum* in 1696, and 7 solutions were published in this journal by Leibniz in 1697, [14], [16]. Therein Jacob Bernoulli anticipated Eulers' idea of reducing the variational problem of an extremum of a functional of the requested extremal function first into a finite number of equidistant discrete problems of infinitesimal calculus for functions, Euler (1744), [17], also using triangular test functions (with value 1 at the considered discrete point and 0 at the neighbouring points) and then performing the transition to infinitely many time steps. Six other solutions of this problem were submitted, among them by Leibniz which will be treated later in this article; but Leibniz did not recognize that the variational calculus was a new important branch of infinitesimal calculus which he liked to dominate in Europe, and he did not contribute further to this new important branch of mathematics.

A second article on the brachistochrone problem was published by Johann Bernoulli with his own solution also in 1697, [16].

2 Snell's Law of Light Refraction and Fermat's Principle of Least Time for the Optical Path Length

Snellius published in 1621 the law of light refraction which was better reasoned later by Huygens with the principle that every point of a wave is the source of a new wave. This law reads

$$\sin \alpha_1 = \frac{c_1 \Delta t}{AB} \; ; \quad \sin \alpha_2 = \frac{c_2 \Delta t}{AB}, (1a) \tag{1}$$

yielding

$$\frac{\sin \alpha_1}{\sin \alpha_2} = \frac{c_1}{c_2} = n \; ; \quad \frac{\sin \alpha_1}{c_1} = \frac{\sin \alpha_2}{c_2} = \text{const}, (1b) \tag{2}$$

n the refraction coefficient.

Fermat established the principle of the light path in minimal time using Cartesian coordinates: $T = s_1/c_1 + s_2/c_2 = n_1/s_1 + n_2/s_2$, $T = n_1 \left[(x-x_1)^2 + y_1^2 \right]^{1/2} + n_2 \left[(x_2-x)^2 + y_2^2 \right]^{1/2}$. The stationarity condition written with the derivative in the later formulation by Newton and Leibniz reads $dt/dx(P_1, P_2; x) = 0 \rightarrow n_1(x-x_1)/s_1 - n_2(x_2-x)/s_2 = 0$ or

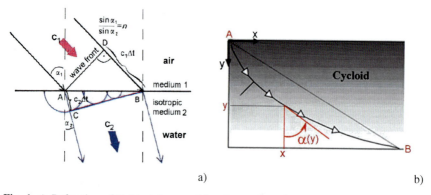

Fig. 1 a) Refraction of light at the transition from a less dense medium (air) to a denser medium (water) with the velocities c_1 and c_2; b) The common cycloid as the light path at least time in a medium with linearly varying density

$$\sin\alpha_1/c_1 = \sin\alpha_2/c_2. \qquad (2)$$

In case of linearly varying density from a more to a less dense medium, Fig. 1b, the light path fulfills the optimality (stationarity) condition

$$T_{AB} = \int ds/c(y)\sqrt{y} = \min; \; dt = ds/c(y), \text{ yelding } \sin\alpha(y)/c(y) = \text{const} \qquad (3)$$

which describes a common cycloid.

Fermat's principle had significant influence on the finding of conservation principles for physical problems in the 17th and 18th century. It will be shown in Sect. 3 that the famous problem of a guided frictionless down-gliding mass due to gravity in shortest time, first posed by Galilei, [15], also has the solution of the common cycloid.

3   The Mechanical Properties of the Catenary Curve

In the 17th century Galilei, Huygens, Leibniz as well as the brothers Bernoulli were searching for the catenary or funicular curve. In 1690 Jacob Bernoulli called for the "solution problematio funicularis" in Acta Eruditorum.

Leibniz discovered the symmetric exponential function as the catenary function

$$y = \frac{a}{2}\left(e^{x/a} + e^{-x/a}\right) = a\cosh(x/a); \quad y(x=0) = a \qquad (4)$$

with the equal normal and curvature radius

$$n(y) = R(y) = y^2/a. \qquad (5)$$

He also gave a representation of the exponential function by the sum of the catenary curve and its derivative.

Furthermore, Leibniz found the catenary curve by a counter clock wise rolling of a parabola on the horizontal axis, Fig. 2a, with the positions $Y, Y', Y'', Y''', \ldots$, where the normals are points of the catenary curve. And finally, Leibniz constructed the logarithmus function from the catenary function, Fig. 2b.

The first term of the power expansion of the catenary function with respect to the parameter f/ℓ, Fig. 2a, yields the quadratic parabola $y = fx(\ell-x)/4\ell^2$. This is the first approximation of the $\cosh(x/a)$-function for small f/ℓ. In this case the vertical line load is constant along the x-axis, whereas the line load of the catenary function caused by dead weight obviously grows from the middle point to the edges.

Another important property of the catenary curve is related to the principle of minimum potential energy of arbitrary connected masses, as outlined in section 1. One gets the function of the hanging catenary curve by postulating the deepest possible position of the gravity centre, and one can show that it has minimal length. Furthermore, the rotational surface of this catenary curve, the catenoid, has minimal surface and is a solution of the Plateau problem.

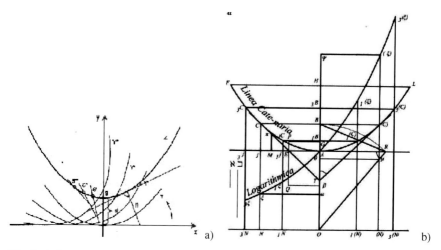

Fig. 2 a) Construction of the funicular function by a rolling parabola; b) construction of the natural logarithmus function from the catenary function

The first scientific application of catenary curves in civil engineering was realised for the restoration of the dome of the St. Peter's Cathedral in Rome which had meridional cracks. Pope Benedikt XIV commissioned the Venecian monks Le Seur, Boscovich and Poleni for a restoration proposal which was submitted in 1742 as the "Parere di tre mathematici", [18], see also Szabo (1987), [19]. They had the ingenious idea to model the dome experimentally by mirroring it with chains and hangings weights (imaging the real dead loads) with respect to the horizontal plane. The meridian of the catenoid surface represents the pure membrane state of internal forces, and the distances of this meridian with respect to the meridian curve of the dome represent the lever arms of the meridian forces, yielding the bending moments in the dome which caused the cracks. Therefore, the restoration was realized by two iron stiffening rings at positions with the largest lever arms.

4 The So-Called Brachistochrone Problem of a Mass Gliding Down Frictionless in Shortest Time

4.1 Galileo Galilei's First Formulation and Approximated Solution

Galilei was the first to pose this famous problem of optimization and variational calculus in his "Discorsi" in 1638, [15]. Not only the position and the value of a function with extremal property is requested but the whole function under the condition of an extremum of an integral of this function and its derivatives within a given domain.

Origins of Mechanical Principles

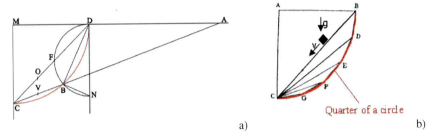

Fig. 3 a) Construction of a polygon point B for a down gliding mass assuming the quarter of a circle as the optimal curve; b) comparative polygons with the quarter of the circle as the approximative optimal solution

Galilei got the experimental results for the gliding times:
$t(BC) > t(BDC) > t(BDEC) > t(BDEFC) > t(BDEFGC)$, with the quarter of a circle as the hull.

About 60 years later the problem was solved analytically with different challenging methods by Johann Bernoulli et al., subsections 4.6 – 4.9, yielding the common cycloid as the solution of the problem.

4.2 Tautochrony or Isochrony Property of the Cycloid

Another access to the extremal properties of the cycloid was provided by Huygens and Newton with the so-called tautochrony or isochrony property, Huygens (1673), [20], as shown in Fig. 4.

With equal gliding times T_{ABo} and $T_{C'Bo}$ for arbitrary starting points A and C' a remarkable property of the cycloid was found. This property is related to Fig. 4b and was also used by Huygens in his famous physical cycloid pendulum, yielding a constant frequency for arbitrary amplitudes, realized by two cycloids on both sides of the pendulum, at which the thread of the pendulum tangentially touches the cycloids, thus reducing the free length of the thread.

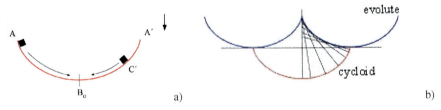

Fig. 4 a) The tautochrony property of the cycloid for a down-gliding mass due to gravity: the times T_{ABo} and $T_{C'Bo}$ are equal; b) The evolute of the cycloid is a congruent cycloid

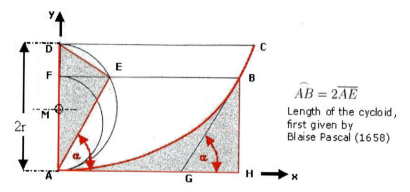

Fig. 5 Proof of the tautochrony of the cycloid using equal area segments

The proof can be found in [20]; it uses the equality of two area segments, Fig. 5,

$$\mathcal{A}_{ADE} = \mathcal{A}_{ABH} \text{ due to } \tan \alpha = HB/GH = AF/FE. \tag{6}$$

4.3 Analysis of the Common Cycloid

For better understanding of the following sections it is helpful to treat first the analysis of the cycloid in a condensed format, Fig. 5 and 6.

The first calculation of the area and the arc length were published by Cavalieri (1629).

Parametric representation:

$$\begin{array}{cccc} \text{coordinates} & & \text{normal and curvature radius} & 1^{\text{st}} \text{ derivative} \\ \begin{aligned} x &= r(\alpha - \sin\alpha) \\ y &= r(1 - \cos\alpha) \end{aligned} & ; & \begin{aligned} \overline{SK} &= n \\ \overline{SM} &= \rho = 2n \end{aligned} & ; & \frac{dy}{dx} = \cot\frac{\alpha}{2} \end{array} \tag{7}$$

Ordinary differential equation of the cycloid, Leibniz (1686):

$$\frac{dy(x)}{dx} = \sqrt{\frac{x}{c-x}}; \, c = 2r; \, n = 2r\sin\frac{\alpha}{2}; \, \rho = 2n = 4r\sin\frac{\alpha}{2}; \, \widehat{OS} = 4r \tag{8}$$

Area of the cycloid, Pascal (1659):

$$2 \cdot \mathcal{A}\, OSR = \int_{x=0}^{x=2\pi r} y \, dx = \int_{0}^{x=2\pi} y\dot{x} \, dt = r^2 \int_{0}^{2\pi} (1 - \cos t)^2 \, dt = 3\pi r^2 \tag{9}$$

Scaled gliding time:

$$T_{OS} = \frac{1}{\sqrt{2g}} = \int_{\alpha_0=0}^{\alpha_S} \left(\frac{\sin^2\alpha + 1 - 2\cos\alpha + \cos^2\alpha}{r(1-\cos\alpha)\sin^2\alpha} \right)^{\frac{1}{2}} r\sin\alpha \, d\alpha = \sqrt{\frac{r}{g}}\pi \tag{10}$$

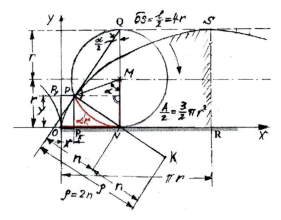

Fig. 6 Cycloid as the rolling or wheel curve on a plane

4.4 Today's Variational Formulation of the Brachistochrone Problem

Due to the singularity at $y = 0$ for $t = 0$, the representation $x = f(y)$ is adequate. The stationarity condition for the wanted extremal function reads, Fig. 7,

$$T_{AB} = \int_{t_A=0}^{t_B} dt(y) = \min, \tag{11}$$

with the velocity, Galilei (1638),

$$v(y) = \sqrt{2gy}. \tag{12}$$

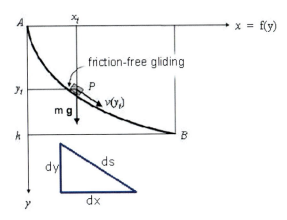

Fig. 7 The brachistochrone problem

With the relations

$$v(y) = \frac{ds(y)}{dt}; \quad dt = \frac{ds(y)}{v(y)}; \quad ds(y) = \sqrt{1+(x'(y))^2}\,dy \qquad (13)$$

the extremal principle for the non-linear implicit time functional reads

$$T_{AB} = \int_{t_A=0}^{t_B} \frac{\sqrt{1+(x'(y))^2}}{\sqrt{2gy}}\,dy \;\to\; \min. \qquad (14)$$

Its 1st variation has to be zero according to the stationarity condition

$$\delta T = \frac{1}{\sqrt{2g}} \int_{y_0}^{y_1} \frac{x'}{\sqrt{y(1+(x')^2)}}\,\delta x'\,dy \stackrel{!}{=} 0 \qquad (15)$$

and yields the ODE of the cycloid. The 2nd variation, the Hesse matrix (analytical tangent), is the basis for a finite element method by using, e.g., linear trial and test functions for the vertical convective coordinate $\eta = y/h$.

$$\delta^2 T = \frac{1}{\sqrt{2g}} \int_{y_0}^{y_1} \delta x' \left[\frac{1}{\sqrt{y(1+(x')^2)}} - \frac{(x')^2}{\sqrt{y(1+(x')^2)^3}} \right] \delta x'\,dy \qquad (16)$$

The positive definit Hesse matrix assigns a minimum of the functional T_{AB}.

It should be remarked that a change of the independent variable, i.e. $y = \tilde{f}(x)$ instead of $x = f(y)$, leads to a variational problem with a more complicated differential equation because of the singularity at the origin.

After these pre-informations from the point of view of the variational calculus of today we turn back to the first solutions in the late 17th century.

4.5 Solutions of the Brachistochrone Problem After the Call of Johann Bernoulli in Acta Eruditorum in 1696

Johann Bernoulli introduced the denotation "brachistochrone" (Greek, means curve of shortest time) for Galilei's problem and called for solutions in one years time in Acta Eruditorum 1696. A total of seven solutions was submitted and published in May 1697 in Acta Eruditorum, [14]; Johann and Jacob Bernoulli had conceived that after the development of infinitesimal calculus this problem required a new branch of analysis. The following 7 solutions were published in [14] and [16], see also Funk (1970), [21], Stein, Wiechmann (2003), [22]:

- by Jacob Bernoulli: the mathematically most important one with the first development of variational calculus;
- two by Johann Bernoulli himself: using ingenious geometrical and analytical insight;

- two by Leibniz: the concept for an approximated discrete solution and a geometric integration of the ODE of the cycloid;
- by Newton anonymously, without proof, provided only in 1724;
- by L'Hôpital and Tschirnhaus: analytical ansatz with incomplete proof.

4.6 Jacob Bernoulli's Ingenious Derivation of the ODE for the Cycloid through a Variational Problem

In anticipation of Euler's idea of piecewise discrete triangular test functions for the derivation of the 1st variation of a functional from 1743, Jacob Bernoulli introduced the same idea for the first time in 1796/97. Starting from the known condition for a minimum or maximum value of a function $y = f(x)$ at a certain point, i.e. $dy/dx = 0$, he first follows this idea for the extremum of the functional $T_{AB} = \int F[y,y']dx = \min$ by dividing the time domain, parametrized by the coordinates $y_A = 0$ and $y_B = 2r$, into a set of equidistant support points $y_i - h$, y_i, $y_i + h$ and choosing triangular test functions for the wanted extremal function, Fig. 8.

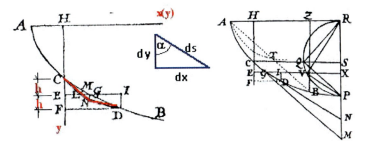

Fig. 8 Jacob Bernoulli's original figures for the variational derivation of the cycloid, published in Acta Eruditorum, May 1697

The obvious discrete stationarity condition for the extremal time consists in the equality of gliding times for the searched extremal curve and the neighboured test curve for all intervals as

$$t_{CG} + t_{GD} \stackrel{!}{=} t_{CL} + t_{LD}, \tag{17}$$

yielding the differential equation of the cycloid for the limit case $h \to 0$ as

$$\frac{ds}{dx} \sim \frac{k}{\sqrt{y}}; \quad \frac{dx}{dy} = \tan\alpha = \sqrt{\frac{y}{k^2 - y}}; \quad k^2 = 2r, \tag{18}$$

where r is the radius of the rolling wheel.

In Euler's derivation of the differential equation for the extremal function of an isoparametric variational problem from 1744, [17], he used the same discrete method with equidistant support points for the minimum problem

$$\int_{x=A}^{Z} F[y(x), y'(x), x] dx = \min, \tag{19}$$

with the required extremal function $y = f(x)$. Euler got the following wellknown ODE for the general problem

$$F_y - \frac{d}{dx} F_{y'} = 0 \quad \text{or} \quad F_x - \frac{d}{dy} F_{y'} = 0 \text{ for } x = f(y), \tag{20}$$

whereas Jacob Bernoulli directly derived the differential equation of the cycloid as the extremal function of the brachistochrone problem.

4.7 Johann Bernoulli's Two Solutions

Johann Bernoulli presented two very tricky solutions, combining geometrical and analytical experience and deep knowledge of function theory. They are both treated in [14], see also [20], [21]. The first solution is got for the substitute problem of the lightway in shortest time through a medium with linearly changing density, Fig. 1b and equ. (3). From the related stationarity condition

$$\frac{\sin \varphi_i(x_i)}{v_i(x_i)} = \text{const.}; \quad i = 1, 2, \ldots, (n+1), \tag{21}$$

where α (in equ. (3)) is replaced by φ and v is the velocity, two coordinate substitutions yield directly the coordinates of the cycloid, equ. (7), in implicit form.

The second solution treats the brachistochrone problem, postulating the stationarity condition as follows: the gliding time for a path increment of the stationary solution must be equal to an infinitesimally varied path increment, Fig. 9, which in principle is the same condition as Jacob Bernoulli's criterion.

The stationarity condition reads

$$\Delta t' - \Delta t = \frac{1}{\sqrt{2g}} \left(\frac{\Delta s(y + \Delta y)}{\sqrt{y + \Delta y}} - \frac{\Delta s(y)}{\sqrt{y}} \right) \stackrel{!}{=} 0$$

$$= \frac{1}{\sqrt{2g}} \left(\frac{(\rho + \Delta \rho) \Delta \varphi}{\sqrt{y + \Delta y}} - \frac{\rho \Delta \varphi}{\sqrt{y}} \right); \quad \Delta y = \Delta \rho \cdot \sin \varphi, \tag{22}$$

with

$$\frac{(\rho + \Delta \rho) \Delta \varphi}{\sqrt{y + \Delta y}} = \frac{\rho + \Delta \rho}{\sqrt{y}} \underbrace{\left(1 - \frac{1}{2} \frac{\Delta y}{y} + \ldots \right)}_{\stackrel{!}{=} 0} \Delta \varphi \tag{23}$$

$$\Delta t' - \Delta t = \frac{1}{\sqrt{2gy}} \left(1 - \frac{\rho \cdot \sin \varphi}{2y} \right) \Delta \rho \Delta \varphi \stackrel{!}{=} 0. \tag{24}$$

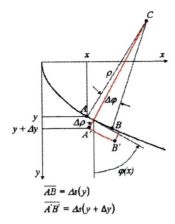

$$\overline{AB} = \Delta s(y)$$
$$\overline{A'B'} = \Delta s(y + \Delta y)$$

Fig. 9 Johann Bernoulli's 2nd solution of the brachistochrone problem, postulating equal incremental time steps for the stationary solution and a neighboured test function

This yields

$$\rho \stackrel{!}{=} \frac{2y}{\sin\varphi} \quad \text{and} \quad n = \frac{y}{\sin\varphi} = \frac{\rho}{2} \quad \text{with} \quad \varphi = \frac{\alpha}{2}, \tag{25}$$

which is a special property of the cycloid. Thus the wanted solution was determined.

4.8 Leibniz' Draft of a Discrete Approximation Called "Tachystoptota"

Leibniz contributed two very different solutions to the brachistochrone problem in [14]. The first one is a discrete geometrical integration of the ODE of the cycloid by means of the so-called "quadratrix", [10], based on his transmutation theorem.

The second one treated here in more detail is a short draft of a direct numerical method which is of considerable interest as a predecessor of a finite element method, Fig. 10.

Leibniz did not present an algorithm for the discrete problem in Fig. 10b, although he was asked for this by Johann Bernoulli in a letter.

Using only two equidistant "finite elements", Fig. 11, the discrete solution is shown by the author in the sequal.

From the continuous minimum problem

$$T_{AB} = \int_A^B dt = \int_{y_A}^{y_B} \frac{ds(y)}{v(y)} = \int_{y_A}^{y_B} \frac{\sqrt{1+(x'(y))^2}}{\sqrt{2gy}} dy = \min_{x(y)} \tag{26}$$

the discrete approximation with one discrete unknown $x_1(y_1)$ reads

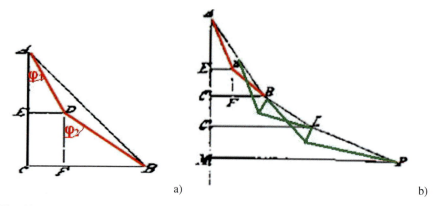

Fig. 10 a) Discrete stationarity condition: the gliding times t_{AB} and t_{ADB} must be equal, similar to Jacob Bernoullis derivation with discrete supports b) Leibniz' draft of a discrete variational method with equidistant support points E, C, C' and triangular (local) test and trial functions between these points

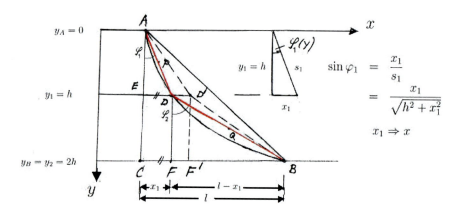

Fig. 11 Discrete variational algorithm for the brachystochrone problem with one unknown nodal position x_1, i.e. two equidistant "elements" with length h

$$T_{AB} = \frac{\overline{AD}}{v_{AD}} + \frac{\overline{DB}}{v_{DB}} = \frac{s_1}{v(h)} + \frac{s_2}{v(2h)} = \frac{\sqrt{h^2+x^2}}{\sqrt{2gh}} + \frac{\sqrt{h^2(l-x)^2}}{\sqrt{2g \cdot 2h}} = \min_{x_h(y)}. \quad (27)$$

The discrete stationarity condition, not given by Leibniz, is

$$\frac{\partial T_{AB}}{\partial x} \stackrel{!}{=} 0 \rightsquigarrow \frac{x}{\sqrt{h^2+x^2}} \cdot \frac{1}{\sqrt{2gh}} - \frac{l-x}{\sqrt{h^2+(l-x)^2}} \cdot \frac{1}{\sqrt{2g2h}} \stackrel{!}{=} 0 \quad (28)$$

$$\rightsquigarrow \frac{\sin\varphi_1}{v_1} - \frac{\sin\varphi_2}{v_2} \stackrel{!}{=} 0, \quad (29)$$

i.e. the same condition as for the optical path length in least time.

This condition, equ. (28), yields the 4th order polynomial

$$f(x) = x^4 - 2lx^3 + x^2(l^2 + h^2) + 2h^2lx - h^2l^2 \stackrel{!}{=} 0 \qquad (30)$$

or with

$$\xi := \frac{x}{l} \; ; \; \left(\frac{h}{l}\right)^2 = \frac{1}{4} \quad \rightsquigarrow \quad f(\xi) = \xi^4 - 2\xi^3 + \frac{5}{4}\xi^2 + \frac{1}{8}\xi - \frac{1}{4} \stackrel{!}{=} 0. \qquad (31)$$

The linear approximation is $x_{1,lin} = 0,5l$, and the exact solution results in

$$x \to x_1 = 0,69l \qquad (32)$$

as the first approximation with only two elements.

4.9 Discrete Variational Formulations by Schellbach

In 1851, Schellbach presented 12 discrete solutions of variational problems in the sense of Leibniz' draft for the brachistochrone problem with various boundary conditions and for related problems in analytical form, also using equidistant support points, Schallbach (1851), [23], Fig.12.

In this paper, entitled "Probleme der Variationsrechnung", Schellbach points out in the introduction: "The reasons for Bernoulli's, Euler's, and Lagrange's methods [for the variational calculus] cannot be clearly understood yet", and: "the variational calculus is the most abstract and most sublime area of all mathematics".

This gives the information that variational calculus and moreover discrete variational calculus was not yet well-known in the mathematical community in the middle of the 19th century.

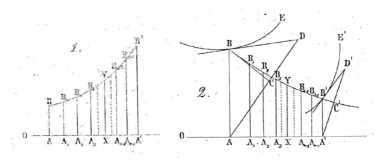

Fig. 12.1 & 12.2 Figures from K. H. Schellbach's discrete formulation and variational setting of the brachystochrone problem with coupled algebraic equations at equidistant points $A, A_1, A_2, \ldots, A'$ in analytical form

In this paper the following discrete problems are treated:

1. Minimum area of a polygon with fixed ends, given length, and extensions (Fig. 12.1)
2. Minimum area of a rotational surface with given meridian arc length and boundary conditions, with extended versions ((Fig. 12.1)
3. Brachistochrone problem with generalized boundary conditions in B and B (Fig. 12.2)
4. Brachistochrone problem in a resisting medium (Fig. 12.2)
5. Problem similar to 3. & 4., but with the condition of largest or smallest final velocity in B' (Fig. 12.2)
6. to 12. Further problems of this type

The numerical calculation of Schellbach's equations yields systems of non-linear algebraic equations for the treated problems. This can be conceived as a first special analytical version of the finite element method.

4.10 Finite Element Method for the Brachistochrone Problem in Today's Fashion

The variational problem of the brachistochrone reads, using the adequate coordinate representation $y = f(x)$, y the vertical coordinate in order to avoid a singularity at the starting point A (Fig. 13)

$$T = \frac{1}{\sqrt{2g}} \int_{y_0}^{y_1} \sqrt{\frac{1+(x')^2}{y}} dy \to \min. \tag{33}$$

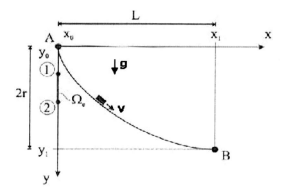

Fig. 13 Finite element analysis of the brachistochrone problem

Origins of Mechanical Principles

The first variation (the stationarity condition) of the time functional follows as

$$\delta T = \frac{1}{\sqrt{2g}} \int_{y_0}^{y_1} \frac{x'}{\sqrt{y(1+(x')^2)}} \delta x' dy \stackrel{!}{=} 0; \quad x' = \frac{dx(y)}{dy} \quad (34)$$

and the second variation (Hesse matrix, analytical tangent) reads

$$\delta^2 T = \frac{1}{\sqrt{2g}} \int_{y_0}^{y_1} \delta \bar{x}' \left[\frac{1}{\sqrt{y(1+(x')^2)}} - \frac{(x')^2}{\sqrt{y(1+(x')^2)^3}} \right] \delta x' dy. \quad (35)$$

The discretization with linear finite elements for the dimensionless vertical coordinate $\eta = y/\ell_e$ as parametrized time variable realizes the original ideas of Jacob Bernoulli and Leibniz, Fig. 14.

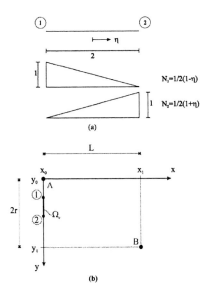

Fig. 14 Linear finite element test and trial shape functions for the discretization of the brachistochrone curve

The linear finite element Ansatz for the horizontal time-dependent coordinate $x(\eta)$, $\eta = f(t)$, reads, Fig. 14,

$$x_h = \sum_{I=1}^{2} N_I(\eta) x_I \quad \forall \, \Omega_e \subset \Omega \quad (36)$$

with the shape functions

$$N_I(\eta) = \frac{1}{2}(1 + \eta_I \eta); \quad -1 \leq \eta \leq 1. \tag{37}$$

The first variation δx_h of x_h reads

$$\delta x_h = \sum_{I=1}^{2} N_I(\eta) \delta x_I \quad \forall \Omega_e \subset \Omega. \tag{38}$$

Inserting this into the global stationarity condition for the total gliding time, (34), now of the discretized nonlinear problem with n_e finite elements, $e = 1, 2, \ldots, n_e$, requires the iterative solution of the nonlinear algebraic equation system, here for the i-th iteration step in matrix notation

$$\delta T_{h,i} = \bigcup_{e=1}^{n_e} \left[\frac{1}{\sqrt{2g}} \int_{y_0}^{y_1} \frac{x'}{\sqrt{y(1+x'^2)}} N_{1,y} dy \middle| \frac{1}{\sqrt{2g}} \int_{y_0}^{y_1} \frac{x'}{\sqrt{y(1+x'^2)}} N_{2,y} dy \right] \times \begin{bmatrix} \delta x_1 \\ \delta x_2 \end{bmatrix}. \tag{39}$$

From (39) the wanted nodal coordinates $x_{h,k}(y_k)$ at equidistant nodal points y_k follow from element-wise numerical integration and then assembling the system. Quadratic convergence is achieved by the use of the Hessian matrix, (35).

The results for regular element refinements up to $n_e=1000$ elements are shown in Fig. 15. About 100 elements are needed for a displacement error in the L_2 norm of less than 1%, executed on a laptop in about 0,1 second.

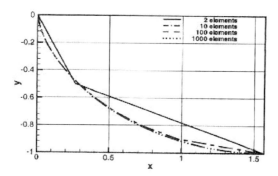

Fig. 15 Finite element solution of the brachistochrone problem with four regularly refined element meshes

5 Concluding Remarks

Important first contributions to variational calculus, based on the infinitesimal calculus by Newton and Leibniz, were discovered and developed at the end of the 17th century with application to conservation principles in mechanics, originating from Torricelli, de Fermat, Huygens, Newton and Leibniz. The brachistochrone problem, first posed and approximately solved by Galilei, was the reason for the request by Johann Bernoulli in 1696 to solve this problem which did not fit into Newton's and

Leibniz's infinitesimal calculus. The seven submitted solutions in one year's time were published as a survey article by Leibniz in Acta Eruditorum in 1697. The most important mathematical solution came from Jacob Bernoulli and anticipated Euler's theory from 1743 to a large extend. Johann Bernoulli took benefit in his two solutions of both, the physical minimum principles, namely the brachistochrone problem of the gliding mass, and the corresponding problem of the light path in shortest time in a medium with linearly changing density according to de Fermat's minimum principle.

Leibniz geometrical integration of the differential equation of the cycloid using a power series of the integrand and his transmutation theorem does not fit directly into the scope of Johann Bernoulli's request from 1696. But his second contribution, a short draft of the discrete "direct variational" approximated solution, was the first pre-version of the finite element method. In total, Leibniz did not recognize the importance of the new calculus, and he did not spend further efforts on this matter.

Also the tautochrony property of the cycloid, investigated by Huygens, Newton and Leibniz, is a remarkable discovery of the 17th century.

This article tries to point out that the invention of analytical geometry by Descartes, the infinitesimal calculus by Newton and Leibniz, Newton's "mechanica rationalis" and furthermore the first achievements in variational calculus in connection with conservation principles in mechanics really characterize the 17th century as the cradle of modern mathematics and natural sciences.

References

1. Bacon, F.: Novum Organum. Original Latin text, England (1620), English translation by Rees, Graham and Maria Wakely. Clarendon, Oxford (2004)
2. Descartes, R.: Discours de la méthode, Leiden, Netherlands (1637)
3. de Spinoza, B.: Tractatus Theologico-Politicus. Anonymous, Amsterdam (1670)
4. de Spinoza, B.: Ethica, ordine geometrico demonstrata. Posthumnous, The Hague (1677)
5. Leibniz, G.W.: La Monadologie (1714), Published in German (1720), in original French language (1840)
6. Leibniz, G.W.: Discours de métaphysique. In: Herring, H. (ed.) Philosophische Bibliothek, vol. 260, pp. 2–95 (1686) (Felix Meiner Verlag, 2. improved edition, Hamburg (1985))
7. Newton, I.: Philosophiae naturalis principia mathematica, London (1687), rev.ed. Cambridge (1713)
8. Hutter, K. (Hrsg.): Die Anfänge der Mechanik. Springer, Berlin (1989)
9. Leibniz, G.W.: Nova methodus pro maximis et minimis. Acta Eruditorum 1684, 466–473 (1684)
10. Leibniz, G.W.: De quadratura arithmetica circuli ellipseos et hyperbolae cujus corollarium est trigonometria sine tabulis. Submitted to the Academie des Sciences, Paris (1676), Translated into German with comments by E. Knobloch, Berlin
11. Leibniz, G.W.: Brevis demonstratio erroris memorabilis Cartesii. Acta Eruditorum 1686, 161–163 (1686)
12. Szabó, I.: Geschichte der mechanischen Prinzipien. Birkhäuser, Basel (1987)
13. Torricelli, E.: De Motu gravium. Opere 2, 105 (1919)

14. Leibniz, G.W.: Communicatio suae pariter, duarumque alienarum ad edendum sibi primum a Dn. Jo. Bernoullio, deinde a Dn. Marchione Hospitalio communicatarum solutionum problematis curvae celerrimi descensus a Dn. Jo Bernoullio geometris publice propositi, una cum solutione sua problematis alterius ab eodem postea propositi. Acta Eruditorum, 201–206 (May 1697)
15. Galilei, G.: Discorsi e dimonstrazioni matematiche, intorno a due nuove scienze, Leiden (1638)
16. Bernoulli, J.: Curvatura radii in diaphanis non uniformibus. Acta Eruditorum, 206–211 (1697)
17. Euler, L.: Methodus inveniendi lineas curvas maximi minimive proprietate gaudens sive solutio problematis isoperimetrici latissimo sensu accepti. Lausanne and Genevae, Opera Omnia, Series I 25 (1744)
18. Le Seur, Jaquier, Boscovich: Parere di tre mattematici sopra i danni che si sono trovati nella Cupola di S. Pietro sul fine dell' Anno 1742, presented by Poleni as Memorie istoriche della Gran Cupola del Tempio Vaticano in 1748, expert opinion by the "tre mattematici" (1742)
19. Szabo, I.: Geschichte der mechanischen Prinzipien und ihrer wichtigsten Anwendungen, 3rd enhanced edn. Birkhäuser, Basel (1987)
20. Huygens, C.: Horologium oscillatorium sive de motu pendularium (1673)
21. Funk, P.: Variationsrechnung und ihre Anwendung in der Technik, 2nd edn. Springer, Berlin (1970)
22. Stein, E., Wiechmann, K.: New insight into optimization and variational problems in the 17th century. Engineering Computations 20, 699–724 (2003)
23. Schellbach, K.H.: Probleme der Variationsrechnung. Crelle's Journal für die Reine und Angewandte Mathematik 41, 293–363 + 1 table (1851)

Principles of Least Action and of Least Constraint

Ekkehard Ramm

Abstract. The present contribution describes the evolution of two major extremum principles in mechanics proposed in the 18th and the first half of the 19th century. In the first part the essay describes the scientific controversy on the Principle of Least Action associated with the name of Maupertuis, certainly one the most documented affairs in the history of sciences. In this scientific dispute at the Prussian Royal Academy of Sciences in Berlin in the early 1750ies the allegation for plagiarism and forgery played a central role. The second part focuses on the Principle of Least Constraint proposed by Gauss in a short article in the *Journal für die reine und angewandte Mathematik* in 1829. Its affinity to the Least Square Method and its relation to d'Alembert's Principle are outlined.

1 Introduction

The search for the existence of extremum principles in nature and technology can be traced back to the ancient times. The scholars were not only driven by scientific observations; often their objective was based on metaphysical arguments. In Figure 1 major figures involved in the development of extremum principles in mechanics are arranged in the respective time course. In the following essay the era of the mid 18th and the early 19th century is selected; in particular we will concentrate on Pierre-Louis Moreau de Maupertuis related to the *Principle of Least Action* in 1744/46 and on Carl Friedrich Gauss who stated the *Principle of Least Constraint* in 1829.

Ekkehard Ramm
Institut für Baustatik und Baudynamik, Universität Stuttgart, Pfaffenwaldring 7,
70550 Stuttgart, Germany
e-mail: ramm@ibb.uni-stuttgart.de

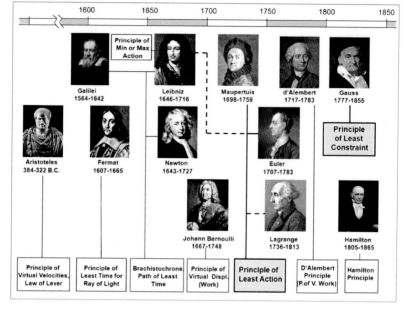

Fig. 1 Evolution of Extremum Principles

2 Principle of Least Action

2.1 Maupertuis and His Principle

The discussion on the Principle of Least Action is one of the most extensive examinations of a scientific controversy. The protagonists (Figure 2) are Maupertuis and Euler on one side and König and Voltaire on the other side. Leibniz' work also

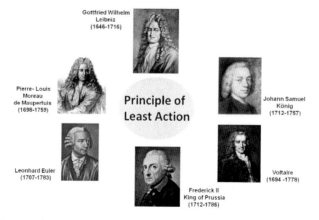

Fig. 2 The Protagonists

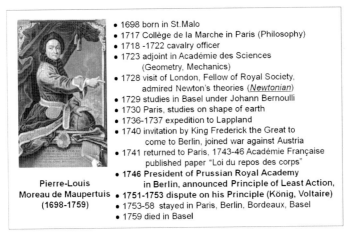

- 1698 born in St.Malo
- 1717 Collège de la Marche in Paris (Philosophy)
- 1718 -1722 cavalry officer
- 1723 adjoint in Académie des Sciences (Geometry, Mechanics)
- 1728 visit of London, Fellow of Royal Society, admired Newton's theories (*Newtonian*)
- 1729 studies in Basel under Johann Bernoulli
- 1730 Paris, studies on shape of earth
- 1736-1737 expedition to Lappland
- 1740 invitation by King Frederick the Great to come to Berlin, joined war against Austria
- 1741 returned to Paris, 1743-46 Académie Française published paper "Loi du repos des corps"
- 1746 President of Prussian Royal Academy in Berlin, announced Principle of Least Action,
- 1751-1753 dispute on his Principle (König, Voltaire)
- 1753-58 stayed in Paris, Berlin, Bordeaux, Basel
- 1759 died in Basel

Pierre-Louis Moreau de Maupertuis (1698-1759)

Fig. 3 Pierre-Louis Moreau de Maupertuis

played a role in the fight along with Frederick the Great, King of Prussia, as a leading authority at that time in Berlin.

The French mathematician Pierre-Louis Moreau de Maupertuis (Figure 3) was very much influenced by the work of Newton. Already in papers to the French Academy of Sciences in 1741 and 1744 he mentioned a principle minimizing a quantity which he called action. After he became the President of the Prussian Academy of Sciences in Berlin, on invitation of King Frederick the Great in 1746, he presented the book "*Les Loix du Movement et du Repos*" (The Laws of Movement and of Rest derived from a Metaphysical Principle) [27], see Figure 4, left. In the introduction he points to his previous work at the Paris Academy in 1744 and adds the remark "*At the end of the same year, Professor Euler published his excellent book Methodus inveniendi lineas curvas maximi minimive proprietate gaudentes*" (A method for finding curved lines enjoying properties of maximum or minimum), see [7], Figure 5. He then continues "*In a Supplement to his book, this illustrious Geometer showed that, in the trajectory of a particle acted on by a central force, the velocity multiplied by the line element of the trajectory is always a minimum*". In other words Maupertuis was aware of maximum and minimum properties however he reduced his further considerations to a minimum principle.

In his "*Les Loix...*" Maupertuis first critized the usual proofs of the existence of God and refers to the fundamental laws of nature. His self-confident statements are remarkable when he writes "*After so many great men have worked on this subject, I almost do not dare to say that I have discovered the universal principle upon which all these laws are based. ... This is the principle of least action, a principle so wise and so worthy of the Supreme Being, and intrinsic to all natural phenomena; one observes it at work not only in every change, but also in every constancy that Nature exhibits*".

Finally he states the general principle as follows (Figure 4, right): "*When a change occurs in Nature, the Quantity of Action necessary for that change is as*

Fig. 4 Maupertuis' Les Loix (1746) [27] and the Principe de la Moindre Action (Principle of Least Action)

small as possible", and defines "*The Quantity of Action is the product of the Mass of Bodies times their velocity and the distance they travel. When a Body is transported from one place to another, the Action is proportional to the Mass of the Body, to its velocity and to the distance over which it is transported*". Expressing the Action A in a formula yields

$$A \sim M \cdot v \cdot \mathrm{d}s \qquad (1)$$

with the mass of a particle M, the velocity v and the transported distance $\mathrm{d}s$. The relation of the Action to the kinetic energy is apparent if the distance $\mathrm{d}s$ is replaced by velocity times the time increment $\mathrm{d}t$.

In the sequel of this definition Maupertuis adds three examples as a proof for the generality of his principle, namely on the laws of motion for inelastic as well as elastic bodies and the law of mechanical equilibrium. In the mentioned Supplement II "*De motu projectorum...*" (On a motion of particles in a non-resistant medium, determined by a Method of maxima and minima), Figure 5, right, Leonard Euler (Figure 6) says "*1. Since all natural phenomena obey a certain maximum or minimum law, there is no doubt that some property must be maximized or minimized in the trajectories of particles acted upon by external forces. However, it does not seem easy to determine which property is minimized from metaphysical principles known a priori. Yet if the trajectories can be determined by a direct method, the property being minimized or maximized by these trajectories can be determined, provided that sufficient care is taken. After considering the effects of external forces and the movements they generate, it seems most consistent with experience to assert that the integrated momentum, i.e. the sum of all momenta contained in the particle's movement, is the minimized quantity*".

Fig. 5 Euler's Methodus Inveniendi Lineas Curvas (1744) [7] and Supplement II

In section 2 of the Additamentum Euler defines mass M, velocity $\sqrt{v}$ (he used $\sqrt{v}$ instead of v), infinitesimal distance ds, the momentum integrated over the distance ds as $M \cdot \mathrm{d}s \cdot \sqrt{v}$ and writes: *"Now I assert that the true trajectory of the moving particle is the trajectory to be described, from among all possible trajectories connecting the same end point, that minimizes $\int M \mathrm{d}s \sqrt{v}$ or since M is constant $\int \mathrm{d}s \sqrt{v}$. Since the velocity $\sqrt{v}$ resulting from the external forces can be calculated a posteriori from the trajectory itself, a method of maxima and minima should suffice to determine the trajectory a priori. The minimized integral can be expressed in terms of the momentum (as above), but also in terms of the living forces* (kinetic energies). *For, given an infinitesimal time* dt *during which the element* ds *is traversed, we have* $\mathrm{d}s = \mathrm{d}t\sqrt{v}$. *Hence* $\int \mathrm{d}s \sqrt{v} = \int v \mathrm{d}t$, *i.e. the true trajectory of a moving particle minimizes the integral over time of its instantaneous living forces* (kinetic energies). *Thus, this minimum principle should appeal both to those who favor momentum for mechanics calculations and to those who favor living forces"*.

Leonhard Euler (1707-1783)

- 1707 born in Basel
- 1720-1723 studies mathematics at University of Basel under Johann I Bernoulli and Jakob Hermann
- 1723/24 study of theology
- 1727 Member of Academy in St. Petersburg, Professors of Physics (1731) and Mathematics (1733)
- 1741 call by Frederick II to Berlin,
- **1741 works on *"Methodus Inveniendi Lineas Curvas..."* published in 1744**
- **1744 Director of class of mathematics of new Prussian Academy of Sciences, treatise on variational methods**
- 1753 takes Maupertuis' part although he must have known the weaknesses and does not claim any priority of the Principle; see *"Dissertatio de Principio Minimae Actiones...."*
- 1766 after disagreement with Frederick II returned to St. Petersburg;
- in following years publishes several fundamental treatises
- 1771 looses his sight
- 1783 died in St. Petersburg

Fig. 6 Leonard Euler

- 1712 born in Büdingen (Hesse) (family expelled from Bern)
- 1730-1734 studied at Basel University mathematics under Johann I & Daniel Bernoulli (on Newton's *principia mathematica*); introduced by J. Hermann to Leibniz philosophy, met Maupertuis
- 1735 studies in Marburg, publ. on math. and mechanics
- 1738 Paris, met Maupertuis, Voltaire et al., went to Bern
- 1744 exiled from Bern (considered too liberal) chair of mathematics and philosophy University Franeker (NL)
- 1749 Member of Prussian Academy on nomination of Maupertuis
- 1750 draft of essay on Principle of Least Action, published in 1751 in Nova Acta Eruditorum directed against Maupertuis: pleads for a *Principle of Minimum or Extremum Energy* and quotes a letter of Leibniz to J. Hermann 1708.
- 1752 left Academy due to its decision: letter is forgery
- 1751 Law of Kinetic Energy
- 1757 died in Zuilenstein, Netherlands

Johann Samuel König (1712-1757)

Fig. 7 Johann Samuel König

Euler realized that for the entire path the action has to be a sum along all segments ds and essentially defined as quantity (velocity defined as v as usual)

$$A = \int Mv \mathrm{d}s = \int Mv^2 \mathrm{d}t \qquad (2)$$

which is equal to the kinetic energy up to the factor $\frac{1}{2}$.

Thus it turned out that Maupertuis' pretension on universality of his principle was wrong, confer [30, 22]; the examples were not well chosen and partially not correct. His claim that the minimization is an economy principle did not hold because some problems lead to a maximum. Chevalier P. D'Arcy reproached Maupertuis for this error; for example the refraction of a concave mirror is based on a maximum property, [30]. Consequently his interpretation of Euler's statements was not correct.

2.2 The Critics

It was his fellow student and friend Johann Samuel König (Figure 7) who became Maupertuis' strongest critic. Two years after he was appointed member of the Berlin Academy, strongly supported by the president, the Swiss lawyer and later mathematician König (1712–1757) passed an essay to Maupertuis who approved it for publication without reading. In his article König pled for a principle of extremal energy from which an extremum for the action can be derived. The essay appeared in Nova Acta Eruditorum 1751 [23]. When Maupertuis read it he became upset because not only the limitation of his principle was questioned but it also claimed that G.W. Leibniz (1646–1716) has given a more precise formulation already in a letter to the Swiss mathematician and theologian Jakob Hermann (1678–1733) in 1707. At the end of the paper [23] he quotes from Leibniz' letter to Hermann [25] (see

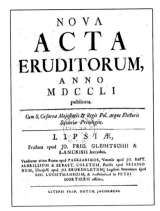

Fig. 8 König's reference letter of Leibniz to Hermann (1708) in Nova Acta Eruditorum (1751) [23]

Figure 8) *"But action is in no way what you think, there the consideration of the time enters; it is the product of the mass and the time*[1]*; or of the time by the living force. I have pointed out that in the modifications of movements, it is usually derived as a Maximum, or a Minimum. From this can be deduced several important propositions; it can be used to determine the curves describing the bodies that are attached to one or several centres".*

According to König the "force vive" (vis viva) has been already introduced by Leibniz expressing $M \cdot v^2$, i.e. it is two times the kinetic energy. It was not his intention though to accuse Maupertuis being a plagiarist or to disparage his merits; he simply wanted to point out the more general formulation of Leibniz (minimum or maximum). Nevertheless it was the beginning of intensive mutual disparagements ending in one of the ugliest of all scientific disputes [19].

The letter König presented was a copy, a fact which would play a key role in the subsequent vehement quarrel. All attempts to find the original letter ended only in further copies; König claimed that he got his copy from the Swiss poet and politician Samuel Henzi in Bern who collected letters of Leibniz among other things. Henzi was decapitated in 1749 because of conspiration and all his original papers were supposed to be burned. The Berlin Academy met in spring 1752 charging König of forgery. It is noteworthy that the meeting was headed by Euler who from the very beginning was a strong supporter of Maupertuis despite his own more rigorous work on the same principle in 1744 (Supplement II, Figure 5); the reason for Euler's reaction is still a matter of conjectures; see e.g. [30, 22, 1]. König's appeal was rejected; he resigned as member of the Academy in summer 1752.

King Frederick used all his authority to support Maupertuis, the President of "his" Academy. It was the time when another figure entered the scene, namely Voltaire (Figure 9), the famous French writer. Voltaire liked satires and polemics

---

[1] The quote contains a misprint; in the first definition of the action time must be replaced by distance and speed.

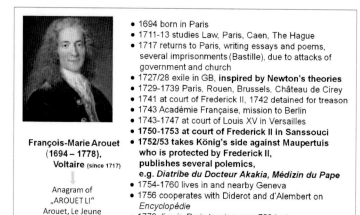

Fig. 9 François-Marie Arouet, called himself Voltaire

and was known for his sharp, malicious, sometimes also witty remarks. Although he was invited by Frederick to come to Berlin he took part of König; he published an anonymous letter "*A reply from an Academician of Berlin to an Academician of Paris*" in fall 1752 defending König (Figure 13, right); he refers to Monsieur Moreau de Maupertuis and says: "*He asserts that in all possible cases, Action is always a Minimum, which has been demonstrated false; and he says he discovered this law of Minimum, what is not less false. Mr. Koenig, as well as other Mathematicians, wrote against this strange assertion, & he cited among other things, a fragment of Leibnitz in which the great man has remarked, that in modifications of movement, the action usually becomes either a Maximum, or a Minimum*". König once more referred to copies of Leibniz' letter in a further pamphlet. Frederick was provoked and defended Maupertuis in a "*Letter of an Academician of Berlin to an Academician in Paris*" (Figure 13, right).

Voltaire published further polemics, the most famous being the "*Diatribe due Docteur Akakia, Médicin du Pape*" (Figure 10, left), which was a defamatory piece of writing [33]. The Greek word Akakia means "without guile". The *Diatribe* printed in Potsdam was burned by order of King Frederick. However since it was also published in Holland, some extra copies appeared in Berlin and were again burned in front of Voltaire's apartment. This was the final break-up with the king; Voltaire left for Leipzig in spring 1753 where he continued attacking Maupertuis.

Several letters of Voltaire followed compiled in "*Histoire du Docteur Akakia et du Natif de St. Malo*" in April 1753 (Figure 10, right) again making snide remarks because St. Malo is the birthplace of Maupertuis. Like the *Diatribe* this pamphlet was a sham presented as a defence but in reality being a backhanded compliment. In the *Histoire* a fictitious peace contract between the President (Maupertuis) and the Professor (König) *Traité de Paix conclu entre M. Le Président et M. Le Professeur, 1st January 1753* [32] was included saying among other things:

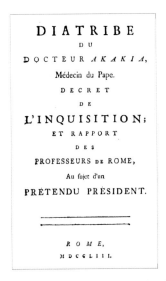 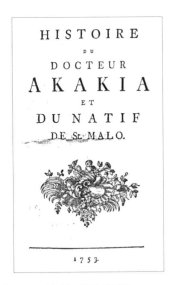

Fig. 10 Voltaire's Diatribe [33] and Histoire du Docteur Akakia [32] (1753)

"In future we promise, not to put the Germans down and admit that the Copernicus's, the Kepler's, the Leibnitz's..., are something, and that we have studied under the Bernoulli's, and shall study again; and that, finally, Professor Euler, who was very anxious to serve us as a lieutenant, is a very great geometer who has supported our principle with formulae which we have been quite unable to understand, but which those who do understand have assured us they are full of genius, like the published works of the professor referred to, our lieutenant", [6].

Thus Voltaire also included Euler in this dispute and holds him up to ridicule. Euler was indeed a vehement advocate of the Academy President despite his own more rigorous exposition of the subject. Dugas refers in [6] to an essay of Euler and quotes: *"This great geometer has not only established the principle more firmly than I had done but his method, more ubiquitous and penetrating than mine, has discovered consequences that I had not obtained. After so many vested interests in the principle itself, he has shown, with the same evidence, that I was the only one to whom the discovery could be attributed"*.

Euler discussed his view of the development of the principle in [8], presented already in 1751 but not printed in the Histoires of the Prussian Academy until February 1753 (Figure 11). He referred to König's role and made a strong statement: *"But there is no one with whom we would be less likely to have dispute than Professor Koenig, who boldly denies that there is in nature such a universal law and pushes this absurdity to the point of mocking the Principle of Conservation, which constitutes the Minimum that nature appoints. In addition, he introduces the great Leibnitz as claiming and explaining that he was far from knowing such a principle. From this we see that Mr. Koenig cannot deny our President the discovery of the principle that he himself considers to be false"*. Finally he said: *"The principle that Mr. de Maupertuis discovered is therefore worthy of the greatest of praise; and without a doubt*

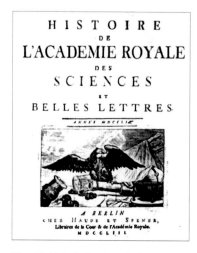

Fig. 11 Euler's Sur le Principe de la Moindre Action 1751/53 [8]

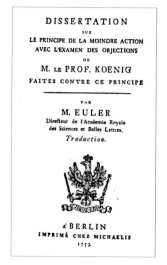

Fig. 12 Bilingual edition of Euler's Dissertatio [9]

it is far superior to all discoveries that have been made in dynamics up until now" (Figure 11, right). This essay was included in the "*Dissertatio de Principio Minimae Actionis...*", a bilingual edition in Latin and French [9] together with an examination of the objections of M. Professor Koenig made against the principle (Figure 12), see also [5].

Shortly thereafter in April 1753 sixteen polemics and letters in dispute were put together in a publication initiated by Voltaire [5]; it was entitled "*Maupertuisiana*" (Figure 13) and printed in a fictitious location Hambourg; among them are the aforementioned "*Reply*" and "*Letter*". The figure on the title page shows Don Quixote

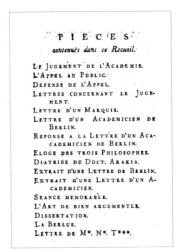

Fig. 13 Maupertuisiana (1753), List of Essays [5]

(Maupertuis) fighting the windmills with a broken lance shouting *"Tremblez"* (tremble); behind him Sancha Panza (Euler) is riding a donkey. On the right side a satyr is depicted saying *"sic itur ad astra"* (thus one reaches the stars), meaning "thus one is blamed forever". Above the picture a line of Virgil's "Aeneis" is added: *"Discite Justitiam, moniti"* (be warned and learn justice).

The *"Traité de Paix"* was also included in the *Maupertuisiana*; in the title page of that volume the first part of a quote from Horace's Satires was added: *"ridiculum acri forties ac melius (plerumque secat res)"* or "ridicule often settles matters of importance better (and with more effect than severity)".

With the publication of the *Maupertuisiana* the quarrel came more or less to an end. Maupertuis was ill, spent a year in Paris and St. Malo, returned to Berlin which he left in 1756. After another year in St. Malo he was accommodated by Johann II Bernoulli in Basel where he died in 1759. S. König being the moral victor in this battle died from a stroke in Holland two years earlier in 1757. Voltaire had still some active and eventful years; he left Berlin in 1753 for Paris, but was banned by Louis X. He then lived in Geneva and in Ferney across the French border. In 1778 he returned for the first time in 20 years to Paris where he died after three months. Euler as the fourth figure in this dispute stayed a couple of years in Berlin, but returned to St. Petersburg in 1766. He was extremely active and published several fundamental treatises. He passed away in 1783.

2.3 On Leibniz' Alleged Letter to Hermann

The claim for forgery was taken up again 140 years later when C.I. Gerhardt, editor of Leibniz' mathematical œvre at the Prussian Academy, reinvestigated the case. The Physicist H. von Helmholtz, who gave a speech on the history of the Principle of Least Action in the Academy in 1887 [20], refers to Gerhardt's remark, that the letter does not fit to the correspondence with Hermann. Leibniz may have postponed the publication because he planned a later application. In a detailed report [16] in 1898 (see minutes for a meeting of the Academy) Gerhardt claims that three of the four letters presented by König are genuine and concludes at the end of his paper: *"From the above explanations it should undoubtedly follow that the letter fragment, which König published, is not made up, and the entire letter has not been foisted. The letter is written by Leibniz. Also with a probability close to certainty it has been proven that it had been directed to Varignon* (Pierre de Varignon (1654–1722), French mathematicien and physicist). *The coincidental letters of the correspondence with Varignon in the Royal Library of Hannover have disappeared as the letters of Leibniz in the respective correspondence with Hermann did"*. The assumption for the authenticity was underlined by another argument in 1913, when W. Kabitz [21] found a further copy of the letter in Gotha among a collection of other genuine letters of Leibniz. This was accepted as a "proof" for the originality of the letter at that time.

Recently Breger has re-examined the case in a remarkably thorough discussion [3] in which he elaborated on eight arguments against the authenticity of the letter. Among those is the argument that Leibniz never applied the notion *"limites"* in the sense as it is used in the letter. Leibniz also often referred to minimum and maximum properties and also mentioned the term action, however combining both these concepts would be a singular event; for further arguments see [3], a paper worth reading. Breger recomments that now it is time to inversely analyse which arguments are speaking in favor of the genuineness. He also mentions that in case of forgery the rather strange reaction of Euler is easier to comprehend; he understood the letter as an action against his own person. A further question namely with respect to the identity of the falsifier still remains open.

2.4 *Application of Principle of Least Action*

As said above the Principle of Least Action has been defined in more rigorous form already by Euler in 1744 as an energy principle. It turned out to be rather a *Principle of Stationary Action*. Lagrange derived the equation of motion based on his newly developed calculus of variations (1760). In 1834/35 Hamilton introduced the Lagrangian function into the variational principle deriving what has been called later the Euler-Lagrange equations. Whereas Euler and Maupertuis concentrated in the Principle of Least Action on the kinetic energy (vis viva, living forces) tacitly assuming constant potential energy the difference between the kinetic and the potential energy entered the Lagrangian in Hamilton's principle. Despite this close relationship the Principle of Least Action has not found its way into the engineering community which mostly refers to the works of Lagrange, d'Alembert and Hamilton. In the 19th century the well known Principles of Minimum of Total Potential Energy (Dirichlet-Green) and Minimum of Total Complementary Energy (Menabrea-Castigliano) became basic theorems in mechanics. The Principles of Virtual Work

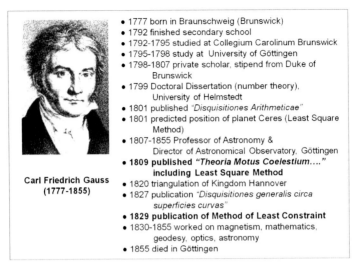

Fig. 14 Carl Friedrich Gauss

and Complementary Work entered the scene in particular because of their generality being applicable also for problems where no potentials exist. They are nowadays together with mixed Variational Principles (Hellinger-Reissner, Fraeijs deVeubeke-Hu-Washizu, etc.) the foundations for the derivation of discretization methods.

Opposite to the engineering community the notion of Principle of Least (Stationary) Action is still present in modern physics. Entire chapters of treatises are devoted to the principle, e.g. [28, 2]; see also the famous Feynman Lectures on Physics [13]. Occasionally they are looked upon as generalization of Hamilton's principle rather than as its predecessor. The Principle has been applied for example in the theory of relativity or in quantum mechanics, see [31].

3 Principle of Least Constraint

3.1 Method of Least Squares

Carl Friedrich Gauss (1777–1855) (Figure 14) applied his method of least squares in 1801 when he determined the elliptical orbit of the astroid Ceres; however he had developed the basis of this method already in 1795 when he was 18 years old. It was not until 1809 that he published the method in the second volume of his book on the Theory of Celestial Bodies [14] (Figure 15). He said *"Our principle which we have made use of since the year 1795 has lately been published by Legendre in the work Nouvelles methodes..."*. It was a further priority argument in the history of mechanics; however Gauss could present his correspondence with colleagues concerning its use much earlier.

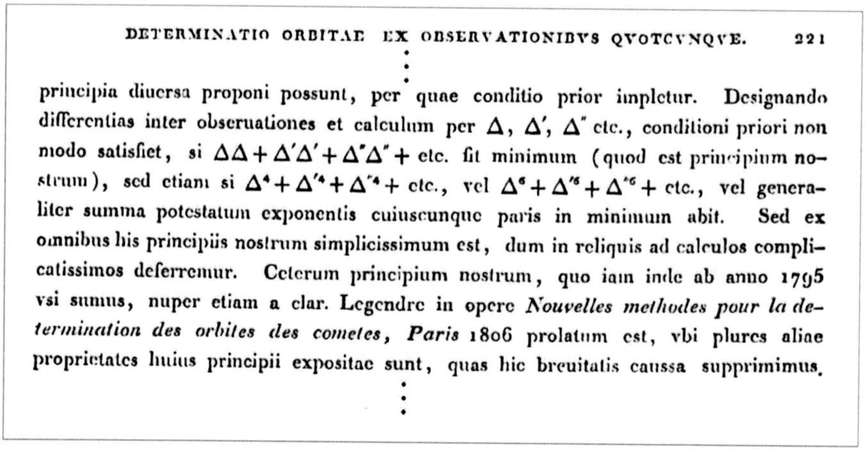

> DETERMINATIO ORBITAE EX OBSERVATIONIBVS QVOTCVNQVE. 221
>
> ⋮
>
> principia diuersa proponi possunt, per quae conditio prior impletur. Designando differentias inter obseruationes et calculum per Δ, Δ', Δ" etc., conditioni priori non modo satisfiet, si ΔΔ + Δ'Δ' + Δ"Δ" + etc. fit minimum (quod est principium nostrum), sed etiam si Δ⁴ + Δ'⁴ + Δ"⁴ + etc., vel Δ⁶ + Δ'⁶ + Δ"⁶ + etc., vel generaliter summa potestatum exponentis cuiuscunque paris in minimum abit. Sed ex omnibus his principiis nostrum simplicissimum est, dum in reliquis ad calculos complicatissimos deferremur. Ceterum principium nostrum, quo iam inde ab anno 1795 vsi sumus, nuper etiam a clar. Legendre in opere *Nouvelles methodes pour la determination des orbites des cometes*, Paris 1806 prolatum est, vbi plures aliae proprietates huius principii expositae sunt, quas hic breuitatis caussa supprimimus.
>
> ⋮

Fig. 15 C.F. Gauss's Method of Least Square (1809), extract from [14]

3.2 Principle of Least Constraint

Twenty years later in 1829 Gauss wrote an essay on "*Über ein neues allgemeines Grundgesetz der Mechanik*" (On a new Fundamental Law of Mechanics) [15]. The paper was published in the 4th issue of *Journal für die reine und angewandte Mathematik (ed. A.L. Crelle)*, a journal still existing today. He describes his new law on four pages that became known as the Principle of Least Constraint (Figure 16), by the way with very little algebra. It says "*The motion of a system of material points... takes place in every moment in maximum accordance with the free movement or under least constraint;...*". He continues "*the measure of constraint, ..., is considered as the sum of products of mass and the square of the deviation to the free motion*".

Applying the usual mathematical notation the acceleration a_i^{free} of the free unconstrained motion is defined by force F_i divided by mass m_i; r is the position vector.

$$\text{free motion} \qquad a_i^{\text{free}} = \frac{F_i}{m_i} \qquad a_i = \ddot{r}_i \qquad (3)$$

$$\text{constrained motion} \qquad a_i \neq a_i^{\text{free}} \qquad (4)$$

The acceleration of the constraint motion is called a_i where kinematic conditions constrain the corresponding material point i. For example a kinematic constraint may be given by a prescribed trajectory. The measure Z for the "constraint" (Zwang) is proportional to the sum of squares of the differences between free and constrained accelerations. Each term in the sum is weighted by mass m_i.

Principles of Least Action and of Least Constraint

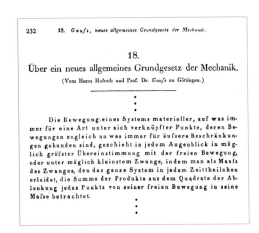

Fig. 16 C.F. Gauss in 1828 and his Paper on the Principle of Least Constraint (1829) [15]

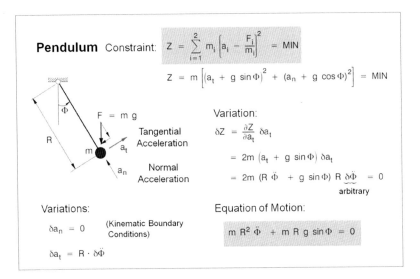

Fig. 17 Application of the Principle of Least Constraint for Pendulum

$$Z \sim \sum_{i=1}^{N} \left(a_i - a_i^{\text{free}} \right)^2 \tag{5}$$

or $\quad Z = \sum_{i=1}^{N} m_i \left(a_i - \frac{F_i}{m_i} \right)^2 = \text{MIN} \quad$ + kinematical conditions, e.g. prescribed displacements $\tag{6}$

According to the principle Z is supposed to be a Minimum. Thus from all possible motions (accelerations) the actual motion leads under given conditions to the least constraint.

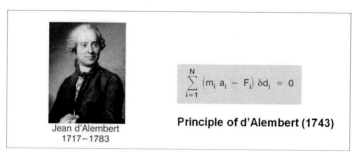

Fig. 18 Principle of d'Alembert

Figure 17 applies the principle to the motion of a pendulum, the circle being the constrained trajectory. As expected the principle yields the equation of motion. This example elucidates that there is a strong relation with d'Alembert's Principle (Jean d'Alembert, 1717-1783). We start from the Principle of Least Constraint and its variation with respect to its free parameters a_i

$$\text{constraint} \qquad Z = \sum_{i=1}^{N} m_i \left(a_i - \underbrace{\frac{F_i}{m_i}}_{\text{fix}} \right)^2 = \text{MIN} \qquad (7)$$

$$\text{variation} \qquad \delta Z = \sum_{i=1}^{N} 2 m_i \left(a_i - \frac{F_i}{m_i} \right) \underbrace{\delta a_i}_{\hat{=}\delta r_i \hat{=} \delta d_i} = 0 \qquad (8)$$

where the position vector and its variation are

$$\text{position vector} \qquad r(t+\mathrm{d}t) = \underbrace{r(t) + v(t)\,\mathrm{d}t}_{\text{fix}} + 1/2\, a(t)\,\mathrm{d}t^2 \qquad (9)$$

$$\text{variation} \qquad \delta r(t+\mathrm{d}t) = 1/2\, \delta a(t)\,\mathrm{d}t^2 \qquad (10)$$

The variation or virtual displacements satisfy the essential boundary conditions. $\delta Z = 0$ directly yields the Principle of d'Alembert (1743) as a variational principle of the equation of motion (Figure 18).

Gauss finishes his paper with the remark (Figure 19) "*It is strange that the free movements, when they cannot withstand the necessary conditions, are modified in the same way as the analyzing mathematician, applying the method of least squares, balances experiences which are based on parameters depending on necessary interactions*", see the comparison in Figure 20.

He adds a final sentence to his paper which says: "*This analogy could be further followed up, but this is currently not my intention*". However he never picked up this matter again.

> Es ist sehr merkwürdig, daſs die freien Bewegungen, wenn sie mit nothwendigen Bedingungen nicht bestehen können, von der Natur gerade auf dieselbe Art modificirt werden, wie der rechnende Mathematiker, nach der Methode der kleinsten Quadrate, Erfahrungen ausgleicht, die sich auf unter einander durch nothwendige Abhängigkeit verknüpfte Gröſsen beziehen. Diese Analogie lieſse sich noch weiter verfolgen, was jedoch gegenwärtig nicht zu meiner Absicht gehört.

Fig. 19 Conclusions of Paper [15]

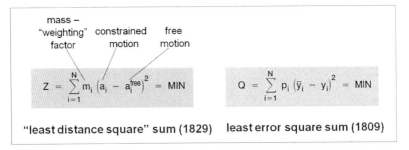

Fig. 20 Principle of Least Constraint and Least Square Method

3.3 Application of Principle of Least Constraint

Gauss's Principle is not very well known although it is mentioned as a fundamental principle in many treatises, e. g. [30, 24, 10, 29], see also [26]; correspondingly it has not been applied too often. Evans and Morriss [10] discuss in detail the application of the Principle for holonomic (constraints depend only on co-ordinates) and nonholonomic constraints (non-integrable constraints on velocity) and conclude *"The correct application of Gauss's principle is limited to arbitrary holonomic constraints and apparently, to nonholonomic constraint functions which are homogeneous functions of the momenta"*. Evans et al. [11] applied the principle in the context of statistical mechanics, see also [4]. As another example the work of Glocker [17, 18] is mentioned where accelerations in rigid multibody systems under set-valued forces are evaluated applying Gauss's principle. Recently Udwadia and co-workers extended the Principle to underdetermined systems [12].

As a final remark the author would like to point out to the possibility of using Gauss's Principle in optimization and design of structures under dynamic excitation where the system ought to be tuned for selected accelerations.

Acknowledgement. The summary of the historical background for two major principles in mechanics can by far not replace any of the excellent rigorous and detailed descriptions on the subject, some of them mentioned in the list of references. This holds in particular for the scientific controversy on the Principle of Least Action. The comprehensive information in the literature is highly appreciated. The author especially acknowledges the permission of the following German libraries for reproducing parts of related historical documents:

- Akademiebibliothek Berlin-Brandenburgische Akademie der Wissenschaften, Berlin (Figures 4, 11);
- Interdisziplinärisches Zentrum für Wissenschafts- und Technikforschung, Bergische Universität Wuppertal (Figure 8);
- Bayrische Staatsbibliothek digital, München (Figures 10 left, 12, 13);
- Sächsische Landesbibliothek - Staats- und Universitätsbibliothek Dresden (SLUB) (Figure 10 right);
- Niedersächsische Staats- und Universitätsbibliothek Göttingen (Figures 15, 16 right, 19);

All images of involved scientists are taken from public domains like Wikipedia, Wikimedia etc. which are invaluable sources. All quoted homepages are dated on January 5, 2013.

References

1. Ariga, N.: Science and Its Public Perception: The Principle of Least Action in Eighteenth-Century Europe. Department of Philosophy and History of Science, Graduate School of Letters, Kyoto University, Japan (2008),
http://www.ariga-kagakushi.info/strage/eastssw2008_article.pdf
2. Berdichevsky, V.: Variational Principles of Continuum Mechanics. Part I Fundamentals. Springer, Berlin (2009)
3. Breger, H.: Über den von Samuel König veröffentlichten Brief zum Prinzip der kleinsten Wirkung (On the Letter published by Samuel König on the Principle of Least Action). In: Hecht, H. (ed.) Pierre Louis Moreau de Maupertuis. Eine Bilanz nach 300 Jahren. Schriftenreihe des Frankreich-Zentrums der Technischen Universität Berlin, pp. 363–381. Verlag Arno Spitz GmbH/Nomos Verlagsgesellschaft, Berlin (1999)
4. Bright, J., Evans, D., Searles, D.: New observations regarding deterministic, time reversible thermostats and Gauss's principles of least constraint. J. Chem Phys. 122(19), 1–8 (2005)
5. Drouet de Maupertuy, J.B.: Maupertuisiana, Collection of pamphlets related to controversy on Principle of Least Actions. Hambourg, Bayrische Staatsbibliothek digital, München, Germany (1753),
http://reader.digitale-sammlungen.de/de/fs2/object/display/bsb10135875_00005.html
6. Dugas, R.: A History of Mechanics. Dover Classics of Science and Mathematics. Dover Publications Mineola, New York (1988)
7. Euler, L.: Methodus inveniendi lineas curvas maximi minimive proprietate gaudentes. Marcum-Michaelem Bousquet & Socios, Lausannæ & Genevæ (1744), Citation from Euler Archive (Mathematical Association of America) E65:
http://www.math.dartmouth.edu/~euler/pages/E065.html,
English Translation of Additamentum II: http://en.wikisource.org/wiki/Methodus_inveniendi/Additamentum_II

8. Euler, L.: Sur le principe de la moindre action (On the Principle of Least Action). In: Histoire de l'Académie Royale des Sciences et des Belles-Lettres de Berlin, p. 199 et seqq. Haude et Spener, Berlin (1751, 1753), Akademiebibliothek Berlin-Brandenburgische Akademie der Wissenschaften, Germany,
http://bibliothek.bbaw.de/bbaw/bibliothek-digital/digitalequellen/schriften/anzeige?band=02-hist/1751,
English Translation: Euler Archive (Mathematical Association of America) E 198 (1753),
http://www.math.dartmouth.edu/~euler/tour/tour_07.html
9. Eulero, L.: Dissertatio de principio minimae actionis una cum examine obiectionum Prof. Koenigii contra hoc principium factarum. Berolini, Ex Officina Michaelis Bayrische Staatsbibliothek digital, München, Germany (1753),
http://www.bsb-muenchen-digital.de/~web/web1008/bsb10080685/images/index.html?digID=bsb10080685&pimage=00010&v=2p&nav=0&l=de,
See also Euler Archive (Mathematical Association of America) E186,
http://www.math.dartmouth.edu/~euler/pages/E186.html
10. Evans, D., Morriss, G.: Statistical Mechanics of Nonequilibrium Liquids, 2nd edn., pp. 33–44. ANU EPress, The Australian National University, Canberra (2007)
11. Evans, D., Hoover, W., Failore, B., Moran, B., Ladd, A.: Nonequilibrium molecular dynamics via Gauss's principle of least contraint. Physical Review A 28, 1016–1021 (1983)
12. Fan, Y., Kabala, R., Natsuyama, H., Udwadia, F.: Reflections on the Gauss Principle of Least Constraint. Journal of Optimization Theory and Applications 127, 475–484 (2005)
13. Feynman, R., Leighton, R., Sands, M.: Lectures on Physics, extended 2nd edn., vol. 2, ch. 19. Addison-Wesley (2005)
14. Gauss, C.F.: Determinatio orbitae observationibus quotcunque quam proxime satisfacientis. In: Theoria Motus Corporum Coelestium in Sectionibus Conicis Solem Ambientium, Göttingen, Liber II, Sectio III, pp. 172–189 (1809); In "'Carl Friedrich Gauss Werke'". Siebenter Band, pp. 236–257, Hrsg. Königl. Gesellschaft der Wissenschaften zu Göttingen in Commission bei B.G. Teubner, Leipzig (1906), Niedersächsische Staats- und Universitätsbibliothek Göttingen, Germany; Signatur: 4 MATH I, 740:4,
http://gdz.sub.uni-goettingen.de/dms/load/img/?PPN=PPN236008730&DMDID=DMDLOG_0017
15. Gauss, C.F.: Über ein neues allgemeines Grundgesetz der Mechanik (On a New Fundamental Law of Mechanics). Journal für die reine und Angewandte Mathematik, herausg. v. CRELLE, Band IV, 232–235 (1829), Niedersächsische Staats- und Universitätsbibliothek Göttingen , Germany; Signatur: 4 MATH I, 3253:7,
http://gdz.sub.uni-goettingen.de/dms/load/img/?PPN=PPN243919689_0004&DMDID=dmdlog23
16. Gerhardt, C.I.: Über die vier Briefe von Leibniz, die Samuel König in dem Appel au public, Leide MDCCLIII, veröffentlicht hat (On the four Letters of Leibniz, which Samuel König published in Appel au public, Leide MDCCLIII). Sitzungsberichte der Königlich Preussischen Akademie der Wissenschaften I, 419–427 (1898)
17. Glocker, C.: The Principles of d'Alembert, Jourdain and Gauss in Nonsmooth Dynamics. Part I: Scleronomic Multibody Systems. ZAMM, Z. Angew. Math. Mech. 78, 21–37 (1998)
18. Glocker, C.: Set-Value Force Laws: Dynamics of Non-Smooth Systems. LNACM. Springer, Heidelberg (2001)

19. Harnack, A.: Geschichte der Königlich Preussischen Akademie der Wissenschaften, Band 1.1: Von der Gründung bis zum Tod Friedrich des Großen (History of Royal Prussian Academy of Sciences, Volume 1.1: From the Foundation until the Death of Frederick the Great), pp. 331–345. Reichsdruckerei, Berlin (1900)
20. Helmholtz, H.: Rede über die Entstehungsgeschichte des Princips der kleinsten Action (Speech on the Evolutionary History of the Principle of Least Action). In: Harnack, A. (ed.) Geschichte der Königlich Preussischen Akademie der Wissenschaften zu Berlin. Band 2 Urkunden und Actenstücke. Akademiebibliothek Berlin-Brandenburgische Akademie der Wissenschaften, Germany, pp. 282–296. Reichsdruckerei, Berlin (1887, 1990),
http://bibliothek.bbaw.de/bbaw/bibliothek-digital/digitalequellen/schriften/anzeige/index_html?band=ak-gesch/harn-2&seite:int=295
21. Kabitz, W.: Über eine in Gotha aufgefundene Abschrift des von S. König in seinem Streite mit Maupertuis und der Akademie veröffentlichten, seinerzeit für unecht erklärten Leibnizbriefes (On a Copy of the Letter of Leibniz, found in Gotha and published by S. König during his Controversy with Maupertuis and the Academy, at that time declared as faked). Sitzungsberichte der Königlich Preussischen Akademie der Wissenschaften II, 632–638 (1913)
22. Knobloch, E.: Das große Spargesetz der Natur: Zur Tragikkomödie zwischen Euler, Voltaire und Maupertuis (The great Law of Conservation in Nature: On the Tragicomedy between Euler, Voltaire and Maupertuis). In: Biegel, G., Klein, A., Sonar, T. (eds.) Leonhard Euler 1707-1783. Mathematiker - Mechaniker - Physiker, pp. 79–89. Braunschweigisches Landesmuseum, Braunschweig (2008)
23. Koenigio, S.: De universali principio aequilibrii et motus, in Vi viva reperto, deque nexu inter Vim vivam et Actionem, utriusque minimo. Dissertatio, Nova Acta Eruditorum, pp. 125–135 (1751), Repertorium deutscher wissenschaftlicher Periodika des 18. Jahrhunderts' of 'Interdisziplinärisches Zentrum für Wissenschafts- und Technikforschung', Bergische Universität Wuppertal, Germany:
http://www.izwtalt.uni-wuppertal.de/repertorium/MS/Acta/NAE1751.pdf#page=128
24. Lanczos, C.: The Variational Principles of Mechanics, 3rd edn., pp. 106–110. University of Toronto Press, Toronto (1966)
25. Leibniz, G.W., quoted by Koenig, S.: Lettre de Mr. de Leibnitz, dout Mr. Koenig a cité le Fragment (Octobre 16, 1707). In: Appel au Public, du Jugement de L'Académie Royal de Berlin, Sur un Fragment de Lettre de Mr. de Leibnitz, cité par Mr. Koenig. 2nd Edition. A Leide de L'Imp. d'Elie Luzac Jun., pp. 166–171 (1753)
26. Linkwitz, K.: Numerical Computation and Principles of Mechanics: C.F. Gauss, the Great Mathematician, Muses Comprehensively on Figures and Formulas. In: Papadrakakis, M., Samartin, A., Oñate, E. (eds.) Proc. IASS-IACM 2000 Colloquium on Computational Methods for Shell and Spatial Structures, Chania, Crete. ISASR, Athens (2000)

27. Moreau de Maupertuis, P.-L.: Les Loix du Mouvement et du Repos déduites d'un Principe Metaphysique. In: Histoire de l'Académie Royal des Sciences et des Belles Lettres, pp. 267–294. Haude, Berlin (1746, 1748), Akademiebibliothek Berlin-Brandenburgische Akademie der Wissenschaften, Germany,
http://bibliothek.bbaw.de/bbaw/bibliothek-digital/digitalequellen/schriften/anzeige/index.html?band=02-hist/1746&seite:int=293, English Translation:
http://en.wikisource.org/wiki/Derivation_of_the_laws_of_motion_and_equilibrium_from_a_metaphysical_principle
28. Pars, L.: A Treatise on Analytical Dynamics. Heinemann, London (1965); reprinted by Ox Bow Press, Woodbridge, CT, USA (1979)
29. Päsler, M.: Prinzipe der Mechanik. Walter de Gruyter & Co., Berlin (1968)
30. Szabo, I.: Geschichte der mechanischen Prinzipien und ihrer wichtigsten Anwendungen (History of Principles of Mechanics and its most important Applications). Birkhäuser, Basel (1979)
31. Taylor, E.F.: Principle of Least Action. MIT (2008),
http://www.eftaylor.com/leastaction.html
32. Voltaire: Histoire du Docteur Akakia et du Natif de St. Malo (Containes Diatribe, Traité de Paix among other documents) (1753), Sächsische Landesbibliothek - Staats- und Universitätsbibliothek Dresden (SLUB), Signatur: SLUB Dresden / Biogr.erud.D.4365,misc.1,
http://digital.slub-dresden.de/werkansicht/dlf/9030/3/cache.off
33. Voltaire: Diatribe du Docteur Akakia, Médecin du Pape, Rome. In: [5], p. 311 et seqq. Bayrische Staatsbibliothek digital, München, Germany (1753),
http://reader.digitale-sammlungen.de/de/fs2/object/display/bsb10135875_00311.html

Lagrange's "Récherches sur la libration de la lune"
– From the Principle of Least Action to Lagrange's Principle

Hartmut Bremer

Abstract. Among the variety of successful eighteenth century contributions, Lagrange's first paper on the moon's libration contains a breakthrough in analytical mechanics. Though it is his very own achievement, Lagrange himself was of course not isolated. Being aware of the results of his predecessors, and being in contact with his scientific contemporaries, he was well prepared for such ground-breaking insight, embedded in the cultural-philosophical surrounding of his time. Behind every named formula and equation there was or is, even in our computer dominated times, always a thinking and feeling human – a fact that we "users", still conneced to our scientific ancestors, should be well aware of.

The historical background, however, is not only fascinating but also yields real benefits: "As far as I studied the basic contributions of our great masters I got insight and understanding which was far beyond everything which could be extracted from secondary sources" says Wilhelm Ostwand (1855–1932).

The formerly tedious search for these sources has become comfortable with the rise of the internet. The corresponding addresses were added to the reference list in order to inspire further reading.

Keywords: Virtual veocities, Principle of Least Action, d'Alembert-Principle, Lagrange-Principle, Méchanique Analytique.

1 Introduction

LAGRANGE's contribution on the "libration of the moon" from 1764 marks a turning point in his research work. JOSEPH LOUIS LAGRANGE was then 28 years old and already scientificly well established. LEONHARD EULER with

---

Hartmut Bremer
Institut für Robotik, Johannes Kepler Universität Linz, Altenberger Straße 69
e-mail: `hartmut.bremer@jku.at`

whom he was in frequent contact was already 57. A vivid impression of the social and cultural events of that time – the young MOZART traveled around as a child prodigy at age eight, the fifteen years old GOETHE was ready to study law at the Leipzig university – a vivid impression is given by MENZEL's famous drawing *Die Tafelrunde Friedrich II von Sanssouci*, printed 1850 (destroyed 1945 in the Flakturm Friedrichshain). Here we see the leading personalities from military and from the Berlin Academy: under the gracious view of the great king, FRANCOIS-MARIE AROUET (VOLTAIRE) speaks to earl FRANCESCO ALGAROTTI right across the table. Everyone is spellbound by VOLTAIRE – except JAMES KEITH (LORD MARISCHALL) (far left) and CHARLES DE BOYER (EARL D'ARGENS) who jokes with JULIEN DE LA METTRIE (far right).

Fig. 1 Adolph von Menzel: Die Tafelrunde Friedrich II in Sanssouci (in part)

ALGAROTTI was well-liked by VOLTAIRE. Probably because of of this: VOLTAIRE loved the women, and ALGAROTTI had written a book *il Newtonianismo per le dame* (Newton for ladies) in 1739, so to say a precursor of the *women–for–technology*-movement. LA METTRIE, in 1748, had also published a monography: *l'homme machine*. His "Machine Man" gained attention, if only, in the robotics community, but not by VOLTAIRE, whose comment "LA METTRIE was a fool, and his profession was to be a fool", with which he characterized him posthumously, was not really polite. Nor was it how he mocked PIERRE-LOUIS DE MAUPERTUIS, the president of the academy, until MAUPERTUIS gave up in 1753 as an embittered man and moved to his friend JOHN II BERNOULLI in Basel 1756. (MAUPERTUIS is not to be seen on MENZEL's drawing but instead his successor MARQUIS D'ARGENS).

Also VOLTAIRE, who had come to Potsdam in 1750, had to leave the court as a consequence of that inglorious affair. He was dishonourly dismissed 1753 by FRIEDRICH. The scene in MENZEL's picture must therefore have been before 1753. Because JAMES KEITH was sent as an ambassador to Paris in 1751, and because the hilarious LA METTRIE died the same year, we may assume that the scene depicted by ADOLF MENZEL took place in 1751. In that year, the fifteen years old LAGRANGE became a professor at the Turin Artillery School (other sources say with 16 or with 19, respectively, after H. SERVUS, [1]).

EULER, only director of the mathematical class, is of course not depicted in MENZEL's painting. (Curious: the year on EULER's grave reads MDCCCXXXVII (1837). He would then have been one hundred and thirty years old. Did the stonecutter make a mistake in the fifth and ninth digit instead of writing MDCCLXXXIII (1783)? The answer: on the rear side of the tombstone one finds the correct data). When EULER 1766 moved back to the great KATHARINA in St. Petersburg, LAGRANGE with thirty years became his successor. One year after FRIEDRICH's death, in 1786, he moved to Paris, towards the french revolution. (He survived in good shape and was even ennobled by NAPOLEON). LAGRANGE had already finished his *Méchanique Analytique* in Berlin.

2 The Award

In 1762, the French Academy posed a question, which LAGRANGE on page six of his contribution summaries in short: "will it be possible to explain, by physical reasons, why the moon always shows us the same face, and how one can determine, by theory and observation, whether this planet moves similar to the earth whose axis performs a precession of the equinoxes and a nutation." And he adds: "I am keen to present the results of my research on that essential matter to it [the academy]. If they do not conform to the opinion of that wise society, they might at least throw a new light on one of the most important phenomena in celestial mechanics." As we know today, LAGRANGE's contribution did conform to the academy's opinion, he received the great award, and his work was published 1764, [2].

On page 10 he praises JOHANN BERNOULLI and his principle of virtual velocities in statics. Its origin reaches far back in time: GUIDO UBALDI [MARCHESE DEL MONTE] (1545-1607) is "denoted the first one who used the priciple of virtual velocities which had already been hinted at by ARISTOTELES and reexamined by LEONARDO DA VINCI" ([3], p.49/50). "The ... principle has

> Tels sont, en abrégé, les points principaux de la Dissertation suivante. L'Académie Royale des Sciences ayant proposé pour le sujet du Prix de l'année prochaine : « Si l'on peut expliquer par quelque raison physique pourquoi la Lune nous présente toujours à peu près la même face; et comment on peut déterminer par les observations et par la théorie si l'axe de cette Planète est sujet à quelque mouvement propre, semblable à celui qu'on connait dans l'axe de la Terre, et qui produit la précession des équinoxes et la nutation »; j'ose lui présenter le fruit de mon travail sur cette importante matière. S'il ne répond pas entièrement aux vues de cette savante Compagnie, au moins servira-t-il à jeter de nouvelles lumières sur un des principaux phénomènes célestes.

Fig. 2 The Prize Question (Page 6 of the "récherches")

been explained by JOHANN BERNOULLI in a letter to VARIGNON dated 1717 ... It hereby has o be emphasized that Johann B. used the term virtual velocity (vitesse virtuelle) for the first time" ([3], p.149. VARIGNON dates that letter to January 26, 1717 [4], p.174).

Afterwards, LAGRANGE speaks about MAUPERTUIS' *principle of rest* (memoirs of the French Academy 1740, [5]) and its generalization by EULER in the memoirs of the Prussian Academy 1751, [6]. From this it follows clearly that it was originally important to LAGRANGE to develop mechanics from the principle of least action, which was one of the most established principles at that time. This is confirmed by his contributions in the Miscellanea Taurinensis (volume II, printed 1762): In *essai d'une nouvelle méthode...* ([7], considering a new method to find the maxima and the minima of undetermined integral forms, German translation by [8]), the Brachistochrone problem is considered, and in *"application de la méthode...*, [9], it is applied to the principle of least action.

LAGRANGE had communicated that new method (*nouvelle méthode*) to EULER already in 1755 (letter dated November 20, [10]) when he was nineteen years old. He later reports in Miscellanea Taurinensis ([8], part II, p.4): "I have to emphasize, because this method needs to vary the same terms in two different ways, in order not to confuse these variations I have introduced a new characteristic δ. Thus, δZ represents a differential with respect to Z, which is not the same as dZ, but which is nevertheless calculated by the same rules, such that, if any $dZ = m\,dx$ exists, one has simultaneously $\delta Z = m\,\delta x$, and this holds for any other equation." (In his *Analytical Mechanics*, LAGRANGE expresses himself similarly, [1], p. 65 and p. 205, resp.).

The field of application of the principle of least action, however, is quite narrow because, as a variational principle, it requires integrability of the

Fig. 3 Coordinate Systems. LAGRANGE intentionally refrains from using drawings: his aim is – contrary to the geometrical methods of that time – to proceed analytically. Nevertheless, a small sketch will be helpful to characterize the problem under consideration: he chooses an inertial frame which is parallel to the ecliptic (γ: first point of Aries); furthermore L: moon ("lune"), T: earth ("terre"), S: sun ("soleil"), and α: a mass element of the moon.

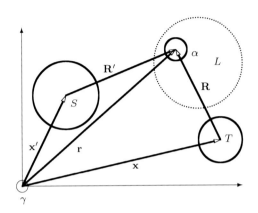

impressed forces, thus being restricted to conservative (potential) forces. In his contribution of 1764, LAGRANGE then consequently applies his calculation rules – and leaves the minimal principles behind.

(The coordinate system is, moreover, shifted to the moon's gravity center, which is, however, of no interest here). In Fig.3, $\mathbf{r} = (X\ Y\ Z)^T$ comprises the components X, Y, Z that are used by LAGRANGE. And now one thing quickly leads to another:

On page 9 we find the equation, marked with (A), which is the core of all following considerations (where $R = |\mathbf{R}|$ und $R' = |\mathbf{R}'|$ according to Fig.3):

$$(A) \quad \frac{1}{dt^2}\int \alpha(d^2X\,\delta X + d^2Y\,\delta Y + d^2Z\,\delta Z) + T\int \frac{\alpha\,\delta R}{R^2} + S\int \frac{\alpha\,\delta R'}{R'^2} = 0.$$

Les quantités δX, δY, δZ, δR, δR' ne sont autre chose que les différentielles des lignes X, Y, Z prises à l'ordinaire et affectées de la caractéristique δ au lieu de la commune d, pour les distinguer des autres différentielles des mêmes lignes qui ont rapport au mouvement réel du corps.

Fig. 4 The Method (Page 9 of the "récherches")

"The terms $\delta X, \delta Y, \delta Z, \delta R, \delta R'$ do not represent anything else than the differentials assigned to δ instead to the common d, in order to distinguish them from the differentials of the same lines which refer to the *real* motions of the bodies".

Calculating the moon problem (p.13–S.61), LAGRANGE uses "Euler angles" ω, ϵ, π as independent variables to express X, Y, Z, R, R' and their variations. The solution process, however, is not of interest here (for more information see e.g. [11]); we are focusing on the method itself. In today's notation we would first refer $(S\alpha/R'^2)\delta R'$ to the variable $\mathbf{r}$, i.e. $(S\alpha/R'^2)(\partial R'/\partial \mathbf{r})\delta \mathbf{r} := d\mathbf{f}_S^{e\,T}\delta\mathbf{r}$, where $d\mathbf{f}_S^e$ is the impressed force exerted by the sun. Combined with the one from the earth into $d\mathbf{f}^e$ and along with dm instead of α, LAGRANGE's equation (A) reads

$$\int_{(S)} (\ddot{\mathbf{r}}\,dm - d\mathbf{f}^e)^T \delta\mathbf{r} = 0 \qquad (1)$$

where (S) denotes the "system" under consideration. The "system" here consists of one single body, the moon, because feedback reactions on the sun and on the earth are not considered. On page 12, however, LAGRANGE resumes:

> **12** RECHERCHES
>
> Au reste le principe de Statique que je viens d'exposer, étant combiné avec le principe de Dynamique donné par M. d'Alembert, constitue une espèce de formule générale qui renferme la solution de tous les Problèmes qui regardent le mouvement des corps. Car on aura toujours une équation semblable à l'équation (A) (Article précédent), et toute la difficulté ne consistera plus qu'à trouver l'expression analytique des forces qu'on suppose agir sur les corps et des lignes suivant lesquelles ces forces agissent, en n'employant dans ces expressions que le plus petit nombre possible de variables indéterminées, de manière que leurs différentielles désignées par le δ soient entièrement indépendantes les unes des autres; après quoi, faisant séparément égaux à zéro les termes qui se trouveront multipliés par chacune des différentielles dont je parle, on aura tout d'un coup autant d'équations particulières qu'il en faudra pour la solution du Problème, comme on le verra dans les Articles qui suivent.

Fig. 5 The Method (Page 12 of the "récherches")

Eventually, the principle in statics which I have just explained, combined with the principle of dynamics, which has been given by mister d'Alembert, will lead to a general formula which comprises the solution of all problems in moving bodies. This is because one will always have an equation similar to (A) (last paragraph), and the whole difficulty will only be to find an analytical expression for the forces that one assumes to act, and their lines of action, using the minimal number of undetermined variables such that those differentials that are expressed by δ will be totally independent one from each other; by separately setting all those terms equal to zero that one finds multiplied with the mentioned differentials, one obtains at once as many individual equations as are necessary to solve the problem, as will be seen in the following articles.

That means in today's notation:

- formulate force laws $d\mathbf{f}^e \in \mathbb{R}^3$,
- introduce minimal variables $\mathbf{q} \in \mathbb{R}^f \Rightarrow \mathbf{r} = \mathbf{r}(\mathbf{q}) \in \mathbb{R}^3$,
- formulate the equations of motion via $\delta \mathbf{r} = (\partial \mathbf{r}/\partial \mathbf{q})\delta \mathbf{q}$, δq_i independent, $i = 1 \cdots f$.

Equation (1) represents the LAGRANGE-Principle "... which up to the time being has been confused with a formulation of D'ALEMBERT's Principle" says KARL HEUN,[12]. (HEUN(1859–1929) studied mathematics and physics in Göttingen, afterwards he was a private lecturer in Munich, teacher in Berlin, and from 1902 on, on the recommendation of FELIX KLEIN(1849–1925), professor for theoretical mechanics in Karlsruhe. One of his co-workers was GEORG HAMEL (1877–1954), habilitation 1903, afterwards Brünn, Aachen and Berlin, since 1938 member of the Prussian and 1953 of the Bavarian Academy). HAMEL confirms: "the *combination* [Fig.5, first line: *étant combiné*] of the two principles, which originally do not have anything in common, has eventually been given by LAGRANGE. It is therefore also called the *"Lagrangean Principle"*, [13]. Though today's contemporaries mostly call eq.(1) the "d'Alembert-Principle in the Lagrangean form" or, lazily, "d'Alembert-Principle", this does neither justice to the matter, nor to LAGRANGE's achievement. LAGRANGE's result is brandnew, and it is a breakthrough in his attempts to free himself from the stranglehold of the least action principle with its restrictive potential forces.

The difference becomes very clear when compared to the original D'ALEMBERT-Principle. For the reason of convenience we translate the German text from [14]:

- *"Decompose the motions (momentums)* **a, b, c** *which are impressed on each body (particle) each into two others* ***a***,**α**; ***b***,**β**; ***c***,**γ** *etc. in such a way that, if only the motions **a,b,c** would have been impressed, the bodies would be able to maintain their motion without hindering each other and, if one would impress on these only the motions* **α,β,γ**, *then the system would stay at rest"*.

Though the forces here are not restricted to potential ones, D'ALEMBERT never got around the problem of decomposing the forces into those which perform motion and those which do not. This makes the application of his principle extremely complicated and even unfeasible for larger systems.

It should furthermore be mentioned that LAGRANGE on page 11 considers the "living forces" (energy conservation) by replacing his δ with the common d (see Fig.6).

The last sentence there is noteworthy: "$V, V', V'', ...$ are the initial values of $v, v', v'', ...$; and this equation yields, as one sees, the conservation of the living forces *in its entire extension*". Here he anticipates, in principle, HELMHOLTZ's results (generalized energy conservation, [15]). At least, however, he demonstrates the superiority of his analytical procedure with a hint on D'ALEMBERT's *traité de dynamique*, "where he [d'Alembert] at the end of his *dynamique* had remarked it [the energy conservation principle]": In the *traité de dynamique* (edition 1758, [16]) we find on twenty pages the energy conservation for various cases in the common, today nearly incomprehensible geometrical representation of his time.

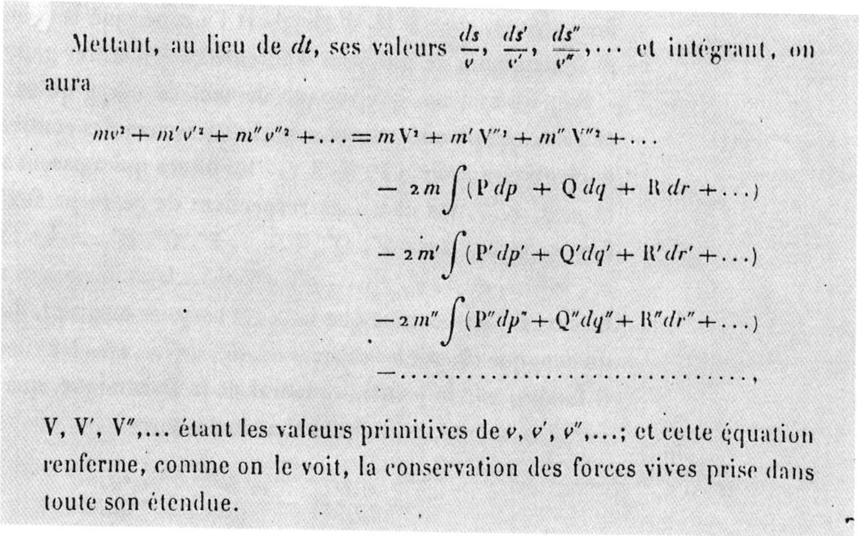

Mettant, au lieu de dt, ses valeurs $\frac{ds}{v}, \frac{ds'}{v'}, \frac{ds''}{v''}, ...$ et intégrant, on aura

$$mv^2 + m'v'^2 + m''v''^2 + ... = mV^2 + m'V'^2 + m''V''^2 + ...$$
$$- 2m \int (P\,dp + Q\,dq + R\,dr + ...)$$
$$- 2m' \int (P'\,dp' + Q'\,dq' + R'\,dr' + ...)$$
$$- 2m'' \int (P''\,dp'' + Q''\,dq'' + R''\,dr'' + ...)$$
$$- ...,$$

$V, V', V'', ...$ étant les valeurs primitives de $v, v', v'', ...$; et cette équation renferme, comme on le voit, la conservation des forces vives prise dans toute son étendue.

Fig. 6 Conservation of Living Forces (Page 11 of the "récherches")

3 Evaluation of the Récherches

Subsequently, LAGRANGE builds his analytical mechanics consequently on the basis of his principle (1). In 1780, he improves the results from 1764 in his *théorie de la libration de la lune*, [17] (here, the LAGRANGE-Equations (of the second kind) appear for the first time), and 1788 he publishes the first edition of his *Méchanique Analytique* in Paris (printed by Desaint (Widow)).

The first German translation followed already nine years later by FRIEDRICH WILHELM AUGUST MURHARD (1779–1853), [18]. (Noteworthy: the first English translation two hundred years later, in the year 1997, by Kluwer).

MURHARD, "mathematician, later journalist and researcher in constitutional law", [19], had aptly been educated by JOHANN MATTHIAS MATSKO (1717–1797). He moved to Göttingen 1795 to continue his studies with KÄSTNER (1719–1800) and LICHTENBERG (1742–1799). (MATSKO himself attained his scientific education from J.A. VON SEGNER in Göttingen and was 1767 appointed to a professorship in Kassel, the hometown of MURHARD. SEGNER became 1755 successor of CHRISTIAN WOLFF in Halle; KÄSTNER overtook his chair in Göttingen and LICHTENBERG, one of his students, became 1770 (1780) professor for mathematics (physics). The famous "Lichtenberg–Figures" caused sensation all over Germany and inspired FLORENS FRIEDRICH CHLADNI (1756–1827), to his well-known "Chladni–Figures", which eventually, in 1909, were mathematically verified and completed by WALTHER RITZ (1878–1909)).

Already 1796, the seventeen years old MURHARD became Magister and gave lectures. In 1797, he was appointed Assessor by the Royal Society of Sciences in Göttingen. The fact that KÄSTNER 1765 had written the first textbook on mechanics in German, [20], may have inspired MURHARD to translate the LAGRANGEan opus.

The LAGRANGEan *Méchanique Analytique* had two further editions (entitled *Mécanique Analytique*, without h) in 1811 (first volume) and 1815, respectively (second volume, posthumously finished by PRONY, GARNIER and LACROIX). It was released 1887 in German by HERMANN SERVUS. A comparison with MURHARD's translation by random samples shows that the latter (subtitled "translated from French with some comments and explanatory additions") has to be called accurate. A typical passage by SERVUS (p.242) is for instance

- Ist ferner s der von dem Körper m in der Zeit t durchlaufene Raum oder Bogen, so ist
$$ds = \sqrt{dx^2 + dy^2 + dz^2} \text{ und } dt = \frac{ds}{u},$$
also wird
$$d^2x\delta x + d^2y\delta y + d^2z\delta z = d(dx\delta x + dy\delta y + dz\delta z) - ds\delta ds$$
und daraus
$$\frac{d^2x}{dt^2}\delta x + \frac{d^2y}{dt^2}\delta y + \frac{d^2z}{dt^2}\delta z = \frac{d(dx\delta x + dy\delta y + dz\delta z)}{dt^2} - \frac{u^2\delta ds}{ds}.$$

MURHARD translates (p.216):

- Es sey nun s der Raum oder der durch den Körper m in der Zeit t beschriebene krummlinige Bogen; alsdenn hat man
$$ds = \sqrt{dx^2 + dy^2 + dz^2} \text{ und } dt = \frac{ds}{u};$$

folglich

$$d^2x\delta x + d^2y\delta y + d^2z\delta z = d(dx\delta x + dy\delta y + dz\delta z) - ds\delta ds;$$

und (*wenn man mit dt^2 überall dividirt und nur beim letzten Gliede $ds\delta ds$ statt dt^2 seinen Werth $\frac{ds^2}{u^2}$ setzt, M.*)

$$\frac{d^2x}{dt^2}\delta x + \frac{d^2y}{dt^2}\delta y + \frac{d^2z}{dt^2}\delta z = \frac{d(dx\delta x + dy\delta y + dz\delta z)}{dt^2} - \frac{u^2\delta ds}{ds}.$$

The underlined passage (with the initial M.) is one of the typical comments by MURHARD: "dividing by dt^2 everywhere except for the last term $ds\delta ds$ where instead of dt^2 its value $\frac{ds^2}{u^2}$ is used" – dispensable, but worthy of an eighteen year young man.

MURHARD was an ardent admirer of the ideas of the French Revolution and was therefore frequently arrested. He was eventually, so to say, pulverized in the political mills of his time and only came free after the German revolution in 1848 via a general amnesty. Together with his brother JOHANN KARL ADAM MURHARD they entailed their properties to the town of Kassel to constitute a research library (namely for constitutional law contributions), the well-known "Murhard–Library". MURHARD's activities in natural sciences are, regrettably, no longer mentioned today.

The systematic evaluation of LAGRANGE's *récherches* in his *analytical mechanics* is impressively demonstrated by its contents (here translated from SERVUS' German edition, page XXVIII, XXIX):

Section I. **On the different principles in dynamics** 183
Section II. **General formula of dynamics for the motions of a system of bodies which are driven by some arbitrary forces** 203
Section III. **General properties of motions, derived from the equations of motion** 211

§ 1. Properties concerning the mass center 211
§ 2. Properties related to surfaces 215
§ 3. Properties of rotations, created by instantaneous forces 222
§ 4. Properties of fixed rotational axes of a free body
    with arbitrary shape 227
§ 5. Properties concerning the living forces 235
§ 6. Properties from the least action principle 241

Here, the central role of the LAGRANGEan principle (the "general formula") becomes evident. From this all subpoints are deduced, not least the principle of least action. (It is remarkable that LAGRANGE derives the gyro equations using minimal coordinates (EULER-angles) as well as with the aid of quasi-coordinates (angular velocities) coming out with EULER's equations (of course he acknowledges EULER's contributions, see p. 547 of SERVUS' translation of the second edition from 1815. This edition, however, had been completed

Lagrange's "Récherches sur la libration de la lune"   55

(from p. 523 on) by the publishers using LAGRANGE's handwritten notes in his personal copy. LAGRANGE, until his death in 1813, had only been able to revise the first volume and about half of the second one).

As mentioned above, LAGRANGE had written the first exemplar of his *méchanique* still in Berlin. But he wanted it to be printed in Paris because he assumed a better print style there. Therefore, he sent a copy to Abbé Marie and commissioned him to look for a publisher. "One can really not believe that he did not find an editor. At long last Desaint accepted the oeuvre, but only under the condition that Lagrange, after a fixed time period, should buy the remaining unsold copies himself. Thus, the monograph appeared in 1788, but with little success; the ideas and theories were too new and the comprehending readers too few," (H. SERVUS, [1], p. XIX, XX).

4 Comprehending Readers Too Few...

Some samples: It already starts with MURHARD. In his "Vorerinnerung" he writes

1797: "LAGRANGE wrote for the thinking mathematician, who will always find something to enjoy, even in the absence of any practical usefulness ... Of course one may not look for benefits or applications in every-day life, it will be useless to calculate even a lever from it ..."

The lever will once more play a role about two hundred years later. As careful as MURHARD's translation was, he probably did not understand its contents so much.

Nine years after that, POINSOT wrote an article, which he repeated word by word in his textbook from 1837, [21]: on page 476 he says

1806: (1837) "... multiplied by the real velocities ... this principle does not leave any trace of infinitesimal motions, which seem strange ... and which leave something obscure in one's mind."

He thus replaces LAGRANGE's variations with the real velocities. A severe mistake! Doing so one obtains the energy conservation principle (as already had been shown by LAGRANGE himself, see Fig.6). By this, all forces that do not perform work are eliminated, not only the constraint forces but also the CORIOLIS–"forces". Did POINSOT really read the *récherches* or the *méchanique*? If so, at least not very careful.

Five years later, JACOBI criticizes the missing proof of force laws in LAGRANGE's contributions (Königsberg lectures [22], Berlin lectures [23]):

1842: (1847) "... because mathematics cannot cook up the characteristics of forces."

In our opinion, force laws fall into the domain of modelling. Asking for a proof of modelling decisions is a really strange request.

In the year 1964, TRUESDELL, that hot-blooded contemporary, offers LAGRANGE a shameful indictment in favour of his protagonist EULER, [24]:

1964: "General principles or notions of mechanics are misunderstood or neglected by Lagrange ... The Mécanique Analytique did not contain many novelties; its contents are mostly from earlier articles of Lagrange himself or from contributions of Euler or other predecessors."

This is almost an accusation of plagiarism! Here – hot-bloodedness aside – TRUESDELL misses the mark by far.

Around twenty years later, THOMAS KANE prevails over LAGRANGE in [25] when he states:

1986: "among the most obscure concepts ... are those marked by the word "virtual". .. They are the closest thing in dynamics to black magic, and they are the heart of Lagrange's approach."

It is needless to comment on that.

Finally, in 1998, the lever surfaces again: H. Pulte, [26], asserts:

1998: "Lagrange's programme – an overambitious exercise. A challenge for numerous mathematicians. They found a better foundation which reduced the principle of virtual velocities to the lever principle which contains Lagrange's principle as a simple corollary."

We are thus back with ARCHIMEDES – 250 years before Christ.

These are, starting with MURHARD, more than 200 years of "too few comprehending readers". Granted: some of LAGRANGE's formulations sound curious, starting already with his variations from 1762. But, of course, mechanics is not completed with LAGRANGE, nobody did ever assert that. "Large parts of mechanics will remain open to further development and some might even be in need of it" (KURT MAGNUS, [29]).

Because $(\partial\Phi/\partial\mathbf{r})d\mathbf{r} = 0$ represents a homogeneous equation, one can just as well write $(\partial\Phi/\partial\mathbf{r})(d\mathbf{r}/dt) = (\partial\Phi/\partial\mathbf{r})\dot{\mathbf{r}} = 0$. Comparison then yields $(\partial\Phi/\partial\mathbf{r})\dot{\mathbf{r}} = 0 \Rightarrow (\partial\Phi/\partial\mathbf{r})\delta\mathbf{r} = 0$. The unloved infinitesimality (POINSOT) disappers, and $\delta\mathbf{r}$ now characterizes a (finite) tangential vector with respect to the constraint surface. It has, however, not to coincide with $\dot{\mathbf{r}}$, see Fig.7. This is precisely the ingenuity of LAGRANGE's approach.

LAGRANGE's principle (1) represents a "raw form" because it still needs an integration $\int_{(S)}$. But if we, for the general case, introduce nonholonomic variables, then we obtain with only few steps a *central equation* [27]

$$\frac{d}{dt}\left[\left(\frac{\partial T}{\partial \dot{\mathbf{s}}}\right)\delta\mathbf{s}\right] - \delta T - \delta W = 0 \tag{2}$$

(T: kinetic energy, δW: virtual work, $\dot{\mathbf{s}} = \mathbf{H}(\mathbf{q})\dot{\mathbf{q}}$: (potentially nonholonomic) minimal velocities, which represent a linear combination of $\dot{\mathbf{q}}$, $\mathbf{q}$: minimal

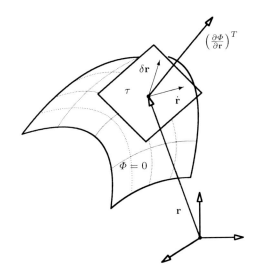

Fig. 7 Constraint Surface. We will probably advance a bit if we do not relate the obscure and black magical "differentials" $\delta \mathbf{r}$ to $d\mathbf{r}$ but to $d\mathbf{r}/dt$: The admissible $\delta \mathbf{r}$ have to fulfill a constraint equation $\Phi = 0$. They are built, according to LAGRANGE, by comparison with the real differentials: $(\partial \Phi/\partial \mathbf{r})d\mathbf{r} = 0 \Rightarrow (\partial \Phi/\partial \mathbf{r})\delta \mathbf{r} = 0$ (*such that, if any $dZ = m\,dx$ exists, one has simultaneously $\delta Z = m\,\delta x$, see above*).

coordinates). From eq.(2) one obtains a considerable body of procedures, just by elementary calculation (– also the EULERian one, Dr. TRUESDELL!).

By the way, as we later realized, HAMEL had published the relation (2) in an equivalent representation already in 1904, [28][1].

5 Conclusion

"To obtain the afore mentioned facts so easily and clearly is mainly due to LAGRANGE himself" states HAMEL, [13], and to this day, LAGRANGE's principle (1) offers an unsurpassed powerful tool. *The short contribution from 1764 marks a milestone in mechanics.*

Acknowledgements. Support of the Austrian Center of Competence in Mechatronics (ACCM) is gratefully acknowledged.

References

1. de Lagrange, J.: Analytische Mechanik, 2nd edn., German by H. Servus. Springer, Berlin (1897)
2. de Lagrange, J.: Récherches sur la libration de la lune. Oeuvres de Lagrange 6 (1873)
3. Rühlmann, M.: Vorträge über Geschichte der Technischen Mechanik. Baumgärtners Buchhandlung, Leipzig (1885)

---

[1] I am grateful to Prof. John Papastavridis, Georgia Tech., for this information. John is the most knowledgeable scientist I have met.

4. de Varignon, P.: Nouvelle Mécanique ou Statique, vol. 2. Claude Jombert, Paris (1725), http://imgbase-scd-ulp.u-strasbg.fr
5. de Maupertuis, P.: Loi du repos des corps. Académie Royale des Sciences, Paris (1740), http://www.academie-sciences.fr
6. Euler, L.: Harmonie entre les principes généraux de repos et de mouvement de M. de Maupertuis. Mémoires de l'Academie des Sciences, Berlin (1751), http://www.math.dartmouth.edu/~euler/
7. de Lagrange, J.: Essai d'une nouvelle méthode pour déterminer les maxima et les minima des formules intégrales indéfinies. Oeuvres de Lagrange 1, 389 (1873), http://gdz.sub.uni-goettingen.de
8. P. Stäckel (ed.): Variationsrechnung. Wissenschaftliche Buchgesellschaft, Darmstadt (1976)
9. de Lagrange, J.: Application de la méthode exposée dans le mémoire précédent a la solution de différents problèmes de dynamique. Oeuvres de Lagrange 1, 419 (1873), http://gdz.sub.uni-goettingen.de
10. de Lagrange, J.: Correspondance. Oeuvres de Lagrange 14, 146–151, 152–154 (1892), http://gdz.sub.uni-goettingen.de
11. Fraser, C.: J.L. Lagrange's Early Contributions to the Principles and Methods of Mechanics. Arch. Hist. Ex. Sci. 28(3), 197–241 (1983)
12. Heun, K.: Ansätze und allgemeine Methoden der Systemmechanik. Enzykl. Math. Wiss. IV(II), 361 (1913)
13. Hamel, G.: Theoretische Mechanik, p. 219. Springer, Heidelberg (1949)
14. Heun, K.: Die Bedeutung des d'Alembertschen Prinzips für starre Systeme und Gelenkmechanismen. Archiv der Math.u.Phys. III(2), 61 (1901)
15. Helmholtz, H.: Über die Erhaltung der Kraft. G. Reimer, Berlin (1849), Ostwald's Klassiker, vol. 1. W. Engelmann, Leipzig (1889)
16. d'Alembert, J.: Traité dynamique, 2nd edn., pp. 252–272. David, Paris (1758), http://fr.wikisource.org
17. de Lagrange, J.: Théorie de la libration de la lune. Oeuvres de Lagrange 5 (1870)
18. de Lagrange, J.: Analytische Mechanik, 1st edn. German by F.W.A. Murhard. Vandenhoeck und Ruprecht, Göttingen (1797), http://gdz.sub.uni-goettingen.de
19. Allgemeine Deutsche Biographie, vol. 23, pp. 62–65. Duncker und Humblot, Leipzig (1886)
20. Kästner, A.G.: Anfangsgründe der höhern Mechanik, 2nd edn. Vandenhoek und Ruprecht, Göttingen (1793)
21. Poinsot, L.: Éleméns de statique, p. 476. Bachellier, Paris (1837)
22. Jacobi, C.G.J.: Vorlesungen über Mechanik, A. Clebsch, ed. G. Reimer, Berlin (1866)
23. Jacobi, C.G.J.: Vorlesungen über analytische Mechanik Berlin 1847/48. In: Pulte, H. (ed.) Dokumente zur Geschichte der Mathematik, vol. 8. Vieweg, Braunschweig (1996)
24. Truesdell, C.: Die Entwicklung des Drallsatzes. ZAMM 44(3/4), 149–159 (1964)
25. Radetsky, P.: The Man who Mastered Motion. Science, 52–60 (May 1986)
26. Pulte, H.: Jacobi's Criticism of Lagrange: The Changing Role of Mathematics in the Foundations of Mechanics. Historia Mathematica 25, 154–184 (1998)

27. Bremer, H.: Über eine Zentralgleichung in der Dynamik. ZAMM 68, 307–311 (1988)
28. Hamel, G.: Über die virtuellen Verschiebungen in der Mechanik. Math. Annalen 59, 416–434 (1904)
29. Magnus, K.: Vom Wandel unserer Auffassung zur Technischen Mechanik. THD-Schriftenreihe 28, 125–136 (1986)

The Development of Analytical Mechanics by Euler, Lagrange and Hamilton – From a Student's Point of View

Maximilian Gerstner, Patrick R. Schmitt, and Paul Steinmann

Abstract. The techniques and methods that were devised during the development of 'Analytical Mechanics' still feature prominently in many branches of Physics, Engineering Sciences and also in various technical applications. In this paper we give an overview of the historical development of this theory by Leonhard Euler (1707-1783), Joseph-Louis Lagrange (1736-1813) and Sir William Rowan Hamilton (1805-1865). The simple example of the pendulum or more correctly the mathematical pendulum serves to illustrate this development. In the conclusion we present a glimpse of the further development this theory has undergone in modern mathematics and physics and discuss some select use cases.

1 Introduction

This paper aims to give a concise overview of the historical development of mathematical modelling in the area of mechanics, especially of 'Analytical Mechanics'. Starting from elementary experiments conducted by Galilei the axiomatization of mechanics was first discussed in Newton's 'Philosophiae naturalis principia mathematica' (1687) [1]. Two papers by Johann Bernoulli and Jacob Hermann concerned with the inverse problem for central forces in 1710 lead the way to a proper 'Analytical Mechanics'.

A first culmination of this development is Euler's systematic application of Leibniz's differential calculus to a small sample mass in his treatise 'Mechanica sive motus scientia analytice exposita' (1736) [2].

Maximilian Gerstner · Paul Steinmann
Chair of Applied Mechanics, University of Erlangen-Nuremberg,
Egerlandstraße 5, D-91058 Erlangen, Germany
e-mail: paul.steinmann@ltm.uni-erlangen.de

Patrick R. Schmitt
Department of Mechanical Engineering,
University of Erlangen-Nuremberg,
Haberstraße 2, D-91058 Erlangen, Germany

Lagrange's 'Mécanique analytique' (1788) [3] then proposes a closed, very mathematical framework in which the dynamics of a system is described solely by a single scalar function - the so-called Lagrangian function or short Lagrangian. Using the theory of variations with classes of admissible functions as developed by Lagrange the equations of motion of the system can be determined via the principle of stationary action as the Euler-Lagrange equations of the variational problem. This approach simplifies many problems because constraints can be dealt with more easily than in Newton's approach. Constraint forces can be either calculated explicitly (Lagrange's equations of the first kind) or be taken into account by choosing suitable generalized coordinates (Lagrange's equations of the second kind).

Hamilton's 'On a General Method in Dynamics' (1834 and 1835) [4, 5] represents a geometrized alternative to Lagrange's formalism by introducing the concept of phase space. Hamilton describes the dynamics of a system with a scalar function called Hamilton's function or short Hamiltonian. The Hamiltonian is a Legendre transform of the Lagrangian replacing the generalized velocities by generalized momenta also known as conjugate momenta. This leads to a reduction of the order of the occuring differential equations from two to one but at the same time doubles the number of equations of motion, now called Hamilton's equations.

To illustrate the merits but also the disadvantages of these different formulations of 'Analytical Mechanics' we will repeatedly consider the simple pendulum (or better mathematical pendulum) as an example.

We will use a modern notaion throughout, e.g. we will write the derivative of a variable x with respect to time t formally as $\frac{d}{dt}x$ and abbreviated this as $\dot{x}$.

2 The Beginnings of the Axiomatization of Mechanics

The very first quantitative experiments especially concerning velocities during free fall and motion on inclined planes performed by the Italian polymath Galileo Galilei (1564-1642) mark the hour of birth of modern mechanics. Galilei was the first modern thinker that realized that the laws of nature are of a mathematical nature. The english physicist, mathematician, astronomer, and natural philosopher Sir Isaac Newton (1642-1727) is the first scholar to introduce a strictly theoretical treatment of mechanics. In his monograph 'Philosophiae naturalis principia mathematica' (1687) [1] Newton formulates the three fundamental laws of motion, also called axioms of mechanics, which we give in a modernized formulation in the following:

1. A body remains at rest or in uniform motion in a straight line unless acted upon by a force. (law of inertia)
2. A body's rate of change of momentum is proportional to the force causing it.
3. When a force acts on a body due to another body then an equal and opposite force acts simultaneously on that other body. (actio = reactio)

In conjunction with the formalization of differential calculus by Newton and Gottfried Wilhelm Leibniz (1646-1716) the well known concept of force and also the

The Development of Analytical Mechanics

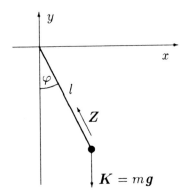

Fig. 1 Model of the planar mathematical pendulum: the mass-less thread or rigid rod has length *l* and can freely rotate about the origin of ordinates. A point mass *m* is attached to its other end.

basic dynamic equation can now be formulated in a mathematically exact way for the very first time

$$\mathbf{F} = m\ddot{\mathbf{r}} \tag{1}$$

and solutions to the thus found differential equations can be stated.

This knowledge and notation are sufficient to describe an elementary example, namely the thread pendulum or mathematical pendulum.

Example: As depicted in fig. 1 the thread pendulum or mathematical pendulum consists of a point mass affixed to one end of a mass-less thread or better yet a mass-less rigid rod which can rotate freely about its other end. The origin of ordinates shall lie in the revolute joint and gravity is assumed to be the only external force $\mathbf{g} = -g\mathbf{e}_y$. Simple geometric considerations (specifically the decomposition of the force of gravity **K** in the direction of the thread and perpendicular to it) yield the well-known (non-linear) equation of motion for the planar mathematical pendulum

$$\ddot{\varphi} + \frac{g}{l}\sin\varphi = 0 \tag{2}$$

Even though a solution to the pendulum problem is found here, at least for small deflections $\varphi \ll 1$, no direct assertions can be made regarding the constraint force acting on the thread.

3 Lagrangian Mechanics

The Italian born mathematician and astronomer Joseph-Louis Lagrange (1736–1813) is able to build upon the preliminary studies concerning the problem of central motion by Johann Bernoulli and Jacob Hermann especially since he is

Fig. 2 Joseph-Louis Lagrange (1736-1813)
source: http://en.wikipedia.org/wiki/File:Langrange_portrait.jpg
(retrieved 2011-10-20, public domain)

also very familiar with the work of Euler, his doctoral adviser. In his treatise 'Mécanique analytique' (1788) [3] Lagrange develops his Lagrangian Mechanics utilizing d'Alembert's principle.

3.1 Lagrange Equations of the First Kind

The main advancement due to Lagrange's equations of the first kind with respect to the previously used Newtonian mechanics is the explicit treatment of constraints. Developments associated with implicit functions, concretely the method of Lagrange multipliers to describe normals to implicitly defined surfaces, make it possible to write down the equations of motion of a system of discrete point masses as Lagrange equations of the first kind:

$$m\ddot{\mathbf{r}} = \mathbf{F} = \mathbf{K} + \mathbf{Z} = \mathbf{K} + \lambda(t)\nabla z(\mathbf{r},t) \qquad (3)$$

where the various variables are to be interpreted as described in table 1.

Table 1 Lagrange equations of the first kind

symbol	meaning
$\mathbf{K}$	impressed forces
$\mathbf{Z}$	constraint forces
$z(\mathbf{r},t)$	constraint
$\lambda(t)$	Lagrange multiplier

Example: At this point let us return to our example system, the thread pendulum. Here the (geometric) constraint $l = const$ can be formulated as follows:

$$z(\mathbf{r},t) = z(\mathbf{r}) = z(x,y) = x^2 + y^2 - l^2 = 0 \tag{4}$$

Taking the gradient of the constraint function yields

$$\nabla z(\mathbf{r},t) = \begin{bmatrix} 2x \\ 2y \end{bmatrix} \tag{5}$$

With this we get for the example of the planar mathematical pendulum the Lagrange equations of the first kind (3) as differential equations of motion of order two in the following form:

$$\begin{bmatrix} m\ddot{x} \\ m\ddot{y} \end{bmatrix} = \begin{bmatrix} 0 \\ -mg \end{bmatrix} + \lambda \begin{bmatrix} 2x \\ 2y \end{bmatrix} \tag{6}$$

In order to determine a solution to this differential equation and especially to concretely calculate the acting constraint forces **Z** we need to determine the Lagrange multiplier λ. The easiest approach is differentiating the constraint twice with respect to time. We have

$$\ddot{z}(\mathbf{r},t) = 2\left[\dot{x}^2 + \dot{y}^2 + x\ddot{x} + y\ddot{y}\right] = 0 \tag{7}$$

Using the equation of motion (6) yields the following expression for the Lagrange multiplier

$$\lambda(t) = \frac{m}{2l^2}\left[gy - [\dot{x}^2 + \dot{y}^2]\right] \tag{8}$$

and therefore the final equations of motion

$$\begin{bmatrix} m\ddot{x} \\ m\ddot{y} \end{bmatrix} = \begin{bmatrix} 0 \\ -mg \end{bmatrix} + \frac{m}{2l^2}\left[gy - [\dot{x}^2 + \dot{y}^2]\right] \begin{bmatrix} 2x \\ 2y \end{bmatrix} \tag{9}$$

respectively the occuring constraint forces

$$\mathbf{Z} = \begin{bmatrix} Z_x \\ Z_y \end{bmatrix} = \frac{m}{l^2}\left[gy - [\dot{x}^2 + \dot{y}^2]\right] \begin{bmatrix} x \\ y \end{bmatrix} \tag{10}$$

The advantage of the Lagrange equations of the first kind is obvious here, namely the explicit expression for the constraint forces (10) given for every moment in time. A clear disadvantage though is the often occuring inherent nonlinearity of the equations of motion (9).

3.2 Lagrange Equations of the Second Kind

The main advantage of Lagrange's equations of the second kind with respect to the equations of the first kind is the incorporation of the constraints that had previously to be dealt with explicitly. A system of N discrete point masses in three dimensional space initially has $3N$ degrees of freedom. If these are restricted by constraints we get a smaller number $n = 3N - \#(\text{non}-\text{redundant constraints})$ of actual degrees of freedom. Using the implicit function theorem we can locally find generalized coordinates q_i, where i ranges over the actual degrees of freedom n of the system. With this we can describe the original coordinates as follows:

$$r_j = r_j(q_1, \ldots, q_n, t) \quad \forall j = 1, \ldots, 3N \tag{11}$$

With (11) we can now also describe the time derivates of the coordinates $\mathbf{r}$ by generalized coordinates $\mathbf{q}$

$$\dot{r}_j = \frac{dr_j}{dt} = \frac{\partial r_j}{\partial t} + \sum_{i=1}^{n} \frac{\partial r_j}{\partial q_i} \dot{q}_i \quad \forall j = 1, \ldots, 3N \tag{12}$$

where the $\dot{q}_i = \frac{dq_i}{dt}$ are called generalized velocities.
In the same way we can write the variations of the r_j as

$$\delta r_j = \sum_{i=1}^{n} \frac{\partial r_j}{\partial q_i} \delta q_i \tag{13}$$

The d'Alembert principle of virtual displacements, named after Jean-Baptiste le Rond commonly called d'Alembert (1717–1783), states that the constraint forces $\mathbf{Z}$ do no virtual work. This leads to the formulation of d'Alembert's principle as employed by Lagrange

$$\sum_{j=1}^{3N} [m_j \ddot{r}_j - K_j] \delta r_j = 0 \tag{14}$$

This solitary scalar equation with the use of (12) and (13) by way of a short calculation leads to the precursors of Lagrange's equations of the second kind

$$\sum_{j=1}^{3N} m_j \ddot{r}_j \delta r_j = \sum_{i=1}^{n} \left[\frac{d}{dt} \left[\frac{\partial}{\partial \dot{q}_i} T \right] - \frac{\partial}{\partial q_i} T \right] \delta q_i \tag{15}$$

and

$$\sum_{j=1}^{3N} K_j \delta r_j = \sum_{i=1}^{n} Q_i \delta q_i \tag{16}$$

where the definition of generalized force Q_i

$$Q_i \stackrel{def}{=} \sum_{j=1}^{3N} K_j \frac{\partial r_j}{\partial q_i} \tag{17}$$

and kinetic energy T

$$T \stackrel{def}{=} \sum_{j=1}^{3N} \frac{m_j}{2} \dot{r}_j^2 \qquad (18)$$

were employed. Because the δq_i are arbitrary and linearly independent the plugging in of above equation into (14) yields the Lagrange equations of the second kind:

$$\frac{d}{dt}\left[\frac{\partial T}{\partial \dot{q}_i}\right] - \frac{\partial T}{\partial q_i} = Q_i \quad \forall i = 1,\dots,n \qquad (19)$$

If the forces K_i can be derived from a velocity-independent potential Π we can write the Lagrange equations of the second kind employing the Lagrangian $L = T - \Pi$ in modern notation as

$$\frac{d}{dt}\left[\frac{\partial L}{\partial \dot{q}_i}\right] - \frac{\partial L}{\partial q_i} = 0 \quad \forall i \qquad (20)$$

Example: We use our example to illustrate the general approach. In our case the constraint can be expressed via a single generalized coordinate $q = \varphi$ due to simple geometric considerations

$$\mathbf{r} = \begin{bmatrix} x \\ y \end{bmatrix} = \begin{bmatrix} l\cos\varphi \\ -l\sin\varphi \end{bmatrix} \Leftrightarrow |\mathbf{r}| = l = const \qquad (21)$$

For the Lagrangian of the thread pendulum comprised of kinetic energy T and potential energy Π we have

$$L(q,\dot{q},t) = L(\varphi,\dot{\varphi}) = T(\dot{\varphi}) - \Pi(\varphi) = \frac{1}{2}ml^2\dot{\varphi}^2 - mgl[1-\cos\varphi] \qquad (22)$$

With this we can state the Lagrange equations of the second kind (20) for the thread pendulum as

$$\frac{d}{dt}\left[\frac{\partial L}{\partial \dot{\varphi}}\right] - \frac{\partial L}{\partial \varphi} = ml^2\ddot{\varphi} + mgl\sin\varphi = 0 \qquad (23)$$

from which the well-known (non-linear) equation of motion of the planar mathematical pendulum follows

$$\ddot{\varphi} + \frac{g}{l}\sin\varphi = 0 \qquad (24)$$

Obviously the Lagrange equations of the second kind directly and automatically yield the equations of motion of the considered system as soon as the Lagrangian is constructed from kinetic and potential energy contributions. In this formalism of the second kind of Lagrange the constraints are eliminated from the calculation of the equations of motion right from the start by choosing suitable generalized coordinates. It needs to be mentioned that choosing generalized coordinates cleverly is far from trivial. Likewise in this context of Lagrange equations of the second kind the explicit determination of acting constraint forces is not easily possible.

4 Hamiltonian Mechanics

The Irish physicist, astronomer and mathematician Sir William Rowan Hamilton (1805-1865) is able to formulate an additional and alternative version of 'Analytical Mechanics' based on the at this time already excellently understood theory of Lagrange. This new formulation is especially elegant in its formulation due to the use of extremal principles as a foundation. Even today it is still of great relevance and not only used in the field of mechanics (refer to conclusion for examples).

Fig. 3 William Rowan Hamilton (1805-1865)
source: http://en.wikipedia.org/wiki/File:
William_Rowan_Hamilton_portrait_oval_combined.png
(retrieved 2011-10-20, public domain)

4.1 Hamilton's Principle

Already in papers concerned with geometrical optics entitled 'Theory of Systems of Rays' (1827) [6] we encounter the so-called Hamilton's function or Hamiltonian for the first time in Hamilton's oeuvre. In the following we will give a short explanation how this Hamiltonian found its way into mechanics. The basis for this is the principle of stationary action as used in the papers 'On a General Method in Dynamics' [4,5] published by Hamilton in 1834 and 1835. Hamilton defines the action or better the action functional $S[\mathbf{q}]$ as follows

$$S[\mathbf{q}] \stackrel{def}{=} \int_{t_1}^{t_2} L(\mathbf{q}(t), \dot{\mathbf{q}}(t), t) \, dt \qquad (25)$$

The Development of Analytical Mechanics

The action depends not only on the start- and end-times t_1 and t_2 and the start- and end-positions $\mathbf{q}_1 = \mathbf{q}(t_1)$ and $\mathbf{q}_2 = \mathbf{q}(t_2)$ but also on the path $\mathbf{q}(t)$ taken in between. Hamilton's principle now states that the action is stationary for physical paths

$$\delta S = 0 \tag{26}$$

Based on work by the french mathematician Pierre Louis Moreau de Maupertuis (1698–1759) who already in 1746 proposed a similar extremal principle to describe natural processes, Euler and Lagrange had shown in the middle of the 18th century that this implies the validity of the so-called Euler-Lagrange equations. The variation of S can be derived by formally replacing $\mathbf{q} \to \mathbf{q} + \delta\mathbf{q}$, expanding the Lagrangian in $\delta\mathbf{q}$ to the first order and then partially integrating with fixed endpoints $\delta\mathbf{q}_1 = \delta\mathbf{q}_2 = \mathbf{0}$. This yields

$$\delta S = \sum_i \int_{t_1}^{t_2} \left[\frac{\partial L}{\partial q_i} - \frac{\mathrm{d}}{\mathrm{d}t} \frac{\partial L}{\partial \dot{q}_i} \right] \delta q_i \, \mathrm{d}t \tag{27}$$

and, because the variations $\delta\mathbf{q}$ were arbitrary except for the endpoints

$$\delta S = 0 \Leftrightarrow \frac{\partial L}{\partial q_i} - \frac{\mathrm{d}}{\mathrm{d}t} \frac{\partial L}{\partial \dot{q}_i} = 0 \quad \forall i \tag{28}$$

With this Hamilton showed that the Lagrange equations of the second kind follow directly from a variational principle, concretely from the variation of the action (25) with respect to the physical path $\mathbf{q}(t)$ with fixed start- and end-points $\mathbf{q}_1$ and $\mathbf{q}_2$ in configuration space. The main step that distinguishes Hamilton's description of mechanics from that of Lagrange is the realization that position $\mathbf{q}$ and momentum $\mathbf{p}$ are mutually conjugate variables. This leads to the formulation of the Hamiltonian and therefore to Hamilton's equations of motion also called Hamilton equations.

4.2 Hamilton Equations

Hamilton's function or short the Hamiltonian is derived from the Lagrangian of a physical system by employing a Legendre-transform. Here the generalized velocities $\dot{\mathbf{q}}$ are eliminated and replaced by the canonical conjugate momenta $\mathbf{p}$. The canonical momenta are defined by Hamilton as

$$p_i = \frac{\partial L}{\partial \dot{q}_i} \quad \forall i \tag{29}$$

For regular Lagrangians, meaning $\det \frac{\partial^2 L}{\partial \dot{q}_i \partial \dot{q}_j} \neq 0$, we get the Hamiltonian as Legendre-transform of the Lagrangian as follows:

$$H(\mathbf{q},\mathbf{p},t) \stackrel{def}{=} \sum_i \dot{q}_i p_i - L(\mathbf{q},\dot{\mathbf{q}},t) \tag{30}$$

In this definition of the Hamiltonian the position **q** and momentum **p** are now regarded as mutually independent variables. Using Hamilton's principle (26) we can derive the Hamilton equations as follows

$$\delta S = \sum_i \int_{t_1}^{t_2} \left[\left[\dot{q}_i - \frac{\partial H}{\partial p_i} \right] \delta p_i - \left[\dot{p}_i + \frac{\partial H}{\partial q_i} \right] \delta q_i \right] dt \qquad (31)$$

Because the (independent) variations $\delta\mathbf{q}$ and $\delta\mathbf{p}$ are again arbitrary except for the start- and end-points Hamilton's equations follow immediately

$$\dot{q}_i = \frac{\partial H}{\partial p_i}, \quad \dot{p}_i = -\frac{\partial H}{\partial q_i} \quad \forall i \qquad (32)$$

Already at first glance substantial differences to the Lagrangian equations of the second kind (20) are clear. Lagrange's equations of the second kind for a system with n degrees of freedom were n ordinary differential equations of second order. But Hamilton's equations (32) are now $2n$ ordinary differential equations of first order in time. This considerable difference can be very useful, especially for numerical integration of the equations of motion.

Furthermore both formulations of 'Analytical Mechanics' are very different with regard to their geometric interpretation. The Lagrangian $L(\mathbf{q},\dot{\mathbf{q}},t)$ is a function on the (extended) tangent bundle of the configuration manifold $\mathcal{M}$, concretely $L : T\mathcal{M} \times \mathbb{R} \to \mathbb{R}$. The Hamiltonian $H(\mathbf{q},\mathbf{p},t)$ however is a function on the (extended) cotangent bundle of the configuration manifold $\mathcal{M}$, symbolically $H : T^*\mathcal{M} \times \mathbb{R} \to \mathbb{R}$. The cotangent bundle of the configuration manifold is usually called the phase-space of the physical system and has inherent geometric properties, e.g. the existence of a symplectic form which the tangent-bundle of a manifold does not usually feature.

Example: In the case of our thread pendulum the Hamiltonian can be directly calculated from the Lagrangian as

$$H(\varphi, p_\varphi, t) = \dot{\varphi} p_\varphi - L = \frac{1}{2} m l^2 \dot{\varphi}^2 + mgl[1 - \cos\varphi] = \frac{1}{2} \frac{p_\varphi^2}{ml^2} + mgl[1 - \cos\varphi] \qquad (33)$$

where the canonical momentum $p_\varphi = \frac{\partial L}{\partial \dot{\varphi}} = ml^2 \dot{\varphi}$ is conjugate to the generalized coordinate φ. We get the Hamilton equations of the thread pendulum as follows

$$\dot{p}_\varphi = -\frac{\partial H}{\partial \varphi} = -mgl\sin\varphi, \quad \dot{\varphi} = \frac{\partial H}{\partial p_\varphi} = \frac{p_\varphi}{ml^2} \qquad (34)$$

Differentiating the second equation of (34) with respect to time and plugging in the first equation we immediately see the equivalence to the equation of motion determined by the Lagrange formalism of the second kind (24).

5 Summary and Generalizations

The preceding discussion has shown how starting from the early days of a still poorly understood mechanics before Newton and Euler an elegant and mathematically sound theory emerged through constant theoretical advancements throughout the 18th and 19th century. This theory is fit to describe all dynamic processes of systems of discrete point masses and also rigid bodies within the usually tangible world. Further development of this theory into classical field theories, e.g. the description of Maxwell's electrodynamics or continuum mechanics by Lagrangian or Hamiltonian formalisms, is another high point. Together with the hereafter described developments in initially pure mathematics the triumph of 'Analytical Mechanics' continues on into the present.

The work of the Norwegian mathematician Marius Sophus Lie (1842–1899) dealing with transformation groups and the associated algebras seemed to his contemporaries quirky and incomprehensible. But combined with ideas from the so-called Erlangen program of Christian Felix Klein (1849–1925) Lie's ideas revolutionized conceptions about and approaches to geometry. No longer the formerly basic building blocks of geometry namely points, lines, etc. are in the spotlight but instead their symmetry properties are the main focus of modern geometry. These symmetries are described by properties of groups, a formerly purely algebraic concept.

The French mathematician Élie Joseph Cartan (1869–1951) developed the theory of so-called Lie-groups and Lie-algebras originally discovered by Lie even further. Together with his contributions to differential geometry and mathematical physics Cartan is chiefly responsible for the current formulation of Hamiltonian mechanics in the framework of symplectic geometry.

In the context of the development of quantum mechanics at the beginning of the 20th century the theoretical physicists Erwin Schrödinger (1887–1961), Werner Heisenberg (1901–1976) und Paul Dirac (1902–1984) follow in the footsteps of their historic forerunners Newton, Euler, Langrange, and Hamilton and lead the theory of 'Analytical Mechanics' to a new golden age. Through the use of novel concepts, first and foremost the formal replacement of clasically real variables like position q and momentum p by abstract operators $\hat{q}$ and $\hat{p}$ that do no longer commute, they succeed in establishing Hamiltonians for formerly poorly understood quantum mechanical systems describing their dynamics correctly. Following these early successes quantum effects in field theories become explainable. Especially the theory of quantum electrodynamics (QED) as developed by Richard Feynman (1918–1988) is able to explain and describe certain effects in a new way and with previously unattainable accuracy. The description of these quantum field theories usually relies on Lagrange densities, the field theoretical counterpart of the Lagrangian for discrete systems. Here equations of motion for the fields are again obtained by means of variational calculus. The action functional, an integral of the Lagrange density, is varied with respect to field degrees of freedom. A milestone in these developments is the so called 'Standard Model of particle physics' which is able to very precisely predict all currently observable effects in the domain of elementary particles. The Lagrange density for the standard model would fill multiple pages of this publication.

References

1. Newton, I.: Philosophiae naturalis principia mathematica. Jussu Societatis Regiae ac Typis Josephi Streater, Londini (1687)
2. Euler, L.: Mechanica sive Motus scientia analytice exposita; instar supplementi ad commentar. Acad. Scient. Imper. Ex typographia Academiae Scientiarum, Petropoli (1736)
3. La Grange, J.L.: Mechanique analytique. Veuve Desaint, Paris (1788)
4. Hamilton, W.R.: Philosophical Transactions of the Royal Society, Part II, 247–308 (1834)
5. Hamilton, W.R.: Philosophical Transactions of the Royal Society, Part I, 95–144 (1835)
6. Hamilton, W., Conway, A., Synge, J.: The mathematical papers of Sir William Rowan Hamilton. Cunningham memoirs, vol. 2. The University press (1931)

Heun and Hamel – Representatives of Mechanics around 1900

Hartmut Bremer

Abstract. At the beginning of the twentieth century, at a time when classical mechanics had long been considered to be complete, two scientists appeared on the scene: KARL HEUN and GEORG HAMEL. We owe them the further development of analytical mechanics, especially in the new field of nonholonomic systems (HAMEL's equations, nowadays mostly referred to as HAMEL-BOLTZMANN-equations). The impact of their research in the twentieth century and up to now is the focus of the present contribution.

Keywords: Analytical Mechanics, Nonholonomic Variables and Constraints, Central Equation, Methodology.

1 Introduction

One quarter century after the founder crisis, in those times characterized by the Jugendstil, European culture experienced a period of exuberant cultural bloom, creative in music, architecture, visual arts. And also in natural sciences: in spring 1899, MAX PLANCK discovered his natural constant h. Although he was, at the beginning, definitely against atomistics at all, he eventually accepted BOLTZMANN's statistical approach. This is because for over more than one year he could not find a theoretical basis for his radiation law. Clearly, no reasonable thinking being would assign physical properties to the mathematical tool itself, but, nevertheless, the procedure opened ad hoc new methods, subsequently pursued enthusiastically. By this, the fate of classical mechanics was sealed ...

Hartmut Bremer
Institut für Robotik,
Johannes Kepler Universität Linz,
Altenberger Straße 69
e-mail: hartmut.bremer@jku.at

"People have dated the completion of that mechanics almost exactly to the turn of the century", says KURT MAGNUS in his inauguration lecture in Munich, February 1967 [1] – and at the same time protests against this claim. Rightly so, because it seems far-fetched that during that brilliant period no "mechanic" would have raised his voice. "With few exceptions they [the physicists] obviously do not take note of those developments which have been accomplished during the last decades". Is KURT MAGNUS allowed to use such harsh words? He is: he studied mathematics, chemistry and *physics* in Göttingen and earned his doctorate under SCHULER and PRANDTL in the year 1937.

Thirty years after MAGNUS' inauguration a book entitled *The Great Physicists* appeared. It comfirms the popular view: After EULER, LAGRANGE, COULOMB, LAPLACE there is no progress in classical mechanics any more. And what is particularly interesting: we learn that history of science has not been established until the first decades after the last world war [2]. Really? Has there not been any history of science before? What about EUGEN DÜHRING [3], ERNST MACH [4], WILHELM OSTWALD [5] or even JOSEPH LAGRANGE [6] whose historical outline had highly been appreciated by POINSOT? History in a specialist book – not a common sight in our days. However, the researcher should be aware of his roots, because without roots there is no growth. In this sense, HEUN always grants history a lot of space, and the same holds for HAMEL who obtained his habilitation 1903 at HEUN's institute in Karlsruhe.

1977 ISTVÁN SZABÓ, one of HAMEL's scholars, wrote a book on the history of the mechanical principles and their most important applications – once more connecting the current subject with its roots [7]. However, he was severely berated for this by our professional historians [8].

Professional historians and philosophers think in other categories than natural scientists do, and sometimes their opinions seem a bit strange. For instance, according to [9], HERTZ in his last contribution wanted to free mechanics from the dubious approaches of NEWTON, LAGRANGE and KELVIN. However, taking HERTZ's book into one's hands one reads: he did not at all want to reject NEWTON's results but, just a contrary, generalize them [10]. What a contradiction! One gets the impression that historians sometimes "interpret" intermediate steps with the aim of a certain desired result. The aim here is presumably to reject determinism and to classify classical mechanis as, if at all, a poor approximation of the true and only real physics. This truth is told in *Triumph and Crisis of Mechanics* [11], advising where the new perspectives of mechanics can be found in an understandable way: *Chaos* – making a new science JAMES GLEICK calls his work and claims: where chaos begins, classical science ends. "A revolution in natural sciences! Natural laws do not longer hold", supposes VOLKER ARZT 1985 in the German television, and 1990 the Bavarian Broadcast simply recommends: Throw out your old equations! Chaos and the butterfly effect – *The Beauty of Fractals*, 1986 praised by PEITGEN and RICHTER, can of course (depending on personal taste) not be denied. But 13 years later co-author PETER H. RICHTER [12] proclaims a surprising message: Chaos as a solution of classical mechanics! Not very surprising, this had already been known to POINCARÉ about one hundred years before. Nevertheless, one is tempted to say: physics has come back to its senses, classical mechanics is saved. So, what happened to it after 1900? Here, first and foremost, KARL HEUN and GEORG HAMEL have to be mentioned.

2 Heun and Hamel

KARL HEUN was born on 3, April 1859 in Wiesbaden. He studied mathematics and philosophy in Halle and Göttingen, earned his doctoral degree 1881 in Göttingen and his habilitation 1886 in Munich, where he stayed for three years as a private lecturer. With his income as a private lecturer he was soon unable to feed his family, meanwhile grown to six persons, and he changed to Berlin as a senior teacher at the first Realschule. He would thus have been lost for our story if he would not have met a group of engineers there. Their problems had grown seemingly insurmountabe, since up to that time the wise practicians had rejected the role of kinetics as irrelevant for their craft, they now where almost helpless when confronted with fast moving steam engines. On the other hand, the LAGRANGEan scientific kinetics engaged only in celestial mechanics and did not care so much about terrestrial applications.

Fig. 1 Karl Heun (1859 – 1929) **Fig. 2** Georg Hamel (1877 – 1954)

Here, KARL HEUN enters the scene: His report *Über die kinetischen Probleme der wissenschaftlichen Technik* [13] (On the kinetic problems in scientific technology, 1900) initiated the development for the next decades, strongly focusing on analytical mechanics. By recommendation of FELIX KLEIN he obtained a professorship for theoretical mechanics in Karlsruhe 1902 where he conducted scietific research for the next twenty years [14].

GEORG HAMEL habilitated 1903 at HEUN's institute. He was born 1877 in Düren (Rheinland), studied mathematics and philosophy in Aachen, Berlin and Göttingen and obtained his doctorate 1901 in Göttingen. Between 1902 and 1905 he was a private lecturer at HEUN's chair, afterwards professor for technical mechanics and mathematics in Brünn, Aachen and Berlin; he retired 1949. In that year he published his final monograph "Theoretische Mechanik" [15], mainly written from memory because he had lost most of his private books and personal writings during the bombing of Berlin at the end of 1943. In 1961, his "Theoretische Mechanik" was reprinted one time only; today it is an exciting adventure to find "the Hamel" second hand. HAMEL died 1954 in Landshut. His profound thinkings have by no means been fully exploited yet.

KARL HEUN started a textbook on mechanics, but, because he was very busy with several tasks (e.g. participating in the encyclopedia of mathematical sciences [16]), he only finished the first volume [17]. 1921 he suffered a stroke. Heartfully cared for by his wife Henriette, he died nine years later.

HAMEL's habilitation thesis was printed 1904, entitled *Die Euler-Lagrangeschen Gleichungen* (The Euler-Lagrange-Equations) [18]. It begins harmless: "I found them when I asked myself: which equations replace the LAGRANGEan ones, when I insert, besides the n position variables $\mathbf{q}$, some other n independent linear combinations of $d\mathbf{q}/dt$? ... To obtain the ... facts so easily and clearly is mainly due to Lagrange himself" says HAMEL.

In this context, HEUN talks about the LAGRANGEan Central Equation. This term cannot be found in any of LAGRANGE's contributions, but in his *théorie de la libration de la lune* from 1780 one finds on page 21 (here in vectorial notation): "il s'ensuit de là que la quantité $d^2\mathbf{r}^T\delta\mathbf{r}$ sera la même chose que celle ci: $d(d\mathbf{r}^T\delta\mathbf{r}) - \delta(d\mathbf{r}^T d\mathbf{r})/2$". This yields, when related to the time element dt and integrated with respect to dm, what HEUN calls the LAGRANGEan Central Equation.

HAMEL raises concerns about this, because the LAGRANGEan identity only holds for $\delta d\mathbf{r} = d\delta\mathbf{r}$, or, in small sketch: if the curves that are created by the different $d-$ and $\delta-$processes are closed. This argumentation, however, might not have been very lucky. At least also great masters in the field like A. LUR'E were misled to the statement *avec l'égalité evidente* ("with the obvious identity"), but here there is nothing obvious about it:

Let $\mathbf{r}$ be the position vector of the considered mass element dm, $\mathbf{q} \in \mathbb{R}^n$ the vector of minimal coordinates of the system and $\dot{\mathbf{s}} = \mathbf{H}(\mathbf{q})\dot{\mathbf{q}} \in \mathbb{R}^n$ the vector of minimal velocities (see above HAMEL: "some n independent linear combinations of $d\mathbf{q}/dt$"), nonholonomic in HERTZ's notation. Writing down the ($\delta d\mathbf{r} - d\delta\mathbf{r}$)-process (and ($\delta d\mathbf{s} - d\delta\mathbf{s}$) analoguously), then ($\delta d\mathbf{q} - d\delta\mathbf{q}$) vanishes when assuming a LAGRANGEan variational approach for $\delta\mathbf{q}$. This would stand to reason (and leads to the fact that $\delta\mathbf{q}$ needs not be infinitesimal – as is important in optimization). By this, only SCHWARZ's integrability condition remains which is fulfilled for $\mathbf{r}$ just from reasons of modelling (but not for the quasi-coordinates $\mathbf{s}$). Then, the above mentioned curves indeed form a closed loop, but this question is out of interest here. However, because $\mathbf{q}$ does not undergo any constraints, one might use arbitrary ansatzes for $\delta\mathbf{q}$. HAMEL keeps this door open which later leads him to LIE's algebra.

Back to LAGRANGE, whom we owe "the opportunity to obtain the aforementioned facts so easily and clearly" [18]. In his *récherches sur la libration de la lune* from 1764 we find the LAGRANGEan principle (virtual work in kinetics) from which one obtains by elementary (maybe sometimes a bit tricky) mathematical operations a relationship that can deservedly be called the *Central Equation of Dynamics*: $d[(\partial T/\partial \dot{\mathbf{s}})\delta\mathbf{s}]/dt - \delta T - \delta W = 0$ [T: kinetic energy, $\delta W = \mathbf{Q}^T \delta\mathbf{s}$: virtual work (accomplished by impressed forces/torques), $\mathbf{Q}$: generalized forces]. From this one can develop a considerable body of analytical procedures just by elementary calculations [19], see Table 1.

The variational form of HAMEL's equations (one line above the Central Equation) contains $\delta\mathbf{s}$ explicitly in the first term while in the second one it is still hidden. One can thus not directly extract the corresponding motion equations. Of course, HAMEL removes this flaw (by his famous coefficients), but, as he puts it, "this is perhaps not always reasonable ... it is sometimes easier to extract $\delta\mathbf{s}$ from the expression $(d\delta\mathbf{s} - \delta d\mathbf{s})/dt$ directly" ([15], p.483). Therefore, we are not going to discuss the HAMEL-coefficients here in detail.

The *Central Equation of Dynamics* can, if at all, only be found in fragments in HAMEL's habilitation thesis [18] or in his last work [15]. But concluding that he was not aware of it is false: A footnote in HEUN [16] refers to one of HAMEL's contributions [20] where according to its title one would not expect it. The reason to later go back to the "raw form" can probably only be interpreted as historical fairness. HAMEL always cites his sientific contemporaries and ancestors (see above: the historical roots), even if he seems to loose the claim for priority: honesty and sincerity are two of HAMEL's distinct features.

3 Outlook

For a final look on HAMEL's equations let us consider a multi body system, which consists of n rigid bodies. The kinetic energy of a single body is $T_i = (\mathbf{p}_i^T \ \mathbf{L}_i^T)\dot{\mathbf{s}}_i/2$ where $\mathbf{p}$: translational momentum, $\mathbf{L}$: angular momentum, $\dot{\mathbf{s}}_i^T = (\mathbf{v}_{si}^T \ \boldsymbol{\omega}_{si}^T)$ corresponding mass center velocities. In the first step, all terms will be represented in a body fixed frame since then the inertia tensor has constant elements. Choosing the mass center velocities as (nonholonomic) intermediate variables one obtains for the directional derivatives the (generalized) momenta themselves: $(\partial T_i/\partial \dot{\mathbf{s}}_i) = (\mathbf{p}_i^T \ \mathbf{L}_i^T)$. The term $(d\delta\mathbf{s} - \delta d\mathbf{s})/dt$

Table 1 The Most Essential Methods in Dynamics

1673	$T + V = H$
1879/99	$\left(\dfrac{\partial G}{\partial \ddot{\mathbf{s}}}\right)^T = \mathbf{Q}$
1788, 1828	$\delta \int_{t_o}^{t_1}(T - V)dt = 0$
1835	$\dot{\mathbf{p}}^T = -\left(\dfrac{\partial H}{\partial \mathbf{q}}\right);\ \dot{\mathbf{q}}^T = \left(\dfrac{\partial H}{\partial \mathbf{p}}\right)$
1780	$\dfrac{d}{dt}\dfrac{\partial T}{\partial \dot{\mathbf{q}}} - \dfrac{\partial T}{\partial \mathbf{q}} - \mathbf{Q}^T = 0$
1962	$\dfrac{1}{m}\dfrac{\partial T^m}{\partial \mathbf{q}^m} - \dfrac{m+1}{m}\dfrac{\partial T}{\partial \mathbf{q}} - \mathbf{Q}^T = 0$
1903	$\left[\dfrac{d}{dt}\dfrac{\partial T}{\partial \dot{\mathbf{q}}} - \dfrac{\partial T}{\partial \mathbf{q}} - \mathbf{Q}^T\right]\left(\dfrac{\partial \dot{\mathbf{q}}}{\partial \dot{\mathbf{s}}}\right) = 0$
1904	$\left[\dfrac{d}{dt}\dfrac{\partial T}{\partial \dot{\mathbf{s}}} - \dfrac{\partial T}{\partial \mathbf{s}} - \mathbf{Q}^T\right]\delta\mathbf{s} + \dfrac{\partial T}{\partial \dot{\mathbf{s}}}\left(\dfrac{d\delta\mathbf{s} - \delta d\mathbf{s}}{dt}\right) = 0$

CENTRAL	$\dfrac{d}{dt}\left[\left(\dfrac{\partial T}{\partial \dot{\mathbf{s}}}\right)\delta\mathbf{s}\right] - \delta T - \delta W = 0$	EQUATION

1988	$\sum_{i=1}^{n}\left\{\left(\dfrac{\partial \mathbf{v}_S}{\partial \dot{\mathbf{s}}}\right)^T(\dot{\mathbf{p}} + \widetilde{\boldsymbol{\omega}}_R\mathbf{p} - \mathbf{f}^e) + \left(\dfrac{\partial \boldsymbol{\omega}_S}{\partial \dot{\mathbf{s}}}\right)^T(\dot{\mathbf{L}} + \widetilde{\boldsymbol{\omega}}_R\mathbf{L} - \mathbf{M}^e)\right\}_i = 0$

{ 1673: HUYGENS, 1879/99: GIBBS/APPEL, 1788/1828: LAGRANGE/HAMILTON, 1835: HAMILTON, 1780: LAGRANGE, 1962: MANGERON&DELEANU, 1903: MAGGI, 1904: HAMEL, 1988: "PROJECTION EQUATION" }

delivers the convective parts where $\widetilde{\boldsymbol{\omega}}_i$ is the (skew symmetric) spin tensor related to the angular velocities $\boldsymbol{\omega}$. Along with $\delta \mathbf{s}_i = (\partial \dot{\mathbf{s}}_i / \partial \dot{\mathbf{s}}) \delta \mathbf{s} = [(\partial \mathbf{v}_{si}/\partial \dot{\mathbf{s}}) \ (\partial \boldsymbol{\omega}_{si}/\partial \dot{\mathbf{s}})] \delta \mathbf{s}$ one obtains at first the motion equations in a body fixed representation, e.g. for the translational part $\delta \mathbf{s}^T \sum (\partial \mathbf{v}_{si}/\partial \dot{\mathbf{s}})^T (\dot{\mathbf{p}} + \widetilde{\boldsymbol{\omega}}_s \mathbf{p} - \mathbf{f}^e)$. Next, unit matrices in the form $\mathbf{A}^T \mathbf{A}$ are introduced. $\mathbf{A}(\mathbf{q})$ is orthonormal and transforms into an abitrary coordinate system which rotates with $\boldsymbol{\omega}_R$. Then, via $\delta \mathbf{s}^T \sum (\partial \mathbf{v}_{si}/\partial \dot{\mathbf{s}})^T \mathbf{A}^T \mathbf{A} (\dot{\mathbf{p}} + \widetilde{\boldsymbol{\omega}}_s \mathbf{p} - \mathbf{f}^e) = \delta \mathbf{s}^T \sum (\partial [\mathbf{A} \mathbf{v}_{si}]/\partial \dot{\mathbf{s}})^T (\dot{\mathbf{p}} + \widetilde{\boldsymbol{\omega}}_R \mathbf{p} - \mathbf{f}^e)$, one obtains the motion equations in an arbitrary reference frame representation R (I: inertial system, R: reference system; all vector quantities in R, index s: mass center). $\delta \mathbf{s}$ is free under variations and therefore removed, yielding the Projection Equation in its final form. (It is remarkable that the functional matrices (or Jacobians) are nothing but the coefficient matrices w.r.t. $\dot{\mathbf{s}}$. These are obtained as a by-product during the calculations. There is no additional calculation required).

Hence, one may add an additional row to the aforementioned table, below the Central Equation, and one gains (the core of) a methodology.

A methodology was obviously not on HAMEL's agenda: after KARL HEUN set the course towards the analytical methods at the turn of the 20th century, and led HAMEL to his generalization of LAGRANGE's equations, he had already enough to do with that. The value of the last expression, the *Projection Equation*, only emerges with the availability of computers which enable the treatment of large systems. Nevertheless, HAMEL had the key already in his hands.

4 Conclusion

The completion of classical mechanics dates back to the turn to the twentieth century? Mechanics is and remains a vivid and inspiring science. *Large parts of mechanics will remain open to further development and some might even be in need of it* (KURT MAGNUS, [21]).

Acknowledgements. Support of the Austrian Center of Competence in Mechatronics (ACCM) is gratefully acknowledged.

References

1. Magnus, K.: Klassische Mechanik im Zeitalter der Raumfahrt? Phys. Blätter 24, 4–12 (1967)
2. von Meyenn, K. (ed.): Die großen Physiker, Band I, p. 8. C.H. Beck, München (1997)
3. Dühring, E.: Kritische Geschichte der allgemeinen Principien der Mechanik. Martin Sändig, Wiesbaden (1887)
4. Mach, E.: Die Mechanik in ihrer Entwicklung. F.A. Brockhaus, Wiesbaden (1921)

5. Ostwald, W.: Große Männer. Akad. Verlagsges, Leipzig (1910)
6. de Lagrange, J.L.: Méchanique analytique. Desaint, Paris (1788), German by H. Servus. Springer, Berlin (1897)
7. Szabo, I.: Geschichte der mechanischen Prinzipien und ihrer wichtigsten Anwendungen. Birkhäuser, Basel (1977)
8. Szabo, I.: Bemerkungen zur Literatur über die Geschichte der Mechanik. Humanismus und Technik 22(3) (1979)
9. von Meyenn, K. (ed.): Die großen Physiker, vol. II, p. 140. C.H. Beck, München (1997)
10. Hertz, H.: Die Prinzipien der Mechanik, p. XXI. J.A. Barth, Leipzig (1894)
11. von Meyenn, K. (ed.): Triumph und Krise der Mechanik, p. 79. Piper, München (1990)
12. Richter, P.H.: Reguläre und chaotische Bewegungen in der klassischen Mechanik. GAMM-Mitteilungen 1, 73–103 (1999)
13. Heun, K.: Die kinetischen Probleme der wissenschaftlichen Technik. DMV-Bericht (1990)
14. Renteln, M.v.: Karl Heun – his life and his scientific work. In: Centennial Workshop on Heun's Equations, Rottach-Egern (1989)
15. Hamel, G.: Theoretische Mechanik. Springer, Heidelberg (1949)
16. Heun, K.: Ansätze und allgemeine Methoden der Systemmechanik. In: Klein, F., Müller, C. (eds.) Enzyklopädie der Mathematischen Wissenschaften, vol. IV(1,2), pp. 357–504. B.G. Teubner, Leipzig (1913)
17. Heun, K.: Lehrbuch der Mechanik, I.Teil Kinematik. Göschen, Leipzig (1906)
18. Hamel, G.: Die Lagrange-Eulerschen Gleichungen der Mechanik. ZAMP, 1–57 (1904)
19. Bremer, H.: Über eine Zentralgleichung in der Dynamik. ZAMM 68, 307–311 (1988)
20. Hamel, G.: Über die virtuellen Verschiebungen in der Mechanik. Math. Annalen 59, 416–434 (1904)
21. Magnus, K.: Vom Wandel unserer Auffassung zur Technischen Mechanik. THD-Schriftenreihe 28, 125–165 (1986)

The Machine of Bohnenberger

Jörg F. Wagner and Andor Trierenberg

Abstract. Johann Gottlieb Friedrich Bohnenberger (1765-1831) was a Professor of physics, mathematics, and astronomy at the University of Tübingen, Germany, as well as the scientific head surveying officer of the Kingdom of Württemberg. He made both major contributions to introducing modern geodesy in Germany and constructed various physical instruments. The "Machine of Bohnenberger" is considered the first gyro with cardanic suspension and forms the precursor of Foucault's Gyroscope of 1852. This article discusses important documents, the historical context, the initial dissemination, and the further development of J.G.F. Bohnenberger's invention.[1]

Keywords: Gyro with cardanic suspension, precession motion of the Earth, inertial sensors.

1 Introduction

Working on a simple experimental proof for the rotational motion of the Earth, the French physicist J.B.L. Foucault (1819-1868) introduced his well-known pendulum. His investigations, however, also concentrated on gyros, because he could not get a real acceptance of the fact that his pendulum sensed only the vertical component of the Earth's rotation rate [1]. A gyro with cardanic suspension, however, is able to measure theoretically the full rate. Although technical reasons

---

Jörg F. Wagner
German SOFIA Institute, University of Stuttgart,
Pfaffenwaldring 29, 70569 Stuttgart, Germany

Andor Trierenberg
Section for History of Science and Technology, University of Stuttgart,
Keplerstr. 17, 70174 Stuttgart, Germany

[1] Revised, updated version of a paper published in the Bulletin of the Scientific Instrument Society 107 (2010): 10-17.

prevented J.B.L. Foucault from building such a gyro device being sufficiently precise, his work later initiated the development of important navigation instruments such as the artificial horizon, the gyrocompass, and finally modern inertial and integrated navigation systems [2].

Nevertheless, J.B.L. Foucault is not the inventor of gyros with cardanic suspension. He was familiar with this mechanical principle since such instruments were especially popular in 19th century France: Prior to the discovery of its potential as a measurement device, this apparatus was already employed in many French schools to demonstrate the precession of the Earth's rotational axis. Explained below, this circumstance traces back to the work of the French mathematicians and physicists P.S. Laplace (1749-1827), S.D. Poisson (1781-1840), and F.J.D. Arago (1786-1853). All three scientists can be linked to the Ecole Polytechnique in Paris, which had already at an early historical stage a gyro with cardanic suspension. Also shown below this instrument was probably one of the original copies invented by Johann Gottlieb Friedrich Bohnenberger (1765-1831), who taught physics, mathematics, and astronomy at the University of Tübingen, Germany. In 1817, this scientist systematically explained the design and use of such an apparatus by means of the drawing in Fig. 1 for the first time [3]. In fact, its main purpose was to demonstrate the precession motion of the Earth's rotational axis.

J.G.F. Bohnenberger called the device simply "Machine". Therefore, the "Machine of Bohnenberger" can be regarded as the basis for J.B.L. Foucault's important work on gyros. After 1852, when Foucault introduced the term "Gyroscope"

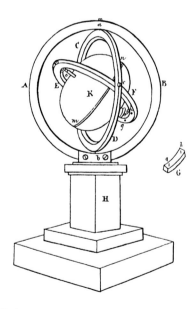

Fig. 1 Drawing of J.G.F. Bohnenberger's original publication of 1817 [3]

for such an instrument being able to observe rotational motions [4], the historically less correct name "Gyroscope of Bohnenberger" was commonly used as well. The same applies to the term "Apparatus of Bohnenberger".

In his publication of 1817, Bohnenberger indicates at the beginning that several copies of the first Machine had been manufactured in Tübingen. At the end of this article, he also names the instrument maker and the price of the device: all instruments were assembled by the "Universitätsmechanicus" Johann Wilhelm Gottlob Buzengeiger (1777-1836) and could be bought for 18 guilders. Nevertheless, all initial specimens of Bohnenberger's Machine seemed to be lost for a long time.

In 2004, at the Kepler Gymnasium of Tübingen, Germany, (a Gymnasium is comparable to a Lyceum or a High School) a copy of the Machine of Bohnenberger was retrieved [5]. Much better than all other samples known to the authors this example matches the historical specifications as far as size, proportions, and construction details are concerned (Fig. 2). A private collector obtained a second copy being identical to the first one at an internet auction in 2010.

The finding of this Machine was the reason for new research into J.G.F. Bohnenberger and his scientific work. Important results are compiled below. To explain the historical background of Bohnenberger's Machine, the next section contains short biographies of J.G.F. Bohnenberger, a well known German geodesist and astronomer of the 19th century and close colleague of C.F. Gauß, as well as of his instrument maker J.W.G. Buzengeiger. A bibliography follows with source criticism on single copies of Bohnenberger's Machine that have been produced by J.W.G. Buzengeiger. The last two sections outline the technical development initiated by J.G.F. Bohnenberger's invention and finish with questions that still need to be clarified. Besides, the article contains a collection of excerpts of letters and papers translated into English to share these with an international public.

Fig. 2 The Instruments retrieved in 2004 (left) and in 2010 (right)

2 Biographies of J.G.F. Bohnenberger and J.W.G. Buzengeiger

Johann Gottlieb Friedrich (von) Bohnenberger (5 June 1765 - 19 April 1831)

Fig. 3 shows a family owned oil painting of J.G.F. Bohnenberger, a portrait done a few months before his death. As a scientist, he was born into a politically and socially agitated time [6]. He saw five sovereigns of the country of Württemberg in southern Germany, all of whom more or less influenced his career. The period from 1789 to 1815 embraces a very facetted development in the history of Württemberg caused by the French Revolution and the Napoleonic Era. A patched duchy became a united Kingdom (in 1806). In 1819, the new state got its own constitution and progressed much more peacefully from then on.

J.G.F. Bohnenberger was the second of three children of the marriage between Gottlieb Christoph Bohnenberger (1733-1807) and Johanne Frederike Schmid (1735-1801). He was born on 5 June 1765 in Simmozheim close to Weil der Stadt, the birthplace of the astronomer Johannes Kepler. His elder brother, whose name he got, had already died at the age of 19 months. His younger brother, Johann Christoph Bohnenberger (1767-1836), later became a fiscal officer at Tübingen. His father C.G. Bohnenberger, a priest in Simmozheim (1761-1784) and Altburg near Calw (1784-1807) in Württemberg had a strong interest in physical instruments, especially electrostatic ones. This passion for sciences and engineering was conveyed to his son at a rather early age. Between 1782 and 1784, J.G.F. Bohnenberger attended the Gymnasium in Stuttgart, the capital of Württemberg. After that, he joined the Tübinger Stift, which is a college run by the Lutheran Church of Württemberg together with the University of Tübingen and which still exists today. Like J. Kepler 200 years ago, he began to study theology there. Among his fellow students was the later famous philosopher G.W.F. Hegel (1770-1831). This education served in parallel to study mathematics and astronomy at the faculty of philosophy at the University of Tübingen.

In 1784, J.G.F. Bohnenberger's parents moved to Altburg, the new parish of his father. Five years later J.G.F. Bohnenberger passed his exam in theology. At his father's request, he became a vicar at Altburg. Tolerated by the Lutheran Church of Württemberg it was Altburg where the first important phase of his scientific work took place [7]. There, one can still find his first observatory having been restored in 2012. J.G.F. Bohnenberger summarized his initial scientific results concerning geodetic instruments and measurement methods in a manuscript about *Astronomic and Trigonometric Contributions to Create a Precise Map of the Duchy of Württemberg* [8]. Through the administration of the Lutheran church this manuscript reached Duke Karl Eugen of Württemberg. He granted Bohnenberger a scholarship for these achievements as well as the prospect of being employed at the University of Tübingen Observatory working together with J.G.F. Bohnenberger's academic teacher Christoph Friedrich Pfleiderer (1736-1821). In addition, he was mandated to map completely the duchy and was granted the privilege for the edition and copyright of the maps [9].

Fig. 3 J.G.F. Bohnenberger (1765-1831) – oil painting of F.S. Stirnbrand, 1831

In the autumn of 1793, J.G.F. Bohnenberger traveled to Altenburg in the duchy of Saxony-Gotha-Altenburg. The local observatory of Altenburg on top of the mountain Seeberg had been built between 1789 and 1792. It was equipped with new instruments from England, and it was headed by the astronomer Baron Franz Xaver von Zach (1754-1832), who instructed the young scientist in using astronomical instruments. At that time, the observatory was considered to be the most modern one in Germany [10]. A letter by F.X. von Zach to the physicist Georg Christoph Lichtenberg (1742-1799) at the University of Göttingen shows that J.G.F. Bohnenberger had probably met the Swiss geodesist Ferdinand Rudolph Hassler (1770-1843) in Altenburg. In the spring of 1794, both young scientists travelled together to Göttingen as F.X. von Zach also remarked [11].

Initially, J.G.F. Bohnenberger worked at the Göttingen Observatory together with the mathematician Abraham Gotthelf Kästner (1719-1800). In the spring semester of 1794, he attended G.C. Lichtenberg's lecture on experimental physics [12]. G.C. Lichtenberg comments on J.G.F. Bohnenberger and F.R. Hassler in a letter [13]: *"When I told you about a buyer, I did not at first really think of the government in Stuttgart but of a rich young Swiss of ample knowledge called Hassler who travels with Master Bohnenberger and who also stays with him here. Namely, as soon as I had told Mister Bohnenberger of the chronometer he immediately said, if it is good, Hassler will certainly buy it! It was only afterwards that he thought of the government of Württemberg or rather himself for he tries to move the government to buy the instrument for his operations (the making of a map of the duchy). If only we could trade this man. Professor Pfaff has reassured me that he is equally good in theory as he is in practice."* In another letter Lichtenberg writes about sound measurements during a thunderstorm [14]: *"I have very good news concerning this thunder, amongst them one of Court Counselor Kästner who was in close proximity of the action. One student stood opposite and was so calm and instructed that he even observed the pendulum. [...] Master Bohnenberger, a son of the electrician, a superb man, mathematician and cold-blooded observer, watched the spot exactly. He may have been 500 feet away."*

The hint that J.G.F. Bohnenberger observed a pendulum sheds a light on the background of his early invention of Kater's Reversible Pendulum. Many years later in a letter to his colleague Heinrich Christian Schumacher (1780-1850), an astronomer from Altona near Hamburg, Bohnenberger wrote about the history of that instrument during his time in Göttingen [15]. The reason for this was a dispute over authorship with the English physicist Henry Kater (1777-1835), who tried to take full credit for the Reversible Pendulum. Later, H. Kater took back his claim to be the first inventor in a letter to H.C. Schumacher [16].

In a letter to the mathematician Carl Friedrich Gauß (1777-1855), J.G.F. Bohnenberger sums up his time in Göttingen in retrospect [17]: *"Many times I have wished to be able to talk to you about many subjects! Perhaps I will decide on a vacation in order to get to know you in person and to see Göttingen again, [...] where I spent the happiest days of my life."*

After his return to Altburg in 1795, J.G.F. Bohnenberger devoted himself again to the work of mapping the duchy of Württemberg. He was employed at the University of Tübingen on 15 January 1796, where he became the head of the Observatory and was commissioned with the operation and maintenance of the physical apparatus. In January 1798, he was appointed Associate Professor. In May of that year, he married Johanna Christine Philipine Luz (1772-1821) from Naißlach near Altburg. The couple had two sons and two daughters. In 1803, J.G.F. Bohnenberger became full Professor. (Until 1818, however, he had no seat or vote in the senate.) At the same time, the young family moved into the eastern wing of the Tübingen Castle right next to the observatory in the northeast tower [5]. Also during that time J.G.F. Bohnenberger got calls to the universities in Freiburg, Germany, St. Petersburg, Russia, and Bologna, Italy. However, he stayed in Tübingen.

J.G.F. Bohnenberger's main scientific area was geodesy based on astronomic observations and trigonometric methods. Sextant and Repetition Theodolite were his most important measuring tools. Already in Göttingen he wrote his first book on surveying [18], his second one from 1826 was path breaking for theoretical geodesy [19]. Along with this scientific work, his research also addressed astronomy, weather observations and meteorological devices, practical mathematics including error analysis, mineralogy and electrical instruments. Besides, it is known that J.G.F. Bohnenberger was a gifted experimenter and systematic lecturer.

To continue his work on cartography, he defined the northeast tower of the Tübingen Castle as the central reference point. In January of 1810, J.G.F. Bohnenberger and his publisher Johann Friedrich Cotta (1764-1832) presented the first main part of the map of Württemberg to King Friedrich I of Württemberg (1754-1816). It was only a short time after Bohnenberger had released the first pages of the map when copies already appeared.

In March 1811, J.G.F. Bohnenberger presented his book *Astronomy* [20] and in December *Basics of higher Analysis* [21] to the censorship department of the government. Both manuscripts were intended to be study material for his lectures.

In October 1816, King Friedrich I died. He left the task of decreeing a constitution for the new Kingdom of Württemberg to his son Wilhelm I, a former student of J.G.F. Bohnenberger. The constitution became finally effective in 1819. One major part in this law was the decree of creating a new land register in order to ensure fair taxation. Contemporary sources, however, show that also another aspect played a role in this: a map of the entire kingdom should enable citizens to identify themselves with the new political as well as social circumstances. In 1819, as the necessary office of land registry was created, J.G.F. Bohnenberger's second phase of his geodetic work began. He was appointed associate scientific advisor to the office with the task of creating the main triangular grid of Württemberg. In addition to this, he had to supervise the work of the other geometers [22].

Already in 1812, J.G.F. Bohnenberger was awarded the Order of Civil Merit of Württemberg and the Minor Cross of the Great Order of the Golden Eagle. With this, he was also given peerage, and his last name changed to "von Bohnenberger". From a scientific point of view, his memberships of the academies of science in Göttingen, Munich, Paris, and Berlin are especially notable honors. J.G.F. Bohnenberger himself valued especially the regular correspondence with colleagues like H.C. Schumacher and C.F. Gauß.

On April 19, 1831, J.G.F. Bohnenberger died of a heart failure. His instrument maker J.W.G. Buzengeiger notified H.C. Schumacher of the death of their common friend [23]: *"Your High Well Born! With this I am giving you the utterly sad news that our revered friend Bohnenberger died on April 19, half past ten in the morning, after a long lasting illness and, according to the doctors', because of [...] a cardiac defect. The local University has lost a lot, a great lot. He was my friend and counselor in everything. [...] By exchanging and combining our ideas, never a thing went wrong when creating a new instrument. I cannot forgive one thing: his splendid thoughts, which he conveyed me and which mostly could not be carried out immediately due to a lack of money and time [...] and which consequently fell into oblivion."*

Johann Wilhelm Gottlob Buzengeiger (25 June 1778 – 26 October 1836)

The last citation characterizes very well the relationship between J.G.F. Bohnenberger and J.W.G. Buzengeiger. J. Broelmann adds [2]: *"An instrument maker did not publish and remained an invisible technician, ingenious but illiterate in status. The mechanic's knowledge of manufacturing remained unknown, and we cannot say why Gauß's mechanic was successful and the one of Zeiss was not. In publications, the mechanics themselves were marginally mentioned at best. J.G.F. Bohnenberger credited the Universitätsmechanicus who had manufactured his Machine very neatly and prettily."*

J.W.G. Buzengeiger grew up in Tübingen. His brother Karl Ignatz Buzengeiger (1771-1835), who excelled mathematically at an early age, studied mathematics in Tübingen under C.F. Pfleiderer together with J.G.F. Bohnenberger and later became Professor at the University of Freiburg, Germany. J.W.G. Buzengeiger could have probably had an equal career if the early death of his mother had not put him under the care of local welfare.

In 1793, J.W.G. Buzengeiger started an apprenticeship as a clock-maker with Johann Jacob Sauter (1743-1803) in Kornwestheim near Stuttgart. J.J. Sauter had been an apprentice and journeyman at the workshop of Philipp Matthäus Hahn (1739-1790) and got his master craftsman's diploma in 1793. Similar to J.G.F. Bohnenberger's father, P.M. Hahn was a priest and also became a very well-known inventor of clocks, scales, and a calculating machine. Around 1799, J.W.G. Buzengeiger had to retire for health reasons from his position as a journeyman at J.J. Sauter's workshop, now in Esslingen near Stuttgart, and moved back to Tübingen into the care of the family of an uncle. After about one year, he went to his brother in Ansbach, near Bayreuth, and worked with the Royal Prussian mechanic Du Mericeau. In the beginning of 1804, J.W.G. Buzengeiger set up his own workshop in Tübingen. Not being a member of a guild, he applied for the privilege of a university citizenship, which was granted to him in February 1804 [24].

It is uncertain whether J.W.G. Buzengeiger and J.G.F. Bohnenberger got to know each other already earlier or if they started their collaboration when working together at the university. As cited above, a congenial relationship between the two men developed. J.W.G. Buzengeiger manufactured the instruments for the university's physical cabinet, he also maintained them, he set up the demonstrations for the lectures in experimental physics, and he tutored the students at their practical work. Apart from these tasks, they had a flourishing Europe-wide trade in instruments, especially telescopes, sextants, barometers, scales, clocks, and electrical devices. For the most part, J.G.F. Bohnenberger acted as the agent for the contracts that J.W.G. Buzengeiger carried out. Reflected by the correspondence with H.C. Schumacher from Altona and with the Swiss mathematician Johann Caspar Horner (1774-1843) from Zurich one can notice, for example, established business relationships with these two scientists over several years [25]. Nevertheless, J.W.G. Buzengeiger applied in a petition to King Friedrich I. for a payment for his services. From then on, he received 150 guilders per annum for his work in assisting in the lectures in experimental physics [26].

After J.G.F. Bohnenberger's death in 1831, J.W.G. Buzengeiger had to rely increasingly on contracts from outside Tübingen. His biggest and probably most faithful customer was H.C. Schumacher. J.W.G. Buzengeiger complained of economic problems to him several times. More and more he had to build instruments on stock and hardly got contracts any more from the local university. Indeed, Johann Gottlieb Nörrenberg (1787-1862) was appointed successor of J.G.F. Bohnenberger in 1832, and it turned out soon that he procured his instruments from Paris.

Around 1834, J.W.G. Buzengeiger suffered from a severe intoxication as some barometer tubes burst while he boiled the mercury inside. He recovered very slowly, and his already feeble constitution confined him even more when working. He died on 26 October 1836.

More details about the work of J.G.F. Bohnenberger and J.W.G. Buzengeiger are published in [27].

3   Historic Documents of the Machine of Bohnenberger

The sources of J.G.F. Bohnenberger and J.W.G. Buzengeiger are quite numerous. However, there are hardly any personal notes. Up to now, the authors could not find any documents like laboratory diaries or construction drawings. Both persons did leave many traces in the files of the University of Tübingen and in various branches of public administration though. In addition, there is the aforementioned correspondence with H.C. Schumacher, J.C. Horner, and C.F. Gauß. Besides, the sources of J.G.F. Bohnenberger are much better in comparison to those of J.W.G. Buzengeiger. Especially, there are additional letters from J.G.F. Bohnenberger to his publisher J.F. Cotta [28] and various documents concerning his appointments as member of the academies of science mentioned above.

It is difficult to classify the letters as several categories apply. On the one hand, they are business correspondence; on the other hand, they also convey personal feelings, messages of other people, and often manuals for the instruments. All these aspects can very well follow each other smoothly in one letter. The traces in the above-mentioned public administration files are similarly diverse.

According to this assessment, the source situation of the Machine of Bohnenberger is both diverse and poor. A copy signed either by J.G.F. Bohnenberger or by J.W.G. Buzengeiger is still missing as the two copies of Fig. 2 do not carry any logo or serial number. In addition, there are no known technical documents of Bohnenberger's Machine, which allow conclusions about the way of the instrument's design or the history of its invention. Therefore, the authors have compiled a chronological list of documents regarding the apparatus. Based on the date of these documents, important facts of the Machine follow in ascending order.

The first notification and oldest description of the Machine of Bohnenberger (at least discovered by the authors) can be found in four letters and in the diary of the geodesist Johann Friedrich Benzenberg (1777-1846) from Düsseldorf in the Rhineland: In 1810, Benzenberg traveled to Switzerland and visited J.G.F. Bohnenberger, C.F. Pfleiderer as well the workshop of J.W.G. Buzengeiger on his way back. His visit lasted from 6 to 10 December 1810 [29]. On 15 December 1810 he sent a letter back to J.C. Horner and simply noted [30]: *"In addition, he* (i.e. J.G.F. Bohnenberger) *has invented a little momentum machine* (in German '*Schwungmaschine*') *that can be used to illustrate several physical problems of Astronomy, e.g. the reason for the backward advancement of the equinoxes."* Moreover, five days earlier, he wrote in his travel report published 1812 as a collection of letters [31]: *"B[ohnenberger] showed me a neat momentum machine, which explains in a vivid way the backward advancement of the points of equinoxes and at the same time illustrates its physical reason. A small wooden Earth, filled with lead, spins around its axis and is, wound up by means of a string, set into rotating motion, like the top boys play with. The poles are attached to a ring that possesses a suspension similar to a compass and can turn freely towards all sides. If you have the Earth spinning, you can move about the entire room with it without a change in direction of its axis of rotation. [...] In order to explain the reason for the retraction of the equinoxes with that machine, a small weight is attached to one side, the intention of which is to change the direction of the Earth axis. This represents the gravitational force of the Sun, which aims also to turn the ring of the Earth mass, put around the equator by the flattening, in order to have the Earth axis stand perpendicularly to its orbit. Now, there is a new phenomenon. The angle of 23.5 degrees, which the equator encloses with the Earth orbit, remains, and the Earth axis circles the point to which it is driven by its weight and which it cannot reach because of the rotation. Now it describes a circle around the poles of the Earth, which it passes through in 25,000 years. The points where the plane of the equator intersects the Earth orbit – the equinoxes – advance as well, and thus the curious phenomenon of the advancement of the equinoxes appears, which was already known to the oldest astronomers and the cause of which could not be given except by Newton until this great geometer explained it by the laws of general gravitation."* On 12 December 1810 he continued: *"I have ordered a momentum machine of Bohnenberger at the local clockmaker Buzengeiger which he delivers in a very orderly manufactured way for the small sum of 12 guilders."*

J.F. Benzenberg also reported about his discussion with C.F. Pfleiderer, who mentioned the Machine as well ([31], letter of 10 December 1810): *"Pfleiderer told me [...] that Segner has already presented an illustration of the momentum machine, which explains the retraction of the equinoxes, and that Bohnenberger would not know this however."* In his diary, J.F. Benzenberg confirms this statement in more detail but without a calendar date [29]: *"Bohnenberger found the momentum machine following the guidance of La Place. Later, Pfleiderer showed him that Segner has the same in his Astronomical Tracts in an imperfect design."*

It is very likely that this remark referred to a textbook [32] by Johann Andreas von Segner (1704-1777) which indeed contains the description and illustration of a device to demonstrate the precession motion of the Earth (Fig. 4). On the other hand, this apparatus lacks a well-defined complete cardanic suspension, which was essential for the further development of the gyroscope. Details about the guidance of P.S. Laplace, however, are still missing. J.F. Benzenberg's and J.G.F. Bohnenberger's explanation of the precession motion of the earth, though, resemble descriptions of P.S. Laplace from 1778 and 1779 [33].

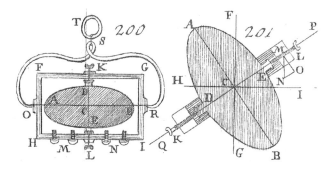

Fig. 4 Apparatus of J.A.v. Segner to demonstrate the precession motion [32]

Obviously, J.F. Benzenberg, a relevant expert, assessed J.G.F. Bohnenberger's invention as new and very important: Less than three weeks later, in his letter of 29 December 1810 from his travel report he told about his effort to introduce the device into the scientific community during a meeting in Frankfurt am Main [31]: *"Yesterday in the evening, we were in the Museum, an amicable foundation, dedicated to arts and sciences. It was the last meeting of the year and was honored by the Prince* (i.e. Karl von Dalberg (1744-1817), Grand-Duke of Frankfurt, appointed by Napoleon Prince Primate of the Confederation of the Rhine) *with his presence. Everybody contributed to the scientific enjoyment. I introduced Bohnenberger's Machine of the Equinoxes to the Prince, which seemed still new to him and which pleased him due to its simplicity."* In a footnote added later J.F. Benzenberg appends: *"During his stay in Paris in 1811, the Prince introduced the gentlemen* (Alexander) *von Humboldt, La Place and La Grange to Bohnenberger's Machine of the Equinoxes, which met with their considerable approval that they recommended it for the education in the Ecole Polytechnique."* Possibly, J.F. Benzenberg gave K. v. Dalberg his copy of the Machine, and K. v. Dalberg donated it the Ecole Polytechnique (see below the article of S.D. Poisson, 1813).

Shortly after J.F. Benzenberg had mentioned the Machine of Bohnenberger, the instrument showed up in a further letter dated 12 May 1811, when J.W.G. Buzengeiger sent a copy to J.C. Horner in Zurich [25].

As mentioned, J.G.F. Bohnenberger wrote the oldest systematic description of his Machine in 1817, and he included Fig. 1. However, the authors discovered an obviously older sketch of the instrument in a fragment of the hand-written manuscript [34] of J.G.F. Bohnenberger's book *Basics of higher Analysis* from 1811 (Fig. 5). The sketch is on the reverse side of the title page, and it was drawn with a graphite or lead pencil. Compared to Fig. 1, it is notable that the weight G, which

Fig. 5 J.G.F. Bohnenberger's sketch of the Machine from 1811 or earlier [34]

effects the precession motion, is missing. Another difference concerns the pedestal, which is not completely depicted; it is round (cf. Fig. 2). It seems as if the page has been cut. Probably J.G.F. Bohnenberger used the sheet twice for lack of paper: firstly, to sketch the Machine, secondly, for the later manuscript. Corresponding to this supposition, the fragment contains no reference of the sketch and no hint has been found up to now in the later published complete edition.

In 1813, i.e. also prior to J.G.F. Bohnenberger's paper from 1817, S.D. Poisson referred to the Machine in a mathematical treatise about gyro dynamics and stated that there was a copy in Paris [35]: *"Il existe ou cabinet de l'Ecole Polytechnique, une machine très-ingénieuse, imaginée par M. Bohnenberger, qui représente parfaitement les diverses circonstances du mouvement que nous considérons. L'auteur la destine à rendre sensible aux yeux le phénomène de la précession des équinoxes."* Using a still well-known notation, the article also includes Euler's Equations and Cardan Angles that apply to Bohnenberger's Machine (Fig. 6).

$$C dr = 0;$$
$$A dq + (A - C) \cdot rp \, dt = a'' gl M dt;$$
$$A dp + (C - A) \cdot rq \, dt = -b'' gl M dt;$$
$$p \, dt = \sin \varphi \cdot \sin \theta \cdot d\psi - \cos \varphi \cdot d\theta;$$
$$q \, dt = \cos \varphi \cdot \sin \theta \cdot d\psi + \sin \varphi \cdot d\theta;$$
$$r \, dt = d\varphi - \cos \theta \cdot d\psi;$$

Fig. 6 Set of equations used by S.D. Poisson to describe the axially symmetric gyro [35]

So, the Machine of Bohnenberger was already well known even before the release of its first systematic description in 1817. J.G.F. Bohnenberger himself writes accordingly [3]: *"I would not have decided to write about a subject which rather belongs to physical astronomy if the machine which I am going to describe here would not present phenomena that are notable enough to deserve the attention of a physicist, aside from its additional use to explain a remarkable movement in our solar system once it is duly set into motion. In addition, there is the request of many who have acquired such a machine that I may give an explicit instruction for the use of it and that I may develop, as far a possible without any calculus, the reasons of its motions."* A detailed explanation of the design and function follows this introduction. Some key sentences are: *"Now, one should attach the small weight, being denoted with G in the figure, on the ring EF near the end f of the axis. This is done by pinning the weight using the tacks g and h, which fit into two holes being placed in the ring and being there denoted with the same letters g and h. As long as the sphere does not turn around its axis, this weight will press down the Ring EF at the side F. [...] If one provides, however, the sphere with a rotational motion by means of the [silk] thread and if one aligns thereupon the [inner] ring in such a way that it is inclined at an arbitrary angle with respect to the horizon and that its weighted side is the lower one, it will be noticed that the angle of inclination of this ring and consequently also of the axis of the sphere remains constant with respect to the horizon whereas the axis no longer stays parallel to itself. It will rather move around very slowly together with the ring CD in a direction that is opposite to the direction of the rotational motion of the sphere."*

Another paper of J.G.F. Bohnenberger with practically the same content but with a slightly modified drawing was released in the following year [36]. With this second publication, the Machine of Bohnenberger became definitely well known. From 1818 onwards, more and more duplicates and variants appeared respectively. Typical examples from 1820s are those of the father of Georg Simon Ohm (1789-1854) [37] and of the instrument makers W. & S. Jones from London [38]. A few years later, the Machine of Bohnenberger was an inherent part of the catalogues of well-recognized instrument manufacturers [39, 40].

In 1827, Emil Nürnberger presented in his essay *"Advancement of the equinoxes"* for the Precursor of the *Brockhaus Encyclopedia* again the connection to France. First, he writes about the astronomical problem and then adds [41]: *"To illustrate the phenomenon, Professor Bohnenberger of Tübingen has invented an apt machine where a weight, which acts on a sphere, symbolizes the attracting force of the Sun and the Moon onto the Earth in a way that if the sphere is simultaneously set in rotating motion its axis will start to describe a small circle around an imagined vertical which is needed for the above given explanation of the advancement of the equinoxes. The mechanic Buzengeiger of Tübingen manufactures such machines for one Karolin, and these have been introduced on Laplace's suggestion at the Ecole Polytechnique due to their usability."*

Obviously, original copies were sold by J.W.G. Buzengeiger for many years. The correspondence between him and H.C. Schumacher shows that also Karl Friedrich Knorre (1801-1883), astronomer in Nikolayev, Ukraine, had ordered in 1832, apart from two barometers, a, *"rotation machine for the explanation of the laws on the rotation of the Earth around its axis"* [42]. In the same letter, J.W.G. Buzengeiger asked H.C. Schumacher whether he already had a Machine. Possibly, he sent him a copy in memory of J.G.F. Bohnenberger. Whether K.F. Knorre received his Machine is uncertain. One year later, J.W.G. Buzengeiger mentioned to H.C. Schumacher that he has received no answer from Nikolayev [43].

In 1833, the precursor of the University of Stuttgart purchased a copy from J.W.G. Buzengeiger for their collection of physical instruments. This example remained in the school's possession for a long time. Its last trace is an entry in an inventory of the University's institute for physics, where it was declared in 1966 to be obsolete [44]. Another copy is recorded in an old inventory list at the University of Tübingen, which was given away in 1857 to an unknown recipient [45].

For the further dissemination of Bohnenberger's Machine it was important that J.B.L. Foucault presented in 1852 his own variant, which he called, as cited above, Gyroscope [4]. Fig. 7 shows that instrument featuring a special low-friction bearing of the outer gimbal by a torsion wire. With this, he caused a debate about Bohnenberger's Machine [46], wherein the physicist Charles Cleophas Person (1801-1884) especially stated [47]: *"L'appareil de Bohnenberger est un petit instrument très portatif qu'on trouve dans tous les cabinets de physique. Il a été introduit en France par M. Arago á qui Bohnenberger en avait envoyé deux exemplaires."*

Fig. 7 J.B.L. Foucault's variant of the Machine of Bohnenberger

In the middle of the 19th century, the gyro with cardanic suspension had also reached America: Bohnenberger's Machine showed up in 1856 in the catalogue of the instrument maker Benjamin Pike Jr., New York [48]. Even before in 1832, the instrument is mentioned in the description of a similar apparatus by Walter R. Johnson, Philadelphia [49]. However, due to a mistake in probably oral tradition, the author of this paper names P.S. Laplace as the inventor.

A more general aspect about the majority of statements in newer literature on the origin of Bohnenberger's Machine should be added: In many encyclopedias as well as in Anglophone textbooks on gyro technology, the year 1810 is given as the time of invention of this instrument (e.g. [50]). However, this is done without giving a direct reference. Furthermore, J.G.F. Bohnenberger is regularly not credited as inventor but inadvertently his father G.C. Bohnenberger (sometimes even if J.G.F. Bohnenberger's article [36] from 1818 is cited), or the completely wrong initials C.A. are used alternatively. This incongruity may have its origin in the circumstance that the discussion on Foucault's Gyroscope skipped from Paris to Scotland: During that debate the Proceedings of the Philosophical Society of Glasgow published a lecture from 1857, which did not contain any initials of Bohnenberger, but cited his article of 1818 [51]. An additional report [52] about this lecture adds the two years 1810 and 1812 of J.F. Benzenberg's book [31].

It was the mathematician and engineer Edward Sang (1805-1890), who probably initiated the debate of Foucault's Gyroscope in Scotland. In a meeting of the Royal Scottish Society of Arts on 14 March 1856, he claimed that he had already described the experiment of proofing the rotational motion of the Earth by Bohnenberger's Machine in 1836. Also he mentioned how this instrument came to Scotland: In 1816 or 1817, F.J.D. Arago gave one of his copies to the mathematician John Playfair (1748-1819), who brought it to Edinburgh [53]. Moreover, E. Sang presented that copy, which did not carry a signature, on that occasion.

Today, the term Machine of Bohnenberger is almost unknown. The reason for that is unclear. The debate of the word Gyroscope and the missing signatures on J.W.G. Buzengeiger's originals may be two reasons. Around 1865 the initial name of the instrument was still usual [54, 55], some decades later the designation "gyro with cardanic suspension" prevailed.

4 Technical Derivatives

The quick dissemination of Bohnenberger's Machine led also to derivatives like the Fessel Gyroscope and the Stréphoscope, which were used in experimental physics [56, 57]. Due to its complex dynamics, the gyro was furthermore considered as a philosophical instrument in the middle of the 19th century [58] and grew more and more in popularity [2]. The first technical applications of the gyro in vehicle engineering and geodesy are connected with the approaching 20th century.

Especially the development of navigation and control instruments holds a multifaceted history [59, 60, 61], which is only outlined here. Promoted by the progress in rotor bearings and drive mechanisms, a famous group of scientists, engineers and technical laymen opened the way to such gyro applications. Specially important are W. von Siemens (1816-1892, entrepreneur, E.O. Schlick (1840-1913, engineer), M.G. van den Bos (theologian), E.A. Sperry (1860-1930, entrepreneur), H. Anschütz-Kaempfe (1872-1931, art historian), J. Boykow (1878-1935, actor, officer), R. Grammel (1889-1964, mathematician), F. Klein (1849-1925, mathematician), A. Sommerfeld (1868-1951, mathematician), N. Ach (1871-1946, psychologist), M. Schuler (1882-1972, engineer), and K. Magnus (1912-2003, physicist). The last five individuals were members of the University of Göttingen, which was a world's leading centre of gyro research at that time.

At the beginning of the 20th century developments of gyro technology followed two directions. The first one aimed at momentum wheels being used to stabilize rolling ships and monorail vehicles. However, the handling of the necessary kinetic energy, which had to be accumulated in the wheel, limited significantly practical applications. Today this kind of technology still exists as significant part of attitude control systems for satellites and space stations.

The other area of developing gyro technology was more successful. It based on J.B.L. Foucault's observation that well-directed restraints of the gimbals of Bohnenberger's Machine lead to specific devices detecting single components of rotary motions: The gyro became an inertial sensor for navigation and vehicle control. An example for the latter use is the first autopilot for aircraft, which was already developed before the First World War by the company of E.A. Sperry [62].

The rivalry between H. Anschütz-Kaempfe and E.A. Sperry regarding the first serviceable gyro compass for ship navigation is legendary. It involved even Albert Einstein (1879-1955) [63]. Two aircraft instruments of their companies, artificial horizon and directional gyro, show specially their origin from Bohnenberger's Machine and are shown in a simple and in a mature variant in Fig. 8 and 9. J.G.F. Bohnenberger himself anticipated already in 1817 the directional gyro [3]:

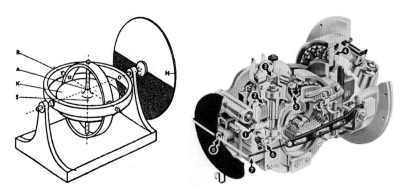

Fig. 8 Artificial horizons: Anschütz-Fliegerhorizont (left), Sperry H.L. 6 (right)

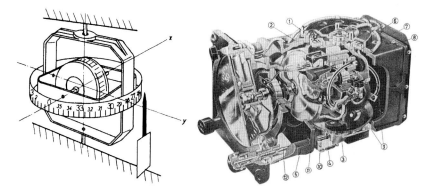

Fig. 9 Directional gyro: Sperry AN 5735-1 (left) and Sperry 16 526-0 (right)

"While the sphere is spinning around its axis, this axis will maintain permanently that direction which was given to it [at the beginning]. This will happen well then if one takes hold of the whole machine at its base H and starts moving it. While carrying around the Machine, one can move in arbitrary directions and with arbitrary velocities, and the axis of the sphere will permanently remain parallel to itself and will permanently stay aligned to north like a magnetic needle if one has, for example, orientated it at the beginning to north."

Between the two world wars and during the cold war, gyro instruments were continuously improved. This included the transition from pure mechanical devices to electromechanical ones (Fig. 8, 9). In addition, the gyro with cardanic suspension was understood as a mechanical calculator, which measures angular rates and calculates Euler Angels thereof. Consequently, further performance improvements aimed at reducing gyro instruments to their sensor property and transferring their mathematical part to electronic computers. As an example, Fig. 10 illustrates the (angular) rate gyro, which results from fixing the outer gimbal of Bohnenberger's Machine and constraining the inner one by a spring and a damper. In the last third

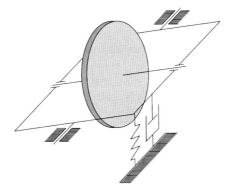

Fig. 10 Principle of the rate gyro

of the 20th century the effort of simplifying gyro sensors mechanically led to the laser and to the fibre-optical gyros as well as to micro-electro-mechanical sensors (MEMS) [64]. The last of these types facilitated the usage even in mass products like care navigation systems and smart phones.

Around 1940, the development of the rocket A4 initiated the development of inertial platforms based on two Machines of Bohnenberger with orthogonal rotor axes and three accelerometers. All these sensors were mounted on a common frame with an own cardanic suspension and allowed the full navigation of a rigid body with six degrees of freedom. Inertial platforms were also subject to mechanical simplifications leading to the replacement of the cardanic suspension of the frame by mathematical algorithms. Today, the combination of such advanced systems, called strap-down-platforms, with satellite navigation receivers form state-of-the-art integrated navigation systems.

To summarize the technical development caused by Bohnenberger's Machine, six chronological phases can be distinguished:

1. Preparative phase – antiquity to 18th
   century invention of the cardanic suspension, usage of tops as a toy
2. Using gyros for demonstrations – 19th century
3. First technical applications, mechanical gyros – ca. 1900-1940
4. Electromechanical gyros and inertial platforms – ca. 1940-1975
5. Electronic and digital systems with gyros – from ca. 1975
6. MEMS gyros for the mass market – from ca. 2000

5   Epilogue

The principle of the gyro with cardanic suspension, which was systematically described by J.G.F. Bohnenberger for the first time, crucially promoted the development of important instruments for navigation, vehicle control and geodesy. Bohnenberger's Machine became quickly part of many physical collections. This is hardly astonishing in view of the various connections J.G.F. Bohnenberger had to his colleagues. On the other hand, he could not survey the importance of his invention, which, as an example, helped considerably to land humans on the moon. For him, the instrument was, in accordance to his modest character, a simple didactic tool for lectures on astronomy. Probably, J.B.L. Foucault did also not anticipate the effect of his contribution to generating a technology of international standing. Therefore, Bohnenberger's Machine is a typical example how unintentional basic research can lead to a big field of technical applications.

The search for some of the initial copies of this instrument clarified interesting historical aspects. Nevertheless, there are still open questions: Although probably eight original copies from J.W.G. Buzengeiger's workshop could be identified in documents, the origin of the two examples of Fig. 2 is still unknown. In addition, there is not yet any certain evidence on how the device was developed or produced. The issue how many copies had been built and who owned one of them will presumably never be completely cleared up. Also, J.G.F. Bohnenberger's contact with P.S. Laplace and F.J.D. Arago is not clarified. Nevertheless, the authors are confident that the research on Bohnenberger's Machine will continue.

References

[1] Tobin, W.: The Life and Science of Léon Foucault. Cambridge Univ. Press, Cambridge (2003)
[2] Broelmann, J.: Intuition und Wissenschaft in der Kreiseltechnik. Deutsches Museum, Munich (2002)
[3] Bohnenberger, J.G.F.: Beschreibung einer Maschine zur Erläuterung der Geseze der Umdrehung der Erde um ihre Axe, und der Veränderung der Lage der lezteren. Tübinger Blätter für Naturwissenschaften und Arzneikunde 3, 72–83 (1817)
[4] Foucault, J.B.L.: Sur les phénomènes d'orientation des corps tournant entrainés par un axe à la surface de la terre. Comptes Rendus Hebd Séances Acad. Sci. 35, 424–427 (1852)
[5] Wagner, J.F., Sorg, H.W., Renz, A.: The Machine of Bohnenberger. Europ. J. Navig. 3(4), 69–77 (2005)
[6] Reist, H.: Johann Gottlieb Friedrich von Bohnenberger, Gedanken zum 200. Geburtstag. Allg. Vermessungsnachr 72, 218–241 (1965)
[7] Baumann, E.: Bohnenberger und Altburg. DVW Mitt, Landesverband Baden-Württemberg 56(1), 77–102 (2009)
[8] Bohnenberger, J.G.F.: Astronomische und Trigonometrische Beyträge zur Verfertigung einer genauen Charte von dem Herzogthum Wirtemberg. State Library of Württemberg. Sig. Cod. Math. 4, 31 (1793)
[9] Bohnenberger, J.G.F.: Report on the negotiations with Duke Friedrich. University Library of Tübingen, Sig. Md 901a (1793-1803)
[10] Stumpf, M., Marold, T.: Zur Geschichte der Sternwarten Gothas. Gothaer Museumsheft, 33–48 (1985)
[11] von Zach to Lichtenberg, March 12, 1794. In: Joost, U., Schöne, A., Hoffmann, J. (eds.) Georg Christoph Lichtenberg, Briefwechsel, Beck, Munich, vol. 4, pp. 229–235 (1992)
[12] Heerde, J.H.: Das Publikum der Physik: Lichtenbergs Hörer, p. 114. Göttingen, Wallstein (2006)

[13] Lichtenberg to Heyne, June 14, 1794. In: Promies, W. (ed.) Georg Christoph Lichtenberg, Schriften und Briefe, Zweitausendeins, Frankfurt/Main, vol. 4, pp. 881–883 (1998)

[14] Lichtenberg to Reimarus, August 18, 1794. In: Promies, W. (ed.) Georg Christoph Lichtenberg, Schriften und Briefe, Zweitausendeins, Frankfurt/Main, vol 4, pp. 895–899 (1998)

[15] Bohnenberger to Schumacher, December 26, 1824. State Library of Berlin, Sig. Nachlass Schumacher

[16] Kater, H., Schumacher, H.C.: Auszug aus einem Schreiben des Capitains Henry Kater an den Herausgeber, London (Junius 7, 1825); Astron Nachr. 4(85), 225–226

[17] Bohnenberger to Gauß, March 18 ,1818. State Library of Lower Saxony. Sig. Cod. MS. Gauß

[18] Bohnenberger, J.G.F.: Anleitung zur geographischen Ortsbestimmung, vorzüglich vermittelst des Spiegelsextanten. Vandenhoek & Ruprecht, Göttingen (1795)

[19] Bohnenberger, J.T.F.: De computandis dimensionibus trigonometricis in superficie terrae sphaeroidica institutis. Eifert, Tübingen (1826)

[20] Bohnenberger, J.G.F.: Astronomie. Cotta, Tübingen (1811)

[21] Bohnenberger, J.G.F.: Anfangsgründe der Höheren Analysis. Cotta, Tübingen (1811)

[22] Kohler, K.: Die Landvermessung im Königreich Württemberg. Cotta, Stuttgart (1858)

[23] Buzengeiger to Schumacher, April 21, 1831. State Library of Berlin, Sig. Nachlass Schumacher

[24] Archives of the University of Tübingen, Sig. 9/6 19-25. The spelling of Du Mericeau is not clear

[25] Letters of Buzengeiger to Horner, Central Library of Zürich, Sig. MS M 5.12

[26] Buzengeiger to King Friedrich I, 9 September 1815. Archives of the State of Baden-Württemberg, Sig. E 202 BÜ 737

[27] Baumann, E. (ed.): DVW Landesverein Baden-Württemberg e.V. Mitteil 57(2) (2010)

[28] Letters Bohnenberger to Cotta. Archives of Literature, Marbach, Sig. Cotta Br

[29] Benzenberg, J.F.: Unpublished works, diary. Heinrich Heine Institute, Düsseldorf (1810)

[30] Vierteljahresschrift der Naturforschenden Gesellschaft in Zürich 22, 126 (1877)

[31] Benzenberg, J.F.: Briefe geschrieben auf einer Reise in die Schweiz im Jahr 1810, vol. 2, pp. 357–359, 387. Schreiner, Düsseldorf (1812)

[32] Segner, J.A.: Astronomische Vorlesungen, vol. 2, pp. 650–653. Curts, Halle (1776)

[33] Laplace, P.S.: Œuvres complètes de Laplace, publiées sous les auspices de l'Académie des Sciences. Gauthier-Villars, Paris (1893)

[34] State Library of Württemberg, Sig. Cod. Math. 4° 64a. It should be mentioned in addition that the covering letter (Archives of the State of Baden-Württemberg, Sig. E 5 BÜ) of the manuscript is dated December 21, 1811 and that probably the book manuscript had originally been handed in together with the covering letter

[35] Poisson, S.D.: Mémoire sur un cas particulier du mouvement de rotation des corps pesans. J. École Polytechn. 9(16), 247–262 (1813)

[36] Bohnenberger, J.G.F.: Beschreibung einer Maschine, welche die Gesetze der Umdrehung der Erde um ihre Axe, und der Veränderung der Lage der Erdaxe zu erläutern dient. Annalen der Physik 30, 60–71 (1818)

[37] May, P.: Georg Simon Ohm – Leben und Wirkung. Mayer, Erlangen (1989)

[38] Simpson, A.D.C.: A Sub-Contractor of W & S Jones Identified. Bulletin Sci. Instrument Soc. 39, 23–27 (1993)

[39] Pixii, père et fils, Catalogue des principaux instruments de physique, chimie, optique, mathématiques et autres a l'usage des sciences. Pixii, Paris (1845)

[40] Lerebours, S.: Catalogue et prix des instruments d'optique, de physique, de chimie, de mathématiques, d'astronomie et de marine. Lerebours et Secretan Paris (1853)

[41] Nürnberger, J.F.: Vorrücken der Nachtgleichen. In: Allgemeine deutsche Real-Encyklopädie für die gebildeten Stände, 7th edn., vol. 11, pp. 786–787. Brockhaus, Leipzig (1827)
[42] Buzengeiger to Schumacher, July 30, 1832. State Library of Berlin, Sig. Nachlass Schumacher
[43] Buzengeiger to Schumacher, July 24, 1833. State Library of Berlin, Sig. Nachlass Schumacher
[44] Archives of the University of Stuttgart, Sig. ZA 640 Inventar D, Inventar Physikalisches Institut 1937 (-1974)
[45] Physikalisches Institut der Universität Tübingen, Inventar der Sternwarte Tübingen
[46] Person, C.C.: Disposition de l'appareil de Bohnenberger pour les différentes latitudes. Comptes Rendus Hebd Séances Acad. Sci. 35, 549–552 (1852)
[47] Person, C.C.: L'appareil de Bohnenberger pour la précision des équinoxes peut servez à constater la rotation de la Terre. Comptes Rendus Hebd Séances Acad. Sci. 35, 417–420 (1852)
[48] Pike, B.: Pike's Illustrated Descriptive Catalogue of Optical, 2nd edn. Mathematical and Philosophical Instruments, p. 142. Pike, New York (1856)
[49] Johnson, W.R.: Description of an Apparatus Called the Rotascope, for Exhibiting Several Phenomena and Illustrating Certain Laws of Rotary Motion. Am. J. Sci. 21, 265–280 (1832)
[50] Yust, W. (ed.): A New Survey of Universal Knowledge, p. 47. Encycl Britannica, Chicago (1956)
[51] Hunt, E.: On Certain Phenomena Connected With Rotatory Motion, the Gyroscope, Precession of the Equinoxes, and Saturn's Rings. Proc. Royal Philos. Soc. Glasgow. 4(1), 130–157 (1859)
[52] Phenomena Connected with Rotatory Motion. Mechanics' Magaz. 67, 587–588 (1857)
[53] Sang, E.: Remarks on the Gyroscope, in Relation to his "Suggestion of a New Experiment Which Would Demonstrate the Rotation of the Earth". Transact. Royal Scottish Soc. Arts 4, 413–420 (1856)
[54] Chevalier, C.: Catalogue explicativ et illustré des instruments de physique expérimentale, d'optique expérimentale, chimie, astronomie, minéralogie, chirugie. Chevalier, Paris (1861)
[55] Gantzer, R.: Über die Rotationsmaschine von Bohnenberger. Ratz, Jena (1867)
[56] Heinen, F.: Über einige Rotations-Apparate, insbesondere den Fessel'schen. Vieweg, Braunschweig (1857)
[57] Gruey, M.L.J.: Le Stréphoscope Universel. Chaix, Paris (1883)
[58] Magnus, K.: Kreisel: Theorie und Anwendungen. Springer, Heidelberg (1971)
[59] Sorg, H.W.: From Serson to Draper. Navigation 23(4), 313–324 (1976)
[60] Wrigley, W.: History of Inertial Navigation. Navigation 24(1), 1–6 (1971)
[61] AGARD, The Anatomy of the Gyroscope, Part I-III, Agardograph No. 313. NATO, Neuilly sur Seine (1988/1990)
[62] Hughes, T.P.: Elmer Sperry: Inventor and Engineer. Johns Hopkins, Baltimore (1971)
[63] Lohmeier, D. (ed.): Einstein, Anschütz und der Kieler Kreiselkompass, 2nd edn. Raytheon Marine, Kiel (2005)
[64] Lawrence, A.: Modern Inertial Technology: Navigation, Guidance, and Control, 2nd edn. Springer, New York (2001)

On the Historical Development of Human Walking Dynamics

Werner Schiehlen

Abstract. Human walking is an interdisciplinary research topic. It started in the Ancient World with early observations and questions in philosophy, it was treated in the Middle Ages with experiments and data collection by physiologists, and in the 20th century models were designed, equations of motion were generated and simulations by multibody dynamics approaches were performed. More recently parameter optimization was used to overcome the problem of muscle overactuation and inverse dynamics methods were introduced. In the first part of the paper the early developments and mechanism models are described while in the second part parameter optimization is applied as an example of recent research results on gait disorder simulations.

Keywords: Early observations, Questions in philosophy, Experiments by physiologists, Multibody dynamics, Parameter optimization, Gait disorder.

1 Introduction

Human walking dynamics is based on a huge body of research results documented by monographs, textbooks and many journals. Human body dynamics is devoted on one hand to the explanation of the complexity of the walking motion of healthy people and to the improvement of the performance of sporting walkers, and on the other hand to the development of medical therapies and mechanical prostheses for handicapped people with gait disorders. Therefore, human walking is an important topic for medical sciences, in particular physiology, and for sports sciences. From a more general point of view biomechanics includes human walking as a research topic, too. Advanced simulation technologies as well as multibody dynamics provide many tools supporting human walking research.

Werner Schiehlen
Institute of Engineering and Computational Mechanics
University of Stuttgart
Pfaffenwaldring 9, 70550 Stuttgart, Germany

This paper is an extended version of the author's summary published in PAMM, Ref. [1]. It is organized as follows. In Chap. 2 early developments in the Ancient World and the Middle Ages are summarized. Chap. 3 is devoted to mechanism models. Chap. 4 presents contributions to parameter optimization while Chap. 5 shows some recent research results on gait disorder simulations.

2   Early Developments and Kinematical Gait Analysis

A first contribution to human walking is due to the Greek philosopher Aristotle (384 – 322 BC). His book written in Greek language is entitled *De Motu Animalium*. This book was translated to English and commented on by the American philosopher Martha C. Nussbaum [2] and it was published by Princeton University Press in 1978. In the Introduction of this book it is stated:

> «We see animals moving around – walking, swimming, flying, creeping. And we ourselves are also moving animals. Why? What role does motion from place to place play in animal lives? What would be an adequate explanation of a particular animal movement? Is there some general account we can give of these phenomena that will hold good for humans and animals alike? With these questions, and with the hope that such a general answer will be found, Aristotle begins the De Motu Animalium.»

Thus, Aristotle made observations of animals and humans, he stated many questions but he did not give scientifically satisfactory answers.

In the Middle Ages a book with the same title *De Motu Animalium* was written by the Italian mathematician Giovanni Alfonso Borelli (1608 – 1679) in Latin language. This book [3] was published after he passed away in 1680 by Angeli Bernabo in Roma, and there have been many editions [4, 5], see Fig.1, reprints [6] and translations [7, 8] later on.

The translation by the Belgian scientist Paul Maquet [8] is based on the 1743 edition [5] by Peter Gosse in The Hague, see Fig. 1. In the Preface of this translation it is stated:

> «De Motu Animalium of Borelli is sometimes referred to in papers on biomechanics. Having acquired the book, Maquet was amazed by the concepts which the figures illustrating the work suggested. ..... De Motu Animalium thus comprises two parts. The first one is divided in 23 chapters and 224 propositions and deals with the movements of limbs and displacement of man and animals. ..... The second part of De Motu Animalium is divided in 22 chapters and 233 propositions. In this part Borelli deals with physiology and analyses the working of viscera considered as machines.»

Chapter 19 of Part 1 is devoted to the walking of bipeds with the propositions illustrated by the Tables X, XI, and XII as shown in Fig. 2. The kinematics of human gait is elaborated, a simple framework model is used to illustrate three steps walking straight ahead, and moving on ice with skates is depicted. The book of Borelli is generally considered as the very first treatise on biomechanics.

On the Historical Development of Human Walking Dynamics

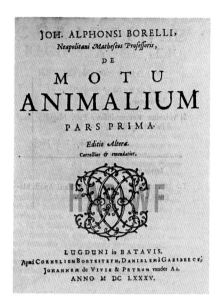

Fig. 1 Printed title page of the 1685 edition [4] by Boutesteyn in Leiden, and engraved title page of the 1743 edition [5] by Peter Gosse in The Hague featuring physico-mechanics of animal and men motion

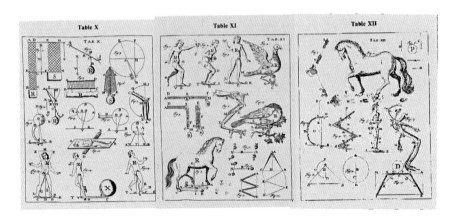

Fig. 2 Kinematics of human gait (Table X, bottom and Table XI, top), simple framework model for three steps walking straight ahead (Table XI, bottom right), and moving on ice with skates (Table XII, top left)

A century later in 1787 the French physician Paul Joseph Barthez (1734 – 1806) published a book entitled *Nouveaux elements de la science de l'homme* which was translated in German by Kurt Sprengel [9]. Barthez dealt mainly with the mechanics of standing, he considered the bending of the spine in different directions and the resulting actions for humans and animals. In particular he pointed out some mistakes of Borelli with respect to the principles of mechanics which have been disclosed by several mathematicians like Varignon, Parent, Pemberton and Hamberger. However, there aren't any figures and formulae in Barthez's book, Fig. 3.

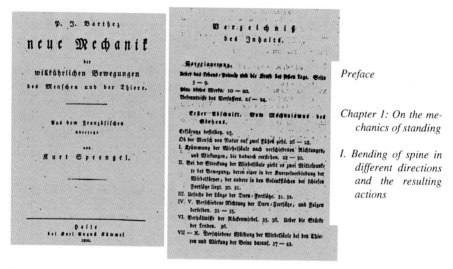

Fig. 3 Title page of the German translation [9] of the French book *Nouveaux elements de la science de l'homme,* and table of contents

The famous book *Mechanik der menschlichen Gehwerkzeuge* was co-authored by two German brothers, the physicist Wilhelm Eduard Weber (1804 – 1891) and the physician Eduard Friedrich Weber (1806 – 1871) and published 1836 by Dieterich in Goettingen. This book [10] characterizes an interdisciplinary approach to the science of human walking and running. The human body has been divided in carrying and supporting parts where the legs are modelled like pendula swinging at the trunk. The book presents detailed measurements, e.g. of the pelvis inclination on living humans and also graphical motion analysis, Fig. 4 and Fig. 5. Nevertheless, formulae haven't been used. Recently, the book has been translated to English, too [11].

Improved motion analysis with selfregistering methods using sophisticated shoes was performed by the German physician Hermann Vierordt (1853 – 1943) and documented in his book *Das Gehen des Menschen in gesunden und kranken Zuständen* [12] published 1887 by Laupp in Tuebingen. Vierordt studied walking in spatial and temporal relations, e.g. for old man and amputee of the right femur with right crutch, Fig. 6.

On the Historical Development of Human Walking Dynamics

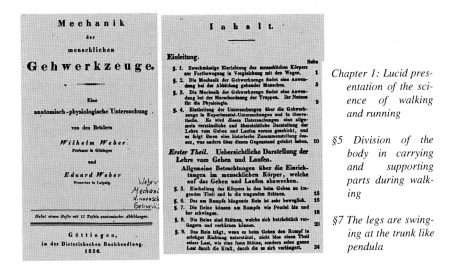

Chapter 1: Lucid presentation of the science of walking and running

§5 Division of the body in carrying and supporting parts during walking

§7 The legs are swinging at the trunk like pendula

Fig. 4 Title page of the German book [10] *Mechanik der menschlichen Gehwerkzeuge* and table of contents

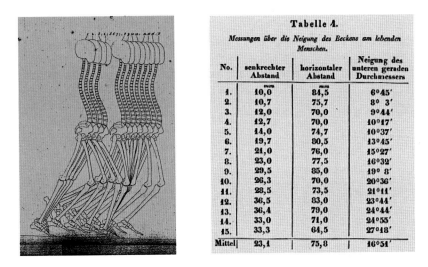

Fig. 5 Graphical gait analysis and measurements of pelvis inclination on living humans from [10]

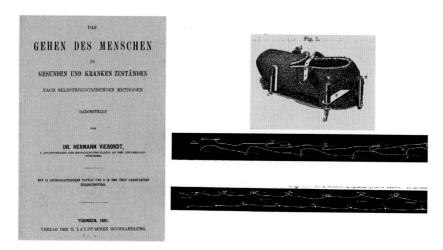

Fig. 6 Title page of the German book [12] *Das Gehen des Menschen in gesunden und kranken Zuständen,* the experimental device and documented experiments for an old man (top) and an amputee of right femur with right crutch (bottom)

3 Mechanism Models and Dynamical Principles

A theoretical foundation for mechanics of the living body was presented by the German physicist and mathematician Otto Fischer (1861 – 1916) in his book *Theoretische Grundlagen für eine Mechanik der lebenden Koerper* [13] published 1906 by Teubner in Leipzig. For the modelling of human walking Fischer used planar three-link systems and spatial n-link systems. The book is organized as follows:

I. General part
    A. The three-link planar joint system
    B. The n-link planar and the spatial joint system
II. Special part
    A. Applications to the mechanics of the human body
    B. Some applications on the motions of machines

Fischer considered a free flying three-link planar system as depicted in Fig. 7 with 5 degrees of freedom. He reduced the complexity of the equations of motion by introducing barycenters (Hauptpunkte) and centers of mass (Schwerpunkte), Fig. 8. The barycenter approach is explained in detail by Wittenburg [14] while Fischer used parallelogram mechanisms to evaluate the system's overall center of mass, Fig. 9. An application to the human body shows Fig.10.

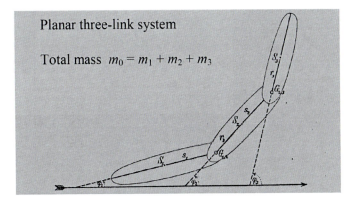

Fig. 7 Free-flying three-link system

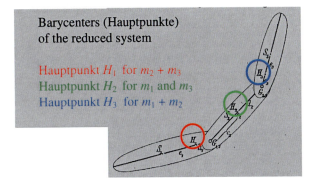

Fig. 8 Barycenters for reducing the complexity of the equations of motion by replacing the body centers of mass

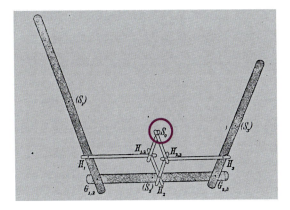

Fig. 9 Fischer parallelograms for the construction of the overall center of mass S_0

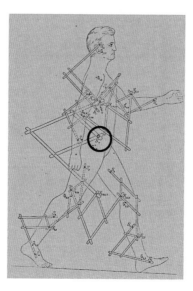

Fig. 10 Construction of the overall center of mass S_0 for a walking human [13]

Fischer used Lagrange's equations to generate the equations of motion. Due to his barycentre approach the equations of motion are very compact, Fig. 11.

$$m_0 \cdot x_0'' = Q_{x_0},$$
$$m_0 \cdot y_0'' = Q_{y_0},$$
$$m_0\{k_1^2 \cdot \varphi_1'' + d_1 c_2 \cos(\varphi_1 - \varphi_2) \cdot \varphi_2'' + d_1 c_3 \cos(\varphi_1 - \varphi_3) \cdot \varphi_3''$$
$$+ d_1 c_2 \sin(\varphi_1 - \varphi_2) \cdot \varphi_2'^2 + d_1 c_3 \sin(\varphi_1 - \varphi_3) \cdot \varphi_3'^2\} = Q_{\varphi_1},$$
$$m_0\{k_2^2 \cdot \varphi_2'' + d_2 c_3 \cos(\varphi_2 - \varphi_3) \cdot \varphi_3'' + c_2 d_1 \cos(\varphi_2 - \varphi_1) \cdot \varphi_1''$$
$$+ d_2 c_3 \sin(\varphi_2 - \varphi_3) \cdot \varphi_3'^2 + c_2 d_1 \sin(\varphi_2 - \varphi_1) \cdot \varphi_1'^2\} = Q_{\varphi_2},$$
$$m_0\{k_3^2 \cdot \varphi_3'' + c_3 d_1 \cos(\varphi_3 - \varphi_1) \cdot \varphi_1'' + c_3 d_2 \cos(\varphi_3 - \varphi_2) \cdot \varphi_2''$$
$$+ c_3 d_1 \sin(\varphi_3 - \varphi_1) \cdot \varphi_1'^2 + c_3 d_2 \sin(\varphi_3 - \varphi_2) \cdot \varphi_2'^2\} = Q_{\varphi_3}.$$

Fig. 11 Equations of motion of a free-flying three-link joint system from [13]

The five generalized coordinates are the translations x_0, y_0 of the overall center of mass, and the rotations φ_1, φ_2, φ_3 of the three links while the five generalized forces Q include the gravitational and muscle forces.

Furthermore, Fischer applied inverse dynamics on a right leg system for the evaluation of the torque generated by the muscles. Thus, Fischer introduced mechanism theory to biomechanics. Some of his important results are summarized as follows:

- The swing motion of the leg is not only a pendulum oscillation as suggested by Weber but also driven by muscles.
- The computed muscle torque does not identify the muscles directly involved in the different phases of the leg motion.
- Fischer suggested only that most probably Iliopsoas, Rectus Femoris and Tibialis Anterior are initiating the swing phase and their contraction disappear after one third of the swing period.

More recently the Russian mechanician and mathematician Vladimir V. Beletsky published a book on *Two Leg Walking – The Model Problems of Dynamics and Control*.[15] in 1984. Beletsky used a planar mechanism model with seven bodies, and proved the existence of limit cycles for the trunk motion, Fig. 12.

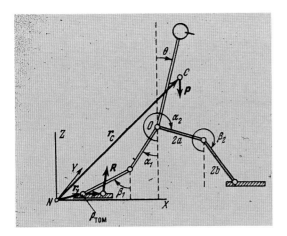 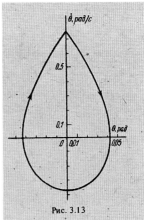

Fig. 12 Model and trunk limit cycle of walking biped from [15]

Complex mechanism models for human walking research are recently designed by multibody system dynamics, see e.g. Silva and Ambrosio [16] or Chenut, Fisette and Samin [17]. For sophisticated models the commercial software AnyBody [18] developed by Rasmussen can used, Fig. 13.

4 Parameter Optimization for Muscle Coordination

The problem of muscle coordination of human walking already stated by Fischer [13] is a central research topic in biomechanics. The review articles by Zajac et al. [19, 20] present a comprehensive introduction to concepts, dynamics, simulations and clinical implications. With respect to optimization and control the contributions by Anderson and Pandy [21, 22] are very essential using global parameter optimization for jumping and human gait. Furthermore, Thelen et al. [23], and Thelen and Anderson [24] proposed the computed muscle control method with

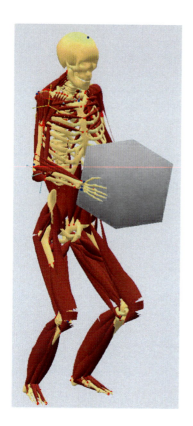 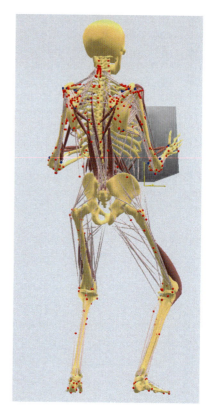

Fig. 13 Commercial model of a human with skeleton and muscles from Anybody Technology [18]

local optimizations to produce empirically reasonable muscle activations. Seth and Pandy [25] use the neuromusculoskeletal tracking method which shows a high computational speed up compared to parameter optimization.

There are several choices for the optimization criteria of human walkers. Healthy walker usually will walk with minimal energy costs per distance travelled what is useful for walkers with orthoses or prostheses, too. For disabled walkers the normal gait aesthetics could be an additional criterion. Elderly or injured walkers may prefer a minimal pain criterion which is related to the joint reaction forces. More general criteria are the least action principle by De Sapio et al. [26] and the min/max criterion by Rasmussen et al. [27].

5   Gait Disorder Simulations Considering Energy and Aesthetics

Mechanism models and parameter optimization are important tools for studies of human walking. These approaches are applied to a gait disorder simulation. First of all a spatial mechanism model is introduced, then the inversion of contraction and activation dynamics are added, the constraints of the parameter optimization are considered and a numerical example is presented. The gait disorder is modeled in the simulation by adding a weight of two kilograms to one foot resulting in an unsymmetrical walking.

The three-dimensional model of the human body used by Garcia-Vallejo and Schiehlen [28] for the parameter optimization approach is composed of 7 moving bodies: a body called HAT representing the pelvis and the trunk, two thighs, two shanks and two feet, see Fig. 14. The thighs are connected at the hips to the HAT by spherical joints, the shanks and thighs are connected by revolute joints representing the knees, and the foot and shanks are connected by revolute joints representing the ankles. This is a simplification compared to other three-dimensional models that can be found in the literature. However, this simplification allows the derivation of the symbolic equations of motion by software Neweul-M^2 [29] for the free flying tree structure composed by the mentioned 7 bodies without any external constraint.

The kinematic tree in Fig. 14 is described by the following position vector of $n_c = 16$ generalized coordinates:

$$q = [x_{I1} \; y_{I1} \; z_{I1} \; \alpha_{I1} \; \beta_{I1} \; \gamma_{I1} \; \alpha_{13} \; \beta_{13} \; \gamma_{13} \; \beta_{34} \; \beta_{45} \; \alpha_{16} \; \beta_{16} \; \gamma_{16} \; \beta_{67} \; \beta_{78}]^T \quad (1)$$

where the subscript I refers to the inertial frame, subscript 1 refers to body HAT, which is in composed of the pelvis and the trunk, subscripts 3 and 6 refer to right and left thighs, respectively, subscripts 4 and 7 refer to right and left shanks, respectively, and subscripts 5 and 8 refer to right and left feet, respectively. When a subscript is written as ij it means a relative motion of body j with respect to body i.

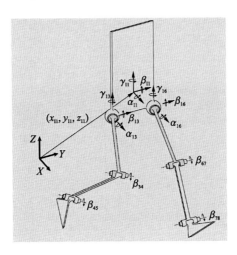

Fig. 14 3D-Model of human body

Once the kinematic tree representing the skeleton is described, the contact of this tree with the ground is added. The contact conditions in the different walking phases must be included by means of unilateral constraints. However, due to the use of an optimization framework in which it is possible to constrain the normal contact forces to be positive only, the contact with the ground can be modeled using the simple bilateral constraints associated to revolute joints. Therefore, the contact forces can be easily added to the model by using Lagrange multipliers as:

$$M(q)\ddot{q} + k(q,\dot{q}) = q_r(q,\dot{q}) + B A f^m + C_{ph}^T \lambda_{ph} \quad (ph = 1, 2, \ldots 8) \tag{2}$$

where $M(q)$ is the $(n_c \times n_c)$-mass matrix of the system, q is the $(n_c \times 1)$-position vector, respectively. k is a $(n_c \times 1)$-vector describing the generalized Coriolis forces, q_r is a $(n_c \times 1)$-vector including generalized gravitational forces, passive generalized moments at the joints due to tissues interacting with the joints and generalized viscous damping torques at the knees and hips according, and $B A f^m$ is a $(n_c \times 1)$-vector that includes the generalized forces exerted by the muscles actuating the model. The $(N_m \times 1)$-vector f^m summarizes the forces generated by a reduced set of Nm muscles included in the model as described in Ref. [26]. Matrix A is the constant $(n_b \times N_m)$-matrix of moment arms and is used to calculate the torques generated by all muscles at the actuated joints, where n_b is the number of actuated joints, and matrix B is a $(n_c \times n_b)$-distribution matrix used to obtain the generalized torques due to the torques at the actuated joints.

Moreover λ_{ph} is the vector of Lagrange multipliers at phase ph of the motion. Note that the previous equation is used together with constraint equations forcing the normal contact forces to be always positive. Moreover, hard impacts will be avoided.

The inversion of the contraction dynamics is needed to obtain the values of the muscle activation a since they are required to evaluate the energy expenditure according to the procedure proposed by Umberger et al. [30]. Once the activations are obtained, using their time derivative it is possible to invert also the activation dynamics so that the neural excitations u are also obtained. The neural excitations are required for two reasons: one is that they are also involved in the calculation of the muscle energy expenditure and the other is that they are involved in some of the nonlinear constraints of the optimization procedure since their values are restricted to the interval [0, 1]. A flow diagram summarizing the inversion process of the contraction and activation dynamics is presented by Ackermann [31] and shown in Fig. 15.

The simulation of human walking motion is now treated as a huge parameter optimization problem. The optimization parameters, also called design variables, are used to reconstruct the muscle force histories and the generalized coordinate histories of a walking cycle as well. Such a set of parameters is found by minimizing a cost function f which is evaluated based on the energetic and aesthetic reasons. Finally, the motion and muscle forces time histories reconstructed from the

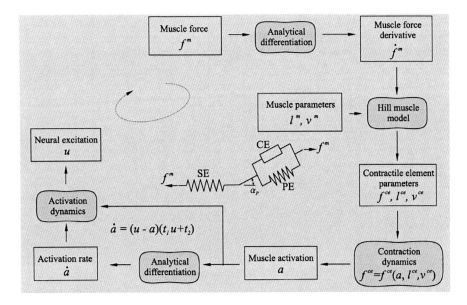

Fig. 15 Flow diagram of the inversion of the activation and contraction dynamics

optimization parameters are asked to fulfil many constraints. The constraints of the constrained optimization problem ensure the fulfilment of the equations of motion of the multibody system, the kinematic constraints as well as other physical and physiological relations.

The complete set of design variables is summarized in vector χ. This vector is itself built from four different vectors as follows:

- A vector q_i, $i = 1, 2, \ldots, nc$, containing all nodal values of the different generalized coordinates.
- A vector f_{mj}, $j = 1, 2, \ldots, N_m$, containing all nodal values of the different muscle forces. Since each muscle force is parameterized as the generalized coordinates are, a similar number of design variables can arise from muscle forces.
- A vector with eight components representing the durations of the eight phases of a walking cycle t_{ph}.
- A vector with geometrical parameters describing the kinematic constraints of the feet on the ground p_g composed of the right and left step lengths L_R and L_L, the right and left orientation angles α_R and α_L of the feet, and the lateral distance L_W between both ankles.

According to the previous explanation, the vector of design variables can be written as:

$$\chi = \begin{bmatrix} q_i^T & f_j^{mT} & t_{ph}^T & p_g^T \end{bmatrix}^T \tag{3}$$

where indices i, j and ph are running from 1 to n_c, N_m and 8, respectively, and g is just a subscript meaning that parameters in p_g are geometrical. In the three-dimensional model presented before the number of coordinates n_c is equal to 16 while the number of muscles N_m is equal to 28. The final value of the cost function f is calculated using the metabolical cost of transportation, E', and the measure of the deviation from normal walking patterns, J_{dev}, as follows

$$f = \omega_E E'/100 + \omega_J J_{dev} \qquad (4)$$

where E' is divided by the factor 100 for balancing of the two terms of the cost function to get comparable numbers, and ω_E and ω_J are two weighting factors.

The procedure suggested by Ackermann [31] and used in this research, too, avoids the forward integration through parameterization of the time histories of the generalized coordinates by using spline polynomials and by searching for their optimum values at certain node positions. Spline functions have many possibilities that can be used to improve the efficiency of the procedure. In fact, it is easy to have access to the derivatives of the parameterized function, avoiding the numerical differentiation used by Ackermann [31]. In addition, the interpolation can be split into two parts: a more computationally expensive one that can be done in a pre-processing stage and the other that is done during the optimization.

As an example, the measured motion as well as the measured ground reaction forces has been used as a reference motion (or normal walking pattern). Two 18 nodes models are considered with one of them having a weight of 2 kg attached to one foot. The attachment of a weight to the ankle results in a higher energy expenditure during walking. The metabolical cost of transportation of the model without the attached weight is 569.56 J/m while 631.14 J/m are obtained when the weight is attached. On the other hand, the deviation from normal walking patterns increased from 3.06 to 3.40 when the weight is attached. Fig. 16 shows the time histories of the two components of the ground reaction forces at one foot. The similarity between the solutions with and without the attached weight is remarkable.

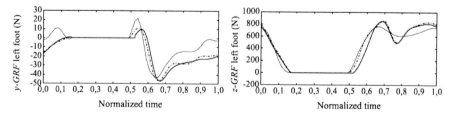

Fig. 16 Time histories of the ground reaction forces experimentally measured (*solid thin line*), simulated with the 18 nodes 3DNS model (*solid thick line*) and with the 18 nodes 3DNS model with one-sided disorder (*dash dot thick line*)

6 Conclusions

Human walking is a fascinating topic attracting scientists from many disciplines. In the Ancient World the philosopher Aristotle made observations and asked questions. In the Middle Ages the physiologists Borelli and Barthez made experiments. the Weber brothers and Vierodt collected data. Later on Fischer, Vaughan, Zajac and Winter developed models for research. More recently the multibody dynamicists Silva, Ambrosio and Fisette dealt with computerized models, equations of motion and simulations. Today parameter optimization is used to overcome the problems of muscle overactuation by inverse dynamics as shown by Umberger and Ackermann. Efficient simulations with 3D-models for gait disorders are presented by Garcia Vallejo and the author. Including the metabolical energy expenditure and the activation and contraction dynamics of muscles in the simulation method allows the possibility to study coordination aspects of human walking and shows a potential application of this approach to the design of rehabilitation therapies and to the design of assistive devices. Efficient methods as are most important for the development of rehabilitation therapies and the design of assistive devices.

References

[1] Schiehlen, W.: PAMM Proc. Appl. Math. Mech. 11, 903–906 (2011)
[2] Nussbaum, M.C.: Aristotle's de motu animalium. Princeton University Press, Princeton (1978)
[3] Borelli, G.A.: De motu animalium. Bernabo, Rome (1680)
[4] Borelli, J.A.: De motu animalium. Boutesteyn, Leiden (1685)
[5] Borelli, J.A.: De motu animalium. Petrum Gosse, The Hague (1743)
[6] Borelli, G.A.: De motu animalium. Kessinger, Whitefish (2009)
[7] Borelli, G.A.: Die Bewegung der Tiere. Akad Verlagsges, Leipzig (German transl. Mengeringhausen, M.) (1927)
[8] Borelli, G.A.: On the Movement of Animals. Springer, Berlin (English transl. Maquet, P.) (1989)
[9] Barthez, P.J.: Neue Mechanik der willkuerlichen Bewegungen des Menschen und der Thiere. Kuemmel, Halle (German transl. Sprengel, K.) (1800)
[10] Weber, W., Weber, E.: Mechanik der menschlichen Gehwerkzeuge. Dieterich, Goettingen (1836)
[11] Weber, W., Weber, E.: Mechanics of the Human Walking Apparatus. Springer, Berlin (English transl. Maquet, P., Furlong, R.) (1992)
[12] Vierordt, H.: Ueber das Gehen des Menschen in gesunden und kranken Zustaenden. Laupp, Tuebingen (1881)
[13] Fischer, O.: Theoretische Grundlagen fuer eine Mechanik der lebenden Koerper. Teubner, Leipzig (1906)
[14] Wittenburg, J.: Dynamics of multibody systems. Springer, Berlin (2008)
[15] Beletsky, V.V.: Two leg walking – The model problems of dynamics and control. Nauka, Moscow (1984)
[16] Silva, M.P.T., Ambrosio, J.A.C.: Kinematic data consistency in the inverse dynamic analysis of biomechanical system. Multibody System Dynamics 8, 219–239 (2002)

[17] Chenut, X., Fisette, P., Samin, J.C.: Recursive formalism with a minimal dynamic parametrization for the identification and simulation of multibody systems – application to the human body. Multibody System Dynamics 8, 117–140 (2002)

[18] AnyBody Technology: Press. AnyBody Modeling System™ (2013), http://www.anybodytech.com (accessed January 25, 2013)

[19] Zajac, F.E., Neptune, R.R., Kautz, S.A.: Biomechanics and muscle coordination of human walking, Part I. Gait Posture 16, 215–232 (2002)

[20] Zajac, F.E., Neptune, R.R., Kautz, S.A.: Biomechanics and muscle coordination of human walking, Part II. Gait Posture 17, 1–17 (2003)

[21] Anderson, F.C., Pandy, M.G.: A dynamic optimization solution for vertical jumping in three dimensions. Computer Meth. Biomechanics Biomedical Engineering 2, 201–231 (1999)

[22] Anderson, F.C., Pandy, M.G.: Dynamic optimization of human walking. J. Biomechanical Eng. 123, 381–390 (2001)

[23] Thelen, D.G., Anderson, F.C., Delp, S.L.: Generating dynamic simulations of movement using computed muscle control. J. Biomechanics 36, 321–328 (2003)

[24] Thelen, D.G., Anderson, F.C.: Using computed muscle control to generate forward dynamic simulations of human walking from experimental data. J. Biomechanics 39, 1107–1115 (2006)

[25] Seth, A., Pandy, M.G.: A neuromusculoskeletal tracking method for estimating individual muscle forces in human movement. J. Biomechanics 40, 356–366 (2007)

[26] De Sapio, V., Khatib, O., Delp, S.: Least action principles and their application to constrained and task-level problems in robotics and biomechanics. Multibody System Dynamics 19, 303–322 (2008)

[27] Rasmussen, J., Damsgaard, M., Voigt, M.: Muscle recruitment by the min/max crierion. J. Biomechanics 34, 409–415 (2001)

[28] Garcia-Vallejo, D., Schiehlen, W.: 3D-Simulation of human walking by parameter optimization. Arch. Appl. Mech. 82, 533–556 (2012)

[29] Kurz, T., Eberhard, P., Henninger, C., Schiehlen, W.: From Neweul to Neweul-M2: symbolical equations of motion for multibody system analysis and synthesis. Multibody System Dynamics 24, 25–41 (2010)

[30] Umberger, B., Gerritsen, K., Martin, P.: A model of human muscle energy expendidure. Computer Meth. Biomechanics Biomedical Eng. 6, 99–111 (2003)

[31] Ackermann, M.: Dynamics and energetics of walking with prostheses. Shaker, Aachen (2007)

Part II
Material Theories of Solid Continua and Solutions of Engineering Problems

On the History of Material Theory – A Critical Review

Albrecht Bertram

Abstract. Material theory in a strict sense is not older than just half a century and from the beginning characterized by fundamental disputes and changes of paradigms. A number of different suggestions have been made and - often - forgotten soon after their publication. In this article the main existing theories shall be briefly described and critically discussed. Some important demands for an adequate theory are raised and applied to these suggestions.

Originally, two totally different lines of approaches existed, namely that of *history functionals* and that of *inner variables* or *state theories*. Since both approaches suffer from fundamental deficiencies, neither of them really achieved global acceptance. And for a long time it remained unclear how the two approaches could be mutually related.

This was changed in 1972 by Noll´s *New Theory of Simple Materials* [19], in which a third approach was suggested that made it possible to compare the two preceding formats. The gain in generality was, however, accompanied by a loss of simplicity. Consequently, this theory has been used only by very few groups within the scientific community.

Since then the majority of the papers in the field of material modeling are pragmatic and do not claim for general validity. However, there remains a clear need for a general theory, as we will finally demonstrate within the context of thermodynamics.

Keywords: Material theory, Constitutive theory, Internal variables, State space.

1 Introduction

During the past several decades, material modeling has become a rich and blossoming branch of research activities. An overwhelming wave of papers,

---

Albrecht Bertram
Institute for Mechanics, Otto-von-Guericke-University Magdeburg,
Universitätsplatz 2, 39106 Magdeburg, Germany

conferences, reports, etc. bear witness to this trend, the peak of which has surely not yet been reached. New materials, new applications, new loading conditions, new computational and experimental facilities, etc. create an almost unlimited demand for material models.

In the presence of this evolution of material models a need for a unifying theory can be stated. One wants to use a theoretical framework for the construction and comparison of particular models or model classes. Surprisingly enough, little research work has been invested in this interesting task during recent decades. This situation, however, has not always been like this. Clifford A. Truesdell (1919 - 2000) remarked in 1993[1]:

> *"It seems to me that in the late 1940s new kinds of continuum mechanics began to be envisioned and explored, different in spirit from the older kinds. (...) Rigorous mathematical analysis based upon consequences of fairly general principles (...) became appealing."*

In fact, from the middle of the last century until the 1980's many attempts have been made to solve this problem. After this period, however, only a few papers have been published in this direction. A general theory was apparently not the focus of the scientific community. A more pragmatic way of extending particular models without placing them within a general framework became habitual thus leaving the situation rather unsatisfying.

With this paper, it is intended to raise the request for a general material theory again. And we will give some good reasons to substantiate this, not only under theoretical aspects, but also under rather practical ones. Further, it is intended to describe the propositions and paradigms which have been suggested in this field, to show their deficiencies, and, thus, to inspire other people to bring this problem closer to a solution.

2   Material Theory

First it should explained what is meant by *Material Theory* in order to prevent the reader from expecting something else. By *Material Theory* we do not have different theories of material models in mind like, e.g., viscoplasticity, micromorphy, or continuum damage theory. What is in fact meant is the theory behind all these examples, or a framework within which one can construct material models like those just mentioned. One may also call it a *(General) Constitutive Theory* or the like.

In this context, the topic of a *Material Theory* can be reduced to a simple question. If a material is elastic then we determine the current stresses by a constitutive function of the current deformation of the body or of a suitable neighborhood of the point under consideration. The question here is: which is the general form of the constitutive equation of a material which is *not* elastic?

---

[1] See http://www.math.cmu.edu/~wn0g/noll/TL.pdf

At first glance this seems to be an easy question. However, we will see in the sequel that the answer is by no means trivial and has not been given in a satisfying way until today.

A theory which deserves the label *Material Theory* is expected to be

- conceptually sound, based on both mathematically and physically clear assumptions or axioms and well defined concepts,
- general in order to include essentially all branches of material modeling,
- practical and applicable.

And we will see when looking at the history of this subject that any theory which does not simultaneously meet all of these three requirements, will soon be a victim of Occam´s razor.

This article is organized as follows. First we want to explain the two original paradigms which exist in the field, namely (*i*) the theory of *history functionals* and (*ii*) the theory of *internal variables*. Then we refer to some attempts to unify them, and finally we describe the state of art and the remaining problems to be solved.

3   History Functionals

In the late 1950's Green, Rivlin (1915 - 2005), Noll, and others made suggestions to give continuum mechanics a rational or even an axiomatic form, like any other branch of mathematics, as they understood mechanics. The demanding aim was expressed under the challenging label *Rational Mechanics*. In the well-known journal *Archive of Rational Mechanics and Analysis*, many papers have been published with such an intention.

In 1965 the famous article *The Non-Linear Field Theories of Mechanics* [26] by Clifford A. Truesdell and Walter Noll appeared in the *Encyclopedia of Physics*. This book soon had an overwhelming impact on continuum mechanics. Here are some important dates of this work:

- January 2nd, 1965 first issue printed 4000 times
- 1971   identical reprint
- 1979   identical reprint
- for a long period unavailable
- 1992   reprint
- 2000   translated into Chinese
- 2004   reprint edited by S. Antman
- 2009   reprint as paperback

On page 56 of this book we find as a starting point of the theory the

Principle of Determinism: *The stress in a body is determined by the history of the motion of that body.*

This short axiom has some important implications like the following ones.

- In mechanics, certain variables are determined by mechanical events.

Perhaps a physicist working in quantum mechanics will not accept this statement. On the other hand, it is the philosophy of any engineer that practically everything in the mechanical world is determined and can be calculated, at least up to a certain degree of precision.

- It is the past that determines the present, whereas the future has no influence on the presence (causality).

This is perhaps the least questionable statement of this principle.

- The authors considered the kinematics (motion of the body) as the independent variables, and the stresses as the dependent ones.

Of course, this is not the only choice. One could also do it vice versa. For a civil engineer it is perhaps more natural to determine the deformations of a building by the given loads, so that the kinematical variables are determined by the dynamical ones. We will, however, see that the principle induces problems in any of these two directions[2]. In fact, the stresses do not determine the deformations, and the deformations not determine the stresses in all cases. This is why one introduces internal constraints.

There are also theories which avoid preferring stresses to strains or vice versa by using constitutive *relations* instead of functionals[3]. However, these concepts are rather complicated to handle and, hence, never really adopted by the community.

Notations. We will denote the dependent variables by σ and the independent ones by ε and call them *observable variables*. One may think of stresses and strains, respectively, but also vice versa, or in the thermodynamical context, by the vector of the caloro-dynamic state and the thermo-kinematic state, respectively. Both variables or sets of variables may be tensors of arbitrary order, i.e., elements of finite dimensional linear spaces with inner product. We will further-on distinguish between *histories* of semi-infinite duration noted as $\varepsilon(\tau)\big|_{\tau=-\infty}^{t}$ for an ε-history with τ being the time parameter, and *processes* of finite duration noted as $\varepsilon(\tau)\big|_{\tau=t_0}^{t}$ starting at some arbitrarily fixed starting time t_0 for an ε-process. Functions of such histories or processes are called *history functionals* and *process functionals*, respectively.

---

[2] As an example one could consider a rigid - perfectly plastic model. This does not allow for a functional dependence of stresses and strains in any direction.

[3] See the method of preparation suggested by Perzyna/Kosiński [24] and Frischmuth/ Kosiński /Perzyna [11].

With the above *principle of determinism* and some other assumptions, Truesdell and Noll come to the conclusion that a history functional like

$$\sigma(t) = F\{\varepsilon(\tau)|_{\tau=-\infty}^{t}\} \qquad (1)$$

would be *"the most general constitutive equation"*. We will examine this demanding title next.

Such *history functionals* or *heredity functionals* were inspired by viscosity or viscoelasticity. Such laws typically consist of initial values weighted by an obliviator or influence function (like an exponential function with negative exponent) and a convolution integral of the deformation process or its rate and the obliviator function. If we consider a typical linear viscoelastic body like the well-known one-dimensional Maxwell model, we obtain for the stresses the integral form

$$\sigma(t) = \sigma(0)\, exp\left(-\frac{E}{D}t\right) + E\int_{0}^{t}\varepsilon(\tau)^{\bullet}\, exp\left(\frac{E}{D}(\tau-t)\right)d\tau \qquad (2)$$

with material constants E and D and initial stress value $\sigma(0)$. Thus, the longer the process lasts, the smaller becomes the influence of the initial values. This effect is called *fading memory*. In the limit, the influence of the initial values has been completely forgotten, and we obtain the history functional

$$\sigma(t) = E\int_{-\infty}^{t}\varepsilon(\tau)^{\bullet}\, exp\left(\frac{E}{D}(\tau-t)\right)d\tau \qquad (3)$$

which is formally simpler than the above integral since it is independent of initial values and initial time. The basic ingredient for this limit is in fact the fading memory property of the material. One has no chance to construct such a history functional for a classical elastoplastic material, since it does not forget the past and, thus, the limit does not exist, as it has been shown by Noll [19].

Such functionals have been extensively used by Green/Rivlin [13], Noll [18], Coleman [6], Wang [28], and many others to construct material theories. Astonishingly enough, the use of these functionals is often mathematically little exact. For example, one will hardly find any regularity conditions on the histories in Truesdell/Noll's book [26].

What are the advantages and disadvantages of the theory of history functionals?

Firstly, an important advantage of them is surely that no new primitive concepts are needed; only the observable variables stresses and strains are used. Secondly, the theory is well justified within viscoelasticity.

However, *"it is both philosophically unacceptable and practically questionable to use semi-infinite histories"* (Noll [19]). Moreover, this format is not general enough since it is essentially limited to fading memory materials and rules out plastic models. This fact, however, is not easy to see.

In *The Non-Linear Field Theories of Mechanics* [26] plasticity is not present at all. The authors claimed that at the time plasticity had not yet gained the state of a mathematical theory. Indeed, the first theories of finite plasticity did not appear before 1965 like, e.g., Green/Naghdi [12], Lee/Liu [16], and Mandel [17].

In 1971 Valanis[4] [27] suggested the *endochronic* theory, which can be considered as an attempt to enlarge the concept of history functionals to rate-independent materials by introducing an artificial time-like parameter. By this ansatz one can give rate-independent materials a form which is similar to that of finite linear viscoelasticity with fading memory. This creates a particular inelastic behavior, which shows effects that are in some ways similar to plasticity. Classical plasticity with elastic ranges, however, can not be brought into this form.

4 Internal Variables

The second format of material theory is concerned with *internal variables, hidden variables, state variables*, or whatever they are called, in contrast to the *observable* variables like σ and ε. It is now difficult to find out who first introduced such variables. This branch was mainly inspired by thermodynamics[5]. We find internal variables already in the early thermodynamical works like those of Eckart [8], [9] who introduces *"certain other variables"* in a more intuitive way. We fully agree to Šilhavý [25]:

> *"Classical thermodynamics uses the concept of state in an informal way, which creates a good deal of confusion in its foundations."*

Percy Williams Bridgman (1882 - 1961) was probably one of the first to try to construct a general concept of state.

> *"Ordinarily the state of a body is characterized by all the measurable properties of the body."* He assumes the *"possibility of an indefinite number of replicas of the original system, all in the same state. Any desired property which determines the state may then be found (...) by making the appropriate measurement on a fresh replica. (...) The "state" is determined by the instantaneous values of certain parameters and their history."*

Although Bridgman's state concept still remains rather vague and lacks of mathematical exactness, some of his ideas later led to concepts like the *method of preparation* and the *minimal state concept* as we will see in the sequel.

In internal variable theories one introduces a (finite) set of tensor-valued variables z as a primitive concept, which enter the constitutive law F (a function, not a functional) for the stresses as additional independent variables

---

[4] See also Haupt [14].

[5] For references to early thermodynamical works see Horstemeyer/Bammann [15] and Maugin/Muschik [20], for references to different state spaces see Muschik/Papenfuss/Ehrentraut [21].

$$\sigma(t) = F(\varepsilon(t), z(t)). \tag{4}$$

For the internal variables an evolution function is needed, which is assumed to be of the form of a process functional of the deformation process

$$z(t) = \mathcal{P}\{\varepsilon(\tau)\big|_{\tau=t_0}^{t}, z(t_0)\} \tag{5}$$

or, simpler, by an evolution function of rate form (first order ODE)

$$z^{\bullet} = E(\varepsilon, z, \varepsilon^{\bullet}) \tag{6}$$

or even in an incrementally linear form

$$z^{\bullet} = e(\varepsilon, z)\, \varepsilon^{\bullet}. \tag{7}$$

The solution of the last two requires initial values at the starting time.

Here all variables are taken at the same time and the same point so that we may suppress the temporal and spatial arguments.

This format is simple and practical, and many specific material models can be brought into it. The important disadvantage of internal variable theories is, however, that the internal variables are not defined but introduced in an ad hoc manner as primitive concepts.[6] In doing so there is hardly any way to assure uniqueness. In fact, in almost all examples one can show that the choice of these variables is rather arbitrary, and so is the structure of the state space.

Apart from non-uniqueness, one can state that not all material models allow for a state space with a finite dimensional linear structure. As a counterexample we could mention history functionals where the histories belong to a functional space of infinite dimension.

In order to give a more concrete example we consider plastic bending of beams[7]. For this purpose, a beam element is modeled by a one-dimensional linear elastic-perfectly plastic law. If we assume Bernoulli's hypothesis, then we can determine the elongation of each fiber by the local curvature of the beam as a linear function of the transversal coordinate x. The stresses in each fiber depend on the deformation process, as usual in plasticity. The resulting moment is the integral over the cross section

$$M = \int_A \sigma(x)\, x\, dA. \tag{8}$$

---

[6] Casey/Naghdi [4] state that *"for a constitutive theory of plasticity it is not enough to assume a constitutive equation for plastic strain (or for its rate). It is necessary, in addition, to provide a prescription for how plastic strain (or even its rate) can be determined from stress and strain measurements, at least in principle."* This statement can also be applied in the same way to any theory of inelasticity other than plasticity.

[7] This example was given by S. Govindjee (2010) in a private communication.

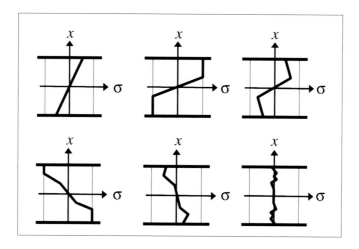

The stress distribution σ(x) is a function of the curvature process and will always be piecewise linear in x. In the Figure a series of cross sections initially, after the first yielding, after reverse yielding, etc. is sketched. The number of possible turning points is countable infinite. Therefore, the bending moment resulting from the past curvature process can not be described by a finite number of variables, in principle.

5   Noll´s New Theory of Simple Materials

For some decades these two paradigms of history functionals and internal variables coexisted, but the relationship between them remained unclear for a long time. This unsatisfying situation ended in 1972, when Noll published his *New Theory of Simple Materials* [19]. As a motivation, Noll claimed that the theory of history functionals

> "*has failed to give an adequate conceptual frame for the mathematical description of such phenomena as plasticity, yield, and hysteresis.*"

Hysteresis could surely be described by such functionals as long as it is not rate-independent, but plasticity (and yield, which is included in plasticity) could in fact not be described in the format of history functionals. So there was good reason to leave the history functionals behind and to create a new theory.

To do this, Noll introduced the following concepts:

- a *process class* as a collection of all possible deformation processes the material could be submitted to starting from some initial state
- a *state space* as a primitive concept
- an *evolution function* for the state in the form of a process functional
- an *output function* for the stresses
- an *output function* for the strains

One can call this theory also a *state variable theory*. Noll's state space is a primitive concept but with a precise mathematical meaning. In a mathematical sense it is uniquely defined, although it may have different representations.

Within this setting, Noll was able to establish an embedding of the theory of *history functionals* within the new theory. According to this, two facts can be stated. Firstly, a representation of a material as a history functional is only possible if it has a fading memory property. And, secondly, the theory of history functionals is a subclass of the internal variable theory if properly constructed ("*semi-elastic*").

Although this theory is rather general and unifies all other ones, its acceptance within the scientific community remained rather limited. This has surely to do with two of its properties. Firstly, it is a rather complicated and demanding construction and, thus, by no means simple. Secondly, the concept of the state space remains rather abstract and difficult to construct. It has hardly any structure. Only a uniform structure on subsets of the state space is introduced in a natural way[8]. It does not possess a linear structure, so nothing can be said about its dimension, neither does a differentiable structure exist a priori which could be used for evolution equations in the form of ODFs like Eq. (6).

It was C. A. Truesdell who wrote us in a personal letter in 1980:

> "Although Noll's paper presenting his "new theory" was dedicated to me, I cannot understand it. It took me ten years to master his "old theory", and now I am much older."

In fact, the paper was conceptually overloaded[9] and difficult to understand by a non-mathematician.

6 Minimal State Space

In the following decades, the community became rather pragmatic with respect to general frameworks. Only few papers or books have been published with the claim of presenting a general constitutive theory. The key problem for this seems to be the construction of the state space. In 1966 and 1968, Emin Turan Onat (1925 - 2000) [22], [23] suggested a construction of the state space in a way which is used in systems theory known there as the *minimal state space*. This concept has been later worked out by Bertram in detail for materials with internal constraints [1] and without constraints [2].

The construction of this state space is rather simple. We assume that we have infinitely many replicas of the material in the same initial situation[10] like

---

[8] See Fabrizio/ Lazzari [10] for the topology of the state space.

[9] In the same paper [20] Noll introduced the intrinsic description which does not make use of any reference placement but instead of concepts of differential geometry, another rather demanding concept for non-mathematicians which has also gained little acceptance in the scientific community.

[10] Which precisely means that all replicas respond to the same ε-process with the identical σ-value.

Bridgman [3] did. Then we can perform certain deformation processes out of this initial situation, the collection of which is called *process class* (in the strain space). And finally it is assumed that at the end of any of these processes we can measure the stresses.

We now define an equivalence relation on the process class. Two processes are considered as equivalent if

(i) both can be continued by the same set of deformation processes.

(ii) the responses (stresses) to all of them are pairwise identical.

Or in other words, two processes are equivalent if no difference in the future behavior can be detected. The equivalence classes according to this equivalence relation define the *state space*.

This state space is unique, although it may allow for different representations. Moreover, it has exactly the right size. If we would drop one bit from it, it would be insufficient and violate determinism. This is why it is named *minimal state space*. And if we would add one bit, it would lead to a redundancy.[11] So it would also deserve the name *maximal state space*.

In the quoted works, the *minimal state space* concept has been related to Noll's new simple materials. There is no principle contradiction with Noll´s *new simple materials*, although the construction of the *minimal state space* is now much simpler. Accordingly, we can at least partly adopt Noll´s concepts. For example, Noll gives a way to construct a topology and a uniformity on state space sections. By the same procedure such a topological structure can also be given to the minimal state space. However, to the best of our knowledge nobody has ever introduced a differentiable structure on this state space in a natural way. This is needed for many purposes like the application of the second law of thermodynamics as it is demonstrated in the next section.

As a résumé, the advantages of this theory are manifold:

- It is very general. In fact there is no class of deterministic models known that does not fit into this format. Moreover, it can also be directly applied to thermodynamics, electrodynamics, and other branches of deterministic sciences.
- No new primitive concepts are needed other than the observable quantities which we called stresses and strains.
- It is based on a derived state concept which is unique and constructed in a physically clear, mathematically exact, and practical way. This state space has precisely the right size.
- Also the evolution equation for the states can be derived.

The main lack of this state concept is that this state space has little mathematical structure. In particular, it does not have a natural linear structure nor

---

[11] This state space coincides with the large state space of Muschik/Papenfuss/Ehrentraut [21]. While within most theories the state space results from the material class under consideration, Muschik/Papenfuss/Ehrentraut start with the state space and work out the material class compatible with it.

a differentiable structure. In examples, this all can be introduced and then it can be of finite or infinite dimension. What we have in mind, however, is a natural way to generally define such structures. And this is still the main open problem.

7 Thermodynamic Consistency

In the sequel we will give an argument for the need of additional mathematical structure of the state space like a differentiable one.

In order to exploit the second law of thermodynamics in the form of the Clausius-Duhem inequality with the entropy η, the mass density ρ, and the heat flux **q**

$$\varphi^{\bullet} + \eta \, \theta^{\bullet} - \frac{1}{\rho} \sigma \cdot \varepsilon^{\bullet} + \frac{\mathbf{q} \cdot \mathbf{g}}{\rho \, \theta} \leq 0 \qquad (9)$$

we have to introduce the free energy as a function of state z and the current independent variables like strain ε, temperature θ, and temperature gradient **g**

$$\varphi(\varepsilon, \theta, \mathbf{g}, z) \qquad (10)$$

so that its time derivative is composed by the partial derivatives

$$\varphi(\varepsilon, \theta, \mathbf{g}, z)^{\bullet} = \partial_{\varepsilon} \varphi \, \varepsilon^{\bullet} + \partial_{\theta} \varphi \, \theta^{\bullet} + \partial_{\mathbf{g}} \varphi \, \mathbf{g}^{\bullet} + \partial_{z} \varphi \, z^{\bullet}. \qquad (11)$$

Then $z^{\bullet}$ must be substituted by the evolution law for the states. Only after that we are able to draw necessary and sufficient conditions for the Clausius-Duhem inequality. However, the last term requires a differentiable structure in the state space.

In Coleman/Gurtin [5] a way to exploit the Clausius-Duhem inequality in this way is given. Since these authors use an evolution equation for the internal variables of the form

$$z^{\bullet} = E(\varepsilon, \theta, \text{grad } \theta, z) \qquad (12)$$

which is less general than Eq. (6), only few material classes like the ideal gases are included, for which they specialize their theory. In particular, plasticity, viscoplasticity, damage, phase changes, etc. are not included since the rates of the independent variables are not included in the list of arguments of the evolution function.

Only for history-functionals with fading memory are the consequences of the Clausius-Duhem inequality given by Coleman [6] in a rather general way[12]. Here, the free energy is derived with respect to thermo-kinematical histories as Frechet-differentials in Hilbert spaces. Besides the usual potentials, a residual dissipation inequality remains, which has to be satisfied.

---

[12] See also Coleman/Owen [7].

It would be desirable to apply this procedure to exploit the CDI for essentially all materials and to establish necessary and sufficient conditions like the potentials and the residual inequality. However, this is only feasible if the state space has a differentiable structure.

8 Résumé

During the last 50 to 60 years many new branches of material theory have been introduced like, e.g., viscoplasticity, continuum damage mechanics, finite plasticity, micromorphic and gradient materials, polar theories, etc. Most authors worked with ad hoc introduced internal variables, a both pragmatic and practical procedure. Nevertheless, a globally accepted general theory or framework for all of these particular material classes is still lacking. Metaphorically speaking, the tree of material theory has gained an overwhelming variety of blossoming branches, while its trunk still remains (conceptually) weak.

If we enlarge our frame and include thermodynamics into our considerations, then we have to state that also here the lack of a general theory leads to undesirable effects like, e.g., that for each and every new constitutive model the thermodynamical consistency has to be proven anew without being able to just refer to a general representation.

We are still far from being able to consider *Material Theory* as a finalized, conceptually sound, and practical theory. Instead, although some progress has been made during its 50 year history, there is still a sizeable piece of work to be done. This article is meant to focus the discussion and stimulate investigations on this important issue.

Acknowledgment. The author would like to sincerely thank Jim Casey, Sanjay Govindjee, Peter Haupt, Arnold Krawietz, Wolfgang Muschik, Miroslav Šilhavý, and Bob Svendsen for stimulating discussions and helpful suggestions during the preparation of this article.

References

[1] Bertram, A.: Material Systems a framework for the description of material behavior. Arch. Rat. Mech. Anal. 80(2), 99–133 (1982)
[2] Bertram, A.: Axiomatische Einführung in die Kontinuumsmechanik. BI Wissenschaftsverlag. Mannheim, Wien, Zürich (1989)
[3] Bridgman, P.W.: The Thermodynamics of plastic deformation and generalized entropy. Rev. Modern Physics 22(1), 56–63 (1950)
[4] Casey, J., Naghdi, P.M.: A prescription for the identification of finite plastic strain. Int. J. Engng. Sci. 30(10), 1257–1278 (1992)
[5] Coleman, B.D., Gurtin, M.E.: Thermodynamics with internal state variables. J. Chemical Physics 47, 597 (1967)
[6] Coleman, B.D.: Thermodynamics of materials with memory. Archive Rat. Mech. Anal. 17(1), 1–46 (1964)

[7] Coleman, B.D., Owen, D.: On the Thermodynamics of materials with memory. Archive Rat. Mech. Anal. 36(4), 245–269 (1970)
[8] Eckart, C.: The thermodynamics of irreversible processes. II. Fluid Mixtures. Physical Rev. 58, 269–275 (1940)
[9] Eckart, C.: The thermodynamics of irreversible processes. IV. The theory of elasticity and anelasticity. Physical Rev. 73(4), 373–382 (1948)
[10] Fabrizio, M., Lazzari, B.: On the notion of state for a material system. Meccanica 14(4), 175–180 (1981)
[11] Frischmuth, K., Kosinski, W., Perzyna, P.: Remarks on mathematical theory of materials. Arch. Mech. 38(1-2), 59–69 (1986)
[12] Green, A.E., Naghdi, P.M.: A general theory of an elastic-plastic continuum. Arch. Rational Mech. Anal. 18(4), 251–281 (1965)
[13] Green, A.E., Rivlin, R.S.: The mechanics of non-linear materials with memory. Archive Rat. Mech. Anal. 1(1), 1–21 (1957)
[14] Haupt, P.: Viskoelastizität und Plastizität. Thermomechanisch konsistente Materialgleichungen. Springer, Berlin (1977)
[15] Horstemeyer, M.F., Bammann, D.J.: Historical review of internal state variable theory of inelasticity. I. J. Plasticity 26, 1310–1334 (2010)
[16] Lee, E.H., Liu, D.T.: Finite-strain elastic-plastic theory with application to plane-wave analysis. J. Appl. Phys. 38(1), 19–27 (1967)
[17] Mandel, J.: Plasticité classique et viscoplasticité. CISM course No. 97. Springer, Wien (1971)
[18] Noll, W.: A mathematical theory of the mechanical behavior of continuous media. Archive Rat. Mech. Anal. 2, 197–226 (1958)
[19] Noll, W.: A new mathematical theory of simple materials. Arch. Rational Mech. Anal. 48, 1–50 (1972)
[20] Maugin, G.A., Muschik, W.: Thermodynamics with Internal Variables, part I and II. J. Non-Equilib. Thermodyn. 19, 217–289 (1994)
[21] Muschik, W., Papenfuss, C., Ehrentraut, H.: A sketch of continuum thermodynamics. J. Non-Newtonian Fluid Mech. 96, 255–290 (2001)
[22] Onat, E.T.: The notion of state and its implications in thermodynamics of inelastic solids. In: Parkus, Sedov (eds.)IUTAM 1966. Irreversible Aspects of Continuum Mechanics, pp. 292–313. Springer, Wien (1968)
[23] Onat, E.T.: Representation of inelastic mechanical behavior by means of state variables. In: Boley (ed.) IUTAM 1968, pp. 213–225. Springer, Berlin (1970)
[24] Perzyna, P., Kosiński, W.: A mathematical theory of materials. Bull. Acad. Polon. Sci., Ser. Sci. Techn. 21(12), 647–654 (1973)
[25] Šilhavý, M.: The Mechanics and Thermodynamics of Continuous Media. Springer, Berlin (1997)
[26] Truesdell, C.A., Noll, W.: The non-linear field theories of mechanics. The Encyclopedia of Physics III(3) (1965)
[27] Valanis, K.C.: A theory of viscoplasticity without a yield surface, part I and II. Archive of Mechanics 23, 517–533 (1971)
[28] Wang, C.C.: Stress relaxation and the principle of fading memory. Arch. Rat. Mech. Anal. 18, 117–126 (1965)

Some Remarks on the History of Plasticity – Heinrich Hencky, a Pioneer of the Early Years

Otto T. Bruhns

Abstract. The history of material equations and hence the development of present material theory as a method to describe the behavior of materials is closely related to the development of continuum theory and associated with the beginning of industrialization towards the end of the 19th century. While on the one hand new concepts such as continuum, stresses and strains, deformable body etc. were introduced by Cauchy, Euler, Leibniz and others and mathematical methods were provided to their description, the pressure of industrialization with the need to ever newer, and likewise also reliably secure, developments has led to the fact that, next to the description of elastic behavior of solid bodies, more appropriate models for the description of plastic or elastic-plastic behavior were introduced. Upon this background, this chapter wants to introduce into the history of plasticity and likewise highlight the contributions of the Darmstadt graduate Heinrich Hencky who started his scientific career there 100 years ago.

Keywords: Plasticity, Prandtl-Reuss-theory, deformation theory, hypoelasticity, large deformation.

1 Introduction

Compared to the history of elasticity the plasticity theory is still relatively young, if its birth - perhaps somewhat arbitrary - is identified with the publication of the results of the French engineer Henri Tresca in 1868 [55][1]. Based on the observations of a series of experiments he had published a hypothesis according to which metals begin to flow when the largest shear stress reaches a critical value. We will come

---

Otto T. Bruhns
Institute of Mechanics, Ruhr-University Bochum,
Universitätsstr. 150, 44780 Bochum, Germany
e-mail: otto.bruhns@rub.de

[1] Reference is also made to a previous publication [54].

back to that point later. At that time Henri Tresca was so highly esteemed that his name was engraved as 3rd out of a total of 72 names on the outside of the first platform of the Eiffel Tower. This birth now dates back almost 150 years.

Despite this relatively short period, the theory of plasticity meanwhile has taken a dynamic development. Thus it would be helpful to differentiate between individual development phases:

1. The origins and basics - development until 1945,
2. The expansion of these basics to approximately 1980,
3. The present status and recent developments.

The first of the above phases is characterized by the pioneering efforts of a few researchers in the rapidly developing industry in Europe. Unfortunately, first results produced in France were hardly noticed. It took about 30 years until in the early 20th Century a small group of - primarily German-speaking - engineers and mathematicians adopted this topic. Among these outstanding persons were the two founding fathers of the Society of Applied Mathematics and Mechanics (Gesellschaft für Angewandte Mathematik und Mechanik - GAMM), Richard Edler von Mises and Ludwig Prandtl, but also Willy (or William) Prager and Heinrich Hencky from Darmstadt. With the specific fate of the latter we want to be concerned more in detail in this contribution[2]. Incidentally, this contribution is to remain limited here to the first of the three phases specified above.

Fig. 1 Richard von Mises (1883-1953) and Ludwig Prandtl (1875-1953)

It is one of the unfortunate facts of the second part of this first phase, that it took place between the two world wars, with all their attendant destruction and distortions. When finally in Germany the Nazi regime came to power, the above mentioned dynamic development soon came to an end; many of the researchers involved were forced to leave their country.

Finally, the following remark appears to be still appropriate: Much of what has been initiated significantly by this – here simply called - "German school" and then,

---

[2] This text comprises a partly extended version of a lecture held at the GAMM Meeting 2012, March 26-30, Darmstadt, Germany.

Some Remarks on the History of Plasticity

Fig. 2 Stepped wheel mechanical calculator, Rheinmetall (Sömmerda)

due to the Nazi regime and its consequences, has been spread all over the world, is today occasionally not appreciated accordingly of its actual achievement. We believe that this attitude does not cope with the partly ground breaking developments of these years. These developments have been made by engineers and applied mathematicians and they could not wait, until the appropriate tools available to us today, were ready. They had to act, i.e.: Find solutions with the tools available at that time. To do this, it is worth remembering: There was no Finite Element Method (FEM), no powerful computers - today assumed as naturally existing - and initially not even simple mechanical calculators, as e.g. the stepped wheel calculator of Rheinmetall, designed in the 1930s (refer to Fig. 2). As a rule, one had to use tables of logarithms[3], as multiplications and divisions were carried out "by hand".

2 The Origins and Basics – Development Until 1945

Let us mentally go back these 150 years to the beginning of industrialization in Europe. In many plants and constructions steel - or as it was called at that time "iron" - is used. To get an idea of the historical context, we want to recall briefly the following data:

- 1811, in Germany, Friedrich Krupp has founded the first cast steel factory in Essen (Ruhr Area),
- 1825, in England, the first public railway is inaugurated (Stockton & Darlington Railway Company, connecting Witton Park and Stockton-on-Tees in north eastern England, 40 km in length),
- 1835, the first railway is opened in Germany (Bavarian Ludwig Railway from Nuremberg to Fürth, 6 km in length).

An essential prerequisite for the operation of these railways was that from

- 1820, first rails were produced by rolling.

---

[3] A typical example is e.g. given with the handbook [1]. These tables contain sequences of mantissas of logarithms - preferably on the basis 10 - of natural numbers.

In this way, the rails until then created in casting processes could be replaced by the much smoother rolled profiles. Of course, one knew even at this early stage about the non-linear behavior of the material in use.

Consider a cylindrical specimen subject to a tensile load F, with length L and cross-sectional area A. Also, consider that for small elongations of this specimen the change of this area is negligible[4]. Then stress and strain may be defined as

$$\sigma = \frac{F}{A}, \qquad \varepsilon = \frac{\Delta L}{L_0} = \frac{L}{L_0} - 1, \qquad (1)$$

where L_0 is the initial length. From a typical tensile test (refer to Fig. 3) it may be observed that up to a specific (yield) point σ_y the stress increases linearly. Beyond this point we observe a general non-linear monotonic increase during loading, and a linearly decreasing behaviour under reverse loading (unloading). Thus, from this process of loading and subsequent unloading it turns out that the total strain reached at final elongation may be split into a reversible part ε_r and an irreversible so-called plastic part ε_i.

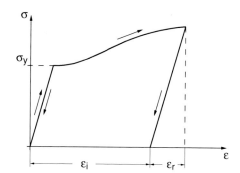

Fig. 3 Stress-strain curve of mild steel during a process of loading and subsequent unloading

What remains is the question how to characterize this behavior. With the above remarks we already have mentioned the two different ways of looking at the behavior of steel in the plastic range:

1. On the one hand, we have for example a vessel of the steam locomotive, which should be tight and safely withstand a given pressure. The necessary strength analyses - eventually carried out – however, were employed on the basis of the elasticity theory developed by Euler (continuum, 1750) and Cauchy (stress concept, 1822). This was admissible because thereby the structure has been assured against the reaching of the yield stress, and even in the case of exceeding this limit a "good-natured" failure - in conjunction with larger in time announcing deformations was observed. We will come back separately to the deviating case of possibly catastrophic failure in stability problems.

    In this first case, the strength properties of the steel are in major focus and accordingly the plastic behavior is interpreted as the behavior of a solid.

---

[4] This assumption does not hold for larger deformations, where the change of cross-sectional area has to be considered. We will come back to that point later.

2. On the other hand, we consider the rolling of rails as a typical forming process. Here, the material, which also then is often heated, is understood as a viscous fluid. Usually, we are interested in the forces that must be applied in such a rolling process. Similar problems might be given by the extrusion or even - on a very different time scale - the tectonic movements in the collision of two tectonic plates.

The general form of a linear elastic relation as the most simple form of a solid material is due to Cauchy[5] [3]

$$\boldsymbol{\sigma} = 2\mu\boldsymbol{\varepsilon} + \lambda \operatorname{tr}(\boldsymbol{\varepsilon})\boldsymbol{I}, \qquad \boldsymbol{\varepsilon} = \frac{1}{2\mu}\left(\boldsymbol{\sigma} - \frac{\lambda}{3\lambda + 2\mu}\operatorname{tr}(\boldsymbol{\sigma})\boldsymbol{I}\right), \qquad (2)$$

if common today's notations are used. Herein $\boldsymbol{\sigma}$ and $\boldsymbol{\varepsilon}$ are stress and strain tensors, λ and μ are the two Lamé's constants and $\boldsymbol{I}$ is a unit tensor. A fully general expression for a relation combining stress and rate of deformation (strain rate) was first given by Poisson [38]

$$\boldsymbol{\sigma} = -p\boldsymbol{I} + 2\mu_v \boldsymbol{d} + \lambda_v \operatorname{tr}(\boldsymbol{d})\boldsymbol{I}. \qquad (3)$$

Accordingly, λ_v and μ_v are the corresponding viscosities, p is the hydrostatic pressure and $\boldsymbol{d}$ the strain rate tensor (stretching), the symmetrical part of the velocity gradient. Special cases of linear viscous fluids were discovered by Navier, de Saint-Venant and Stokes [36; 47; 52][6]

$$\boldsymbol{\sigma} = -p\boldsymbol{I} + 2\mu_v \left[\boldsymbol{d} - \frac{1}{3}\operatorname{tr}(\boldsymbol{d})\boldsymbol{I}\right], \qquad \boldsymbol{d} - \frac{1}{3}\operatorname{tr}(\boldsymbol{d})\boldsymbol{I} = \frac{1}{2\mu_v}(\boldsymbol{\sigma} + p\boldsymbol{I}), \qquad (4)$$

From the very beginning the development of plasticity has been in this conflict: Does the body under consideration behave more solid-like or like a fluid - and does this possibly also depend on the specific task to be solved?

We now return to Henri Tresca. Tresca wanted to know whether a simple criterion can be specified for achieving the flow state of his material. On the basis of numerous experiments on various metallic materials he concluded[7], that "in the plastic state of the solid, the largest shear stress has a fixed value" (in commonly used today's notations):

$$|\tau|_{\max} = \frac{1}{2}(\sigma_1 - \sigma_3) = k, \qquad (5)$$

with σ_1 the largest and σ_3 the smallest value of the principal stresses, and k the shear yield limit. This was the first yield condition. Using this condition the already 73-year-old Barré de Saint-Venant (1797-1886) [48] 1870 has presented his "five equations of hydro- stereodynamics" for the problem of plane deformations starting from the above described material behavior as a viscous fluid. In addition to balance

---

[5] Although the anagram "ut tensio sic vis" of Hooke [22] may be interpreted as a first (uni-axial) step in this direction.

[6] Where in these cases $\mu_v = $ const. and in addition Navier adopted an incompressible material with $\operatorname{tr}(\boldsymbol{d}) = 0$, whereas the latter two introduced $3\lambda_v + 2\mu_v - 0$. Refer to [56].

[7] He probably may have resorted also to earlier works by Coulomb.

Fig. 4 Barré de Saint-Venant (1797-1886) and Maurice Lévy (1838-1910)

equations, the (assumed) incompressibility, as well as Tresca's yield criterion (5), this was a relationship of the form

$$\frac{d_{xx} - d_{yy}}{d_{xy}} = \frac{\sigma_{xx} - \sigma_{yy}}{\sigma_{xy}}. \tag{6}$$

In the same year, his student Maurice Lévy (1838-1910) [28] has transferred this representation to the general spatial problem. For this we can write

$$\boldsymbol{d} - \frac{1}{3}\mathrm{tr}(\boldsymbol{d})\boldsymbol{I} = c\boldsymbol{\tau}, \qquad \boldsymbol{\tau} = \boldsymbol{\sigma} - \frac{1}{3}\mathrm{tr}(\boldsymbol{\sigma})\boldsymbol{I}, \tag{7}$$

with $\boldsymbol{\tau}$ the deviatoric stresses, and c is a simple proportionality.

Thus, the basics of a simple flow theory were established. Because of the associated mathematical difficulties[8] this theory, however, did not find any application. It took another 30 years before these ideas were taken up again. Some spectacular cases of damage as a result of stability failure may have contributed to this. Namely, unlike the above-mentioned strength analysis after exceeding the yield point, in the case of a stability failure, e.g. the buckling of a simply supported beam (Euler problem), the known elastic solution

$$\sigma_{crit} = EI\frac{\pi^2}{Al^2}, \qquad E = \mu\frac{3\lambda + 2\mu}{\lambda + \mu} \tag{8}$$

may lead to an unsafe solution - if plastic deformations occur. Herein the modulus of elasticity E occurs as the slope of the stress-strain curve. Corresponding spectacular failure cases then may have led to the need that already before the turn of the century modifications of the buckling load of elastic-plastic beams were discussed, where - albeit very simplified - the material properties in the plastic range were taken into account (refer to [4; 5; 24]). Incidentally, this discussion about the correct value of

---

[8] The interested reader will recognize that, formerly, it was almost impossible to solve systems of partial differential equations.

reduction of the modulus of elasticity in the above relationship (8)[9] - as a result of non-linearity of the problem, i.e., due to the fact that here the bifurcation load and the stability limit no longer coincide as in the elastic range - took another 50 years. This solution was finally delivered by Shanley [50; 51].

To simplify the flow theory outlined above, in 1913 v. Mises [31] replaced Tresca's yield condition (5) by

$$(\tau_1)^2 + (\tau_2)^2 + (\tau_3)^2 = 2k^2 \tag{9}$$

with τ_i the principal deviator stresses. The meaning of this condition and its relationship with a strength hypothesis already indicated by M.T. Huber[10] in 1904 [23] has been explained by Heinrich Hencky first 1924 [14]. The deformation law

$$\boldsymbol{d} - \frac{1}{3}\mathrm{tr}(\boldsymbol{d})\boldsymbol{I} = c\,\boldsymbol{\tau}$$

of v. Mises has indeed taken over from de Saint-Venant and Lévy $(7)_1$. In the addressed work Hencky also specifies the until today used geometrical interpretation of the v. Mises and the Tresca yield conditions as surfaces in the 9-dimensional space of stresses (refer to Fig. 5)[11].

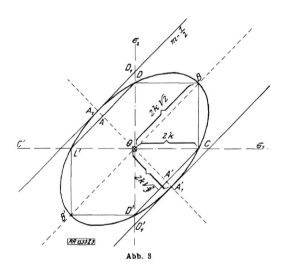

Fig. 5 Yield conditions according to Tresca and v. Mises, refer to Hencky, 1924 [14]. Copyright Wiley-VCH Verlag GmbH & Co. KGaA. Reproduced with permission.

---

[9] One might only contemplate the classical confusion regarding whether the tangent-modulus or the reduced- modulus theory gives an adequate description of the critical load, see e.g. [25; 6; 27]

[10] Refer to footnote 3 on page 327 of the Hencky paper [14].

[11] Figure 5 with a slight modification has been taken from the Hencky paper [14].

This system of partial differential equations at the time has been considered unsolvable – with a few pathological exceptions. In contrast, however, was the wish to develop practically manageable procedures that allowed to regain the progress of the material description in the calculation procedure and design rules. Thus, meaningful and yet reasonable simplifications had to be made. Today, we would say: There was a need for developing simplifying "models". In this sense, also the simplifications introduced by Prandtl [43] to consider the continuum as a) ideal-plastic or b) elastic-plastic body are to be understood. A further simplification is pursued in 1921 and 1923 by Prandtl [44] and Hencky [13], emerging from the following question: Do, possibly, exist special cases such that with the help of balance equations (equilibrium conditions) and the additional yield condition alone, solutions to the given problem can be found?

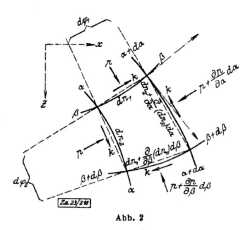

Fig. 6 Orthogonal families of α and β-slip lines, refer to Hencky, 1923 [13]. Copyright Wiley-VCH Verlag GmbH & Co. KGaA. Reproduced with permission.

For the general spatial problem with its 4 conditions and 6 unknown stress components this obviously would be not the case. If we restrict our considerations, however, to axially symmetric problems, we note that only 3 unknown stress components are facing 3 conditions. The focus of the forthcoming work is therefore on the solution of such "statically determinate" cases. For problems of plane deformations, the yield conditions of Tresca and von Mises (5) and (9) coincide[12]. Based on a system of hyperbolic differential equations then two orthogonal families of characteristics can be introduced as geometrical places of directions of the principal shear stresses. Following suggestions from Prandtl [44] and Nádai [34] these curves according to their meaning in plastic behavior will be interpreted by Hencky [13] in 1923 as "slip-lines" (α– and β-lines, see Fig. 6). Inclined at 45° to this family, we

---

[12] We note that this is not the case for problems of plane stress.

find the directions of the largest tensile and compressive stresses[13]. Between these families, there exist certain phenomena which Hencky summarizes in 3 theorems. That same year, Prandtl [45] as well as Carathéodory and Schmidt [2] adopt these thoughts and complement them by graphical solution methods and numerous additional statements. 1925 v. Mises [32] finally summarizes this development: "... from (.) immediately follow the beautiful differential geometrical properties of the slip-lines discovered by Mr. Hencky."

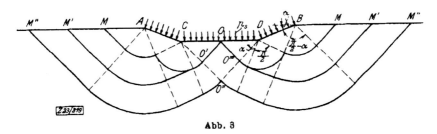

Fig. 7 Half plane with prismatic indentation, refer to Hencky, 1923 [13]. Copyright Wiley-VCH Verlag GmbH & Co. KGaA. Reproduced with permission.

In the above work Hencky, as v. Mises elsewhere, indicates the limits of this approach which are given by the respective boundary conditions: Ideally here only stresses should be prescribed. However, this is generally not the case.

Thus, the slip line theory is developed. Besides v. Mises, Prandtl and Hencky especially Hilda Geiringer and Willy Prager [8] contribute to the spreading of this theory.

Who was actually this Heinrich Hencky[14] from (at the time) Delft, who together with v. Mises and Prandtl has published numerous contributions to the development of the theory of plasticity in the ZaMM[15] (Zeitschrift für angewandte Mathematik und Mechanik), at that time the leading journal of mechanics?

Heinrich Hencky was born in Ansbach (Bavaria), Germany on the 2nd of November 1885 as elder son of a Bavarian school administrator. After the early death of his father and after finishing the senior high school in Speyer, he studies civil engineering at the TH (Technische Hochschule) in Munich from 1904-1908; at the same college his 4 years younger brother Karl Georg follows him, studying mechanical engineering (1908-1912). After his military service in 1908/09 with the 3rd Pioneer Battalion in Munich and a short activity as engineer for the Alsacian Railway in Strasbourg (1910-1912), in 1912 he moves to the Grand Ducal TH (Großherzogliche Technische Hochschule) in Darmstadt as "design engineer and assistant for engineering sciences". In 1913 he receives his doctorate of engineering from the

---

[13] The construction of a typical field of slip-lines is shown in Fig. 7 for a prismatic indentation into a half-plane.

[14] We here also refer to the sensitive article of Tanner and Tanner [53] where especially Hencky's pioneering work on rheology is underlined.

[15] Now ZAMM.

Fig. 8 Heinrich Hencky (1885-1951) and Boris G. Galerkin (1871-1945)

TH Darmstadt with a thesis on the numerical calculation of stresses in thin plates [10].

After completing his doctorate he seeks for a new position in the field of railway engineering[16] and in 1914 moves to Kharkov (Kharkiv) in the Ukraine, an emerging industrial and commercial center. The beginning of World War I and the revolutionary upheavals in Russia procure him his first unpleasant experience. His career is abruptly terminated. Like many others, he is interned 1915-1918 in the Urals region. Although during this time he met his later wife Alexandra Yuditskaya, this period must have had a decisive impact. After the war ended the Hencky family is sent back to Germany. His attempt to restart as test engineer in Warnemünde (on a project to develop a seaplane) fails. As a result of the general demobilization, the project had to be stopped.

1919 he is again "assistant for engineering sciences" at the TH Darmstadt. During this time he turns to the treatment of stability problems of elastic structures. Within this new subject he also wrote his habilitation treatise [11]. In 1920 he moves to the mechanical department of the TH of Dresden, first as assistant, and after acceptance of his habilitation in 1921 as lecturer. He is striving for a professorship. If one may believe the rumors, this, however, failed due to "not convincing achievements in teaching". Perhaps, his occasional reticence or shyness may have been an obstacle on this way. In this Dresden period he also reflects on possible relationships between philosophy and the description of nature [12]. Anyway, in 1922 he moves to the Technical University of Delft (at the chair of Cornelis B. Biezeno) - hoping to get a permanent professorship there. After all he is at the age of 37 and has a wife and two children.

In Delft Hencky probably wrote his most significant works - we come back to that in a moment. This, however, seems to have no influence on his position. In Delft, he all the time remains lecturer. It seems only too understandable that this situation must have affected his relationship with Biezeno. Therefore, in 1929 he leaves Delft with his family towards the USA. Biezeno indeed believes that it was merely a temporary research period; Hencky, however, is looking for a permanent job (he is now at the age of 44). In the summer of 1930, he then becomes associate professor at

---

[16] In those days one of the most attractive branches in civil engineering.

MIT in Cambridge, Massachusetts. This position, again, is not permanent. In 1932 the MIT is reorganized, his former advocate has died and Hencky is no longer employed. Here again, if we can trust the sources, he was regarded as "too theoretical" in a department of mechanical engineering, mainly interested in practical problems. In the following years he tried to survive as a consultant - interrupted by a short job at Lafayette College in Easton, Penn., with E.C. Bingham.

What should he do? In Germany, he could not find any kind of work, as in 1935 an offer arrived from Boris Galerkin, whom he knows from his Delft period. After careful consideration – in fact he probably had no choice – 1936 he accepts this proposition as a professor of engineering mechanics at Kharkov Polytechnic Institute and later at the Institute of Mechanics of the Lomonosov University of Moscow with A.A. Ilyushin[17]. Although he now has an adequate position, in his heart he must have felt as a prisoner. Whether he expressed this accordingly and therefore is "fallen from grace" can not be ascertained anymore. In any case, 1938, he has to leave the Soviet Union with his family within 24 hours.

Back in Germany, he gets help from his brother Karl Georg, who meanwhile was extraordinary professor (apl. Professor) at RWTH Aachen and now holds a senior position at the IG Farben in Ludwigshafen, to find a position at MAN company in Gustavsburg. Here he remains, suspiciously observed by the "security service" (SD), the intelligence agency of the SS, but under the protection of his supervisor at MAN (Dr. Richard Reinhardt) until the end of the war and then until his retirement in 1950. On the 6.07.1951, he died in a climbing accident at the age of 65.

Back to historical development. In the above mentioned paper from 1924 [14], where Hencky explained the v. Mises yield condition and interpreted it in a today still usual way as a hypothesis, according to which the boundary to plasticity is described by the elastic shear strain energy, he also developed a simple constitutive law for elastic as well as elastoplastic behavior, specifying a relationship between stresses and strains. As he mentions in his introduction, he hereby revisits an approach by Haar and v. Kármán [9], whose meaning seems to be not properly recognized.

For this purpose he assumes the specific complementary energy

$$A = \frac{1}{4\mu} \left(\sigma_{ik}\sigma_{ik} - \frac{9\lambda}{2+\lambda} \sigma_m^2 \right) = \frac{1}{4\mu} \left(S_2 - \frac{\lambda}{2+\lambda} S_1^2 \right), \tag{10}$$

with $S_1 = \text{tr}(\boldsymbol{\sigma})$, $S_2 = \text{tr}(\boldsymbol{\sigma}^2)$ and the limit of elastic behavior, which here is specified with the condition

$$\Phi = T_2 - 2k^2, \qquad T_2 = \text{tr}(\boldsymbol{\tau}^2) = S_2 - \frac{1}{3} S_1^2. \tag{11}$$

---

[17] Unfortunately, informations about Hencky's second stay in the Soviet Union are very rare. This is certainly due to the secrecy of those days. It is, however, believed that with the help of Galerkin Hencky was hired to improve the Soviet Union lightweight (airplane) construction. His deformation theory could contribute a lot to this matter, and it is known that Ilyushin later was very much impressed by this theory.

Hencky now solves the variational problem by seeking with

$$W = A + Lu + Mv + Nw + \varphi\Phi \tag{12}$$

for an extremum of $\int W \, dV$. L, M and N herein are the balance equations as auxiliary conditions. Thus, the bounding (yielding) condition $\Phi = 0$ is multiplied by a local function φ and then added to the elastic complementary energy. As a result, he receives

$$\boldsymbol{\varepsilon} - \frac{1}{3}\mathrm{tr}(\boldsymbol{\varepsilon})\boldsymbol{I} = \frac{1+\varphi}{2\mu}\boldsymbol{\tau}, \quad \mathrm{tr}(\boldsymbol{\varepsilon}) = \frac{1}{3K}\mathrm{tr}(\boldsymbol{\sigma}), \quad 3K = 3\lambda + 2\mu \tag{13}$$

a model very similar to the elasticity law (2)$_2$, where φ is a still undetermined Lagrange parameter, and K is the bulk modulus. For $\varphi = 0$, this law changes into that of the elastic material. The compression is purely elastic and at plastic behavior the shear modulus μ is reduced by $(1+\varphi)$, the material thus becomes "softer".

Thus, for the first time it is possible to formulate a constitutive law to describe elastoplastic behavior. This formulation later referred to as "deformation theory" is rapidly accepted, even if it soon meets its limits: A neutral change of stresses, as it for instance occurs in non-proportional loading, cannot be reflected. For many so-called "proportional" problems, however, it represents not only the first but also a very simple method. It should be noted that Hencky in this development has assumed that a body under increasing load will be deformed first elastically and then plastically after having reached the yield point. In the interior of the structure, however, still remains a so-called "elastic core". This corresponds to the above mentioned concept of plastic behavior as that of a solid material.

In the case that the considered body, under further load increase, merges into a "free flow", Hencky in a paper [16] of 1925 reverts to Lévy's approach (7), relating the stresses with strain rates, thus describing plastic flow as a behavior of a fluid.

Apparently independent[18] from a work by L. Prandtl [43], where already elastic deformations have been considered in a plane problem, the Hungarian András (Endre) Reuss in 1930 connects this Saint-Venant/Lévy approach (7) with the description of elastic behavior. For this purpose, like Hencky, he emanates from the v. Mises yield condition and obtains on a comparable way a constitutive law

$$\boldsymbol{d} - \frac{1}{3}\mathrm{tr}(\boldsymbol{d})\boldsymbol{I} = \frac{1}{2\mu}\dot{\boldsymbol{\tau}} + \lambda\boldsymbol{\tau}, \tag{14}$$

that with λ still contains a yet undetermined function, and

$$\mathrm{tr}(\boldsymbol{d}) = \frac{1}{3K}\mathrm{tr}(\dot{\boldsymbol{\sigma}}) \tag{15}$$

for its (purely elastic) compressible part. Thus, the basic version of the nowadays commonly used Prandtl-Reuss theory is introduced.

---

[18] We refer to a footnote of A. Reuss in [46] where it is stated that while elaborating his 1930 paper (Berücksichtigung der elastischen Formänderung in der Plastizitätstheorie, Z. angew. Math. Mech. 10, 266–274, 1930), the lecture of Prandtl was not known to him.

Fig. 9 Hilda Geiringer-v. Mises (1893-1973) and Willy Prager (1903-1980)

Many generalizations of this theory are given shortly later. So H. Geiringer and W. Prager [8] in addition to the yield condition $\Phi = 0$ introduce a second condition $H = 0$ as flow potential. The self-evident case $\Phi = H$ then is called an associated theory and as with Hencky's deformation theory a normality rule can be derived from the potential property

$$\boldsymbol{d}_p = \lambda \frac{1}{2} \frac{\partial \Phi}{\partial \boldsymbol{\tau}}. \tag{16}$$

For the general case of elastoplastic deformations, the yet undetermined parameter λ can be eliminated with the aid of the so-called condition of consistency, e.g. $\dot{\Phi} = 0$. This idea stems from H. Geiringer [7].

Also, the hardening of the material is taken into account by modifying the yield condition. First steps towards the "isotropic" hardening adapt the function k of the uniaxial tensile test for corresponding experiments and arrive at, e.g. the following proposals:

$$k = k_0 + F(J_1), \quad J_1 = \sqrt{\int \boldsymbol{\sigma} : \boldsymbol{d}_p \, dt}, \tag{17}$$

$$k = k_0 + F(J_2), \quad J_2 = \int \sqrt{\operatorname{tr}(\boldsymbol{d}_p^2)} \, dt, \tag{18}$$

where eq. (17) is introduced by Schmidt [49], and eq. (18) by Odqvist [37].

Approaches to the description of the "kinematic" hardening are derived from Reuss and Prager [39][19] in the form

$$\Phi = (\boldsymbol{\tau} - c\boldsymbol{\varepsilon}_p) : (\boldsymbol{\tau} - c\boldsymbol{\varepsilon}_p) - 2k^2 \tag{19}$$

and allow to take into account the Bauschinger effect. Then, the first evolution equation for this kinematic hardening originates from Melan [30]

$$\Phi = (\boldsymbol{\tau} - \boldsymbol{\alpha}) : (\boldsymbol{\tau} - \boldsymbol{\alpha}) - 2k^2, \quad \dot{\boldsymbol{\alpha}} = C\boldsymbol{d}_p. \tag{20}$$

---

[19] We refer to footnote 4 on page 79 of [39].

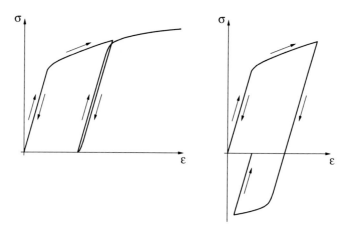

Fig. 10 Tensile test, cyclic loading with reloading in different directions: same direction (left Figure) and reverse loading expressing the Bauschinger effect (right Figure)

The debate as to what might be the "better" theory, starts of course immediately and drags on until the 1950's, 1960's[20]. At the beginning, it even seems as if the deformation theory of H. Hencky and A. Nádai[21] has advantages compared to the Prandtl-Reuss theory. K. Hohenemser [21] and W. Prager [40] for example indicate that in comparison with numerous experiments the deformation theory can reflect these results better. Moreover, in the aforementioned dispute about the "correct" buckling load in the range of plastic deformations, this deformation theory at first provides the "better" results. Finally, one also must recognize that in these pre-war years, e.g. in aircraft construction, the deformation theory simply prevails in practical applications because of its easy handling.

It is noticeable that until the 1930's in the mathematical representation of the so far considered contributions, we find Cartesian coordinates, small deformations, and all relations written in detailed component notation. In his 1925 paper [15], Hencky is one of the first to introduce the tensor analysis into mechanics, "to avoid the maze of formulas, that so far has prevented from calculating finite deformations..." as he says. To this end, in this work he also for the first time introduces convective coordinates. Maybe he is thus too far ahead of the times.

Hencky now turns increasingly to the issues of finite deformations and thus in [17; 18; 19] introduces first of all a logarithmic strain measure[22], which is the only one to allow for a correct superposition and, moreover, the only one capable to allow for a physically meaningful description of a total compression[23],

---

[20] We e.g. refer to Prager [41] and the discussion between B. Budiansky and W. Prager in [42].

[21] A. Nádai with his book [35] has very much contributed to its publicity.

[22] This strain is sometimes called natural strain or - simply - logarithmic strain. Here, we will use the term Hencky strain ***h***.

[23] Such a compression to "zero length" can be accompanied - think of an elastic body - only with an infinitely large force.

$$\boldsymbol{h} = \frac{1}{2}\ln\boldsymbol{b}, \quad \boldsymbol{b} = \boldsymbol{F}\boldsymbol{F}^{\mathrm{T}}. \tag{21}$$

$$\boldsymbol{h} = \boldsymbol{e}^{(0)} = \frac{1}{2}\ln\boldsymbol{b} = \frac{1}{2}\sum_{\sigma=1}^{n}(\ln\chi_\sigma)\boldsymbol{b}_\sigma. \tag{22}$$

where $\boldsymbol{F}$ is the deformation gradient and χ_σ and $\boldsymbol{b}_\sigma$ are n distinct eigenvalues and eigenprojections[24], respectively, of the left Cauchy-Green tensor $\boldsymbol{b}$[25]. Moreover, $\boldsymbol{e}^{(0)}$ is an Eulerian strain of the family of the Seth-Hill-Doyle-Ericksen strains ($m = 0$)

$$\boldsymbol{e}^{(m)} = \frac{1}{2m}(\boldsymbol{b}^m - \boldsymbol{I}). \tag{23}$$

Hencky introduces the Lagrangian and the Eulerian descriptions and discusses in this context the importance of time derivatives occurring in the relevant constitutive laws. For the Lagrangian analysis these will be understood as material time derivatives. In an Eulerian description, which he prefers for physical reasons, he notes that the time derivatives must be independent of the respective rigid-body rotation - or simply objective in today's notation. In [18] he therefore replaces the specified time derivative of the stress tensor by

$$\dot{\boldsymbol{\sigma}} \quad \Rightarrow \quad \overset{\circ}{\boldsymbol{\sigma}} = \dot{\boldsymbol{\sigma}} + \boldsymbol{\sigma}\boldsymbol{w} - \boldsymbol{w}\boldsymbol{\sigma}. \tag{24}$$

Herein, the spin tensor (vorticity) $\boldsymbol{w}$ has been introduced as skew-symmetrical part of the velocity gradient $\boldsymbol{l}$

$$\boldsymbol{w} = \frac{1}{2}(\boldsymbol{l} - \boldsymbol{l}^{\mathrm{T}}), \quad \boldsymbol{d} = \frac{1}{2}(\boldsymbol{l} + \boldsymbol{l}^{\mathrm{T}}), \quad \boldsymbol{l} = \dot{\boldsymbol{F}}\boldsymbol{F}^{-1}. \tag{25}$$

Unfortunately, the original work by Hencky [18] contains a small error. Instead of the above spin tensor he used an alternative definition, which differs by a minus sign. Due to this deviation, however, his derivative (24) looses its objectivity[26].

We immediately recognize the so-called Jaumann derivative, previously already discussed in this context by M.S. Zaremba [60] and G. Jaumann [26].[27]

---

[24] In numerous discussions Hencky is confronted with the difficulties of this measure for problems of non-principal axes. He then always refers to a possible transformation into principal axes. He also notes that this approach is quite common among technologists (refer e.g. to P. Ludwik, [29]). Nevertheless, this measure does not prevail. Knowing the today's computational possibilities, however, we should overcome this position and always use the logarithmic Hencky strain measure in the description of finite deformations.

[25] In today's notations.

[26] This may explain why proposition (24) is criticized by Truesdell in [56; 57], refer also to [53].

[27] It is not delivered whether Hencky knew these works. Rather not, we suspect. Otherwise he certainly would have quoted them. The today occasionally widespread habit, to quote only the works of the own school, has come up later.

But even this - from today's perspective - important statement is ignored. Until much later–after the end of World War II - these ideas are taken up again, e.g. by Oldroyd, Hill, Rivlin, Truesdell, and others. W. Prager e.g. in a work of 1960 refers to Jaumann: "Jaumann's work does not seem to be well known: the definition ... is frequently used in the recent literature without reference to Jaumann".

The perception that even in this accordingly corrected Prandtl-Reuss law still errors can occur, is generally attributed to C.A. Truesdell. For purely elastic processes, e.g. during unloading of a plastically deformed material, the constitutive law formulated in rates of stresses and deformation (which not until later is denoted hypoelastic by Truesdell [57]) does not have the properties of an elastic body. This can be shown easily if a circular process is calculated and thereby at the end an accumulated dissipation is observed. But even this can already be read in the work of G. Jaumann, who then denotes his rate material a material with fluxes (Fluxionen). He states that the behavior of this rate material is indeed not elastic and only in the limit of infinitesimal deformations changes to that of an elastic body. The full solution of this "defect" is given only in recent times by introducing the so-called logarithmic rate [58] to replace the rate of stress in the Prandtl-Reuss law. This allows for an exact integration of these rate equations to reflect elastic behavior [59].

Of course, one could ask where actually the contributions of materials science and materials physics remain, because naturally plastic behavior results primarily from gliding as a consequence of dislocation movement. In the 1920's, Bridgman had managed to produce single crystals. Subsequently, Taylor and Elam, Schmid and Sachs conducted experiments on such mono-crystalline metals. Sachs (1928) and later Taylor (1938) also have determined yield stresses for polycrystalline material by averaging the values of single crystals. As a result of the works of Egon Orowan, Michael Polanyi and Geoffrey Ingram Taylor, 1934 may be considered the birth year of the dislocation theory. The fundamental work of Johannes Martinus Burgers followed in 1939.

And there is also a work by v. Mises [33] on the yield behaviour of single crystals and a work by Hencky from his time at MIT [20], which begins as follows: "The behavior of metals in the inelastic states cannot be explained by theories which do not assume a microstructure." These pioneering works contain a proposal for the description of polycrystalline metals based on a statistical method.

In this first phase, however, such considerations had no further influence on the development of the plasticity theory. Yet the numerical methods and the associated high-performance computers were missing. The plasticity theory was a purely phenomenological one and remained so until the 70's and 80's of the last century.

At the beginning, we stated that research in the theory of plasticity theory took place primarily in German-speaking countries, which among other things can be detected from the fact that the Journal of applied Mathematics and Mechanics ZaMM was the leading journal of mechanics at the time. Certainly, the question comes to our mind, how this situation has changed due to political changes in Germany. Consider the year 1938: Most of the aforementioned persons were Jewish or simply of Jewish descents and as such no longer safe in Germany. Von Mises, 1933 has emigrated to Turkey - although he initially as a highly decorated veteran of World War

I and respected test pilot thought to be safe from reprisals. Via Istanbul, in 1939, he then attended Harvard University, Cambridge.

Prager has studied at the TH in Darmstadt, made his PhD in 1926, and remained there as lecturer (1927-1929). In 1929 he became lecturer in Göttingen, and then in 1932 – as at the time youngest professor in Germany – he went to Karlsruhe. 1933, he also emigrated to Turkey as a professor of theoretical mechanics in Istanbul. In 1940, he left Istanbul for Brown University, Providence.

Hilda Geiringer was assistant to von Mises. In 1927, she habilitated in Berlin where she was a lecturer - incidentally after a short marriage as a single child's mother. After Hitler came to power, however, she was dismissed and 1934 followed von Mises to Turkey - with a small detour over Brussels - and later in the United States, where in 1944 she became a professor at Wheaton College, Norton. At that time Hencky was already in the Soviet Union and was close to his deportation.

The Hungarian Arpád Nádai (1883-1963) and Theodore v. Kármán (1881-1963) both have studied first at the Technical University of Budapest. Nádai then went to the Technical University of Berlin and received his PhD in 1911; in 1918 he went to Göttingen to Prandtl and became a professor there. In 1927 he finally became the successor of Stepan Timoshenko at Westinghouse, Philadelphia. Von Kármán 1906 went to Göttingen to Ludwig Prandtl and Felix Klein, habilitated there, and in 1913 became professor at the RWTH Aachen. He also should be dismissed from his office, however, anticipating his release, 1934 he emigrated to the United States at Caltech, Pasadena.

The third Hungarian was by the way András Reuss (1900-1968), but also Orowan (1902-1989) and Polanyi (1891-1976) were Hungarian, who made their PhD in Berlin and Karlsruhe. Reuss never had left his home country, Orowan and Polanyi went in time via Birmingham at the MIT and to Manchester, respectively.

The hitherto fruitful research in the surrounding of the two nuclei Richard von Mises and Ludwig Prandtl had so definitively lost some of its most important figures; on the horizon, World War II announced itself in the course of which large parts of Europe and Asia should sink into wrack and ruin.

References

[1] Bruhns, C.: Neues logarithmisch-trigonometrisches Handbuch auf sieben Decimalen, 13 edn. Bernhard Tauchnitz, Leipzig (1920)

[2] Carathéodory, C., Schmidt, E.: Über die Hencky-Prandtlschen Kurven. Z. angew. Math. Mech. 3, 468–475 (1923)

[3] Cauchy, A.L.: Recherches sur l'équilibre et le mouvement intérieur des corps solides ou fluides, élastiques ou non élastiques. Bull. Soc. Philomath., 9–13 (1823)

[4] Considère, A.: Résistance des pièces comprimées. Congr. Int. des Procédés de Constr. 3, 371 (1889)

[5] Engesser, F.: über die Knickfestigkeit gerader Stäbe. Z. Arch. Ing-Ver. 35, 455 (1889)

[6] Engesser, F.: Über Knickfragen. Schweiz. Bauzeitung 26, 24–26 (1895)

[7] Geiringer, H.: Fondements mathématiques de la théorie des corps plastiques isotropes. Mém. Sci. Math., Gauthier-Villars, Paris 86, 19–22 (1937)

[8] Geiringer, H., Prager, W.: Mechanik isotroper Körper im plastischen Zustand. Ergebnisse der Exakten Naturwissenschaften 13, 310–363 (1934)

[9] Haar, A., von Kármán, T.: Zur Theorie der Spannungszustände in plastischen und sandartigen Medien. Nachr. Ges. Wiss. Göttingen, Math.-Phys. Kl, 204–218 (1909)

[10] Hencky, H.: Der Spannungszustand in rechteckigen Platten. Oldenbourg, München und Berlin (1913)

[11] Hencky, H.: Über die angenäherte Lösung von Stabilitätsproblemen im Raum mittels der elastischen Gelenkkette. Der Eisenbau 11, 437–452 (1921)

[12] Hencky, H.: Über die Beziehungen der Philosophie des 'Als Ob' zur mathematischen Naturbeschreibung. Annalen der Philosophie 3(2), 236–245 (1923)

[13] Hencky, H.: Über einige statisch bestimmte Fälle des Gleichgewichts in plastischen Körpern. Z. Angew. Math. Mech. 3, 241–251 (1923)

[14] Hencky, H.: Zur Theorie plastischer Deformationen und der hierdurch im Material hervorgerufenen Nachspannungen. Z. Angew. Math. Mech. 4, 323–334 (1924)

[15] Hencky, H.: Die Bewegungsgleichungen beim nichtstationären Fließen plastischer Massen. Z. Angew. Math. Mech. 5, 144–146 (1925)

[16] Hencky, H.: Über langsame stationäre Strömungen in plastischen Massen mit Rücksicht auf die Vorgänge beim Walzen, Pressen und Ziehen von Metallen. Z. Angew. Math. Mech. 5, 115–124 (1925)

[17] Hencky, H.: Über die Form des Elastizitätsgesetzes bei ideal elastischen Stoffen. Z. Techn. Phys. 9, 215–220, 457 (1928)

[18] Hencky, H.: Das Superpositionsgesetz eines endlich deformierten relaxationsfähigen elastischen Kontinuums und seine Bedeutung für eine exakte Ableitung der Gleichungen für die zähe Flüssigkeit in der Eulerschen Form. Annalen der Physik. 5, 617–630 (1929)

[19] Hencky, H.: The law of elasticity for isotropic and quasi-isotropic substances by finite deformations. J. Rheol. 2, 169–176 (1931)

[20] Hencky, H.: On a simple model explaining the hardening effect in poly-crystalline metals. J. Rheol. 3, 30–36 (1932)

[21] Hohenemser, K.: Elastisch-bildsame Verformungen statisch unbestimmter Stabwerke. Ing.-Archiv. 2, 472–482 (1931)

[22] Hooke, R.: Lectures de potentia restitutiva, or of spring explaining the power of springing bodies, London (1678)

[23] Huber, M.T.: Właściwa praca odkształcenia jako miara wytężenia materiału. Czasopismo Techniczne 22, 34–40, 49–50, 61–62, 80–81 (1904)

[24] Jasinski, F.: Recherches sur la flexion des pièces comprimées. Ann. Ponts Chaussées 8, 233, 654 (1894)

[25] Jasinski, F.: Noch ein Wort zu den "Knickfragen". Schweiz. Bauzeitung 25, 172–175 (1895)

[26] Jaumann, G.: Geschlossenes System physikalischer und chemischer Differentialgesetze. Sitzber. Akad. Wiss. Wien, Abt. Iia. 120, 385–530 (1911)

[27] von Kármán, T.: Untersuchungen über Knickfestigkeit. Mitt. VDI, p. 81 (1910)

[28] Lévy, M.: Mémoire sur les équations générales des mouvements intérieurs des corps solides ductiles au delà des limites où l'élasticité pourrait les ramener à leur premier état. C. R. Acad. Sci., Paris 70, 1323–1325 (1870)

[29] Ludwik, P.: Elemente der technologischen Mechanik. Springer, Berlin (1909)

[30] Melan, E.: Zur Plastizität des räumlichen Kontinuums. Ing.-Archiv 9, 116–126 (1938)

[31] von Mises, R.: Mechanik der festen Körper im plastisch-deformablen Zustand. Nachr. Ges. Wiss. Göttingen, Math.-Phys. Kl, 582–592 (1913)
[32] von Mises, R.: Bemerkungen zur Formulierung des mathematischen Problems der Plastizitätstheorie. Z. Angew. Math. Mech. 5, 147–149 (1925)
[33] von Mises, R.: Mechanik der plastischen Formänderung von Kristallen. Z. Angew. Math. Mech. 8, 161–185 (1928)
[34] Nádai, A.: Versuche über die plastischen Formänderungen von keilförmigen Körpern aus Flußeisen. Z. Angew. Math. Mech. 1, 20–28 (1921)
[35] Nádai, A.: Plasticity, a Mechanics of the Plastic State of Matter. McGraw-Hill, New York (1931)
[36] Navier, L.: Sur les lois des mouvements des fluides, en ayant égard à l'adhésion des molécules. Ann. de Chimie 19, 244–260 (1821)
[37] Odqvist, F.K.G.: Die Verfestigung von flußeisenähnlichen Körpern. Z. Angew. Math. Mech. 13, 360–363 (1933)
[38] Poisson, S.D.: Mémoire sur les équations générales de l'équilibre et du mouvement des corps solides élastiques et des fluides. J. École Poly. 13 (1831)
[39] Prager, W.: Der Einfluß der Verformung auf die Fließbedingung zähplastischer Körper. Z. Angew. Math. Mech. 15, 76–80 (1935)
[40] Prager, W.: Strain hardening under combined stresses. J. Appl. Phys. 16, 837–840 (1945)
[41] Prager, W.: Theory of plastic flow versus theory of plastic deformation. J. Appl. Phys. 19, 540–543 (1948)
[42] Prager, W.: A new method of analyzing stresses and strains in work-hardening plastic solids. ASME J. Appl. Mech. 78, 493–496, 79, 481–484 (1956)
[43] Prandtl, L.: Spannungsverteilung in plastischen Körpern. In: Proc. 1st Int. Congr. Appl. Mech., Delft, pp. 43–46 (1924)
[44] Prandtl, L.: Über die Eindringungsfestigkeit (Härte) plastischer Baustoffe und die Festigkeit von Schneiden. Z. Angew. Math. Mech. 1, 15–20 (1921)
[45] Prandtl, L.: Anwendungsbeispiele zu einem Henckyschen Satz "uber das plastische Gleichgewicht. Z. Angew. Math. Mech. 3, 401–406 (1923)
[46] Reuss, A.: Fließpotential oder Gleitebenen? Z. Angew. Math. Mech. 12, 15–24 (1932)
[47] de Saint-Venant, B.: Note à joindre au mémoire sur la dynamique des fluides, présenté le 14 avril 1834. C. R. Acad. Sci. 17, 1240–1243 (1843)
[48] de Saint-Venant, B.: Sur l'établissement des équations des mouvements intérieurs operes dans les corps solides ductiles au delà des limites où l'élasticité pourrait les ramener à leur premier état. C. R. Acad. Sci. 70, 473–480 (1870)
[49] Schmidt, R.: über den Zusammenhang von Spannungen und Formänderungen im Verfestigungsgebiet. Ing.-Archiv 3, 216–235 (1932)
[50] Shanley, F.R.: The column paradox. J. Aeronaut. Sci. 13, 678 (1946)
[51] Shanley, F.R.: Inelastic column theory. J. Aeronaut. Sci. 14, 261–268 (1947)
[52] Stokes, G.G.: On the theories of the internal friction of fluids in motion, and of the equilibrium and motion of elastic solids. Trans. Cambr. Phil. Soc. 8, 287–319 (1845)
[53] Tanner, R.I., Tanner, E.: Heinrich Hencky: a rheological pioneer. Rheologica Acta 42, 93–101 (2003)
[54] Tresca, H.: Mémoire sur lécoulement des corps solides soumis à des fortes pressions. C. R. Acad. Sci. 59, 754–758 (1864)

[55] Tresca, H.: Mémoire sur lécoulement des corps solides. Mém. Pres. Par. Div. Sav. 18, 733–799 (1868)
[56] Truesdell, C.: The mechanical foundations of elasticity and fluid dynamics. J. Rational Mech. Anal. 1, 125–300, 2, 595–616, 3, 801 (1952)
[57] Truesdell, C.: Hypo-elasticity. J. Rational Mech. Anal. 4, 83–133 (1955)
[58] Xiao, H., Bruhns, O.T., Meyers, A.: Logarithmic strain, logarithmic spin and logarithmic rate. Acta Mechanica 124, 89–105 (1997)
[59] Xiao, H., Bruhns, O.T., Meyers, A.: Hypo-elasticity model based upon the logarithmic stress rate. J. Elasticity 47, 51–68 (1997)
[60] Zaremba, S.: Sur une forme perfectionnée de la théorie de la relaxation. Bull. Intern. Acad. Sci. Cracovie, 595–614 (1903)

Prandtl-Tomlinson Model: A Simple Model Which Made History

Valentin L. Popov* and J.A.T. Gray

Abstract. One of the most popular models widely used in nanotribology as the basis for many investigations of frictional mechanisms on the atomic scale is the so-called Tomlinson model, consisting of a point mass driven over a periodic potential. The name "Tomlinson model" is, however, historically incorrect: The paper by Tomlinson from the year 1929 which is often cited in this context did not, in fact, contain the model known as the "Tomlinson model" and suggests, instead, an adhesive contribution to friction. In reality, it was Ludwig Prandtl who suggested this model in 1928 to describe the plastic deformation in crystals and dry friction. Staying in line with some other researchers, we call this model the "Prandtl-Tomlinson model," although the model could simply and rightly be dubbed the "Prandtl model." The original paper by Ludwig Prandtl was written in German and was not accessible for a long time for the largest part of the international tribological community. The present paper is a historical introduction to the English translation of the classical paper by Ludwig Prandtl, which was published in 2012. It gives a short review of the model as well as its properties, applications, and extensions from a contemporary point of view.

Keywords: Prandtl-Tomlinson model, atomic scale friction, superlubricity, velocity dependence, temperature dependence, dislocations.

1 Introduction

Throughout history, there have been contributions which have forever changed science. The discoveries of Newton or Einstein, of course, are among these. However, there are also such discoveries in more specialized fields, theories or models

---

V.L. Popov · J.A.T. Gray
Berlin University of Technology, 10623 Berlin, Germany
e-mail: v.popov@tu-berlin.de
* Corresponding author.

which have developed into an integral part of science. In contact mechanics and friction, these include the Herzian contact problem or the lubrication theory of Petrov-Reynolds. This chapter is dedicated to one of the contributions from Ludwig Prandtl, a model, which he suggested and investigated. This model played a prominent role in frictional physics, especially in nanotribology.

The name Prandtl is associated most often with fluid mechanics. It is easy to believe that the publication [1] was written by him. Less known, however, is that he was also responsible for several meaningful contributions to the physics and mechanics of plastic deformations, friction, and fracture mechanics (e.g., [2]). In 1928, Ludwig Prandtl suggested a simple model for describing plastic deformation in crystals [3]. He considered the process of plastic deformation and abstracted it stepwise to a model in which a stage with many elastically coupled "atoms" is moved along a periodic potential. To start with, Prandtl simplifies it even further and considers only one-dimensional movement of a point mass being dragged in a periodic potential by means of a spring with a constant velocity v_0 and being damped proportional to velocity (Fig. 1)[1]:

$$m\ddot{x} = k(v_0 t - x) - \eta \dot{x} - N \sin(2\pi x / a). \qquad (1)$$

Here, x is the coordinate of the body, m its mass, k the stiffness of the pulling spring, η the damping coefficient, N the amplitude of the periodic force, and a the spatial period of the potential. In the case of a very soft spring, the motion over one or several periods does not change the spring force F, which can be considered to be constant. In this case, the model can be simplified to the form

$$m\ddot{x} = F - \eta \dot{x} - N_0 \sin(2\pi x / a). \qquad (2)$$

This is illustrated in Fig. 2. After analyzing the one-dimensional model (1), Prandtl generalizes it back to a system with many independent "atoms" coupled to a stage as well as to an ensemble of "atoms" with different coupling stiffnesses, and then even to an ensemble of "systems," which from a macroscopical point of view simulates a multi-contact situation.

Over many years, the basic models (1) and (2) have been referred to as the "Tomlinson model" and the paper [5] by Tomlinson from 1929 has been cited in this connection. However, the paper by Tomlinson did not contain the above model. It is now difficult to ascertain who made this historical error and also who corrected it. We believe, however, that we have Martin Müser to thank for this historical rectification. Together with two prominent figures in nanotribology, Michael Urbakh and Mark Robbins, he published a fundamental paper in 2003 [6], in which the above mentioned model was termed the "Prandtl-Tomlinson Model." Since then, this name has become ever more frequently used.

---

[1] Prandtl visualized this system with a mechanical model consisting of a wave-like surface upon which a heavy roller rolls back and forth. The elastic force is realized by the springs whose ends are fastened to a gliding stage.

Fig. 1 The Prandtl model: A point mass dragged in a periodic potential

Fig. 2 Simplified Prandtl model in the case of a very soft spring (modeled here by a constant force)

The model from Prandtl describes many fundamental properties of dry friction. Actually, we must apply a minimum force to the body so that a macroscopic movement can even begin. This minimum force is none other than the macroscopic force of static friction. If the body is in motion and the force reduced, then the body will generally continue to move, even with a smaller force than the force of static friction, because it already possesses a part of the necessary energy due to its inertia. Macroscopically, this means that the kinetic friction can be smaller than that of the static friction, which is a frequently recurring characteristic of dry friction. The force of static friction in the model described by Eq. (1) is equal to N.

The success of the model, variations and generalizations of which are investigated in innumerable publications and are drawn on to interpret many tribological processes, is due to the fact that it is a simplistic model that accounts for two of the most important fundamental properties of an arbitrary frictional system. It describes a body being acted upon by a periodic conservative force with an average value of zero in combination with a dissipating force which is proportional to velocity. Without the conservative force, no static friction can exist. Without damping, no macroscopic sliding frictional force can exist. These two essential properties are present in the PT-model. In this sense, the PT-model is the simplest usable model of a tribological system. It is interesting to note that the PT-model is a restatement and further simplification of the view of Coulomb about the "interlocking" of surfaces as the origin of friction.

The PT-model was designed for understanding plasticity [3], but its extensions and variations are widely used for understanding processes of various physical natures, including dislocations in crystals (Frenkel-Kontorova model [7],[8]), atomic force microscopy [9], solid-state friction [6], control of friction by chemical and mechanical means as well as in the design of nano-drives [10],[11], and handling of single molecules [12].

In his original paper, Ludwig Prandtl considered not only the simplest deterministic form of the model, but also the influence of thermal fluctuations. He was the first "tribologist" who came to the conclusion that thermal fluctuations should lead to a logarithmic dependency of the frictional force on velocity.

The rest of this paper is organized as follows: In Sect. 2 we describe the main findings of the paper by Prandtl, in Sect. 3 we discuss further well known extensions and applications of the Prandtl model. We would like to stress that this paper is not based on a deep historical review and all following discussions make no claims of being complete.

2 Results of Prandtl

Prandtl notes, first, that the movement of the mass point does not follow the movement of stage continuously, provided that the pulling spring is soft enough. For small pulling velocities, this condition of *elastic instability* reads

$$k < N \frac{2\pi}{a}. \tag{3}$$

In this case, a hysteresis is observed when the stage is moved over large distances back and forth (Fig. 3a). This is, according to Prandtl, the physical reason for plastic hysteresis. Since the 1980s, it has been known in the nanotribological community that the existence of elastic instabilities is indeed the necessary condition for a finite static frictional force in atomic scale friction [9],[13]. An example of the atomic stick-slip motion and the hysteresis is shown in Fig. 4. If the stiffness k is larger than the critical value, the average spring force (interpreted as a macroscopic force of friction) becomes identically zero (Fig. 3b). This property has been proven exactly for an arbitrary periodic potential [14]. This model has been frequently used in nanotribolgy, above all to describe the movement of the tip of an atomic force microscope along a smooth surface. This is really almost the Prandtl model to the letter, and everything that one may predict with the model can be measured. Elastic instability occurs at either sufficiently small stiffness or at sufficiently large amplitudes of the potential. The latter may be varied by adjusting the pressure that the tip of the atomic force microscope exerts on the substrate (Fig. 5), which Prandtl already realized at the end of his paper. Thereby, one can clearly see that there are actually certain conditions for which the average friction force practically disappears completely. This can be clearly seen in the hysteresis curves, which can be described very well with the simplest Prandtl model for all but small fluctuations and confirmed experimentally (Fig. 4).

In recent years, a mysterious effect has been discovered in that the movement of the tip exhibits a defined stick-slip characteristic although the average friction is very small. Even this possibility is fundamentally contained in the Prandtl model, as shown in the graphics below. The physical reason is that the mass loses stability at one position, is accelerated by the spring, and then becomes stuck once again in state in which the spring is compressed (Fig. 6b). In this case, there exists nearly no dissipation, and thereby, only little associated friction.

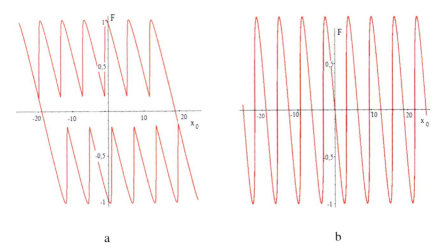

Fig. 3 Dependence of the force on the position x_0 of the slider: (a) $k = 0.2 \cdot k_c$. In this case, a pronounced atomic "stick-slip" and hysteresis are seen, the average "friction force" is non-zero; (b) $k = k_c$. In this case, the tangential force depends continually on the coordinate and the average frictional force is zero, there is no hysteresis.

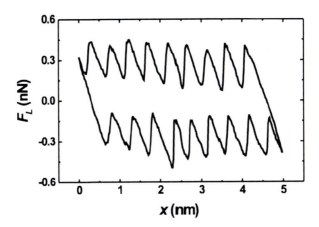

Fig. 4 Friction loop formed by two scan lines measured on a NaCl(100) surface forward and backward, respectively at $F_N = 0.65\ nN$ and $v = 25\ nm/s$. Source: [17]

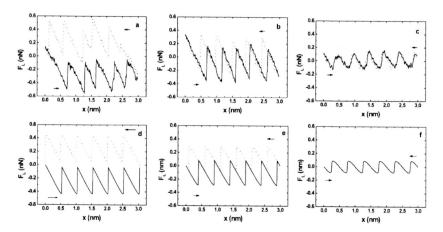

Fig. 5 Experimental data from [19] (a) –(c) Measurements of the lateral force acting on the tip sliding forward and backward in (100) direction over the NaCl(001) surface with decreasing normal force. (d) –(f) Corresponding numerical results from the Tomlinson model with suitable parameters. For small values of the normal force, the hysteresis loop enclosed between the forward and the backward scan vanishes; i.e., there is no more dissipation within this model.

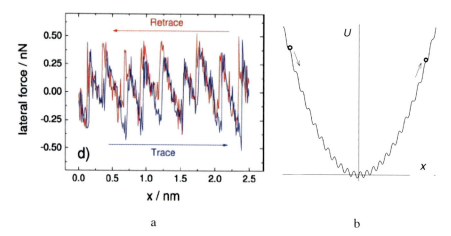

Fig. 6 Stick-Slip without hysteresis: (a) experimental data for a scan on a graphite surface (Source: [20]) an (b) theoretical explanation

One of the surprising discoveries of nanotriblogy was the effect of super lubrication. This effect is a direct result of the fundamental properties of the Prandtl model. If the transversal stiffness is above a critical value and if several "atoms" are arranged in such a way that they have a non-compatible period relative to the opposite periodic potential (Fig. 7), then the spatial average is the same as the temporal average. Therefore, the tangential force at *every* point in time is zero. Moving the stage leads to a redistribution of the atomic structure but not to a change in the total energy. The macroscopic friction force disappears.

In literature on nanotribology, this effect is known as "superlubricity" [15]. Superlubricity can be observed either if the crystal lattices of two contacting bodies are incommensurate or if the lattices are rotated with respect to one another. In all situations, with the exception of discrete compatible orientation, the frictional force is extremely low. This is commonly illustrated by the "egg-carton model" (Fig. 8) and is also seen in experiment (Fig. 9 and Ref. [13]). Note that the similarity of the term superlubricity to the terms superconductivity and superfluidity is misleading, because it refers only to the *static* friction force. The interaction with electrons and phonons will generally lead to a final, velocity dependent frictional force. However, in dielectric materials at very low temperatures, an effect similar to true superfluidity may take place, which in [16] was called *superslipperiness*.

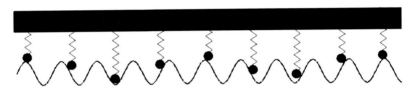

Fig. 7 In the case of a supercritical stiffness, the frictional force between incommensurate bodies vanishes

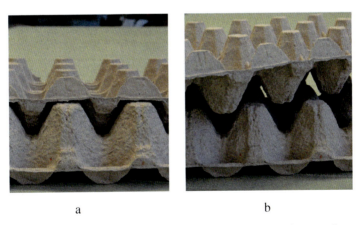

Fig. 8 (a) The surfaces are commensurate – high friction (b) the surfaces are incommensurate – low friction

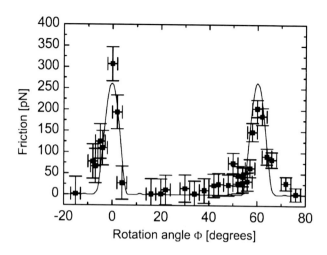

Fig. 9 Experimental data from [13]: Average friction force versus rotation angle of the graphite sample around an axis normal to the sample surface. Two narrow peaks of high friction were observed at 0 and 60°, respectively. Between these peaks a wide angular range with ultralow friction close to the detection limit of the instrument was found.

If the loading is non-monotonous, then the exact state of the system at any moment of time depends on its prehistory. In the case of an ensemble of particular systems with different parameters, this state can be very complicated. Prandtl poses the question of whether it is possible to restore the *virgin state* in which all the "atoms" are again in their initial non-stressed positions. With very simple arguments, he shows that if the stage starts to oscillate with large amplitude and the amplitude then decreases *slowly*, then each "atom" in the ensemble finally comes to the neutral, non-stressed position and the system returns to the virgin state. He compares this result with demagnetization through slowly decreasing oscillating magnetic fields, first studied by E. Madelung [18].

Prandtl further considers the properties of the model at finite temperatures. The principle idea exploited by Prandtl is to calculate back and forth mass currents due to thermal fluctuations. If the height of the periodic potential in (1) is U_0 and the spatial period a, then in a first approximation, a spring force F will change the heights of the "left" and "right" potential barriers and they will become $U_1 = U_0 + Fa/2$, $U_2 = U_0 - Fa/2$, thus, leading to a total sliding velocity $\dot{v}$ of

$$\dot{v} = C\left(e^{-(U_0 - Fa/2)/k_B T} - e^{-(U_0 + Fa/2)/k_B T}\right), \quad (4)$$

where k_B is the Boltzmann constant and T the absolute temperature. For *medium* forces satisfying the condition

$$kT \ll Fa/2 \ll U_0, \tag{5}$$

the back transitions in (4) can be neglected and we have $\dot{v} = Ce^{-(U_0 - Fa/2)/k_B T}$. Solving this equation for F

$$F = 2U_0/a + (2k_B T/a)\ln(\dot{v}/C) \tag{6}$$

gives the famous logarithmic dependency of the frictional force on velocity [6], which is found both at the atomic scale [17] and at the macro scale [21]. Note that the coefficient of the logarithmic term is proportional to the absolute temperature. For extremely small forces

$$Fa/2 \ll kT, \tag{7}$$

the velocity is a linear function of the force

$$v = Ce^{-U_0/k_B T}\left(e^{Fa/2k_B T} - e^{-Fa/2k_B T}\right) \approx Ce^{-U_0/k_B T}\frac{Fa}{k_B T}, \tag{8}$$

as it should be from general thermodynamic principles. The exact solution of (4) with respect to F has the form

$$F = a(T) \cdot \operatorname{asinh}(v/v_c), \tag{9}$$

which has been confirmed recently by direct molecular dynamic simulations [22]. Particularly, for very small velocities, Prandtl comes to the "noteworthy result" that the force is proportional to the deformation rate. Thus, a resistance of the same type as the drag in a fluid exists. In this viscous region he comes to the conclusion that the viscosity is an exponential function of temperature and pressure (as shown in experiments by Bridgeman [23]).

For larger forces in the region $Fa/2 \approx U_0$, a more detailed analysis is needed. In this case, the back transitions can be neglected. If the initial potential is approximated as

$$U = U_0\left[\frac{3}{2}\left(\frac{2x}{a}\right) - 2\left(\frac{2x}{a}\right)^3\right] \tag{10}$$

(preserving the distance $a/2$ between the minimum and the adjacent maximum and the height U_0 of the potential barrier), then in the presence of a constant force, the potential energy will be

$$U = U_0\left[\frac{3}{2}\left(\frac{2x}{a}\right) - 2\left(\frac{2x}{a}\right)^3\right] - Fx. \tag{11}$$

The distance between the minimum and the maximum is equal to

$$\tilde{a} = \frac{a}{2}\left(1 - \frac{1}{3}\frac{Fa}{U_0}\right)^{1/2} \qquad (12)$$

and the energy difference is

$$\Delta U = U_0\left(1 - \frac{1}{3}\frac{Fa}{U_0}\right)^{3/2}. \qquad (13)$$

Therefore, for the macroscopic sliding velocity, we obtain

$$v \approx C e^{-\Delta U/T} = C \exp\left(-U_0\left(1 - \frac{1}{3}\frac{Fa}{U_0}\right)^{3/2} \Big/ k_B T\right). \qquad (14)$$

Solving for F under the condition $\ln(v/C) < 0$ gives

$$F = \frac{3U_0}{a}\left[1 - \left(\frac{k_B T}{U_0}\right)^{2/3}\left|\ln\frac{v}{C}\right|^{2/3}\right]. \qquad (15)$$

Thus, the Prandtl model predicts the following temperature dependence of the force of friction

$$F = C_1 - C_2 T^{2/3}. \qquad (16)$$

This dependency, first found by Prandtl, has been rediscovered in the context of atomic force microscopy [24] and has been observed experimentally [25],[26]. It is interesting to note that the determination of the thermally activated sliding velocity formally coincides with many problems in chemical kinetics. The results of Prandtl, therefore, also find a place in these areas of application [12].

All of the above mentioned results of the paper have been obtained by Prandtl in a much more rigorous way as briefly summarized above. As already mentioned, Ludwig Prandtl considered an ensemble of systems and solved a kinetic equation for them in a very similar way to how it is done by contemporary researchers [24]. Eq. (1) in his paper, which in contemporary notations would read

$$\frac{dP(x,t)}{dt} = \frac{1}{\tau}\left((1 - P(x,t))e^{-\frac{U_2(x)}{k_B T}} - P(x,t)e^{-\frac{U_1(x)}{k_B T}}\right) \qquad (17)$$

(with $P(x,t)$ being the probability for the atoms to take the position x in the potential relief), is nothing other than the kinetic equation for the ensemble of

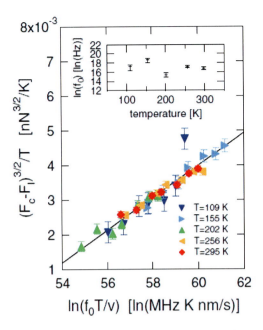

Fig. 10 Source: [26]: Force by scanning the highly oriented pyrolytic graphite (HOPG) as a function of velocity shows the dependence predicted by (15)

systems and the following treatment is essentially equivalent to solving the Fokker-Planck equation. The derivations around Eq. (20) of Prandtl's paper give, in the end, essentially the same results as were found by Kramers in his classical paper on the kinetics of chemical reactions [32]. However, one must admit that the calculations by Prandtl and his terminology are very difficult to follow. This may be one of the reasons why his paper has not had the impact to the development of science that it deserves. Now, it is of course possible to directly solve Eq. (1) with additional stochastic forces modeling thermal fluctuations. A very good review of such "molecular dynamic" investigations of the PT-model can be found in an excellent recent paper by Martin Müser [22]. Prandtl was concentrated on the study of quasi-static movements and creep, so he did not investigate the dynamic properties of the model. An overview of basic dynamic properties is given in [14].

It is interesting to note that the entire paper by Prandtl is devoted to a discussion of plasticity. Only at the end of his paper, does he come to a discussion of friction and notes that "...our conceptual model is also suitable for the treatment of kinetic friction between solid bodies." In this connection, he cites experimental results on the logarithmic dependence of the frictional force on the sliding velocity and states that "the law that frictional force increases with the logarithm of velocity is confirmed very well in a large range of velocities." He finally speculates on the reasons for the approximate proportionality of the frictional force to the normal force and suggests two physical reasons for this: On the one hand, the effective contact area of both bodies increases with the normal force; and on the other hand, through the increase in surface pressure, the molecules of both bodies are brought into closer proximity, thus, a force field with a larger wave amplitude is present. These dependencies of the local potential amplitude on the normal force have in the meantime been investigated experimentally [27] and theoretically and are indeed considered to be one of the reasons for the validity of Coulomb's law of friction [28].

3 Applications and Extensions of the Prandtl-Tomlinson Model

1. The initial application of the model was the understanding of plastic deformation and deformation creep at finite temperatures.

2. Historically, the main application of the Prandtl model has been found in nanotribology, especially for the understanding of experiments with atomic force microscopes (for review see [9]).

3. The equation of type (2) – rewritten in another notation –

$$\left(\frac{\hbar C}{2e}\right)\ddot{\phi} = j - \left(\frac{\hbar}{2eR}\right)\dot{\phi} - j_0 \sin\phi, \qquad (18)$$

describes the dynamics of a single Josephsohn contact [29],[33]. In Eq. (18), $\hbar$ is Planck's constant, ϕ is the phase difference between contacting superconductors, C and R are the respective contact capacity and Ohmic resistance, e is the elementary charge, and j_0 is the maximum contact current. The mathematical equivalence of the two problems means that the effects seen in Josephsohn's contacts must be observed in tribological systems whose microscopic model is described by Eq. (2). One of the effects is in the modification of the current-voltage characteristic of a Josephsohn's contact undergoing an external periodic perturbation. The modification is in the appearance of plateaus where the voltage is rigorously constant (modern quantum voltage standard is based on this effect). An analogy to this effect in terms of tribology is the appearance of a plateau of a constant average velocity of a body in the presence of a periodic oscillating force. These plateaus play a key role in the further discussion of nanomachines.

4. One further extension of the simple PT-model is to consider a number of "atoms" which are elastically coupled with both the moving stage and with each other (Fig. 11). This model has been proposed by Frenkel and Kontorova [7],[8] to describe dislocation-based plastic deformation in crystals.

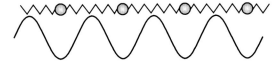

Fig. 11 Kontorova-Frenkel model

Considering the series of elastically coupled atoms as an elastic continuum, one gets the following equation of motion of the system

$$\frac{\partial^2 u}{\partial t^2} = c^2 \frac{\partial^2 u}{\partial x^2} - \alpha \frac{\partial u}{\partial t} - R\sin\left(\frac{2\pi u}{a}\right) + f, \tag{19}$$

where u is the tangential displacement of a mass point with the initial coordinate x. α is a damping constant, f is the force density acting homogeneously on all mass points (not shown in Fig. 3), c is the sound velocity in the elastically coupled chain (without interactions with the periodic potential and the stage) and R is a constant characterizing the amplitude of the periodic potential. Eq. (19) is also known as the Sine-Gordon equation. Its static solutions correspond to dislocations in a crystal. Moving solutions of the form $u = u(x - vt)$ lead to the equation

$$\left(c^2 - v^2\right)\frac{\partial^2 u}{\partial x^2} + \alpha v \frac{\partial u}{\partial x} = R\sin\frac{2\pi u}{a} - f, \tag{20}$$

which is an equation precisely of the same form as the simplified PT-equation (2) (only the time should be replaced by x and the coordinate by u).

5. We consider an automobile that is power by a single-cylinder internal combustion engine and drives downhill at an angle of α to the horizontal plane. Due to the fact that the fuel injection and combustion is coupled to a distinct phase angle of the piston and, therefore, to the angle of the crank shaft, one can consider the moment that the motor exerts on the axle before the gearbox as a function of the angle θ: $M^* = M^*(\theta)$. We can write this function as

$$M^*(\theta) = M_1 + M_2 \sin\theta. \tag{21}$$

If a gearbox with a gear ratio of n is connected between the motor and the wheel axle, then the moment exerted on the axle is equal to $M = nM^*(\varphi/n)$. Neglecting the moment of inertia of the wheels, one obtains the following equation of motion for the automobile:

$$\ddot{\varphi} + \frac{\gamma}{m}\dot{\varphi} = \left(\frac{g}{r}\sin\alpha + \frac{nM_1}{mr^2}\right) + \frac{nM_2}{mr^2}\sin(\varphi/n), \tag{22}$$

where φ is the angle of rotation of the wheel, r the radius of the wheels, m the mass of the automobile, and g the gravitational acceleration. Thus, the simplest case of engine braking is also described by the PT-model!

6. Most of the ways of generating directed motion of molecular objects discussed in literature are based on the interaction between a driven object and an inhomogeneous, commonly periodically structured substrate. The latter can be either asymmetric or symmetric. In the former case, the directed motion can be only unidirectional; it is fixed by the interaction between the substrate and the object following the "ratchet-and-pawl" principle (see, e.g.,[31]). In the latter case, the direction of motion is originally not fixed and is determined dynamically. An example of this kind of a dynamically driven engine was given in [10] and [11]. The models utilized in these works are nothing but a generalized PT-model with oscillating forces.

7. Very thin layers of liquid between crystalline solids have a tendency to undergo layering transitions [34]. Under these boundary lubrication conditions, the fluid takes on the properties of the solid state. The sliding in this case can also be described by a generalized PT-model [35].

8. Further generalizations are very numerous and include, for example, the substitution of a periodic potential by a stochastic one, thus, producing a random PT-model [30]. This allows the transformation of the viscous friction at the nano-scale to a Coulomb-like friction at the macro-scale to be considered by means of a sort of renormalization technique.

4 Conclusion

Ludwig Prandtl is known foremost for his contributions to hydrodynamics. However, like many other accomplished scientists, he provided important contributions to scientific areas which were not the main field of his activities. One of these contributions is the paper on the theory of plasticity analyzed with what we now call the Prandtl-Tomlinson model. Prandtl investigated problems which still remain a subject of scientific investigation until today. The model has played a very prominent role in the history of science, especially tribology, which can be compared with other classical models, such as the contact theory by H. Hertz or the lubrication theory by Petrov and Reynolds.

References

[1] Prandtl, L.: Über Flüssigkeitsbewegung bei sehr kleiner Reibung, Verhandlungen III, p. 484. Intern. Math. Kongress, Heidelberg (1904)

[2] Prandtl, L.: Über die Härte plastischer Körper, Nachrichten Göttinger Akad. Wiss. (1920)

[3] Prandtl, L.: Ein Gedankenmodell zur kinetischen Theorie der festen Körper. ZAMM 8, 85–106 (1928)

[4] Popov, V.L., Gray, J.A.T.: ZAMM. Z. Angew. Math. Mech. 92(9), 683–708 (2012), doi:10.1002/zamm.201200097
[5] Tomlinson, G.A.: A molecular theory of friction. Phil. Mag. 7, 905–939 (1929)
[6] Müser, M.H., Urbakh, M., Robbins, M.O.: Statistical mechanics of static and low-velocity kinetic friction. In: Prigogine, I., Rice, S.A. (eds.) Advances in Chemical Physics, vol. 126, pp. 187–272 (2003)
[7] Kontorova, T.A., Frenkel, Y.I.: К теории пластической деформации и двойникования. I. (On the theory of plastic deformation and crystal twinning). Zh. Éksp. Teor. Fiz. 8(1), 89–95 (1938)
[8] Kontorova, T.A., Frenkel, Y.I.: Z. Éksp. Teor. Fiz. 8(12), 1340–1348 (1938)
[9] Meyer, E., Overney, R.M., Dransfeld, K., Gyalog, T.: Nanoscience: Friction and Rheology on the Nano-meter Scale. World Scientific, Singapore (1998)
[10] Porto, M., Urbakh, M., Klafter, J.: Atomic scale engines: cars and wheels. Phys. Rev. Lett. 84, 6058–6061 (2000)
[11] Popov, V.L.: Nanomachines: Methods of induce a directed motion at nanoscale. Physical Review E 68, 026608 (2003)
[12] Dudko, O.K., Hummer, G., Szabo, A.: Intrinsic rates and activation free energies from single-molecule pulling experiments. Phys. Rev. Lett. 96, 108101 (2006)
[13] Dienwiebel, M., et al.: Superlubricity of graphite. Phys. Rev. Lett. 92, 126101 (2004)
[14] Popov, V.L.: Contact Mechanics and Friction. Physical Principles and Applications. Springer (2011)
[15] Hölscher, H., Schirmeisen, A., Schwarz, U.D.: Principles of atomic friction: from sticking atoms to superlubric sliding. Phil. Trans. R. Soc. A 366, 1383–1404 (2008)
[16] Popov, V.L.: Superslipperiness at low temperatures: quantum mechanical aspects of solid state friction. Phys. Rev. Lett. 83, 1632–1635 (1999)
[17] Gnecco, E., Bennewitz, R., Gyalog, T., Loppacher, C., Bammerlin, M., Meyer, E., Guntherodt, H.J.: Velocity dependence of atomic friction. Phys. Rev. Lett. 84, 1172–1175 (2000)
[18] Madelung, E.: Über Magnetisierung durch schnell verlaufende Ströme, Göttingen 1905. Ann. d. Phys. 17, 861 (1905)
[19] Socoliuc, A., Bennewitz, R., Gnecco, E., Meyer, E.: Transition from Stick-Slip to Continuous Sliding in Atomic Friction: Entering a New Regime of Ultralow Friction. Phys. Rev. Lett. 92(13), 134301 (2004)
[20] Dietzel, D., Mönninghoff, T., Herding, C., et al.: Frictional duality of metallic nanoparticles: Influence of particle morphology, orientation, and air exposure. Phys. Rev. B 82, 035401 (2010)
[21] Dieterich, J.H.: Modeling of rock friction. 1. Experimental results and constitutive equations. J. Geophys. Res. 84(B5), 2161 (1979)
[22] Müser, M.H.: Velocity dependence of kinetic friction in the Prandtl-Tomlinson model. Phys. Rev. B 84, 125419 (2011)
[23] Bridgeman, P.W.: Proc. of the Amer. Acad. of Arts and Sciences 61, 57 (1926)
[24] Dudko, O.K., Filippov, A., Klafter, J., Urbakh, M.: Dynamic force spectroscopy: a Fokker–Planck approach. Chemical Physics Lett. 352, 499–504 (2002)
[25] Sang, Y., Dube, M., Grant, M.: Phys. Rev. Lett. 87, 174301 (2001)
[26] Jansen, L., Hölscher, H., Fuchs, H., Schirmeisen, A.: Temperature dependence of atomic-scale stick-slip friction. Phys. Rev. Lett. 104, 256101 (2010)
[27] Gnecco, E., Bennewitz, R., Meyer, E.: Transition from stick-slip to continuous sliding in atomic friction: Entering a new regime of ultralow friction. Phys. Rev. Lett. 92, 134301 (2004)

[28] Cheng, S.F., Luan, B.Q., Robbins, M.O.: Contact and friction of nanoasperities: Effects of adsorbed monolayers. Phys. Rev. E 81, 16102 (2010)
[29] Barone, A., Esposito, F., Magee, C.J., Scott, A.C.: Theory and applications of the sine-gordon equation. La Rivista del Nuovo Cimento 1(2), 227–267 (1971)
[30] Filippov, A.E., Popov, V.L.: Fractal Tomlinson model for mesoscopic friction: From microscopic velocity-dependent damping to macroscopic Coulomb friction. Physical Review E 75, 027103 (2007)
[31] Reimann, P.: Brownian motors: noisy transport far from equilibrium. Physics Reports 361, 57–265 (2002)
[32] Kramers, H.A.: Brownian motion in a field of force and the diffusion model of chemical reactions. Physica (Utrecht) 7, 284–304 (1940)
[33] Barone, A., Paterno, G.: Physics and Applications of the Josephson Effect. Wiley & Sons (1982)
[34] Thompson, P.A., Grest, G.S., Robbins, M.O.: Phase transitions and universal in confined films. Physical Review Letters 68, 3448–3451 (1992)
[35] Popov, V.L.: A theory of the transition from static to kinetic friction in boundary lubrication layers. Solid State Commun. 115, 369–373 (2000)

A Historical View on Shakedown Theory

Dieter Weichert and Alan Ponter

Abstract. Plastic design started in the early 20th century with the arrival of steel constructions in civil engineering. The objective was to determine the load carrying capacity in particular of steel bridges and steel skeleton buildings beyond the elastic limit. The related studies were first focused on monotonically increasing, "dead" loading. From this point of view they were directly related to the ancient question of determining the load carrying capacity of masonry construction like domes of churches.

It was in the extension of these studies that the problem of plastic design under variable loads came into the picture. Martin Grüning was the first to be attracted by the beneficial effect of limited plastic deformation in redundant elements in hyperstatic structures and opened the door to the fascinating theory of shakedown.

Keywords: Shakedown Theory, Plasticity.

1 From the Beginnings to the Formulation of the Classical Theorems of Shakedown Theory

The usual path to solve problems in solid mechanics today starts with an appropriate formulation of the set of differential equations governing kinematics, balance laws and material behaviour. In particular in the non-linear case, generally the evolutionary or rate formulation of the problem is chosen, which is then reformulated in weak form. This opens the way to approximation methods by introducing appropriate test functions. Solutions are then constructed in discrete form by a

---

Dieter Weichert
Institute of General Mechanics, RWTH-Aachen University,
Templergraben 64, 52056 Aachen, Germany

Alan Ponter
Department of Engineering, University of Leicester,
Leicester, LE 7RH, U.K.

cascade of systems of linear algebraic equations.-The finite element method most prominently represents this strategy, which we may call "classical" in what follows.

This highly successful methodology relies on several important assumptions: The loading history is deterministically known, test functions and material law formulation are sufficient smooth and, above all, means for solving a cascade of very large systems of linear algebraic equations are available. - Non-linear material behaviour such as plasticity renders the problem within this strategy more complicated than linear elasticity which is a simple linear mapping of the space of stresses onto the space of strains. As well-known example, the elastic-perfectly plastic material model rather badly fits into this scheme.

Going back a century, there was no question of adequate computational tools nor did there even exist a general mathematical framework for non-linear solid mechanics. Nevertheless, the question whether a mechanical structure may fail or not under variable loads is as old as engineering activity itself. It is therefore very difficult to pinpoint the instant when mathematical models were first used to predict with rational arguments if a structure will resist the acting forces or collapse. Edoardo Benvenuto [1] gives an excellent view on early developments, going back to antiquity. - We choose more or less arbitrarily the prominent and representative "Poleni's problem" as the starting point for the rational treatment of limit states of structures. By his study published in 1748 [2] G. Poleni answered the question, whether the dome of St. Peter's would collapse due to cracks that had formed on its top? He solved this problem, that would today be classified as a problem of limit analysis with unilateral material behaviour, semi-empirically: he used Hooke's analogy between the "hanging chain problem" and the "arch-problem" (*"as hangs the flexible line, so but inverted will stand the rigid arch"* cited from J. Heyman [3]) which needed to translate the plane problem of the arch to a shell problem (for details see J. Heyman's paper on Poleni's problem already quoted before [3]).

This typical limit-analysis problem does not involve the characteristic features of a shakedown problem which are variable loads, material ductility and residual stresses. However, it illustrates several important historical aspects:

(i) Long before mechanical engineering had gained significant importance, in civil engineering safe design, that means avoiding collapse of masonry structures, was of crucial importance.
(ii) Irreversible material behaviour is present through cracks and pseudo-hinges.
(iii) Elastic behaviour is neglected.

In the beginning of the 20th century, steel was increasingly used to construct bridges and buildings. Very soon engineers found that elastic design, commonly used at that time, was excessively conservative and plastic design became an issue for this type of structures in the 2nd decade. In particular methods to determine the load carrying capacity of beams and trusses under monotonically increasing loads were investigated both from experimental and theoretical point of view (Maier-Leibnitz in 1928 [4], Schaim in 1930 [5], Fritzsche 1931 [6]). But as early as 1926, Martin Grüning (Fig.1) discussed the influence of plasticity in hyperstatic truss systems under repeated loading: He observed the beneficial effect of redundant elements in case of repeated loading: Redistribution of stresses caused by plastic deformation occurring in first loading cycles may be such that in subsequent loading, the members of the structure may be operating solely in the elastic domain for higher load levels than determined if the initial elastic limit would be taken as reference. This might be considered as the first appearance, yet in embryonic form, of the idea of shakedown. Grüning published his results shortly after his famous work "Die Statik des ebenen Tragwerks" [7] in 1925 (still available today) in a 30-pages booklet entitled "Die Tragfähigkeit statisch unbestimmter Tragwerke aus Stahl bei beliebig häufig wiederholter Belastung" [8]: The decisive sentence in German:

"Überschreiten die Spannungen in n Stäben eines n-fach statisch unbestimmten Fachwerks, die als Überzählige eines stabilen (statisch bestimmten) Systems aufgefaßt werden können, infolge einer Belastung die Elastizitätsgrenze, so gehen sie unter hinreichend häufig wiederholter Be-und Entlastung in und unter Umständen unter die Elastizitätsgrenze zurück, sofern die Spannung in keinem Stabe des stabilen Systems sich über die Elastizitätsgrenze hebt".

("Do the stresses in n bars of a n-times over-determined bar-system, which can be considered as redundant elements of a stable (statically determined) system, exceed the elastic limit due to some loading, so they go, under sufficiently often repeated loading and unloading, back to, or under circumstances below the elastic limit, in case that the stress in none of the bars of the stable system goes beyond the elastic limit")

Fig. 1 (Bernhard) Martin Grüning (10.12.1869 – 30.6.1932) was a civil engineer, had several positions in German administration before he became in 1918 professor in Statics and Steel Construction (Eisenbau) at TH Hannover and in 1923 professor in Statics (Baustatik) at TH Wien. The above cited papers were published under his affiliation to TH Hannover.

We see the vicinity to the limit-load problem of civil engineering on one hand, but also the differences that are decisive for the different strands that take shakedown-theory and limit-analysis: Metal ductility combined with elasticity, at the origin of residual stresses, which are introduced as key element and can only play their role in repeated loading, in contrast to monotonically increasing loading in limit analysis that does not need to introduce elasticity and residual stresses in order to determine limit loads.

It was Hans Bleich (Fig.2) from TH Wien who picked up and generalised in 1932 Grünings findings [9], who's studies concerned only locally stationary repeated loads. This restriction was removed by Bleich, who also introduced the notion of "Selbstspannung", nowadays commonly used in its English half-adopted equivalent "eigenstress". One may assume that Bleich's work was inspired by Grüning not only through his publication but also by his presence in Vienna, but this is not confirmed. - With Bleich's work, we come already close to today's notion of shakedown theory.

Fig. 2 Hans Heinrich Bleich (24.3.1909 – 8.2.1985) was, like Grüning, civil engineer and had graduated in 1931 in Structural Engineering from TH Wien. In 1939, Bleich left Austria for England and moved in 1945 to the USA, where he continued his career first in industry and then as highly appreciated and prominent professor at Columbia University's school of Engineering.

Hans Bleich, together with his father Friedrich Bleich had been in close contact with Ernst Melan (Fig.3), publishing together in 1927 the book "Die gewöhnlichen und partiellen Differentialgleichungen der Baustatik" [10]. It might be this contact, but also the contact to Grüning, at that time professor in Vienna, that had inspired Melan to his path-breaking works in 1936 and 1938 in which he, rightly referring to Grüning and Bleich, formulated the general lower bound shakedown theorem in his key contribution "Der Spannungszustand eines Mises-Hencky'schen Kontinuums bei veränderlicher Belastung" [11]:

„*Wir verzeichnen somit folgendes Ergebnis: Unter den gemachten Voraussetzungen besitzt das Integral (15) stets einen nicht negativen Wert. Dieser Wert bleibt unverändert, wenn das Material an keiner Stelle des Körpers fließt oder wenn bleibende Dehnungen auftreten, die keine Zwangsspannungen hervorrufen. Fließt aber das Material, so nimmt J wegen $\dot{J} < 0$ immer nur ab*" *(page 86, lines 6-12)*

("We note therefore the following result: under the given assumptions, the integral (15) has always a non-negative value. This value remains unchanged, if the material does not flow in any point of the body or if permanent strains occur, that do not cause enforced stresses. If however the material flows, then the integral J decreases monotonically because of $\dot{J} < 0$")

Fig. 3 Ernst Melan (16.11.1890 – 10.12.1963) was Civil Engineer graduated from the German Technical University of Prague from where he also obtained his Dr-degree in 1917. In the same place he became extraordinary professor after holding 1916 and 1923 different positions in Austrian administration and industry. In the same place he became extraordinary professor after holding between 1916 and 1923 different positions in Austrian administration and industry.

It may surprise that, in particular, Melan's powerful general theorem from 1938 remained without resonance in the scientific community for many years. However, one must not forget the atrocious situation in Germany and, after WWII had started, in the entire world, at that time. Many brilliant German-Jewish scientists in mechanics had to struggle for their lives and to find new places to live for themselves and their families, if they had the chance to survive. In addition, the relevant papers had been published in the German language, which was evidently not very popular in these and the coming years. It was Prager (Fig.4), then at Brown University, who came back to the problem in his contribution "Problem Types in the Theory of Perfectly Plastic Materials" [12], presented at the Symposium on Plasticity held at Brown University in 1948. There, he refers to the less far reaching formulation of the lower bound theorem for bar-systems by Melan [13]. It was also Prager, at that time funded by the US Office of Naval Research, who introduced the denomination "Shakedown", which is known in shipyards to describe the process of accommodation of parts in new ships due to dynamic loading by engine vibration after putting them into operation for the first time (O. Mahrenholtz, private communication).

Fig. 4 William Prager (23.5.1903 – 16.3.1980) graduated as engineer from the Technical University Darmstadt, where he also obtained his Dr-degree. He then held positions in Göttingen and Karlsruhe (as youngest professor in Germany at that time) before being forced to leave Germany after the Nazi-regime came into power. After moving to Turkey he was appointed professor at Brown University in the USA, where, except for a short period at the University of California San Diego, he continued his brilliant career until retirement.

Brown University with its scientifically particularly fertile atmosphere at that time can be considered as a kind of cradle for the further development of shakedown theory. Besides Prager, also should be mentioned Paul Southworth Symonds (20.8.1916 – 28.3.2005) (Fig.5) and B.G. Neal as particularly active in the field of

Fig. 5 P.S. Symonds

determination of limit loads for frame structures under proportional and variable loads. [14-19]. Special importance has their paper "Recent Progress in the Plastic Methods of Structural Analysis" [17] as it anticipates the general upper bound theorem of shakedown. In his contribution [20], Paul Symonds gives an enlightening personal insight into the developments of shakedown theory at that time at Brown University and Cambridge University.

Generally speaking, in the U.K. to our knowledge, it was basically the group by John Baker working in structural plasticity in the same period and their interest was mainly limit analysis (see also the multi-volume work "The Steel Skeleton", by Baker, Horne and Heyman [21]).

Fig. 6 This photograph was taken at Brown University in April 1960 on the occasion of the Second Plasticity Symposium organized by the US Office of Naval Research.

W.T. Koiter (Fig.7), who had been visitor to Brown University at that time, has the merit of not only formulating the upper bound theorem of shakedown in a general form in 1956 in his paper "A new general theorem on shakedown of elastic-plastic structures" [23], but also to revalue shakedown theory as a whole through his fundamental work "General Theorems for Elastic-Plastic Solids" [24] that summarises pretty well the state of the art at that time in a very accessible manner. We stress however that the contribution by Paul Symonds concerning the upper bound theorem should not be undervalued [25] as it has been in the past. He formulated in fact the first time an upper bound approach in shakedown theory, not in general form, as Koiter did, but for frames. Symonds also greatly simplified the proof of Melan's lower bound theorem [15] in the general case in a form adopted by Koiter [24].

Fig. 7 Warner Tjardus Koiter (16.6.1915 – 2.9.1997), graduated as mechanical engineer from Delft University of Technology in 1936. He worked at the Dutch National Aeronautical Research Institute (NLL) in Amsterdam, the Government Patent Office and the Government Civil Aviation Office. In 1949, he was appointed Professor of Applied Mechanics in Delft where he stayed until his retirement in 1979. His contributions to shell theory and plasticity influenced significantly the development of modern mechanical sciences.

One can say, that with Koiter's formulation of the lower and the upper bound theorems of shakedown theory in the context of continuum mechanics the first chapter of development was closed, leaving evidently many questions open.

2   Theoretical Extensions and Development of Numerical Methods

Open questions left by pioneers are also opportunities for interesting research subjects for the younger generation. In case of shakedown theory, the interest was double: On one side, the theory had an evidently high practical potential and on the other side, basic assumptions were rather restrictive and the application to industrial problems not easy. One can grossly summarize this to two major questions:

(i)   How can these powerful theorems be applied in practical engineering?
(ii)   How to get rid of the very coercive assumptions on which the classical proofs of shakedown theory were based?

Many of the young scientists who had been on visit at Brown University in the 50 and 60ties had been inspired by the strong plasticity group and carried on their research in this field after returning home. They tried to find answers to these questions. Among them, Giulio Maier *(*8.3.1931)* (Fig.8), back from Brown University where he had been as visiting scholar in 1964, continued his work in Milano, Italy, focusing together with his co-workers on the problems of non-associated flow rules [26], geometrical non-linearities [27] and dynamic effects [28]. Non-associated flow rules are particularly important for frictional materials such as soils, masonry, but also, more recently investigated, porous and heterogeneous materials [33-35]. Also, accounting for the progress in numerical methods and in

Fig. 8 G. Maier

view of industrial applications, Maier adapted shakedown theory to the so-called linear programming methods [26]. His very rich scientific oeuvre in the field of shakedown theory, partly referred to in this chapter (ref. [26-41]) was initially stimulated by problems in civil engineering but triggered many research activities in the field of limit analysis in general with greatest impact worldwide, not only in Italy. To mention, from Italian side, are in this context the contributions e.g. by L. Corradi, R. Contro, F. Genna, A. Corigliano, U. Perego, and in recent years, V. Carvelli and G. Cocchetti, in the same group as Giulio Maier.

Fig. 9 A. Ponter

Equally back from Brown University where he stayed 1964-1965, first to Cambridge and later to Leicester, Alan Ponter *(*13.2.1940)* (Fig.9) as applied mathematician interested in problems of mechanics engineering design extended the lower bound theorem to creep behaviour of materials. Due to the explicit time-dependence, absent in plasticity, this problem was (and still is) particularly challenging. He showed that if the cycle time is regarded as a variable, solutions for very short and very long cycle times possessed bounding properties for intermediate finite cycle times. For short cycle times, the residual stress remained constant and hence behaviour could be related to shakedown, resulting in the Creep Modified Shakedown Limit [43-46]. He also derived displacement bounds in shakedown conditions [47] and more general bounding theorems in plasticity [48,49]. These works were carried out in cooperation with Fred Leckie, John Martin and others and influenced both high temperature design and life assessment methods considerably. We note that Martin had also been at Brown University and spent a year at Leicester with Ponter while he wrote his book [42]. Bree's complete solution for a particular problem involving thermal and loading of a tube [50] was an important impulse for the design community to grasp the complete range of possible behaviour, including shakedown, reverse plasticity and ratchetting. This decisively paved the way for the acceptance of shakedown limit solutions as a basis for design in the nuclear design community.

Ponter was then attracted by the problem of mechanical parts in power plants operating at elevated temperatures, related to the European Fast Reactor-project, which was later cancelled. This involved extending shakedown theory to the evaluation of the ratchet limit, and this was achieved, in an approximate way, for the characteristic problems of Fast Reactor design [51-53]. Methods for obtaining the rate of ratchet strain growth in excess of shakedown were also obtained [54,55]. He wrote together with Sami Karadeniz, Keith Carter and Alan Cocks four reports for the EC and many papers, including both creep deformation and creep rupture effects. The final EC report [56], describes the background to a design code for thermal loading as part of the European Fast Reactor Project.

In more recent years, Ponter focused on rolling contact problems [57], composites [58,59] and the development of methods of how to solve most efficiently shakedown and related problems by numerical methods. He introduced the so-called "Linear Matching Method" [60,61] and implemented it successfully into commercial software for design purposes [62,63]. As well as shakedown limits, ratchet limits in excess of shakedown are also obtained. This work continues at the University of Strathclyde by a research group led by H. F. Chen.

Variable thermal loads was also the field David Aronovich Gokhfeld *(31.7.1919 - 14.3.2004)* (Fig.10) from the South Ural University, USSR, was deeply involved in. He studied in the beginning of the 60ties of the last century ratchetting of me-

Fig. 10 D. Gokhfeld

chanical parts in furnaces and observed that under moving thermal loads even without any mechanical loading large deformations may occur that render the considered part unusable [64]. These studies, always joining theoretical and experimental work, were extended to other types of elements such as turbine blades, parts in nuclear reactors and pressure vessels in chemical processing. Aware of the work of Ponter, Williams and Leckie on creep, he and his co-workers developed an alternative way of taking creep into account by adjusting appropriately the yield limit of the material. - This idea resembles the concept of "sanctu-

ary of elasticity" by Nayroles and Weichert, put forward many years after in another context [168]. - Gokhfeld, O.F.Cherniavsky and co-workers published more than 100 papers in the field, almost all of them in Russian language [64-66] and therefore largely unknown outside of the Russian speaking scientific community. Special mention deserves the book "Theory of shakedown and strain accumulation under thermal cycling" from 1980 in English which is still today an invaluable source of information. – The work from this group is today successfully continued in a modern computational environment by A. Cherniavsky and co-workers, with new fields of application like hydrogenated metals, crack forming and use of traveling heat sources for controlled metal forming.

Another question, not addressed by classical theory, was the problem of dynamic (inertia) effects. Here also, in a different way than in case of viscoplasticity, time enters the formulation of the problem explicitly. This subject was first addressed first by Giulio Ceradini from Rome in his theorem on dynamic shakedown in 1969 [68], still of great importance in particular in earthquake engineering.

Fig. 11 T.M. Huber

To come back to Italy, Castrenze Polizzotto from Palermo who came from a more mathematically than engineering-driven background, contributed to the problems of variational formulations, extended classes of material behaviour, dynamics and bounding theorems [69-78]. His work is continued by the groups of Guido Borino, Paolo Fuschi, Aurora Pisano and others.

Particularly important for the development of shakedown theory was research conducted at the Polish Academy of Sciences: There was a long Polish tradition since the times of Tytus Maksymilian Huber *(4.1.1872 – 9.12.1950)* (Fig.11) in the field of Plasticity in Poland that was continued after WWII in particular related to limit states of plastic structures and phenomenological modeling of plastic behavior. Wacław Olszak (Fig.12) had been the initiator of developing the research team in plasticity theory and applications in engineering including both structural mechanics and technology of metal processing. In 1955 he initiated weekly scientific seminars at the Polish Academy of Sciences. The principles of limit state analysis, metal forming, yield conditions, anisotropy, inhomogeneity and flow rules were the main topics of discussion and research.

Fig. 12 W. Olszak

In 1957 a set of lectures was presented by Koiter, who discussed his kinematic approach to shakedown and the upper bound theorem illustrated by some examples. Olszak, who made his PhD in Vienna Polytechnic in 1935, knew well Melan and was familiar with his theorem on shakedown. It can be supposed that this was the main inspiration to develop more intensive study of cyclic loading and shakedown as fundamental for application in structural mechanics. In the sequel, Antoni Sawczuk *(16.1.1927 – 27.5.1984)* (Fig.13), primarily working on limit analysis of plates and shells wrote several papers on application of kinematic theorem to specify upper bounds on load amplitudes.

Fig. 13 A. Sawczuk

It was however Jan Andrej König *(16.5.1937 – 8.12.1990)* (Fig.14), his doctoral student, who contributed most significantly to the advancement of shakedown theory by a large number of theoretical papers on hardening material behavior [80], thermal problems [81], bounding methods [82], structures [83] and numerical methods [84]. His book "Shakedown of Elastic-Plastic Structures" [85] from 1987 is still of great actuality and a prime reference; it is concise and easy to read.

Fig. 14 A. König

Among the great number of Polish scientists working successfully in the field of shakedown between roughly 1960 and 2000, we only mention A. Borkowski, M. Kleiber, Z. Mróz, A. Sawczuk, M. Janas, St. Dorosz, A. Siemaszko, S. Pycko, J. Skrzypek, B. Skoczeń, J. Orkisz, J. Zwoliński, B. Bielawski, among others [86-110]. The book by Michal Życzkowski *(12.4.1930 – 24.5.2006)* (Fig.15) "Combined Loadings in the Theory of Plasticity" [79] gives an excellent account on research on plastic structures with more than 3000 entries of references.

It was also König who brought shakedown theory back to Germany: In the beginning of the 80ties, Oskar Mahrenholtz (*17.5.1931) (Fig.16), then at Hannover University, was involved in studies on failure of zirconium tubes under variable thermal and mechanical loads, which is a typical problem from nuclear power engineering. König, at that time visiting scholar in Hannover, suggested solving this problem by applying the upper bound shakedown theorem and carrying out experiments simultaneously.

Fig. 15 M. Życzkowski

Fig. 16 O. Mahrenholtz

The experimental and numerical results they obtained were in quite good agreement [111-113]. Until today, in particular the experimental results from this study are used as benchmark. – Some years later, König stayed again at Hannover University, this time with Erwin Stein (*5.7.1931) (Fig.17). There he initiated a series of studies focusing on the development of numerical shakedown analysis involving material hardening, cracks and structural optimization [113-122]. These works were successfully continued by Stein, Zhang, Mahnken, Wiechmann and others in the following years. Of particular interest is the "reduced base"-technique, developed by Stein. This technique reduces significantly the numerical efforts to construct numerical solutions and is used in industrial applications.

It should be noted that in the German community of civil engineers, to which Stein belongs, the application of shakedown theory was even in the 90ties far from being commonly accepted and was discussed quite controversially [123].

Fig. 17 E. Stein

Another strand of development, the initial boundary value problem of plasticity, was the entry point for Dieter Weichert (* 5.3.1948) (Fig.18) to discover indirectly shakedown theory through the influence of Pawel Rafalski [124], another Polish scholar from Warsaw, visiting Bochum University in the beginning of the 80ties. Weichert studied in the sequel first the problem of geometrical non-linearities in the context of shakedown theory in the framework of continuum mechanics [125], investigated then the problem of generalized material laws according to the standard material model by [126] with applications to thin-walled structures. Through his stay at the American University of Beirut, he started to work with Lutfi Raad and others on problems in pavement mechanics [127-131]. Later, at Lille University in France, Weichert initiated and carried out a number of studies on numerical methods, dynamic shakedown and the problem of shakedown including material damage and cracked bodies [132-137]. He continued this work back to Germany at Aachen University of Technology with applications to composites [138-140] and with the aim to apply shakedown theory to large scale industrial problems [141-146]. Weichert and his co-workers concentrated on the lower bound theorem and used in the beginning stress-based numerical approaches, which delivered very good results, but which are difficult to combine with displacement-based commercial finite element codes. Today, his group works primarily with displacement-based approaches.

Fig. 18 D. Weichert

Fig. 19 B.F. de Veubeke

Force and stress-based methods have been intensively developed at Liège University, starting in the 60ties. There was a strong group around Baudouin Fraeijs de Veubeke *(3.8.1917 – 16.9.1976)* (Fig.19), Ch. Massonnet, G. Sanders, C. Fleury, mainly involved in general mechanics, limit analysis of plates and shells, and optimization. M. Save and G. Guerlement continued the work at the Polytechnic School of Mons.

Later, in particular Patrick Morelle and Nguyen Dang Hung applied the tools that had been developed in the innovative scientific environment of Liège University to shakedown analysis [147-152]. Their and their co-workers efforts were aimed at the numerical exploitation of duality principles. Manfred Staat from Jülich Research Center continued successfully this work in recent years [153-157].

The path of force methods has been followed in an original manner by Kostas Spiliopoulos, with application to frame structures, based on graph theory and linear programming [158-160].

Géry de Saxcé, who had also started his career at Liège University [161] before moving first to the Polytechnic School of Mons and then to Lille University, introduced the so-called bi-potential theory [162-163] as generalization of Fenchel's inequality, opening new doors to take into account more complex, friction-type material laws in limit analysis and shakedown theory [164]. This novel approach has a high potential and is far from being fully exploited at the time being.

The research by de Saxcé is therefore linked to the scientific tradition of the Belgian group, but also to the French, mathematically inclined community of mechanics, strongly involved in the 70s and 80s of the last century in the development and application of Convex Analysis which is based on the Fenchel inequality and so carries further the idea of classical potential-based principles. Convex Analysis had been developed in the 70s by T. Rockafellar in the context of operational research and by J.J. Moreau [165] in the context of mechanics. B. Nayroles together with O. Debordes, both at that time in Marseille, applied this to the shakedown problem [166,167] and contributed essentially to the strengthening of the mathematical basis of the theory. In this context, the important work by Quoc-Son Nguyen has to be mentioned, who contributed strongly to the understanding of the effect of hardening from mathematical point of view [169-171]. Very fruitful because well suited to extend the classical theorems to larger classes of material behavior was the introduction of the so-called Standard Material Model by B. Halphen and Nguyen Quoc Son [172] as had already been shown by Mandel in 1976 [173]. Radenkovic's work on non-associated flow rules from 1961 [174], although basically related to limit analysis played an important role in the sequel also in shakedown theory and should be mentioned in this place.

Independently and application oriented, Joseph Zarka and his group developed the so-called Simplified Method [175,176], particularly useful for applications involving alternating plasticity and fatigue problems in mechanical engineering as has been shown by Geneviève Inglebert and her co-workers [177]).

Coming back to typical problems of civil engineering (which, as mentioned in the beginning, in some sense has triggered limit- and shakedown analysis), there were some important but for long time spared-off areas, which are mechanics of soils, foundations and pavements. Here, just as in case of concrete and reinforced concrete structures, the complexity of the material behavior and the difficulty to develop realistic material models that fit the framework of shakedown theory are important obstacles. In particular for pavements, other effects like rutting, crack development, moisture, freezing and thawing cycles are very important aside of plasticity as to their long-time behavior; in case of foundations, mostly the fluctuation of loads is by far less important than gravitational "dead loads".

Apart from the groups mentioned before, there is a tradition in this field of research in Australia and New Zealand. Pioneering work on pavements goes back to John Robert Booker *(24.7.1942 – 13.1.1998)* (Fig.20) and R.W. Sharp from Sydney University in 1984 [178], basing their approach on the particular stress pattern that develops in the rolling contact on roads.

Fig. 20 J.R. Booker

This work was followed by others, like Scott Sloan *(* 2.7.1954)* (Fig.21) from Newcastle University and his group [179-181], Ian Collins from Auckland University New Zealand [182,183,131]. In his later work, there was a link to Ponter, Weichert and Raad through the fact that Mostapha Boulbibane, a former PhD-student of Weichert, had been active in all three groups. But also shakedown of structures such as bridges has been widely investigated in Australia: It was Paul Grundy from Monash University who started as early as 1969 to study in many papers the shakedown behavior of mechanical elements, in particular linked to bridge constructions [184,185]. His work is continued by Francis Tin Loi and his coworkers at the University of New South Wales. - It is from Australia, Newcastle University that from Sloans group Hai-Sui Yu brought back to the U.K. at Nottingham University shakedown. Yu and his co-workers concentrate on the numerical application of Melan's theorem to compute lower bounds to the shakedown limit for rolling contact problems for Mohr-Coulomb type yield conditions for road pavement design [186-190].

Fig. 21 S. Sloan

We can however have another way to look at the historical evolution of shakedown analysis, detached from the individual researcher and research groups and their connections and relations: The onset was the observation that residual stresses due to plasticity in redundant elements of hyperstatic bar-structures are beneficial for their survival under variable loads, what differentiates shakedown theory essentially from limit load theory. Application did not appear for long time due to the lack of means how to translate the theory to calculation methods. First applications appear in special types of structures like beams, plates and shells, where by appropriate assumptions and semi-analytical methods the complexity of the problem can be reduced drastically. As the theory of plasticity is genuinely linked to metals, the fields of application were on the side of mechanical engineering pressure vessels and pipes, on the side of civil engineering steel frame structures. This first "bifurcation" was not methodological, but naturally imposed by engineering practice. In the sequel, on both sides, application-driven theoretical extensions were carried out: More complex material models for metals, for concrete and for soil-like materials, material damage and cracks, temperature influence, geometrical non-linearity are the major strands, accompanied by the development of appropriate and more rigorous mathematical foundations, such as the proper formulation of the theorems as optimization problems. The breakthrough to modern engineering however is due to the tremendous development of numerical methods and computer technology: Discretization of structures of almost arbitrary shape connected with fast linear solvers and highly performing optimization algorithms render the theorems of shakedown today easily applicable in practical engineering.

And it is the modern formulation of the theory which makes that the differences between shakedown theory and theory of limit analysis in practice almost vanishes: Limit analysis became a particular case of shakedown analysis and today both are subsumed under the notion of "Direct Methods".

3 Final Remarks

To keep "the beauty of shakedown theory" evoked by Giulio Maier, can be understood as a warning against theoretical overstretching. The beauty of the theory is obvious: without information about the path of loading in an arbitrarily complex loading space, one predicts, if a dissipative mechanical system will fail or not. No need to walk step-by-step through the evolution of a system, which is not only cumbersome but in many cases simply impossible, because the loading history of the considered element is unknown.

What sounds a bit like a miracle, is, from mathematical point of view, due to convexity of the potentials involved and the existence of subgradients describing the evolution of the material. This reflects the physical features of small geometrical transformations and stable material behavior, as particular case of the validity of the second law of thermodynamics in form of the Clausius-Duhem inequality.

The price to pay for this miracle is loss of information about the evolution of local quantities during the process and, most important, about deformation, although certain bounding properties have been proven. This price comes along with rather harsh assumptions about material behavior, in many cases too rough for sophisticated investigations and, if one applies the theorems directly, a rather complex resolution methodology. And here is the risk of overstretching: If the evolution of a system inherently depends on the evolution of local quantities, it becomes very tricky, if not impossible, to find adequate theoretical extensions of shakedown theory. So, to find the border line of usefulness is an important issue.

But this is true for any kind of modeling and we have to keep in mind the issue of our efforts: confronted to a concrete problem, one has to decide about the adequate methodology to solve it. And in this sense, there is plenty of room for the further development and application of shakedown theory.

The authors of this short and far from being comprehensive historical view wish to thank particularly O. Mahrenholtz, E. Stein, G. Maier, Z. Mróz and O.F. Cherniavsky for their advice and valuable information on their personal historical witness.

References

[1] Benvenuto, E.: An Introduction to the History of Structural Mechanics, Part I, Statics and Resistance of Solids. Springer (1991)
[2] Poleni, G.: Memorie istoriche della gran cupola del Tempio Vaticano, Padova (1748)
[3] Heyman, J.: Poleni's problem. Proc. Instn Civ. Engrs, Part 1 84, 737–759 (1988)
[4] Maier-Leibnitz, H.: Beitrag zur Frage der tatsächlichen Tragfähigkeit einfacher, durchlaufender Balkenträger aus Baustahl St. 37 und Holz, Bautechnik, 6. Jg. Heft 1, 11–14 und Heft 2, 27–31 (1928)
[5] Schaim, J.H.: Der durchlaufende Träger unter Berücksichtigung der Plastizität, Stahlbau (1930)

[6] Fritzsche, J.: Die Tragfähigkeit von Balken aus Baustahl bei beliebig oft wiederholter Belastung. Bauingenieur 12, 827 (1931)

[7] Grüning, M.: Die Statik des ebenen Tragwerks. Springer, Berlin (1925)

[8] Grüning, M.: Die Tragfähigkeit statisch unbestimmter Tragwerke aus Stahl bei beliebig häufig wiederholter Belastung. Springer, Berlin (1926)

[9] Bleich, H.: Über die Bemessung statisch unbestimmter Stahltragwerke unter Berücksichtigung des elastisch-plastischen Verhaltens des Baustoffes. Der Bauingenieur, Heft 19/20, 261–267 (1932)

[10] Bleich, F., Melan, E.: Die gewöhnlichen und partiellen Differentialgleichungen der Baustatik. Springer (1927)

[11] Melan, E.: Der Spannungszustand eines Mises-Hencky'schen Kontinuums bei veränderlicher Belastung, Sitzungsberichte, vol. 147, pp. 73–87. Akademie der Wissenschaften, Wien (1938)

[12] Prager, W.: Problem Types in the Theory of Perfectly Plastic Materials. Journal of the Aeronautical Sciences, 337–341 (June 1948)

[13] Melan, E.: Theorie unbestimmter Systeme aus idealplastischem Baustoff. Sitzungsberichte, Akademie der Wissenschaften 145, 195–218 (1936)

[14] Symonds, P.S., Prager, W.: Elastic-plastic analysis of structures subjected to loads varying arbitrarily between prescribed limits. J. Appl. Mech. 17, 315–324 (1950)

[15] Symonds, P.S.: Shakedown in Continuous Media. J. Appl. Mech. 17, 85–89 (1951)

[16] Neal, B.G.: Plastic-Collapse and Shake-Down Theorems for Structures of Strain Hardening Material. J. Aero. Sci. 17, 297–307 (1950)

[17] Symonds, P.S., Neal, B.G.: Recent progress in the plastic methods of structural analysis. J. Franklin Inst. 252(6), 469–492 (1951)

[18] Symonds, P.S., Neal, B.G.: The Calculation of Failure Loads on Plane Frames unbbder Arbitrary Loading Programs. J. Inst. Civil Engrs. 35, 41–61 (1951)

[19] Neal, B.G., Symonds, P.S.: A Method for Calculating the Failure Load for a Framed Structure Subjected to Fluctuating Loads. J. Inst. Civil Engrs. 35, 186 (1951)

[20] Symonds, P.S.: An early upper bound method for shakedown. In: Weichert, D., Maier, G. (eds.) Inelastic Analysis of Structures under Variable Loads – Theory and Engineering Applications. Series, Solid Mechanics and its Applications, vol. 83, pp. 1–9. Kluwer Academic Publishers, Dordrecht (2000)

[21] Baker, J., Horne, M.R., Heyman, J.: The steel Skeleton, vol. II, ch. 9. Cambridge University Press (1956)

[22] Horne, M.R.: The Effect of Variable Repeated Loads in the Plastic Theory of Structures. Research, Engineering Structures Supplement, Colston Papers II, 141 (1949)

[23] Koiter, W.T.: A new general theorem on shakedown of elastic-plastic structures. Proc. Kon. Ne. Ak. Wet. B59, 24–34 (1956)

[24] Koiter, W.T.: General theorems for elastic-plastic structures. In: Sneddon, I.N., Hill, R. (eds.) Progress in Solid Mechanics, ch. IV, pp. 165–221. North-Holland Publ. Co., Amsterdam (1960)

[25] Symonds, P.S.: Basic Theorems in the Plastic Theory of Structures. J. Aero. Sci. 17, 669–670 (1950)

[26] Maier, G.: Shakedown theory in perfect elastoplasticity with associated and non-associated flow-laws, a finite element, linear programming approach. Meccanica 4(3), 250–260 (1969)

[27] Maier, G.: A shakedown matrix theory allowing for workhardening and second-order geometric effects. In: Sawczuk, A. (ed.) International Symposium, Warsaw, August 30-September 2. Foundations of Plasticity, vol. 1, pp. 417–433. Noordhoff, Leyden (1972, 1973)

[28] Corradi, L., Maier, G.: Dynamic non-shakedown theorem for elastic perfectly-plastic continua. Journal of the Mechanics and Physics of Solids 22, 401–413 (1974)

[29] König, A., Maier, G.: Shakedown analysis of elastoplastic structures, a review of recent developments. Nuclear Engineering and Design 66, 81–95 (1981)

[30] Maier, G.: A generalization to nonlinear hardening of the first shakedown theorem for discrete elastic-plastic structures. Rendic. Acc. Naz. dei Lincei, Serie Ottava, 161–174 (1988)

[31] Maier, G., Novati, G.: Dynamic shakedown and bounding theory for a class of nonlinear hardening discrete structural models. Int. J. of Plasticity 6(5), 551–572 (1990)

[32] Maier, G., Pan, L., Perego, U.: Geometric effects on shakedown and ratchetting of axisymmetric cylindrical shells subjected to variable thermal loading. Engineering Structures 15(6), 453–466 (1993)

[33] Corigliano, A., Maier, G., Pycko, S.: Dynamic shakedown analysis and bounds for elastoplastic structures with nonassociative, internal variable constitutive laws. Int. J. Solids and Structures 32(21), 3145–3166 (1995)

[34] Cocchetti, G., Maier, G.: Static shakedown theorems in piecewise linearized poro-plasticity. Archive of Applied Mechanics 68, 651–661 (1998)

[35] Carvelli, V., Maier, G., Taliercio, A.: Shakedown analysis of periodic heterogeneous materials by a kinematic approach. Journal of Mechanical Engineering 50(4), 229–240 (1999)

[36] Carvelli, V., Cen, Z.Z., Liu, Y., Maier, G.: Shakedown analysis of defective pressure vessels by a kinematic approach. Archive of Applied Mechanics 69, 751–764 (1999)

[37] Cocchetti, G., Maier, G.: A shakedown theorem in poroplastic dynamics. Rend. Mat., Accademia Nazionale dei Lincei, s. 9 13, 43–53 (2002)

[38] Maier, G., Cocchetti, G.: Fundamentals of direct methods in poroplasticity. In: Weichert, D., Maier, G. (eds.) Inelastic Behaviour of Structures Under Variable Repeated Loads, Direct Analysis Methods, CISM, pp. 91–113. Springer, Wien (2002)

[39] Maier, G., Carvelli, V.: A kinematic method for shakedown and limit analysis of periodic composites. In: Weichert, D., Maier, G. (eds.) Inelastic Behaviour of Structures Under Variable Repeated Loads, Direct Analysis Methods, CISM, pp. 115–132. Springer, Wien (2002)

[40] Maier, G., Pastor, J., Ponter, A.R.S.: Direct Methods in Limit and Shakedown Analysis. In: De Borst, R., Mang, H.A. (eds.) Numerical and Computational Methods, vol. 3. Elsevier-Pergamon, Amsterdam (2003); Milne, I., Ritchie, R.O., Karihaloo, B. (eds.) Comprehensive Structural Integrity

[41] Cocchetti, G., Maier, G.: Elastic-plastic and limit-state analyses of frames with softening plastic-hinge models by mathematical programming. Int. J. Solids and Structures 40, 7219–7244 (2003)

[42] Martin, J.: Plasticity, Fundamentals and General Results. MIT Press (1975)

[43] Ponter, A.R.S., Williams, J.J.: Work bounds and the associated deformation of cyclically loaded creeping structures. ASME, J. Appl. Mech. 40, 921–927 (1973)

[44] Ponter, A.R.S.: On the relationship between plastic shakedown and the repeated loading of creeping structures. Trans ASME, J. Appl. Mech., Series E 38, 437–440 (1971)

[45] Ponter, A.R.S.: Deformation, displacement and work bounds for structures in a state of creep and subject to variable loading. Trans. ASME, J. Appl. Mech, 39, Series E, 953–959 (1972)
[46] Ponter, A.R.S.: On the creep modified shakedown limit. In: Ponter, A.R.S., Hayhurst, D.R. (eds.) 3rd IUTAM Symposium on "Creep in Structures", Leicester University (September 1980), pp. 264–278. Springer (1981)
[47] Ponter, A.R.S.: An upper bound on the small displacement of elastic-plastic structures. Trans. ASME, J. Appl. Mech. 39, Series E, 959–964 (1972)
[48] Ponter, A.R.S.: General displacement and work bounds for dynamically loaded bodies. J. Mech. Phys. Solids 23, 151–163 (1975)
[49] Ponter, A.R.S.: A general shakedown theorem for inelastic materials. In: Proc. SMiRT-3, London, paper L5/2 (1975)
[50] Bree, J.: Elasto-plastic behavior of thin tubes subjected to internal pressure and intermittent high-heat fluxes with applications to Fast Reactor Fuel Elements. J. Strain Analysis, 2(3), 226–238 (1967)
[51] Ponter, A.R.S., Karadeniz, S.: An extended shakedown theory for structures that suffer cyclic thermal loading, Part I: Theory. Trans. ASME, J. Appl. Mech. 52, 877–882 (1985)
[52] Ponter, A.R.S., Karadeniz, S.: An extended shakedown theory for structures that suffer cyclic thermal loading, Part II: Applications. Trans. ASME, J. of Appl. Mech. 52, 883–889 (1985)
[53] Ponter, A.R.S., Carter, K.F.: The ratchetting of shells subjected to severe thermal loading. In: Tooth, A.S., Spence, J. (eds.) Applied Solid Mechanics - 2, pp. 303–320. Elsevier Applied Science (1987)
[54] Ponter, A.R.S., Cocks, A.C.F.: The incremental strain growth of an elastic-plastic body loaded in excess of the shakedown limit. Trans. ASME, J. Appl. Mech. 51(3), 465–469 (1984)
[55] Ponter, A.R.S., Cocks, A.C.F.: The incremental strain growth of elastic-plastic bodies subjected to high levels of cyclic thermal loading. Trans. ASME J. Appl. Mech. 51(3), 470–474 (1984)
[56] Ponter, A.R.S., Karadeniz, S., Carter, K.F.: The computation of shakedown limits for structural components subjected to variable thermal loading - Brussels Diagrams. Directorate General for Science, Research and Development, Office for Official Publications of the E.C., Report EUR12686EN, Brussels, 170 pages (1990)
[57] Ponter, A.R.S., Chen, H.F., Chiavarella, M., Specchia, G.: Shakedown analysis for rolling and sliding contact problems. Int. J. Sol. Struct. 43, 4201–4219 (2001)
[58] Ponter, A.R.S., Leckie, F.A.: Bounding Properties of Metal Matrix composites subjected to Cyclic loading. Jn. Mech. Phys. Solids 46, 697–717 (1998)
[59] Ponter, A.R.S., Leckie, F.A.: On the behaviour of metal matrix composites subjected to cyclic thermal loading. Jn. Mech. Phys. Solids 46, 2183–2199 (1998)
[60] Ponter, A.R.S., Engelhardt, M.: Shakedown limits for a general yield condition. European Journal of Mechanics, A/Solids 19, 423–445 (2000)
[61] Ponter, A.R.S., Chen, H.F.: A Minimum theorem for cyclic load in excess of shakedown, with applications to the evaluation of a ratchet limit. Euro. Jn. Mech., A/Solids 20, 539–553 (2001)
[62] Chen, H.F., Ponter, A.R.S., Ainsworth, R.A.: The Linear Matching Method ap-plied to the high temperature life assessment of structures, Part 1. Assessments involving constant residual stress fields. Int. J. Pressure Vessels and Piping 83, 123–135 (2006)

[63] Chen, H.F., Ponter, A.R.S., Ainsworth, R.A.: The Linear Matching Method applied to the high temperature life assessment of structures, Part 2. Assessments beyond shakedown involving changing residual stress fields. Int. J. Pressure Vessels and Piping 83, 136–147 (2006)

[64] Gokhfeld, D.A.: On the possibility of increase of plastic deformation due to thermal cycling effects. Calculations of Strength (7) (1961) (in Russian)

[65] Gokhfeld, D.A.: Bearing capacity of structures under thermal cycles. Mashinostroenie publ., Moscow, p. 259 (1970) (in Russian)

[66] Gokhfeld, D.A., Cherniavsky, O.F.: Theory of shakedown and strain accumulation under thermal cyclinges. In: Proceedings of the All-USSR. Symp. on Low-cycle Fatigue at Elevated Temperatures, Chelyabinsk, vol. (3), pp. C3–C31 (1974)

[67] Gokhfeld, D.A., FCherniavsky, O.: Limit analysis of structures at thermal cycling. Sijthoff and Noordhoff. Int. Publ. Alphen aan den Rijn, The Netherlands – Rockville, USA (1980)

[68] G. Ceradini, Sull' adattamento dei corpi elasto-plastic isoggetti ad azioni dinamiche. Giornaledel Genio Civile 415, 239–258 (1969)

[69] Polizzotto, C.: Workhardening adaptation of rigid-plastic structures. Meccanica 10, 280–288 (1975)

[70] Polizzotto, C.: A unified approach to quasi-static shakedown problems for elasto-plastic solids with piecewise linear yield surfaces. Meccanica 13, 109–120 (1978)

[71] Polizzotto, C.: On workhardening adaptation of discrete structures under dynamic loadings. Arch. Mech. Stos. 32, 81–99 (1980)

[72] Polizzotto, C.: A unified treatment of shakedown theory and related bounding techniques. S.M. Arch. 7, 19–75 (1982)

[73] Polizzotto, C.: A convergent bounding principle for a class of elastoplastic strain-hardening solids. Int. J. Plasticity 2, 357–370 (1986)

[74] Polizzotto, C.: On the condition to prevent plastic shakedown of structures ASME. J. Appl. Mech. I, II(60), 15–25, 318–330 (1993)

[75] Polizzotto, C., Borino, G.: Shakedown and steady state responses of elastic-plastic solids in large displacements. Int. J. Sol. Struct. 33, 3415–3437 (1996)

[76] Polizzotto, C., Borino, G., Caddemi, S., Fuschi, P.: Shakedown problems for material models with internal variables. Int. J. Mech. Sci. 35, 787–801 (1993)

[77] Polizzotto, C., Borino, G., Fuschi, P.: An extended shakedown theory for elastic-damaged models. Eur. J. Mech. - A/Solids 15, 825–858 (1996)

[78] Polizzotto, C., Borino, G., Fuschi, P.: Weak forms of shakedown for elastic-plastic structures exhibiting ductile damage. Meccanica 36, 49–66 (2001)

[79] Życzkowski, M.: Combined Loadings in the Theory of Plasticity. Polish-Scientific Publ. (1981)

[80] König, J.A.: Shakedown of strainhardening structures. 1st Canad. Cong. Appl. Mech., Quebec. (1967)

[81] König, J.A.: A shakedown theorem for temperature dependent elastic moduli. Bull. Acad. Polon. Sci. Sér. Sci. Tech. 17, 161–165 (1969)

[82] König, J.A.: Deflection bounding at shakedown under thermal and mechanical loadings. In: Second SMiRT Conf., Berlin, paper L7/3 (1973)

[83] König, J.A.: A method of shakedown analysis of frames and arches. Int. J. Sol. Struct., 327–344 (1971)

[84] König, J.A.: Shakedown deflections, a finite element approach. Teoret. I Priloż. Meh. 3, 65–69 (1972)

[85] König, J.A.: Shakedown of Elastic-Plastic Structures. Elsevier, Amsterdam (1987)

[86] Borkowski, A.: Analysis of Skeletal Structural Systems in the Elastic, and Plastic Range. PWN and Elsevier, Warsaw (1988)
[87] Borkowski, A., Kleiber, M.: On a numerical approach to shakedown analysis of structures. Comput. Methods Appl. Mech. Engng. 22, 101 (1980)
[88] Dorosz, S.: An improved upper bound to maximum deflections of elasticplastic structures at shakedown. J. Struct. Mech. 6, 267–287 (1978)
[89] Dorosz, S., König, J.A., Sawczuk, A., Biegus, A., Kowal, Z., Seidel, W.: Deflections of elastic-plastic hyperstatic beams under cyclic loading. Arch. Mech. 33, 611–624 (1981)
[90] Dorosz, S., König, J.A.: Iterative method of evaluation of elastic-plastic deflections of hyperelastic structures. Ing. Archiv. 55, 202–222 (1985)
[91] Janas, M., Pycko, S., Zwoliński, J.: A min-max procedure for the shakedown analysis of skeletal structures. Int. Journal of Mechanical Sciences 37, 629–649 (1995)
[92] Janas, M., König, J.A.: A cylindrical tank response as an example of shakedown of non-Clapeyronian systems. Arch. Mech. 43, 49–56 (1991)
[93] Kleiber, M., König, J.A.: Incremental shakedown analysis in the case of thermal effects. J. Num. Methods Engng. 20, 15–67 (1984)
[94] König, A., Kleiber, M.: On a new method of shakedown analysis. Bull. Acad. Pol.-Sci., Ser. Sci. Technol. 26, 165 (1978)
[95] König, J.A., Siemaszko, A.: Strainhardening effects in shakedown process. Ing.-Archiv 58, 58–66 (1988)
[96] König, J.A., Maier, G.: Shakedown of elastoplastic structures, a review of recent developments. Nuccl. Eng. Design 66, 81–95 (1981)
[97] König, J.A., Pycko, S.: Shakedown analysis in the case of imposed displacements. Mech., Teor. Stos. 28, 101–108 (1990)
[98] Mróz, Z.: On the theory of steady plastic cycles in structures. In: Proc. 1st SMIRT, L6, 489501 (1971)
[99] Orkisz, J., Orringer, O., Holowinski, M., Pazdanowski, M., Cecot, W.: Discrete analysis of actual residual stress resulting from cyclic loadings. Comput. Struct. 35, 397 (1990)
[100] Pycko, S., Mróz, Z.: Alternative approach to shakedown as a solution of min-max problem. Acta Mechanica 93, 205 (1992)
[101] Pycko, S., König, J.A.: Elastic-plastic structures subjected to variable repeated imposed displacements and mechanical loads. Int. J. Plasticity 8, 603–618 (1992)
[102] Sawczuk, A.: Evaluation of upper bounds to shakedown loads of shells. J. Mech. Phys. Solids 17, 291–301 (1969)
[103] Sawczuk, A.: Shakedown analysis of elastic-plastic structures. Nucl. Eng. Design 28, 121–136 (1974)
[104] Sawczuk, A.: Mechanics and Plasticity of Structures. Ellis-Horwood/PWN, Chichester/Warsaw (1989)
[105] Siemaszko, A., Mróz, Z.: Sensitivity of plastic optimal structures to imperfections and non-linear geometrical effects. Structural Opimization 3, 99–105 (1991)
[106] Siemaszko, A., König, J.A.: Analysis of stability of incremental collapse of skeletal structures. J. Struct. Mech. 13, 301–321 (1985)
[107] Siemaszko, A., König, J.A.: Shakedown optimisation accounting for non-linear geometrical effects. ZAMM 71, 294–296 (1991)
[108] Zwoliński, J., Bielawski, G.: An optimal selection of residual stress for determination limit and shakedown multiplier. In: Proc. of Conf. on Comp'. Meth. struct. Mecł., Jadwisin, p. a59 (1987) (in Polish)

[109] Skoczeń, B., Skrzypek, J., Bielski, J.: Shakedown and inadaptation mechanisms of bellows subject to constant pressure and cyclic axial forces. Mech. Struct. Mach. 20, 119 (1992)

[110] Skoczeń, B., Skrzypek, J.: Inadaptation mechanisms in bellows subject to sustained pressure and cyclic axial loadings in terms of finite deformations. In: Mróz, Z., Weichert, D., Dorosz, S. (eds.) Inelastic Behaviour of Structures under Variable Loads, pp. 341–361. Kluwer Academic Publishers (1995)

[111] Mahrenholtz, O., Leers, K., König, J.A.: Shakedown of tubes: a theoretical analysis and experimental investigations. In: Reid, S.R. (ed.) Metal Forming and Impact Mechanics, W. Johnson Commemorative Volume, pp. 155–172. Pergamon Press (1984)

[112] Leers, K.: Experimentelle und theoretische Shakedownuntersuchung an Rohren. VDI-Verlag, Düsseldorf (1985)

[113] Leers, K., Klie, W., König, J.A., Mahrenholtz, O.: Experimental investigations of shakedown of tubes. In: Sawczuk, A., Bianchi, G. (eds.) Plasticity Today, pp. 259–275. Elevier Appl. Sci. Publ., London (1985)

[114] Stein, E., Zhang, G., Mahnken, R., König, J.A.: Micromechanical Modeling and Computation of Shakedown with Nonlinear Kinematic Hardening including examples for 2-D problems. In: Proc. CSME Mechanical Engineering Forum, Toronto, pp. 425–430 (1990)

[115] Stein, E., Zhang, G., König, J.A.: Shakedown with Nonlinear Hardening including Structural Computation using Finite Element Method. Int. J. Plasticity 8, 1–31 (1992)

[116] Stein, E., Zhang, G., Mahnken, R.: Shakedown Analysis for Perfectly Plastic and Kinematic Hardening Materials. In: Stein, E. (ed.) Progress in Computational Analysis of Inelastic Structures, pp. 175–244. Springer (1993)

[117] Stein, E., Zhang, G., Huang, Y.: Modeling and computation of shakedown problems for nonlinear hardening materials. Comput. Methods Appl. Mech. Engrg. 103, 247–272 (1993)

[118] Huang, Y., Stein, E.: Shakedown of a cracked body consisting of kinematic hardening material. Engineering Fracture Mechanics 54, 107–112 (1996)

[119] Huang, Y., Stein, E.: Prediction of the fatigue threshold for a cracked body by using shakedown theory. Fatigue & Fracture of Engineering Materials & Structures 18(3), 363–370 (1995)

[120] Huang, Y., Stein, E.: Shakedown of a CT-specimen with St52-steel: Experimental, analytical and numerical investigations. Journal of Strain Analysis 30(4), 283–289 (1995)

[121] Wiechmann, K., Barthold, F.-J., Stein, E.: Optimization of elasto-plastic structures using the finite element method. In: 2nd World Congress of Structural and Multidisciplinary Optimisation, pp. 1013–1018 (1997)

[122] Wiechmann, K., Barthold, F.-J., Stein, E.: Shape Optmization under Shakedown Constraints. In: Weichert, D., Maier, G. (eds.) Inelastic Analysis of Structures under Variable Loads, pp. 49–68. Kluwer Academic Publishers (2000)

[123] Stein, E.: Private communication

[124] Rafalski, P.: Minimun Principles in Plasticity, Mitteilungen aus dem Institut für Mechanik Bochum. Band 13 (1978)

[125] Weichert, D.: On the influence of geometrical nonlinearities on the shakedown of elastic-plastic structures. Int. J. Plast. 2(2), 135–148 (1986)

[126] Weichert, D., Groß-Weege, J.: The numerical assessment of elastic-plastic sheets under variable mechanical and thermal loads using a simplified two-surface yield-condition. Int. J. Mech. Sci. 30(10), 757–767 (1989)
[127] Raad, L., Weichert, D., Haidar, A.: Analysis of full-depth asphalt concrete pavements using shakedown-theory. Transactions of the Transportation Research Board (NRC), Transportation Research Record 1227, 53–65 (1989)
[128] Raad, L., Weichert, D., Najim, W.: Stability of multilayer systems under repeated loads. Transactions of Transportation Research Board (NRC), Transportation Research Record 1207, Pavement Design, 181–186 (1988)
[129] Weichert, D., Raad, L.: Extension of the static shakedown-theorem to a certain class of materials with variable elastic coefficients. Mech. Res. Comm. 19(6), 511–517 (1992)
[130] Boulbibane, M., Weichert, D., Raad, L.: Numerical application of shakedown theory to pavements with anisotropic layer properties, Paper No. 99-0342. Journal of the Transportation Research Board (NRC), Transportation Research Record 1687, 75–81 (1999)
[131] Boulbibane, M., Collins, I.F., Weichert, D., Raad, L.: Shakedown analysis of anisotropic asphalt concrete pavements with clay subgrade. Geotech. J. 37, 882–889 (2000)
[132] Hachemi, A., Weichert, D.: An extension of the static shakedown-theorem to a certain class of damaging inelastic material. Arch. of Mech. 44(5-6), 491–498 (1992)
[133] Hachemi, A., Weichert, D.: Application of shakedown theorems to damaging inelastic material under mechanical and thermal loads. Int. J. Mech. Sci. 39(9), 1067–1076 (1997)
[134] Belouchrani, M., Weichert, D.: An extension of the static shakedown theorem to inelastic cracked structures. Int. J. Mech. Sci. 41, 163–177 (1999)
[135] Belouchrani, M.A., Weichert, D., Hachemi, A.: Fatigue threshold computation by shakedown theory. Mech. Res. Comm. 27(3), 287–293 (2000)
[136] Hachemi, A., Weichert, D.: Numerical shakedown analysis of damaged structures. Comp. Meth. Appl. Mech. Engng. 160, 57–70 (1998)
[137] Hamadouche, M.A., Weichert, D.: Application of shakedown theory to soil dynamics. Mech. Res. Comm. 26(5), 565–574 (1999)
[138] Weichert, D., Schwabe, F., Hachemi, A.: Composite design by shakedown analysis for low cycle fatigue service conditions. In: Khan, A.S., Zhang, H., Yuan, Y. (eds.) Proceedings of the 8th Int. Symp. on Plasticity and its Current Applications (PLASTICITY 2000), held at Whistler, Canada, July 16-20, pp. 505–507. Neat Press (2000)
[139] Weichert, D., Hachemi, A.: Shakedown- and Limit Analysis of periodic composites. Journal of Theoretical and Applied Mechanics 40(1), 273–289 (2002)
[140] Hachemi, A., Weichert, D.: On the problem of interfacial damage in fibre-reinforced composites under variable loads. Mech. Res. Comm. 32, 15–23 (2005)
[141] Hachemi, A., Mouhtamid, S.D.: Weichert, Progress in shakedown analysis with applications to composites. Archives of Applied Mechanics, 1–11 (2005)
[142] Weichert, D., Hachemi, A.: A shakedown approach to the problem of damage of fiber-reinforced composites. In: Sadowski, T., (ed.) Proc. IUTAM-Symposium on "Multiscale Modelling of Damage and Fracture Processes in Composite Materials", KazimierzDolny, Poland, May 23-27, Solid Mechanics and its Applications, vol. 135, pp. 41–48. Springer (2006)

[143] Nguyen, A.D., Hachemi, A., Weichert, D.: Application of the interior-point method to shakedown analysis of pavements. Int. J. Numer. Meth. Engng. 75, 414–439 (2008)
[144] Weichert, D., Hachemi, A.: Recent advances in lower bound shakedown analysis. In: Proceedings of Pressure Vessel and Piping Conference 2009, Prague, Czech Republic, July 26-30. ASME (2009)
[145] Simon, J.-W., Chen, M., Weichert, D.: Shakedown analysis combined with the problem of heat conduction. In: Proceedings of the ASME 2010 Pressure Vessels & Piping Division/K-PVP Conference, PVP 2010, Bellevue, Washington, USA, July 18-22 (2010)
[146] Simon, J.-W., Weichert, D.: An improved Interior-Point Algorithm for Large-Scale Shakedown Analysis PAMM. Proc. Appl. Math. Mech. 10, 223–224 (2010)
[147] Morelle, P.: Numerical shakedown analysis of axisymmetric sandwich shells: An upper bound formulation. Int. J. Num. Meth. 23(11), 2071–2088 (1986)
[148] Hung, N.-D., König, J.A.: Finite element formulation for shakedown problems using a yield criterion of the mean. Computer in Applied Mechanics and Engineering 8(2), 179–192 (1976)
[149] Hung, N.-D., Palgen, L.: Shakedown analysis by displacement method and equilibrium finite element. Transactions of the CSME 6(1), 32–39 (1980)
[150] Hung, N.-D.: Shakedown analysis by finite element method and linear programming techniques. Journal de Mécanique Appliquée 2(4), 587–599 (1983)
[151] Hung, N.-D., Yan, A.-M.: Direct finite element kinematical approaches in limit and shakedown analysis of shells and elbows. In: Inelastic Analysis of Structures under Variable Loads, Theory and Engineering Applications, pp. 233–254. Kluwer Academic Publishers (2000)
[152] Hung, N.-D., Vu, D.K.: Primal-dual algorithm for shakedown analysis of structures. Computer Methods in Applied Mechanics and Engineering 193(42-44), 4663–4674 (2004)
[153] Staat, M., Heitzer, M., Yan, A.M., Vu, D., Hung, N.-D., Voldoire, F., Lahousse, A.: Limit Analysis of Defects (In collaboration with, Berichte des Forschungszentrum Jülich 3746, Jül-3746) (2000)
[154] Vu, D., Staat, M., Tran, I.T.: Analysis of pressure equipment by application of the primal-dual theory of shakedown. Commun. Numer. Meth. Engng. 23(3), 213–225 (2007)
[155] Vu, D., Staat, M.: Shakedown analysis of structures made of materials with temperature-dependent yield stress. International Journal of Solids and Structures 44(13), 4524–4540 (2007)
[156] Heitzer, M., Staat, M.: Reliability analysis of elasto-plastic structures under variable loads. In: Weichert, D., Maier, G. (eds.) Inelastic Analysis of Structures under Variable Loads: Theory and Engineering Applications, pp. 269–288. Kluwer, Academic Press, Dordrecht (2000)
[157] Tran, T.N., Phạm, P.T., Vu, D.K., Staat, M.: Reliability analysis of ine-lastic shell structures under variable loads. In: Weichert, D., Ponter, A.R.S. (eds.) Limit States of Materials and Structures: Direct Methods, pp. 135–156. Springer, Netherlands (2009)
[158] Spiliopoulos, K.V.: On the automation of the force method in the optimal plastic design of frames. Comp. Meth. Appl. Mech. Eng. 141, 141–156 (1997)
[159] Spiliopoulos, K.V.: A fully automatic force method for the optimal shakedown design of frames. Comp. Mech. 23, 299–307 (1999)

[160] Spiliopoulos, K.V.: Force Method-based procedures in the limit equilibrium analysis of framed structures. In: Weichert, D., Ponter, A.R.S. (eds.) Limit States of Materials and Structures: Direct Methods, pp. 233–252. Springer, Netherlands (2009)

[161] de Saxcé, G., Massonet, C., Morelle, P.: Discussion of paper Plastic collapse, shakedown and hysteresis. Int. J. Struct. Eng. ASCE 112(9), 2177–2183 (1987); S.A. Grualnik, S. Singh, T. Erber (eds.)

[162] de Saxcé, G.: Une generalization de l'inégalité de Fenchel et ses applications aux lois constitutives. C.R. Acad. Sci. t.314. série II, 125–129 (1992)

[163] Bodoville, G., de Saxcé, G.: Plasticity with non linear kinematic hardening: modeling and shakedown analysis b the bipotential approach. Int. J. Plast. 17(1), 21–46 (2001)

[164] Bousshine, L., Chaaba, A., de Saxcé, G.: A new approach to shakedown analysis for non-standard elastoplastic material by the bipotential. Int. J. Plast. 19(5), 583–598 (2003)

[165] Moreau, J.J.: On unilateral constraints, friction and plasticity, in: new variational techniques in Mathematical Physics. CIME Course, pp. 173–322. Springer (1974)

[166] Debordes, O., Nayroles, B.: Sur la théorie et le calcul à l'adaptation des structures élasto-plastiques. J. Mécanique 20, 1–54 (1976)

[167] Nayroles, B.: Tendences récentes et perspectives à moyen terme en élastoplastici-té asymptotique des constructions. In: Congrès Français de Mécanique, Grenoble, France (1977)

[168] Nayroles, B., Weichert, La, D.: notion de sanctuaire d'élasticité et l'adaptation des structures. C.R. Acad. Sci. Paris 316, Série II, 1493–1498 (1993)

[169] Nguyen, Q.-S.: Extension des théorèmes d'adaptation et d'unicité en écrouissage non linéaire. C.R. Acad. Sc. 282, 755–758 (1976)

[170] Nguyen, Q.-S.: Min-Max Duality and shakedown theorems in plasticity. In: Alart, P., Maisonneuve, O., Rockafellar, R.T. (eds.) Nonsmooth Mechanics and Analysis, Theoretical and Numerical Advances, ch. 8. Springer (2006)

[171] Nguyen, Q.-S., Pham, D.: On shakedown theorems in hardening plasticity. C.R. Acad. Sci. 329, 307–314 (2001)

[172] Halphen, B.: Q-S Nguyen, Sur les matériaux standard généralisés. J. Mécanique 14, 1–37 (1975)

[173] Mandel, J.: Adaptation d'une structure plastique écrouissable. Mech. Res. Comm. 3, 251–256 (1976)

[174] Radenkovic, D.: Théorèmes limites pour un materiau de Coulomb à dilatation non-standardisée. Comptes Rendus de l'Académie des Sciences Paris 252, 4103–4104 (1961)

[175] Zarka, J., Frelat, J., Inglebert, G., Kasmaï-Navidi, P.: A new approach to inelastic analysis of structures, CADLM edition, France (1989)

[176] Inglebert, G., Zarka, J.: On a simplified inelastic analysis of structures. Nucl. Eng. 57, 333–368 (1980)

[177] Hassine, T., Inglebert, G., Pons, M.: shakedown and damage analysis applied to rocket machines. In: Weichert, D., Maier, G. (eds.) Inelastic Analysis of Structures under Variable Loads: Theory and Engineering Applications, pp. 255–2267. Kluwer, Academic Press, Dordrecht (2000)

[178] Sharp, R.W., Booker, J.R.: Shakedown of pavements under moving surface loads. Journal of Transportation Engng. ASCE 110(1), 1–14 (1984)

[179] Krabbenhøft, K., Lyamin, A.V., Sloan, S.W.: Bounds to shakedown loads for a class of deviatoric plasticity models. Comp. Mech. 39(6), 879–888 (2007)

[180] Krabbenhøft, K., Lyamin, A.V., Sloan, S.W.: Shakedown of a cohesive-frictional half-space subjected to rolling and sliding contact. Int. J. Solids and Structures 44(11-12), 3998–4008 (2007)
[181] Zhao, J.D., Sloan, S.W., Lyamin, A.V., Krabbenhøft, K.: Bounds for shakedown of cohesive-frictional materials under moving surface loads. Int. J. Solids and Structures 45(11-12), 3290–3312 (2008)
[182] Collins, I.F., Wang, A., Saunders, L.: Shakedown Theory and the design of unbound pavements. Road and Transport Research 2, 28–39 (1993)
[183] Collins, I.F., Wang, A., Saunders, L.: Shakedown in layered pavements under moving surface loads. Int. J. Numerical and Analytical Methods in Geomechanics 17, 165–174 (1993)
[184] Grundy, P.: Shakedown of Bars in Bending and Tension. J. Eng. Mech. Div. ASCE 95(EM3), 519–529 (1969)
[185] Alwis, W.A.M., Grundy, P.: Shakedown Analysis of Plates. Int. J. Mech. Sci. 27(1/2), 71–82 (1985)
[186] Shiau, S.H., Yu, H.S.: Finite element method for shakedown analysis of pavements. In: 16th Australasian Conference on the Mechanics of Structures and Materials, pp. 17–22 (1999)
[187] Shiau, S.H., Yu, H.S.: Shakedown analysis of flexible pavements. In: The John Booker Memorial Symposium, pp. 643–653 (2000)
[188] Yu, H.S.: Foreward: shakedown theory for pavement analysis. International Journal of Road Materials and Pavement Design 6(1), 7–9 (2005)
[189] Li, H.X., Yu, H.S.: A non-linear programming approach to kinematic shakedown analysis of composite materials. International Journal of Numerical Methods in Engineering 66(1), 117–146 (2006)
[190] Yu, H.S., Wang, J.: Three-dimensional shakedown solutions for cohesive-frictional materials under moving surface loads. Int. J. Sol. Struct. 49(26), 3797–3807 (2013)

Some Remarks on the History of Fracture Mechanics

Dietmar Gross

Abstract. A short survey on the history of Fracture Mechanics is presented, supplemented by some remarks. Subsequently, the thoughts on fracture of two early scientists who are not widely known in the fracture mechanics community are discussed in more detail. Bertram Hopkinson had a clear picture of the stresses ahead of a crack and motivated most probably C.E. Inglis to solve the elliptical-hole problem. Ludwig Prandtl was the first to use the cohesive zone model and who investigated a time dependent fracture problem.

1   A Short History of Fracture Mechanics

Fracture Mechanics is a relatively young scientific branch of solid mechanics. Commonly its beginnings are associated with the work of A.A. Griffith on the theory of rupture [8], [9]. However, regarded as father of modern Fracture Mechanics is G.R. Irwin who introduced in the fifties of the last century new concepts, complemented the theory and made it applicable to real engineering problems [15], [16]. Even though Fracture Mechanics is young, a considerable number of publications exist about its history, see for example [1], [27], [30], [19], [41], [3], [5]. The number increases when the more general field of Failure Mechanics is incorporated, see e.g. [33], [32], [2]. As an example of early studies on the history of Failure Mechanics, Fig.1 shows failure patterns of cast iron under pressure and bending from the frontispiece of I. Todhunter & K. Pearson [33], published in 1886. One simple reason for the huge amount of material is that Fracture was from the beginnings of Mechanics in the focus of interest of the early scientists, because Fracture and Failure influences our daily life considerably. Therefore, the reader cannot expect to find in this paper something historically absolutely new. Who is interested in details is referred in the original literature. A good starting point are the works of Hans-Peter

---

Dietmar Gross
Division of Solid Mechanics, TU Darmstadt, Germany
e-mail: gross@mechanik.tu-darmstadt.de

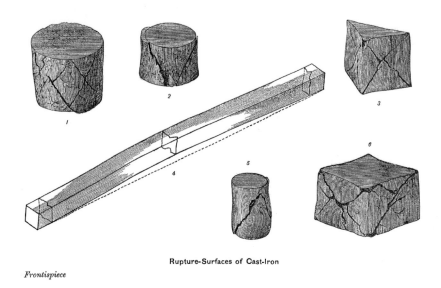

Fig. 1 Frontispiece of I. Todhunter & K. Pearson, The History of the Theory of Elasticity and the Strength of Materials, Vol. 1 [33], published in 1886

Rossmanith and Brian Cotterell [27], [3], authors who have contributed most to the history of Fracture Mechanics.

Fracture is an everyday companion of mankind since his existence and the human race learned very fast to take advantage of it. Techniques like stone knapping to manufacture hand axes date back more than 2 million years, Fig. 2. And making a cup from a skull by well-directed fractures was a relatively easy task. Other examples are the many monuments from the ancient world until the present time. They prove the art of stone cutters shaping brittle materials by fractures using knowledge based on experience.

Fig. 2 a) Acheulean hand axe from Kent, b) skull cup from Gough's cave, 14000 BCE, Credt: Natural History Museum, London

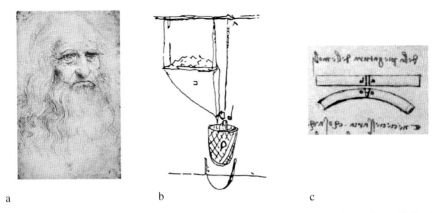

Fig. 3 a) Leonardo da Vinci, b) Leonardo's fracture test setup, reprinted from [21], c) Leonardo's sketch on bending, Codex Madrid page 84 verso [34]

Evidence of a scientific consideration of fracture can be found in the early Renaissance. It is well known that many inventions were anticipated by Leonardo da Vinci (1452–1519). In Mechanics, for example, he stated scaling laws for the bending strength and columns. Long before Jacob Bernoulli (1655–1705) but without any formulas he had a correct picture about the deformation kinematics of beams [25], Fig. 3. He also was the first who described a fracture test for metal wires in his notes. In the Codex Atlanticus, folio 222, a sketch of the test with a detailed description can be found which is worth to be quoted (for the translation see [21] or [32]): *The object of this test is to find the load an iron wire can carry. Attach an iron wire 2 braccia long* (remark: 1 braccia = approx. 60 cm) *to something that will firmly support it, then attach a basket or any similar container to the wire and feed into the basket some fine sand through a small hole placed at the end of a hopper. A spring is fixed so that it will close the hole as soon as the wire breaks. The basket is not upset while falling, since it falls through a very short distance. The weight of the sand and the location of of fracture of the wire are to be recorded. The test is repeated several times to check the results. Then a wire of one-half the previous length is tested and the additional weight it carries is recorded; the a wire of one-fourth length is tested and so forth, noting each time the ultimate strength and the location of the fracture.* It is interesting to note that Leonardo knew that the strength of a metal wire increases with decreasing length. This size effect is the result of the decreasing number of defects (e.g. deviations of the cross section) which were clearly visible in metal wires at that time.

Next, Galileo Galilei (1564–1642) must be mentioned. He is regarded not only as the founder of modern Mechanics as we understand it but he also established a seminal way of scientific thinking. He delivered many contributions to mechanics: well known is the so-called Dialog [7] which led to the famous court case of the Roman Catholic Church. More interesting regarding fracture are the Discorsi [6] which Galilei has written during his house arrest in Arcetri. On the "second day"

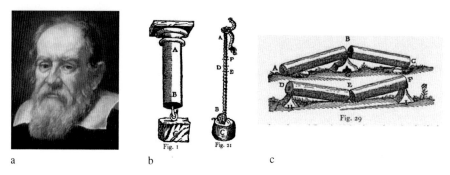

a   b   c

Fig. 4 a) Galileo Galilei, b) Galilei's tensile test setup, c) Galilei's sketch on bending fracture [6]

a discussion of the three participating persons, Salviati, Sagredo and Simplicio, is described about the fracture strength of columns, beams, ropes etc.. Salviati, i.e. Galilei, comes to the correct conclusion that the fracture force of a column under tension is proportional to the area of the cross section, Fig 4. He also reflected about the fracture of beams, which led him to the conclusion that the bending moment is the crucial loading measure. In this context he derived a scaling law for the bending strength of a beam, stating that it scales with the 3rd power of its height. As we know, this is not correct.

Edme Mariotte (1620–1684) was a scientist who was engaged, primarily experimentally, in different fields, Fig. 5. Shortly after l'Académie des Sciences was founded he became one of its first members. Well known from Mariotte are the Boyle-Mariotte gas law or the discovery of the eye's blind spot. When he was in charge of the pipe line design for the fountains of the palace in Versailles, he

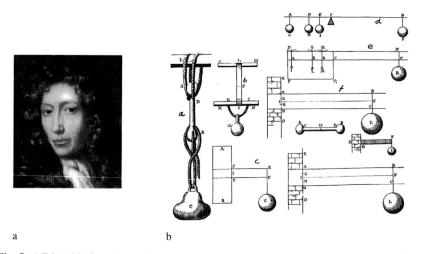

a   b

Fig. 5 a) Edme Mariotte, b) Mariotte's tensile and bending test setup, reprinted from [32]

became interested in the bending strength of beams. In this context he investigated experimentally the fracture load of wooden rods and beams and found that Galilei's theory gives exaggerated values. Having found the linear elastic law at the same time as Robert Hooke, he developed his own theory of elastic rods and beams under bending. He probably was the first who proposed a failure criterion stating that fracture occurs when the elongation of a rod under tension exceeds a certain limit.

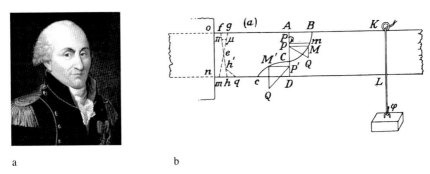

Fig. 6 a) Charles Augustin de Coulomb, b) Coulomb's sketch on bending, reprinted from [32]

Charles Augustin de Coulomb (1763–1806), Fig. 6, was educated as an engineer in the army and then sent for nine years to Martinique. There and later back in France he was in charge of different fortification works. In this context he made many experiments with soil and sandstone. In this way he found that the shear strength of soil or sandstone depends on the normal pressure, and he formulated this finding in the form that today is called Coulomb's law. His experiments indicated also that in certain cases the ultimate shear strength is equal to the ultimate strength in tension. Furthermore, he formulated the correct scaling law for the bending strength of beams. It should be emphasized that in the Renaissance until the end of the 18th century quantities like stress and strain have not yet been clearly defined. Scientists after Robert Hooke and Edme Mariotte in most cases argued by using the one-dimensional concept of forces acting on deformable fibers. Regarding fracture and failure this period can be called the period of scaling laws.

Beginning in the early 19th century, the theory of elasticity has been developed. Without any completeness some leading scientists from the French school shall be

Fig. 7 a) Claude-Louis Navier, b) Augustin-Louis Cauchy, c) Adhémar J.C. Barré de Saint Venant

Fig. 8 a) Gabriel Lamé, b) William J.M. Rankine, c) Eduard Tresca, d) Eugenio Beltrami, e) Otto Mohr

mentioned. Claude-Louis Navier (1785–1836), Fig. 7, was one prominent representative of this school. He has written a famous book on Theory of Elasticity and somewhat later a book on Strength of Materials. In his first book on theory of elasticity, Navier did not yet use the notion of stress. This term was later introduced by Augustin-Louis Cauchy (1789–1857), who also formulated the three-dimensional stress–strain law. The final form of the linear theory was given by Adhémar Jean Claude Barré de Saint Venant (1797–1886) who was recognized as the highest authority in Elasticity at that time. It was due to his strong influence that engineers until the end of the 19th century in most cases used the Maximum Strain Hypothesis as a design criterion and not the Maximum Principal Stress Hypothesis. In the field of Elasticity Theory many other scientists should be mentioned, but we will restrict our attention to failure and fracture. Here the fact is important that hand in hand with the development of the theory of elasticity, criteria for failure and fracture have been developed. Already mentioned was Coulomb's Shear Criterion and the Maximum Strain Criterion of Saint Venant. The Maximum Stress Criterion was proposed independently by Gabriel Lamé (1795–1870) and William John Macquorn Rankine (1820–1872) for mostly brittle material behavior, while the Maximum Shear Criterion of Henri Éduard Tresca (1814–1885) and later in 1900 by James J. Guest [10] was thought to be applicable for plastic flow, see [20]. It should be emphasized that until the end of the 19th century there was not yet a clear differentiation between failure by brittle fracture and failure by plastic flow. Otto Mohr (1835–1913) tried in 1882 to construct a general failure criterion which graphically is given by the envelope of the Mohr's circles at critical stress states. As Mohr-Coulomb Criterion it is still in use e.g. in Geomechanics. Mentioned shall also be the criterion of Eugenio Beltrami (1835–1900) which associates the limit state of the material with a certain critical strain energy density. Well known and commonly used in Plasticity is the Maximum Distortion Energy Criterion of Huber, von Mises and Hencky (Maksymilian Tytus Huber, 1872–1950, Richard von Mises, 1883–1953, Heinrich Hencky, 1885–1951). It is worth to be mentioned that this criterion has been proposed much earlier by J.C. Maxwell [20]. Finally, it shall be noted that J.J. Guest [10] was probably the first who differentiated clearly brittle fracture from failure by plastic yielding.

The disadvantage of the mentioned fracture or failure criteria is that they predict failure at a material point when a certain critical stress or strain state is attained.

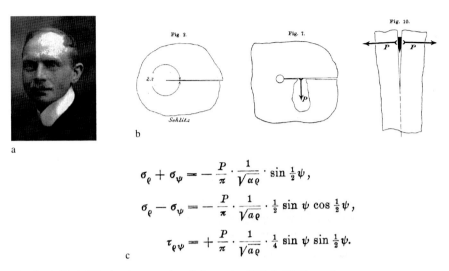

Fig. 9 a) Karl Wieghardt, reproduced from the ICF & ESIS conference announcement: Wieghardt & Irwin Centenary Conference, TU Vienna, March 1-2, 2007, b) sketches from Wieghardt's paper [38], c) crack-tip field from Wieghardt's paper [38]

Usually they do not take into account the kinematics of failure and thermodynamic principles as e.g. the energy balance of a failure process. And they do not take into account the fact that real materials always have defects. The first one who tried to overcome this disadvantage was Karl F.D. Wieghardt (1874–1924), Fig. 9. He was a PhD student of Felix Klein in Göttingen and became in 1906 a Professor of Mechanics and Mathematics in Braunschweig, 1907 in Hannover, 1911 in Vienna, and finally 1920 in Dresden. He was undoubtedly the first who solved exactly a real crack problem. In his publication *Über das Spalten und Zerreißen elastischer Körper* [38], written during his short time in Braunschweig, he treated the elastic problem of a semi-infinite crack in an infinite sheet loaded by a single force at a crack face (for an English translation of the paper see [26], [39]). He determined the exact solution for the singular crack tip field and recognized that the traditional failure criteria cannot predict failure of real cracks. To overcome this shortage, he proposed to introduce an additional characteristic length parameter, i.e. he anticipated a two-parameter fracture concept by stating that fracture does not occur on account of a critical stress in a single point, but rather when the stress level in a small area becomes critical. He also discussed criteria for the crack growth direction and proposed to take the maximum circumferential stress for that. Though he had the essential building blocks for an applicable fracture concept in his hands, he did not pursue this problem further. Engineers and scientist at that time might have considered Wieghardt's work as irrelevant for practical purposes or they did not know it at all. In any case, the time had not yet come for Fracture Mechanics and Wieghardt's paper was forgotten for more than 50 years.

It was Alan A. Griffith (1893–1963), Fig. 10, who set the first cornerstone of a fracture theory of cracks. In 1920 he published his famous paper *The Phenomena of Rupture and Flow in Solids* [8] where he introduced the surface energy of the crack into the energy balance and formulated the energetic fracture concept. Griffith studied Mechanical Engineering at the University of Liverpool and joined in 1915 the Royal Aircraft Establishment at Farnborough. Here, in cooperation with Geoffrey I. Taylor, well-known from fluid mechanics and dislocation theory, he applied Prandtl's soap-film method to the stress analysis of components. It might be that the close contact with G.I. Taylor and with the experimental method, relying on the surface tension of the soap film, brought him to the fundamental idea to associate the crack surface with a surface tension and surface energy, respectively. In his paper [8] he pointed out that defects like scratches on a glass surface reduce the fracture strength considerably and that there is a size effect which cannot be explained by traditional failure hypotheses. Using his energy approach, see Fig. 10b, he derived the famous relation

$$\sigma_c = \sqrt{\frac{2E\gamma}{\pi c}} \qquad (1)$$

for the fracture stress σ_c of an infinite plate under plane stress conditions containing a straight crack of length $2c$ with E being Young's modulus and γ the specific surface energy. To calculate the change of strain energy due to the presence of the crack, Griffith used the solution of C.E. Inglis for an elliptical hole in a plate under tension [14]. It has been noted several times that this calculation is erroneous and that it is not fully clear how Griffith came to his results (there is also a respective note from the publisher at the end of his paper [8]), see e.g. [40], [31], [13], [28], [18]. In a second paper he revised his equation for the fracture stress [9]. But compared with his seminal idea the mentioned error must be regarded as insignificant. Though Griffith's work was recognized and generally accepted by the scientific community, it had no impact on engineering practice. One reason for that was the big difference between theoretically predicted and in fracture experiments measured values of the specific surface energy γ. Another problem was the difficulty to determine the change of strain energy due to crack growth $\partial W/\partial c$ in typical engineering structures. Like Wieghardt before, Griffith did not pursue the fracture problem further.

The condition that the crack may extend is

$$\frac{\partial}{\partial c}(W - U) = 0,$$

b

a

Fig. 10 a) Alan A. Griffith, b) reproduction of the essential statement from Griffith's paper [8]. Here W, U and $2c$ are the strain energy, surface energy and crack length, respectively.

Fig. 11 Waloddi Weibull, photo by Sam C. Saunders

Another milestone was set by Waloddi Weibull (1887-1979), Fig. 11, who formulated in 1939 the statistical theory of fracture [35], [36]. With relatively simple mathematical effort he clarified the volume dependence of strength parameters of a brittle material containing internal defects but no macroscopic crack. A detailed information on the physical nature of the defects is not required. He also explained the influence of the stress state like tension, bending or torsion on the strength. Since his papers have been recognized at that time only by a small group of scientists, Weibull summarized his theory 12 years later in another publication [37] which attracted wide attention. Today, Weibull's statistical fracture concept is a standard tool for the design of components made from brittle materials like ceramics.

The actual breakthrough of Fracture Mechanics was achieved by George Rankine Irwin (1907-1998), Fig. 12. After having received the PhD in physics in 1937 Irwin accepted a job at the US Naval Research Laboratory in Washington (for details of his career and contributions see e.g. [29], [41], [4]). Here he was from the beginning confronted with static and dynamic failure problems. At the end of the forties, he became deeper interested in the macroscopic characterization of fracture. To overcome one drawback of Griffith's theory he proposed in 1948 to replace the specific surface energy γ by the specific effective fracture surface energy $\gamma_{eff} = \gamma + \gamma_p$ which incorporates the work γ_p of inelastic (plastic) deformation at the crack tip during crack growth. The second important step was made in 1957 by his publication *Analysis of stresses and strains near the end of a crack traversing a plate* [15]. Using examples of recently solved crack problems, Irwin pointed out that, independent of the specific loading situation, the singular crack tip field in any case has the structure

Fig. 12 George R. Irwin

$$\sigma_{ij} = \frac{\sqrt{E\mathscr{G}}}{\sqrt{2\pi r}} f_{ij}(\theta) \qquad (2)$$

which was found already 50 years ago by Karl Wieghardt, but was forgotten. Here E is Young's modulus (slightly different in plane strain and plane stress), r, θ are polar coordinates and $f_{ij}(\theta)$ are universal functions. The quantity $\sqrt{E\mathscr{G}}$ was later denoted by K and called *stress intensity factor*. Irwin further showed that $\alpha\mathscr{G}$ is the loss of strain energy during a small crack extension α, i.e. that $\mathscr{G}$ is the *energy release rate* and that it can be regarded as *generalized force* or, as it is called today, *crack extension force*. Since crack growth sets in only when $\mathscr{G}$ attains a critical value $\mathscr{G}_c$, the simple relation $\mathscr{G}_c = K_c^2/E = 2\gamma$ holds. This implies that for linear elastic bodies, the energy approach of Griffith and crack initiation criteria based on K or $\mathscr{G}$ essentially are equivalent. In subsequent works, Irwin and his collaborators completed the theory and proved its applicability to brittle and quasibrittle fracture, see e.g. [16], [17]. Because of its clarity and simple structure, Irwin's fracture concept found rapidly entrance into practical applications and is meanwhile firmly established. In the early sixties the first concepts for an elastic-plastic fracture mechanics were proposed and a rapid development set in. First steps towards an integration of damage mechanics and micromechanics into fracture mechanics have been attempted in the eighties. Nowadays, fracture mechanics is still a lively field of experimental, theoretical and numerical research. It is not an overstatement to say that G.R. Irwin's ideas, improvements and contributions were most essential for the birth of fracture mechanics as a science and an engineering tool. For this reason he is frequently called the "father of fracture mechanics".

2 Bertram Hopkinson Thoughts on Fracture

Bertram Hopkinson (1874-1918), Fig. 13, after having finished his studies at Cambridge in 1876, trained as a patent lawyer until he started in business as a consulting and design engineer, among others in the field of electric tramways. When he was 29 years old, having already a considerable professional reputation, he was appointed as Professor of Mechanism and Applied Mechanics at the University of Cambridge. Hopkinson's manifold interests were directed to the science of explosions, impact, dynamic properties of materials, combustion engines and many more. Well-known is the Split-Hopkinson Bar suggested by him in 1914 as a way to measure dynamic material properties under high-velocity impact.

In 1910 Hopkinson delivered a remarkable lecture on *Brittleness and Ductility* to the Sheffield Society of Engineers and Metallurgists which sheds some light on the knowledge regarding failure and fracture at that time [11]. In the beginning, Hopkinson discusses the typical failure modes, namely brittle failure by tearing (rupture) and ductile failure by sliding, and he points out that the failure behavior depends strongly on the stress state. To characterize the limit states, onset of tearing and onset of sliding, the material parameters "tenacity" (introduced by Rankine) and "rigidity" are used. He then states that the available experimental investigations of tenacity and the comparison of respective data with the failure criteria of, e.g., Lamé,

Fig. 13 Bertram Hopkinson (left), Charles Inglis (right)

Rankine or St. Venant leave questions open. Regarding the "rigidity", i.e. the onset of plastic flow, his review of the state of art comes to a similar conclusion. Though the mentioned experiments of J.J. Guest [10] indicate that the elastic limit is attained when the maximum shear stress becomes critical, it remains unclear to what extent the accompanied normal stress affects this limiting shearing stress. It shall be noted that Hopkinson, as other scientists at that time, did not yet classify the different material types. Metals, glass and stone are not differentiated but mentioned in the same context.

Subsequently, Hopkinson addresses the influence of cracks on the failure behavior. Mentioning first that scratches at glass surfaces reduce its strength considerably he states *"As far as I am aware the distribution of stress in the neighborhood of a crack has not been investigated by mathematicians with the completeness which its importance deserves."* and further *"It may be inferred from the elastic theory that if the crack have an absolutely sharp edge, the stress actually at the edge will be infinitely great"* and finally, concluding that real cracks are not mathematically sharp edged from the fact that the failure stress is limited *"What is wanted is an investigation of the distribution of stress, near a thin lens-shaped cavity with slightly rounded edges, say, of the shape of a very flat spheroid .."*. Here two remarks are necessary. First, Hopkinson did not know the already published paper of K. Wieghardt. Nevertheless, as a mathematically well trained engineer he knew intuitively that at a sharp crack a stress singularity must be present. Secondly, his last statement was most probably not only a wish but rather an order – an order to Charles Edward

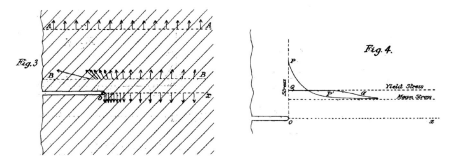

Fig. 14 Sketches of stress distribution near a crack under tension from Hopkinson's paper [11]

Inglis (1875-1952), Fig. 13, whom he had appointed in 1903 as Lecturer and whose mathematical abilities he knew very well. Inglis completed this order in 1913 when he presented the already mentioned solution for an elliptical hole in a plate under tension [14]. One of its important results is that the maximum stress at the vertex of a flat ellipse under tension is proportional to $2\sqrt{c/\rho}$ where ρ is the curvature radius, i.e. it can be expected that the fracture stress scales with $1/\sqrt{c}$.

In what follows, Hopkinson discusses the stress distribution near the crack (notch) tip and the influence of yielding on it, see Fig. 14. His picture of the stresses is qualitatively correct apart from his assumption that the stress state at the tip is practically the same as in a test sample under uniaxial tension. He concludes that on account of the high local stress at the crack tip an ideally brittle material will obviously fail by tear at a much lower mean stress than a piece without a crack. More interesting is the case that the material is able to yield. The elastic stress distribution then *"will be redistributed more or less in a manner shown by the curve QQ'* and further *"The areas under the two stress curves PP' and QQ' are the same, since there is no change in the total load"*, see Fig. 14. He then states, that *"It is however clear that, generally speaking, the amount of slide near the ends of the crack may be of excess amount, though it will extend over a small area only"* and *"This severe local deformation implies hardening and renders the material locally brittle"*. These sentences express what today is generally accepted and proven by stress analysis of elastic and elastic-plastic fracture mechanics. It is worth mentioning that, 50 years later, G.R. Irwin used exactly the same considerations to obtain an approximation for the extension of the plastic zone ahead of a crack tip which is well known in fracture mechanics. Hopkinson, without any formulas, had a qualitative clear picture of what is going on near the crack tip and why a ductile material containing a macroscopic crack shows a quasi-brittle fracture behavior.

When C.E. Inglis presented his paper [14], Hopkinson in the discussion pointed out its importance and underlined that most failures in engineering structures originate in a crack which might be started from the presence of hard particles, impurities and/or cavities, see [12]. In addition, he emphasized the importance of taking into account the influence of the ductility and he encouraged C.E. Inglis carrying out a respective analysis. But for such an elastic-plastic stress analysis the time had not yet come.

3 Prandtl's Contribution to Fracture Mechanics

Ludwig Prandtl (1875-1953), Fig. 16, after having studied Mechanical Engineering and finished his doctorate, worked for a short time in industry. In 1901 he was appointed as Professor of Fluid Mechanics at the University Hannover, 3 years later as Professor of Technical Physics and since 1907 as Professor of Applied Mechanics at the University Göttingen. L. Prandtl is well-known for his research in Fluid and Gas Dynamics but also in fields like Plasticity, Elasticity or Meteorology. Already mentioned was the soap-film technique, proposed by him in 1903, to experimentally solve problems from torsion theory.

Fig. 15 Ludwig Prandtl

In 1928 Prandtl developed a model to describe the delayed elastic response and the rate dependence of the yield stress of plastic materials [23]. It is widely used in nanotribology and known as Prandtl-Tomlinson-Model, see [22]. Five years later he used the same ideas to model time and rate effects of fracture processes, [24] (for an English translation by W.G. Knauss see http://resolver.caltech.edu/Caltech AUTHORS:20110622-132946088). While in plasticity a tangential relative sliding of two bodies and a periodic force field is assumed, now, for fracture, a normal separation of the bodies and an attractive force field with a single equilibrium position is considered. To make the analysis as simple as possible, Prandtl replaced the bodies, i.e. the elastic half-domains, by beams and simplified the cohesive law by a linear traction-separation relationship, Fig. 16. As result, a fracture stress σ is obtained for the quasi-static case which differs from Griffith's result mainly on account of the model simplifications. But both results can be brought into good agreement when the hight of the beams is sufficiently increased.

Prandtl proceeds in his paper with a discussion of time effects due to the statistical character of thermal oscillations of the particles. For a constant remote stress the

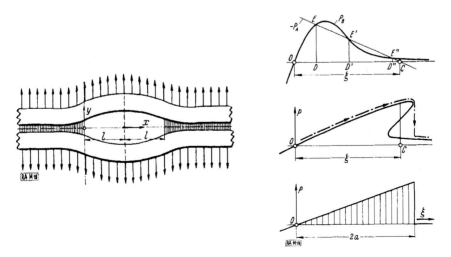

Fig. 16 Sketches of crack configuration and the cohesive law from Prandtl's paper [24]

model leads to continuous crack growth with a velocity, depending on the material, that might be negligibly small. Prandtl emphasizes that the model is conceived for a material with regular structure like a crystal. Finally he mentions that for macroscopic materials with an irregular, inhomogeneous structure the model could be improved by statistical superposition of different force distributions. Such a model also should be able to catch hysteresis effects on account of repeated loading.

Prandtl was only once engaged in a problem of fracture. It is all the more remarkable that his model was the first cohesive zone model that has been published in fracture mechanics and that he was the first who tried to understand time dependent fracture.

Acknowledgements. I would like to thank my colleagues Werner Hauger, Ralf Müller and Baixiang Xu for valuable remarks.

References

[1] Barsom, J.M.: Fracture Mechanics Retrospective. ASTM, Philadelphia (1987)
[2] Benvenuto, E.: An Introduction to the History of Structural Mechanics, Parts 1,2. Springer, New York (1991)
[3] Cotterell, B.: Fracture and Life. Imperial College Press, London (2010)
[4] Dally, J., Irwin, G.R.: Memorial Tributes: National Academy of Engineering 10, 146–153 (2002)
[5] Erdogan, F.: Fracture mechanics. Int. J. of Solids and Structures 37, 171–183 (2000)
[6] Galilei, G.: Discorsi e Dimonstrazioni Mathematiche, 'The New Sciences' English translation. The Macmillan Co., New York (1933)
[7] Galilei, G.: Dialog über die beiden hauptsächlichsten Weltsysteme. Teubner, Stuttgart (1982)
[8] Griffith, A.: The Phenomena of Rupture and Flow in Solids. Phil. Trans. Roy. Soc. London, A 221, 163–198 (1920)
[9] Griffith, A.: The Theory of Rupture. In: Proc. of the Int. Congress for Appl., pp. 55–63 (Delft 1924)
[10] Guest, J.: On the strength of ductile materials under combined stress. Philosophical Magazine Series 5 50, 69–132 (1900)
[11] Hopkinson, B.: Brittleness and Ductility. In: Ewing, A., Larmor, J. (eds.) The Scientific Papers of Bertram Hopkinson, Cambridge (1921)
[12] Hopkinson, B.: On Holes an Cracks in Plates. In: Ewing, A., Larmor, J. (eds.) The Scientific Papers of Bertram Hopkinson, Cambridge (1921)
[13] Horowitz, C., Pines, B.: Beitrag zur Griffithschen Bruchtheorie. Zeitschrift für Physik 47, 904–911 (1928)
[14] Inglis, C.: Stresses in a Plate due the Presence of Cracks and Sharp Corners. Trans. Roy. Inst. Naval Architects 45th Session, 219–241 (1913)
[15] Irwin, G.: Analysis of Stresses and Strains Near the End of a Crack Traversing a Plate. Journal of Applied Mechanics 24, 361–364 (1957)
[16] Irwin, G.: Fracture, in Encyclopedia of Physics. In: Flügge, S. (ed.) Elasticity and Plasticity, vol. IV, pp. 551–590 (1958)

[17] Irwin, G., Paris, P.: Fundamental aspects of crack growth and fracture, in Fracture: an Advanced treatise. In: Liebowitz, H. (ed.) Engineering Fundamentals and Environmental Effects, pp. 1–46 (1971)
[18] Munz, D.: The Griffith equation (2011) (unpublished manuscript)
[19] Ohnami, M.: Fracture and Society. Ohmsha Ltd., Tokyo (1992)
[20] Osakada, K.: History of plasticity and metal forming analysis. Journal of Materials Processing Technology 210, 1436–1454 (2010)
[21] Parsons, W.: Engineers and Engineering in the Renaissance. Williams & Wilkins Co., Baltimore (1939)
[22] Popov, V., Gray, J.: Prandtl-Tomlinson Model: History and applications in friction, plasticity and nanotechnologies. ZAMM, Z. Angew. Math. Mech. 92, 683–708 (2012)
[23] Prandtl, L.: Ein Gedankenmodell zur kinetischen Theorie der festen Körper. Ztschr. f. Angew. Math. und Mech. 8, 85–106 (1928)
[24] Prandtl, L.: Ein Gedankenmodell für den Zerreiss vorgang spröder Körper. Ztschr. f. Angew. Math. und Mech. 13, 129–133 (1933)
[25] Reti, L.: The unknown Leonardo. McGraw-Hill, New York (1974)
[26] Rossmanith, H.: An introduction to K. Wieghardt's historical paper 'On splitting and cracking of elastic bodies'. Fatigue and Fract. Engn. Mater. Struct. 18, 1367–1369 (1995)
[27] Rossmanith, H.: Fracture Research in Retrospect. Balkema, Rotterdam (1997)
[28] Rossmanith, H.: The struggle for recognition of engineering fracture mechanics. In: Rossmanith, H.P. (ed.) Fracture Research in Retrospect, Rotterdam (1997)
[29] Rossmanith, H.: Georg Rankin Irwin - The Father of Fracture Mechanics 1907-1998. FRAGBLAST - Int. J. of Blasting and Fragmentation 2, 123–141 (1998)
[30] Rumpf, H.: About the History of Physics of Fracture Phenomena. Chemie-Ingenieur-Technik 31, 697–705 (1959) (in German)
[31] Smekal, A.: Technische Festigkeit und molekulare Festigkeit. Die Naturwissenschaften 10, 799–804 (1922)
[32] Timoshenko, S.: History of Strength of Materials. McGraw-Hill, New York (1953)
[33] Todhunter, I., Pearson, K.: A History of the Theory of Elasticity and of the Strength of Materials, vol.1, 2. University Press, Cambridge (1886, 1893)
[34] da Vinci, L.: Codex Madrid I: The Madrid Codices: National Library Madrid, Fascimile Edition of Codex Madrid I. McGraw-Hill, Baltimore (1974)
[35] Weibull, W.: A statistical theory of the strength of materials. Ingeniörsvetenskapsakademiens Handlingar 151 (1939)
[36] Weibull, W.: The phenomenon of rupture in solids. Ingeniörsvetenskapsakademiens Handlingar 153 (1939)
[37] Weibull, W.: A Statistical Distribution Function of Wide Applicability. J. of Appl. Mech. 18, 293–297 (1951)
[38] Wieghardt, K.: Über das Spalten und Zerreißen elastischer Körper. Zeitschr. für Mathematik und Physik 55, 60–103 (1907)
[39] Wieghardt, K.: On splitting and cracking of elastic bodies (translated by H.P. Rossmanith). Fatigue and Fract. Engn. Mater. Struct. 18, 1371–1405 (1995)
[40] Wolf, K.: Zur Bruchtheorie von A. Griffith. Zeitschr. für angewandte Mathematik und Mechanik 3, 107–112 (1922)
[41] Yarema, S.: History of Fracture Mechanics, On the contribution of G.R. Irwin to Fracture Mechanics. Material Science 31, 617–623 (1995)

Porous Media in the Light of History

Wolfgang Ehlers

Abstract. At the end of the 18th century, serious problems in dike constructions in Northern Germany and the need to understand coupled solid-water problems initiated first attempts to describe porous media. Many attempts followed until a sound Theory of Porous Media was born on the basis of continuum mechanics of multi-component materials with multi-physical properties.

The present article roughly describes the development of the Theory of Porous Media from its origins to contemporary applications, thus presenting porous media in the light of history.

Keywords: Theory of Porous Media, Mixture Theories, Volumetric Homogenization, Geomechanical, Environmental, and Biomechanical Problems.

1 What is Porous Media Mechanics

Many problems in nature and, as well, in our designed environment are based on water seeping into soil while yielding a destabilization of slopes and embankments through buoyancy forces. This has always led to failure situations with dramatic results for the local population. Especially, heavy rainfall events cause landslides, dam failure and other disasters. As a result, people have always been interested in explaining the reasons for upcoming catastrophes in order to find measures to prevent disastrous situations.

In civil engineering, soil-water problems are nowadays described on the basis of the Theory of Porous Media (TPM) composed of the Theory of Mixtures (TM) and the concept of volume fractions. The TPM is a macroscopic continuum-mechanical approach proceeding from a virtual homogenization over volumetric microstructures of the considered aggregate. In case of a soil-water aggregate, the TPM approach proceeds from a ternary model of a soil constituent, the pore water

Wolfgang Ehlers
Institute of Applied Mechanics,
University of Stuttgart, Pfaffenwaldring 7, 70569 Stuttgart, Germany

and the pore gas, such that the overall domain does not contain any vacant space. Furthermore, the TPM is rigorous in the sense that it proceeds from superimposed continua with internal interactions and individual states of motion. As a result, each constituent is governed by an individual set of balance equations for mass, momentum, moment of momentum and energy. The balance equations of one component are coupled to the balance equations of all other components of the aggregate by so-called production terms of mass, momentum, moment of momentum and energy. For details, cf. the articles by the author [1, 2] and his scientific teacher Reint de Boer [3] or the basic work by Truesdell [4], Bowen [5-7] and other authors cited therein.

To set an example of a geomechanically-based porous-media problem, consider a heavy rainfall event acting on a railway dam, cf. Figure 1 (right). The water content varies from red (empty) directly beneath a nearly impermeable layer and fully saturated (blue) on top of this layer and at the bottom of the displayed zone which is understood as the groundwater level. As a result of the rainfall, the water accumulates on top of the impermeable layer and leaks at the airside of the slope, while the soil is subjected to buoyancy forces. As is seen from Figure 1 (left), this leads to an accumulation of plastic soil deformations in a thin shearing zone with a shape of a double-cranked shell embedded in the three-dimensional (3-d) soil.

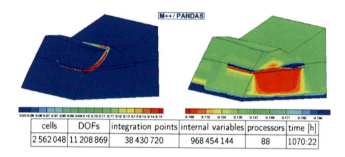

cells	DOFs	integration points	internal variables	processors	time [h]
2 562 048	11 208 869	38 430 720	968 454 144	88	1070:22

Fig. 1 Detail of a railroad dam, shear zone development (left) as a result of a heavy rainfall event (right)

This example has been computed as a triphasic aggregate of an elasto-plastic soil saturated by two pore fluids, the pore water and the pore gas. The numerical treatment proceeds from a monolithic computation by the finite-element method (FEM), where the momentum balance of the overall aggregate corresponds to the solid displacement vector, and the mass balances of the fluids are governed by the effective water and gas pressures. As can be seen from the Table included in Figure 1, the computation has been carried out by 2.5 million elements on 88 processors.

To make computations like this possible, it took a long time to build up the theoretical basis and numerical possibilities on modern computers.

2 The Early Days

In its roots, the Theory of Porous Media dates back to the end of the 18th century, when Reinhard Woltman (*December 28, 1757; † April 20, 1837), the director of hydraulic engineering (Direktor der Strom- und Uferwerke und Leiter des gesamten Wasserbaus) of the city of Hamburg, discovered the concept of volume fractions as the ratio of the volumetric portions of the soil and the pore water compared to the volume of the overall dike as significant components of any dike construction [8]. Based on this concept, cf. Figure 2, he was also able to conclude to the partial densities of mud as a mixture of soil and water.

Comparable ideas have been set by Achille Ernest Oscar Joseph Delesse (*February 3, 1817; † March 24, 1881). In his early career as a mining engineer, he had the problem to distinguish between the portions of the minerals in a mine. From a seam, he could observe the area fraction of the minerals, but was that equivalent to the volume fractions?

Fig. 2 The concept of volume fractions and the weightiness of mud components, taken from [8]

By intensive studies, Delesse [9] found out by statistical investigation of various slices of mineral conglomerates that area fractions and volume fractions are equivalent. In a modern setting, this leads to

$$n^\alpha = dv^\alpha/dv = da^\alpha/da$$

stating that the volume fraction n^α is obtained by relating the local volume or the local area element of the αth constituent to the overall volume element dv or

the overall area element da. After his period as a mining engineer, Delesse became a renown scientist, when he was appointed as a professor for geology and mineralogy at Besançon (1845-1850) and later as a professor for geology at the Sorbonne in Paris (1850-1864). In 1864, he became a professor for agriculture at the École des Mines, where he finally was appointed as the inspector-general of mines in 1878.

In the same area of time, Henry Philibert Gaspard Darcy (*June 10, 1803; † January 3, 1858) was working as a hydraulic engineer at the city of Dijon (1834-1840), where he became the chief engineer for the Cote d'Or in 1840. As a result of health problems, he asked for an early retirement in 1850. Nevertheless, he published his famous law

$$n^F \mathbf{w}_F = -k^F \operatorname{grad} h$$

in 1856 [10] after his retirement. In Darcy's law, which can also be found by modern approaches within the Theory of Porous Media, $n^F \mathbf{w}_F$ is the filter velocity given as the product of the seepage velocity $\mathbf{w}_F$ and the fluid volume fraction n^F, k^F is the Darcy permeabilty or the hydraulic conductivity, respectively, and $\operatorname{grad} h$ is the pressure head. From a modern point of view, Darcy's law can be found by combination of a constitutive equation for the direct momentum production term and the momentum balance of the liquid component of a binary system of solid and fluid. Furthermore, Darcy's law only proves to hold in case of quasi-static situations, when a lingering hydraulic flow motivates a neglect of acceleration terms and when, furthermore, the frictional fluid stresses can be neglected by arguments of dimensional analysis in comparison with the friction included in the constitutive assumption for the momentum production, cf. [1]. However, Darcy's law is widely used today in hydraulic engineering as a given constitutive equation without the consideration of its domain of validity.

Apart of the fields of mining and hydraulic engineering, porous media systems appear in various application areas. Following this, it is not astonishing that, apart from fluid flow in porous media, also diffusion problems had to be studied. The first one who empirically investigated these problems was Adolf Eugen Fick (*September 3, 1829; † August 21, 1901) who started to study mathematics but changed to medicine, later on. Fick's laws were published in 1855 [11], shortly before he became an adjunct professor in 1856 and a full professor for physiology in 1862 at the university of Zürich. Fick's first law states that the concentration flow of a species in a mixture of two components is proportional to its concentration gradient. Inserting this finding into the mass conservation equation yields Fick's famous second law

$$\frac{\partial c}{\partial t} = D \operatorname{div} \operatorname{grad} c.$$

In this equation, c is the concentration, D the diffusion coefficient and $\operatorname{div} \operatorname{grad}(\cdot)$ the *Laplace*an.

IV. *Ueber Diffusion; von Dr. Adolf Fick,*
Prosector in Zürich.

Die Hydrodiffusion durch Membranen dürfte billig nicht blofs als einer der Elementarfactoren des organischen Lebens sondern auch als ein an sich höchst interessanter physikalischer Vorgang weit mehr Aufmerksamkeit der Physiker in Anspruch nehmen als ihr bisher zu Theil geworden ist. Wir besitzen nämlich eigentlich erst vier Untersuchungen, von Brücke¹), Jolly²), Ludwig³) und Cloetta⁴) über diesen Gegenstand, die seine Erkenntnifs um einen Schritt weiter gefördert haben. Vielleicht ist der Grund dieser spärlichen Bearbeitung zum Theil in der grofsen Schwierigkeit zu suchen, auf diesem Felde genaue quantitative Versuche anzustellen. Und in der That ist diese so grofs, dafs es mir trotz andauernder Bemühungen noch nicht hat gelingen wollen, den Streit der Theorien zu

Fig. 3 Fick's original work [11] on diffusion problems

All investigations and findings by Woltman, Delesse, Darcy and Fick have been based on experimental observations and conclusions from other scientific

laws. The first who studied diffusion behavior in the sense of continuum mechanics was Josef Stefan (Jožef Štefan) (*March 24, 1835; † January 7, 1893). Stefan enhanced Fick's diffusion laws by also investigating mixtures of three components and extended his findings to the diffusion of gases across porous walls. In 1871, he described the gas diffusion through rigid membranes [12]. The relation of the effective (free) gas pressure compared to the partial pressure

of the pore gas in the porous wall was described on the basis of the porosity of the porous solid. Therewith, Stefan was the first who included the concept of volume fractions into a continuum theory for porous media.

Apart of the early pioneers of porous media theories, a lot of other scientists contributed to the success of the Theory of Porous Media. In Austria, Gustav Jaumann (*April 18, 1863; † July 21, 1924) published on continuum mechanics of

complex systems. Based on prior work by Gibbs, he was the first to make extensively use of the tensor calculus, which, in 1911, was called dyade calculation. His contribution to continuum mechanics, however, sometimes difficult to read, does not only include the "Jaumann derivative" which is still in use in various plasticity approaches. Instead, his pioneering work also presents a phenomenological continuum theory for a closed system of physical and chemical differential equations for a mixture of an arbitrary amount of chemical components including chemical reactions and electromagnetic effects [13].

From a scientific point of view, Jaumann's work can still be considered as modern, and he can definitely be called the pioneer of continuum mechanics and mixture theories as the bearing pillar of the modern Theory of Porous Media.

Merging continuum mechanics and thermodynamics, important contributions have been made by Rudolf Julius Emanuel Clausius (*January 2, 1822; † August 24, 1888) whose famous articles "Ueber die bewegende Kraft der Wärme ..." [14] and "Ueber die veränderte Form des zweiten Hauptsatzes ..." [15] established the basis of irreversible thermodynamical processes and their later impulse on continuum theories. In a series of articles published between 1876 and 1878 by Josiah Willard Gibbs (*February 11, 1839; † April 28, 1903) with the concluding title "On the Equilibrium of Heterogeneous Substances", Gibbs used thermodynamical methods to interpret chemo-physical phenomena, for example, by use of the phase rule, later called after him. Gibbs can furthermore not only be considered as the father of vector calculus but, together with Hermann Ludwig Ferdinand von Helmholtz (*August 31, 1821; † September 8, 1894), also as the establisher of the entire field of physical chemistry, which is also part of modern porous media approaches, for example, in the domain of phase transitions and electro-chemical reactions.

The Scottish scientist William John Macquorn Rankine (*July 5, 1820; † December 24, 1872) contributed like Duhem, Reynolds, Planck and others to the evolution of thermodynamics, but Rankine was furthermore involved in the field of soil mechanics by offering, like Coulomb, a method for the computation of the earth pressure [16]. Furthermore, his maximum normal-stress yield criterion is well known in the field of plasticity.

3 The Period of Geomechanics

At the beginning of the 20th century, geotechnical constructions had to be considered that needed a precise analysis of soil systems composed of soil particles and pore water. Especially, dam-construction and foundation problems such as the computation of settlements of tall buildings had to be considered. In the framework of these engineering-dominated problems, geotechnical experts like Terzaghi and Biot came into play.

Karl von Terzaghi (Karl Anton Terzaghi Edler von Pontenuovo) (*October 2, 1883; † October 15, 1963) was a civil engineer with a deep interest in geotechnical problems. In 1912, he visited various dam construction sites in the USA. Ob-

viously, he was aware of the complexity of soil as a binary medium of solid grains and water, but he was completely unaware of any theoretical description of porous-media problems. Therefore, he tried himself to solve the problem, when he published on soil mechanics, for example, in his paper "Die Berechnung der Durchlässigkeitsziffer des Tones aus dem Verlauf der hydrodynamischen Spannungserscheinungen" [17] of 1923 and in the books "Erdbaumechanik auf

bodenphysikalischer Grundlage" [18] and "Theorie der Setzung von Tonschichten" [19], the latter together with Otto Karl Fröhlich. However, Terzaghi was not a mathematician nor was he very much interested in struggling with theories such that he did not find continuum-mechanically based approaches satisfying the standard that has earlier been found, for example, by Stefan and Jaumann. As an engineer, Terzaghi always tried to combine theory and practice. Thus, it was not astonishing that Terzaghi's work led to oppositions. Especially, when Terzaghi was a professor for hydraulic engineering at the Technical University of Vienna (1929-1938), Paul Fillunger (*June 25, 1883; † March 7, 1937), who was a professor for applied mechanics at the same university since 1923, became his major scientific opponent.

In 1913, Fillunger published a first paper on buoyancy forces in gravity dams [20]. With this work, he presented a masterpiece, when he considered the problem as a binary medium of two interacting continua, soil and water. From this point of view, Fillunger can be regarded as the pioneer of the modern Theory of Porous Media, a conclusion that has been drawn by de Boer in his book on porous media [3]. However, and this is the tragedy of Fillunger's work, his buoyancy equation included a mistake by presenting the buoyancy force as a linear function of the difference between the volume and the surface porosity, cf. [3]. This mistake was recognized by Terzaghi. Fillunger tried to justify his result – the onset of a severe conflict between the opponents that has been carried out both personally and scientifically. When both worked on capillary forces in porous media, it came to a final conflict. In his book "The engineer and the scandal" [21], de Boer portrayed the whole story that ended with Fillunger's suicide in 1937, after a scientific commission of the Technical University of Vienna came to the conclusion that Fillunger was wrong.

$$(7) \quad \begin{cases} \frac{\partial v_1}{\partial t} + v_1 \frac{\partial v_1}{\partial z} = \frac{1}{\varrho_1}\left(-Z - \frac{\partial p_1}{\partial z}\right), \\ \frac{\partial v_2}{\partial t} + v_2 \frac{\partial v_2}{\partial z} = \frac{1}{\varrho_2}\left(Z - \frac{\partial p_2}{\partial z}\right), \\ \frac{\partial \varrho_1}{\partial t} + \frac{\partial (\varrho_1 v_1)}{\partial z} = 0, \\ \frac{\partial \varrho_2}{\partial t} + \frac{\partial (\varrho_2 v_2)}{\partial z} = 0. \end{cases}$$

Fig. 4 Fillunger's basic equations of a binary aggregate taken from [28]

Fillunger's and Terzaghi's basic work on geotechnically-based porous media has been continued by a variety of scientists, in the early days by the Belgium Maurice Anthony Biot (*May 25, 1905; † September 12, 1985) and by the Austrian Gerhard Heinrich (*April 18, 1902, † 1983) who later also published with

Kurt Desoyer. While Biot, who worked as a research associate with Theodore von Kármán at Caltech, was a follower of Terzaghi, Heinrich exclusively used the ideas of Fillunger when he started to publish on the settlement of clay layers in 1938 [22]. This dissociation of the porous-media society is, by the way, still active. While the procedure of Terzaghi and Terzaghi and Fröhlich is, from a modern point of view, more or less unsatisfactory, Fillunger's approach is still modern, because he started with the balance equations of two overlaying constituents, soil and water, and treated this aggregate in the sense of a mixture with immiscible but interacting constituents, cf. Figure 4, and not, like Terzaghi, from an intuitive basis. Biot followed Terzaghi's basic ideas in his first papers of 1935 and 1941 [23, 24] and published intuitively-based treatises. Nevertheless, Biot's work, especially his famous papers of 1955 and 1956 [25-27], are still highly cited and in use whenever young researchers are looking for basic material when they want to solve porous-media problems. Unfortunately, Fillunger's work got nearly lost. This might be a result of his early suicide. His final work, "Erdbaumechanik?" of 1936 [28], a polemic reaction on Terzaghi's work, has only been published posthumously by Fillunger's son Erwin in 1937. Like Fillunger's work, also the papers by Heinrich and Desoyer, for example, [22, 29], got lost in the scientific jungle. Only Reint de Boer recovered these articles during his visits in Vienna in 1987 and later. A first publication reviewing the history of porous media has been published in 1988 by de Boer and the author [30].

Fig. 5 Reint de Boer at the age of 65

4   The Onset of the Modern Era

The modern era started with the recovery of continuum mechanics at the beginning of the 1950th, when the US-American scientist Clifford Ambrose Truesdell III (*February 18, 1919; † January 14, 2000) entered the stage. He held a BSc in Mathematics and Physics, a MSc in Mathematics and an additional graduation in mechanics. Truesdell is truly the originator of "Rational Mechanics", a topic, on which he publishes a huge amount of articles and books. Later, he also became a historian for natural sciences and mathematics. However, in his early days, Truesdell visited Europe learning French, German and Italian. He also improved his Latin and Greek. As a result, he did not only publish in English, but also in

Fig. 6 Professor Truesdell and his wife Charlotte in their Guilford home (Wikipedia)

European languages, for example, in Italian [31, 32]. Truesdell's work originated the modern view on continuum mechanics and thermodynamics [31] including mixture theories [32]. Truesdell was the son of a wealthy family and turned out to exhibit a baroque personality. This fact can also nicely be seen from Figure 6 exhibiting Truesdell and his wife in their home stylishly dressed and surrounded by precious paintings.

Considering earlier work on continuum mechanics by Jaumann [13], it is astonishing that 46 years elapsed until continuum mechanics was brought to light again. This might be a result of a technical view on mechanics and the need to use mechanics as a design tool. However, in those days without powerful computers (the civil engineer Konrad Zuse, 1910-1995, just had found a basic computation machine without great efficiency), people were more interested in simple numerical computation methods than in complex continuum-mechanical equations and constitutive relations resulting in unsolvable systems of partial differential equations.

With his early work, Truesdell became attractive for a variety of young researchers with theoretical interest such as Richard Toupin, Walter Noll and others. Especially, the publications with Toupin and Noll [33, 34] both included in the famous "Handbuch der Physik" nearly contain the complete continuum-mechanical knowledge of the 1950th and early 1960th.

The origin of the modern Theory of Mixtures as the scientific basis of the Theory of Porous Media must be dated as of 1957, when Truesdell [31] presented his local balances of mass, momentum and energy for arbitrarily constituted mixtures. These balances have been constructed in analogy to the standard balance equations for closed systems of single-component materials. In contrast to the standard balance equations, they include convenient production terms for the description of the coupling mechanisms between the mixture constituents. Thus, Truesdell described a closed mixture system where the single components behave like open systems. An elaborated version of these equations is also part of the "Classical Field Theories" [33] published with Toupin in "Handbuch der Physik", cf. [33, Sections 158, 159, 215, 243, 254, 255, 259, 295]. The basic procedure of constructing balance equations for multi-component systems such as mixtures explained Truesdell with the aid of his so-called metaphysical principles which he raised to a thermodynamical principle in 1969 [35], cf. [4, p. 221].

> 1. *All properties of the mixture must be mathematical consequences of properties of the constituents.*
> 2. *So as to describe the motion of a constituent, we may in imagination isolate it from the rest of the mixture, provided we allow properly for the actions of the other constituents upon it.*
> 3. *The motion of the mixture is governed by the same equations as is a single body.*

However, in Truesdell's description of mixtures, there was no relation for a balance of moment of momentum for the mixture constituents [33, Section 215]. Furthermore, an entropy inequality was also missing, although the entropy principle of Clausius was part of the description of standard single-phasic materials [33, Sections 245-258] for a long time. Without raising their hypothesis to a principle, Truesdell and Toupin surmised that the entropy inequality of heterogeneous media would basically be the same as that of a single-component medium. However, this assumption turned out to be wrong.

In 1964, Kelly [36] generalized Truesdell's mixture theory by the inclusion of electromagnetic effects towards a multi-physical and multi-component mixture. Moreover, Kelly additionally included jump conditions across a discontinuity surface. However, his essential merit is the derivation of balance equations for multi-component systems on the basis of a so-called fundamental balance law. For details, also compare the articles by the author [1, 2]. By use of this fundamental concept, Kelly terminated the uncertainties in deriving balance equations for multi-component systems, especially, when a balance equation of the overall aggregate has to be computed from the sum of the balance equations of the constituents. In addition, Kelly formulated angular-momentum balances for the components and allowed for a moment-of-momentum production term. As a result, the partial stress tensors of the components turned out to be non-symmetric, although the overall stress obtained from the partial stresses was symmetric. With Kelly's publication, the continuum-mechanical frame of mixtures was built apart from the formulation of a sound version of the entropy inequality.

At this point, one might only hypothesize, why Truesdell and Toupin refrained from formulating an angular-momentum balance for mixtures. The inclusion of an angular-momentum production yielding non-symmetric partial stresses in the sense of the Cosserat brothers [37, 38] is somehow contradictory to the fact that the whole system has symmetric stresses in the sense of a standard Cauchy continuum. To elucidate this point, consider a liquid-saturated porous solid, where a homogenization over the solid and fluid microstructures yields the constituent balance equation of mixture theories. If both the solid and the fluid are Cauchy continua on their microstructures, the homogenization will result in partial continua with symmetric stresses and vanishing angular-momentum productions. As a result, no angular-momentum productions are necessary. From a formal point of view, the introduction of these terms can be seen, in principle, as a possibility to homogenize the construction of general balance equations.

Based on Truesdell's mixture theory, Adkins [39-41] and Green and Adkins [42] developed first purely mechanically motivated approaches for mixtures of fluids and for a mixture of a single fluid and an elastic solid. Although these models have been subjected to certain invariance criteria obtained from the "principle of objectivity", a thermodynamic investigation of the constitutive equations for mixtures was missing and had only been introduced to standard continua by Coleman and Noll [43].

With respect to the procedure of Green and Adkins [42], when they formulated constitutive equations for mixtures, Truesdell and Noll commented in "The Non-Linear Field Theories of Mechanics" [34, p. 541]:

"The foregoing summary makes clear that the rational theory of diffusion is in its infancy, with many possibilities awaiting search. The uncertainty as to even the overall pattern for the right constitutive equations, ... , suggest that some basic principle of the subject remains to be discovered. ... The "missing principle", surely, is a proper generalization of the Clausius-Duhem inequality."

First attempts to treat the constitutive theory of mixtures in the frame of thermodynamics have been worked out by Eringen and Ingram [44] and by Green and Naghdi [45]. For this purpose, it was necessary to transfer the entropy principle from single to heterogeneous media. The completely different procedure of Eringen and Ingram and of Green and Naghdi, respectively, reveals how difficult it was in those days to find the right formulation of the principle. While Eringen and Ingram assumed that each constituent of a mixture with different constituent temperatures is associated with an individual entropy inequality, Green and Naghdi made the point that there could only be one single entropy inequality for the whole medium. Also the basic procedure by Green and Naghdi in formulating only one single entropy inequality for the whole mixture proved to be right, while the assumption of Eringen and Ingram cannot be accepted from a modern point of view, their result, which they postulated again in 1967 [46], was wrong. Bowen and Wiese later pointed out in 1969 [47] that the theory of Green and Naghdi was lacking the free-energy transport produced by the diffusion process.

The application of Green and Naghdi's basic version of the entropy inequality for mixtures quickly proved to be too specific to be used as a basis for a general theory of mixtures. In 1967, Bowen [48] observed also with respect to his first own attempts to formulate constitutive equations for mixtures

> "that this procedure led to the result that, in equilibrium, the partial free energy density of a given constituent is independent of the deformations of the other constituents in the mixture. ... Such independence fails to be confirmed by experiments on fluid mixtures."

Bowen believed that the mistake of the existing theories could be seen in the fact that they have been based on partial stresses instead of chemical potentials as  they are used in chemistry and that, as a result, the existing theories could only describe ideal mixtures. Therefore, he formulated his version of a mixture theory on the basis of so-called tensors of chemical potentials, thus extending the notion "chemical potential", originally introduced by Gibbs as a scalar. The entropy inequality formulated by Bowen later on proved as the first fully correct version of an entropy inequality for mixtures. This inequality is basically equivalent to that of single continua, except for the fact that Bowen introduced, in addition to the usual heat flux, an energy flux motivated by the diffusion process [48, inequality (3.3)]. The chemical-potential tensors and their relation to the partial stresses could be found through constitutive equations. Green and Naghdi criticized Bowen's work by the argument that the introduction of tensors of chemical potentials instead of partial stresses would not lead to a basically different theory; an argument that has been accepted by Bowen in 1969 [47].

Concerning the development of the entropy inequality, Truesdell developed in 1968 his own version based on the assumption that the entropy inequality of the mixture must be obtained as the sum of the entropy inequality of the components [32] in analogy to the classical continuum mechanics of single-component materials. Truesdell's result was identical with Bowen's entropy postulate. As a result, the commonly accepted form of the entropy inequality of mixtures was called Bowen-Truesdell form.

Lots of papers followed. Around 1970, the basis of a general "Theory of Mixtures" was found. Insofar, it is not astonishing that Eringen's collection on *Continuum Physics* in four volumes, published between 1971 and 1976, contained Bowen's famous *Theory of Mixtures* [5], the basis for any modern approach to porous media. On the basis of mixture theories, it was Bowen himself who published on incompressible and compressible porous media models in 1980 and 1982 [6, 7], where he extended the "Theory of Mixtures" by the concept of volume fractions.

With Bowen's articles, porous media theories was split in two directions, the one following the Terzaghi-Biot line and the other following Bowen's line which can be seen as following the old ideas of Fillunger and Heinrich, although Bowen did not take notice of their articles.

Given a sound frame for the "Theory of Porous Media" does not necessarily imply that one could find constitutive equations for various porous-media problems. Especially, the application of the five basic principles for construction of constitutive equations "determinism", "equipresence", "local action", "material frame indifference" and "dissipation" formulated by Truesdell, Noll and Coleman between 1949 and 1963 led to some confusion, when the components of the mixture were treated like standard first-grade materials. It has been found by Müller in 1968 [49] that mixture components always have to be treated as materials of second grade, if one does not want to describe only simple mixtures. In particular, Müller's finding includes, for example, that, if a solid skeleton is part of a mixture, the constitutive setting does not only depend on the solid deformation gradient $\mathbf{F}_S$ alone but also on the second deformation gradient $\text{Grad}_S \mathbf{F}_S$ following the skeleton motion. Although this concept is conductive, it leads to considerable confusion when complicated aggregates are investigated.

To avoid these problems, the author introduced in 1989 [50] the concept of phase separation based on the idea that a component of a porous medium like the solid skeleton is only part of the multi-component aggregate after homogenization over its microstructure. Since the deformation of a single solid is only governed by $\mathbf{F}_S$ corresponding to the first Piola-Kirchhoff stress $\mathbf{P}^S$ in the sense of energetically conjugate variables and not by $\text{Grad}_S \mathbf{F}_S$ as far as there is no corresponding stress gradient, this concept includes the assumption that a standard constituent of the porous-media aggregate is governed by standard variables. Only the interaction terms like the direct momentum production, which can be interpreted as the local interaction force between the constituents, obviously depend on the sum of the constitutive variables of the interacting components. However, when real mixtures are concerned, Müller's concept has to be fully applied, also compare [2], where solids with pore-fluid mixtures are investigated.

5 Porous Media Today

In recent time, geotechnical applications like the investigation of the deformation and stability behavior of fully and partially fluid-saturated soil constructions, such as dikes, embankments, railroad dams, or foundation and settlement problems have to be computed on the basis of the Theory of Porous Media. But not only in civil and environmental engineering, porous media problems occur. Also in mechanical engineering, porous media come into play, for example, when foamed materials have to be considered or when smart materials like electro-active polymers have to be computed. Figure 7 shows a smart gripper controlled by the application of electric potentials. The result of this computation as well as all other computations presented in this article has been carried out by use of PANDAS[1], a finite-element solver for the monolithic computation of strongly volumetrically coupled problems.

---

[1] **P**orous Media **A**daptive **N**on-linear Finite-Element solver based on **D**ifferential-**A**lgebraic **S**ystems.

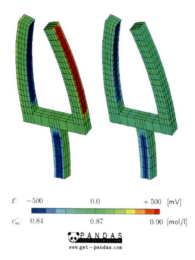

Fig. 7 Motion of a gripper made from an electro-active material (hydrogel), left: electrical potential; right: anion concentration

Another field, where porous media play a dominant role, is the whole area of biomechanics. For example, if one is interested in the investigation and the description of the overall human, one must be aware that living biological tissue contains more than 90% of fluids, interstitial fluid and blood. To set an example, Figure 8 exemplarily exhibits the deformation of an intervertebral disk of the human lumbar spine during flexion.

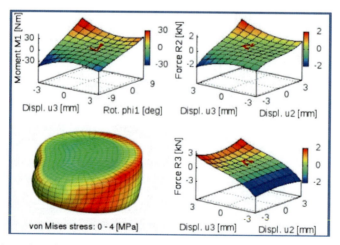

Fig. 8 Deformation of an intervertebral disk and bending moment and forces in vertical and horizontal direction during flexion

6 Final Remarks

The examples show that porous-media mechanics is a very interesting field with an increasing importance for the solution of complex coupled problems including solid mechanics, fluid mechanics, thermodynamics, computational mechanics, etc. The present historical view on this field can obviously only highlight a particular segment of the overall history of porous-media research. Further impressions can be found in [51] (in German) and in various articles by Reint de Boer on the early days of this topic.

Finally, I must confess that I am convinced that the field of porous media is and will be a very important matter of research for interesting and future-oriented investigations.

Acknowledgement. All included portraits are taken from Wikipedia.

References

[1] Ehlers, W.: Foundations of multiphasic and porous materials. In: Ehlers, W., Bluhm, J. (eds.) Porous Media: Theory, Experiments and Numerical Applications, pp. 3–86. Springer, Berlin (2002)

[2] Ehlers, W.: Challenges of porous media models in geo- and biomechanical engineering including electro-chemically active polymers and gels. International Journal of Advances in Engineering Sciences and Applied Mathematic 1, 1–24 (2009)

[3] de Boer, R.: Theory of Porous Media. Springer, Berlin (2000)

[4] Truesdell, C.: Thermodynamics of diffusion. In: Truesdell, C. (ed.) Rational Thermodynamics, 2nd edn., pp. 219–236. Springer, New York (1984)

[5] Bowen, R.M.: Theory of Mixtures. In: Eringen, A.C. (ed.) Continuum Physics, vol. III, pp. 1–127. Academic Press, New York (1976)

[6] Bowen, R.M.: Incompressible porous media models by use of the theory of mixtures. International Journal of Engineering Science 18, 1129–1148 (1980)

[7] Bowen, R.M.: Compressible porous media models by use of the theory of mixtures. International Journal of Engineering Science 20, 697–735 (1982)

[8] Woltman, R.: Beyträge zur Hydraulischen Architektur, Dritter Band, Dietrich, Göttingen (1794)

[9] Delesse, A.: Procédé mécanique pour déterminer la composition des roches. Annales des Mines 4. séries 13, 379–388 (1848)

[10] Darcy, H.: Les fontaines publiques de la ville de Dijon, Dalmont, Paris (1856)

[11] Fick, A.: Ueber Diffusion. Annalen der Physik und Chemie 94, 59–86 (1855)

[12] J. Stefan: Über das Gleichgewicht und die Bewegung, insbesondere die Diffusion von Gasgemengen. Sitzungsberichte der Kaiserlichen Akademie der Wissenschaften (Wien), mathematisch-naturwissenschaftliche Klasse. Abteilung IIa 63, 63–124 (1871)

[13] Jaumann, G.: Geschlossenes System physikalischer und chemischer Differerentialgesetze. Sitzungsberichte der Kaiserlichen Akademie der Wissenschaften (Wien), mathematisch-naturwissenschaftliche Klasse. Abteilung IIa 120, 385–530 (1911)

[14] Clausius, R.: Ueber die bewegende Kraft der Wärme und die Gesetze, welche sich daraus für die Wärmelehre selbst ableiten lassen. Poggendorfs Annalen der Physik und Chemie 79, 368–397 (1850)
[15] Clausius, R.: Ueber die veränderte Form des zweiten Hauptsatzes der mechanischen Wärmetheorie. Annalen der Physik und Chemie 18, 481–506 (1854)
[16] Rankine, W.: On the stability of loose earth. Philosophical Transactions of the Royal Society London 147, 9–27 (1857)
[17] von Terzaghi, K.: Die Berechnung der Durchlässigkeitsziffer des Tones aus dem Verlauf der hydrodynamischen Spannungserscheinungen. Sitzungsberichte der Akademie der Wissenschaften (Wien), mathematisch-naturwissenschaftliche Klasse. Abteilung IIa 132, 125–138 (1923)
[18] von Terzaghi, K.: Erdbaumechanik auf bodenphysikalischer Grundlage. Franz Deuticke, Leipzig-Wien (1925)
[19] von Terzaghi, K., Fröhlich, O.K.: Theorie der Setzung von Tonschichten. Franz Deuticke, Leipzig-Wien (1936)
[20] Fillunger, P.: Der Auftrieb in Talsperren. Österreichische Wochenschrift für den öffentlichen Baudienst 19, 532–556, 567–570 (1913)
[21] de Boer, R.: The Engineer and the Scandal – A Piece of Science History. Springer, Berlin (2004)
[22] Heinrich, G.: Wissenschaftliche Grundlagen der Theorie der Setzung von Tonschichten. Wasserkraft und Wasserwirtschaf 33, 5–10 (1938)
[23] Biot, M.A.: Le problème de la consolidation des matiéres argileuses sous une charge. Annales de la Société scientifique de Bruxelles 55(B), 110–113 (1935)
[24] Biot, M.A.: General theory of three-dimensional consolidation. Journal of Applied Physics 12, 155–164 (1941)
[25] Biot, M.A.: Theory of elasticity and consolidation for a porous anisotropic solid. Journal of Applied Physics 26, 182–185 (1955)
[26] Biot, M.A.: General solutions of the equations of elasticity and consolidation for a porous material. Journal of Applied Mechanic 23, 91–96 (1956)
[27] Biot, M.A.: Theory of propagation of elastic waves in a fluid-saturated porous solid, I. Low frequency range. Journal of the Acoustical Society of America 28, 168–178 (1956)
[28] Fillunger, P.: Erdbaumechanik? Selbstverlag des Verfassers, Wien (1936)
[29] Heinrich, G., Desoyer, K.: Theorie dreidimensionaler Setzungsvorgänge in Tonschichten. Ingenieur-Archiv 30, 225–253 (1961)
[30] de Boer, R., Ehlers, W.: A Historical Review of the Formulation of Porous Media Theories. Acta Mechanic 74, 1–8 (1988)
[31] Truesdell, C.A.: Sulle basi della termomeccanica. Rendiconti Lincei 22, 33–38 (1957)
[32] Truesdell, C.A.: Sulle basi della termodinamica delle miscele. Rendiconti Lincei 44, 381–383 (1968)
[33] Truesdell, C.A., Toupin, R.A.: The classical field theories. In: Flügge, S. (ed.) Handbuch der Physik, vol. III(1), pp. 226–902. Springer, Berlin (1960)
[34] Truesdell, C.A., Noll, W.: The nonlinear field theories of mechanics. In: Flügge, S. (ed.) Handbuch der Physik, vol. III(3), Springer, Berlin (1965)
[35] Truesdell, C.A.: Rational Thermodynamics. McGraw-Hill, New York (1969)
[36] Kelly, P.D.: A reacting continuum. International Journal of Engineering Science 2, 129–153 (1964)
[37] Cosserat, E., Cosserat, F.: Sur la mécanique générale. Comptes Rendus de l'Académie des Sciences 145, 1139–1142 (1907)

[38] Cosserat, E., Cosserat, F.: Théorie des Corps Déformables. A. Hermann & Fils, Paris (1909)
[39] Adkins, J.E.: Non-linear diffusion, I. Diffusion and flow of mixtures of fluids. Philosophical Transactions of the Royal Society of London, Series A 255, 607–633 (1963)
[40] Adkins, J.E.: Non-linear diffusion, II. Constitutive equations for mixtures of isotropic fluids. Philosophical Transactions of the Royal Society of London, Series A 255, 635–648 (1963)
[41] Adkins, J.E.: Diffusion of fluids through aelotropic highly elastic solids. Archive for Rational Mechanics and Analysis 15, 222–234 (1964)
[42] Green, A.E., Adkins, J.E.: A contribution to the theory of non-linear diffusion. Archive for Rational Mechanics and Analysis 15, 235–246 (1964)
[43] Coleman, B.D., Noll, W.: The thermodynamics of elastic materials with heat conductionand viscosity. Archive for Rational Mechanics and Analysis 13, 167–178 (1963)
[44] Eringen, A.C., Ingram, J.D.: A continuum theory of chemically reacting media. International Journal of Engineering Science 3, 197–212 (1965)
[45] Green, A.E., Naghdi, P.M.: A dynamical theory of interacting continua. International Journal of Engineering Science 3, 231–241 (1965)
[46] Green, A.E., Naghdi, P.M.: A theory of mixtures. Archive for Rational Mechanics and Analysis 24, 243–263 (1967)
[47] Bowen, R.M., Wiese, J.C.: Diffusion in mixtures of elastic materials. International Journal of Engineering Scienc 7, 689–722 (1969)
[48] Bowen, R.M.: Toward a thermodynamics and mechanics of mixtures. Archive for Rational Mechanics and Analysis 24, 370–403 (1967)
[49] Müller, I.: A thermodynamic theory of mixtures of fluids. Archive for Rational Mechanics and Analysis 28, 1–39 (1968)
[50] Ehlers, W.: On thermodynamics of elasto-plastic porous media. Archives of Mechanics 41, 73–93 (1989)
[51] Ehlers, W.: Poröse Medien – ein kontinuumsmechanisches Modell auf der Basis der Mischungstheorie, Kapitel 2: Zur Geschichte der Mischungstheorie und der Theorie poröser Medien. Report No. II-21 of the Institute of Applied Mechanics, University of Stuttgart 2011 (Reproduction of "Forschungsberichte aus dem Fachbereich Bauwesen 47, Universität-GH-Essen 1989")

Parameter Identification in Continuum Mechanics: From Hand-Fitting to Stochastic Modelling

Rolf Mahnken

Abstract. After a brief review on the development of three mathematical models in the history of mechanics, we outline today's role of parameter identification within the process of model building. On this basis, parameter identification is illustrated as a direct method by hand-fitting for some simple mathematical structures from the early stages of mathematical modeling. The beginning of so called "advanced constitutive modeling" in the sixties of the twentieth century rendered parameter identification as a least-squares problem. For its solution evolution strategies and more efficient optimization algorithms have been introduced by several researchers. After addressing these developments, we will also outline some newer aspects such as inhomogeneous field problems and stochastic methods.

1 Model Building in the History of Mechanics

1.1 Introduction

In the earliest time understanding and interpretation of physical phenomena was based on experience and judgement. Examples are empirical rules by the Egyptians to build their temples and pyramids and the universal preference of semicircular arches with comparatively small span by the Romans, although unfavourable with respect to weight, [27]. Even in the fifteenth century, when Leonardo da Vinci elaborated on the strength of structural materials, craftsmen continued to fix the dimensions of structural elements according to rules of experience, [27]. Also, artillerymen and gunners continued to predict the trajectory of a cannonball in the spirit of Aristotle teaching. Mathematical modeling played hardly any role in those times.

Rolf Mahnken
University of Paderborn, Chair of Engineering Mechanics (LTM), Warburger Str. 100,
D-33098 Paderborn, Germany
e-mail: mahnken@ltm.upb.de

From the seventeenth century onwards experimental and theoretical methods were merged as a basis for reliable model building. The former produces experimental data by observation and measurement. The latter uses the branches of mathematics and logic for predictive results. Repeated attempts in both disciplines and its interaction renders the process of model building schematically illustrated as:

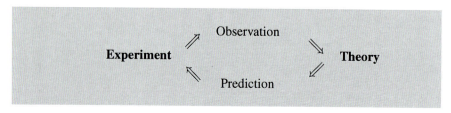

As an ultimate goal of the above process the gap between the outcome of both disciplines should be closed. However, this can became a laborious and lengthy task, as illustrated by the following three examples from the history of mechanics.

1.2 Experiment and Theory on Newton's Law of Gravity

One of the most famous achievements of classical mechanics is Newton's law of universal gravitation. He derived it from empirical observations by what he called induction and formulated it in his work *Philosophie Naturalis Principia Mathematica* ("the Principia"), first published on 5 July 1687, as follows: "I deduced that the forces which keep the planets in their orbs must [be] reciprocally as the squares of their distances ....", as illustrated in Fig. 1. Newton himself said that the fall of an apple from a tree inspired him to formulate his theory of gravitation. But there is no confirmation by himself on the more popular version of the fall on his head.

In modern notation the law reads

$$F = G\frac{Mm}{r^2}, \qquad (1)$$

where F is the force between the masses, G is the gravitational constant, M is the first mass, m is the second mass, and r is the distance between the centers of the masses.

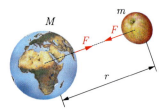

Fig. 1 Newton's law of gravity

However, Newton could not apply his law to predict the trajectories of any planets, since he was unable to determine the gravitational constant G in Eq. (1). This left his law as a hypothesis with no verification by experiments. However, any critics he retorted with saying: "Hypotheses non fingo", ("I do not invent hypothesis"), [24].

The determination of the gravitational constant G based on experiments – shortly: parameter identification – took place 111 years after the publication of Newton's "Principia" by the British scientist Henry Cavendish in 1798.

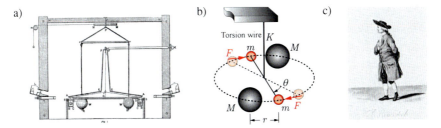

Fig. 2 Torsion balance instrument by (Mitchell-) Cavendish (1798), a) Cross section of laboratory, b) Schematic of torsion balance apparatus (reproduced from [31]), c) H.Cavendish (1731 - 1810)

The principle of the (Mitchell-)Cavendish experiment is illustrated in Fig. 2. Fig. 2.a shows a cross section of the laboratory, Fig. 2.b provides a schematic illustration of the torsion balance apparatus An extensive description of the experimental setup and the theoretical conception is described by Cavendish himself in [5]. In the first place he was interested on the average *density of the earth*, which he expressed as a specific gravity, or a ratio of the earth's density to the density of water. Additionally, as a side result he succeeded in determination of G as

$$G = \frac{F \cdot r^2}{m \cdot M} = 6.6673 \, 10^{-11} \frac{m^3}{kg \cdot s^2}. \qquad (2)$$

In this way, by experimental determination Cavendish was able to close the gap between theory and experiment in Newton's law of universal gravitation, which till today constitutes a firm part of classical mechanics.

1.3 Galilei's Experiments and Theories on Strength of Materials

In his famous book *Two New Sciences* Galileo Galilei contributed on experimental and theoretical investigations for structural elements. His work on the new science field *strength of materials* includes the two experiments illustrated in Fig. 3. As an experimental finding for the tension test he stated, that the strength of a bar is proportional to its cross-sectional area and is independent of the length, [27], p.11, which is well established today. Galileo also was the first to perform experiments on the strength of a cantilever beam, however, with conclusions not accepted by today's knowledge. For instance, he postulated a uniform distribution of the "resistance" (in today's term the stress) over the cross section BA. Furthermore, Galileo did not distinguish an elastic from a plastic regime prior to failure. As a result Galileo's theory gives a value three times larger than a breaking load on the basis of Hooke's law. As a further deficiency, Galileo's theory did not differentiate between materials, and therefore no material constants were involved.

Despite the shortcomings, Galileo inspired various ensuing research activities, e.g. by Mariotte (1620-1684), Leibniz (1646-1716), Jacob Bernoulli (1654-1705)

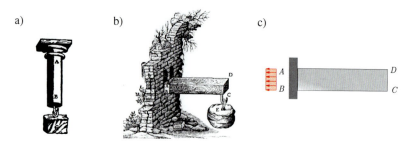

Fig. 3 Strength experiments by Galilei: a) tensile test, b) cantilever beam, c) assumption on "resistance"

and in particular Euler (1707-1783), who firstly introduced Young's modulus as a material constant in 1727. Nowadays, by means of limit load analysis plastic methods of structural analysis are common practice engineering, see e.g. [13].

1.4 Experiments and Theories on the Elastic Constant Controversy

Another historical example, where theory firstly exhibited a broad gap to experiment is given by the elastic constant controversy in the eighteenth and nineteenth century. The early stages of the theory of elasticity included a hypothesis regarding the behavior of the molecular structure of elastic bodies. Navier assumed in 1824, that an ideal elastic body consists of molecules between which forces appear on deformation. These forces are proportional to changes in the distances between the molecules and act in the directions of lines joining them, [27]. As a consequence he obtained equations of equilibrium of a particle for an isotropic, linear elastic material by using only *one* elastic constant. The next important step in the theory of elasticity was made by Cauchy (1789-1857) with the introduction of strains and stresses. Although, in the beginning he introduced two constants, he became a supporter of a one-constant theory, [27]. Concurrently, Poisson (1781-1840) claimed that for simple tension of a prismatic bar, the axial elongation ε must be accompanied by a lateral contraction with factor $\mu = 0.25$. The concept that elastic properties of an isotropic body can completely be defined by one constant (say the modulus in tension E) was generally accepted in the early stages of the development of elasticity. According to [27] well known scientists like Navier, Cauchy, Poisson, Lamé and Clapeyron, all agreed to this.

However, further theoretical and experimental investigations deviated from the one-constant theory. In 1859 Lamé published a book on elasticity, where he no longer favors the use of Navier's derivation involving molecular structures and molecular forces of elasticity, [27, p.118]. Furthermore, Kirchhoff used circular steel cantilevers in his experiments, so that bending and torsion were produced simultaneously. From these tests Kirchhoff found that Poisson's ratio for steel is 0.294 and for brass he obtained the value 0.387, [27, p.222]. These theoretical and experimental results were incompatible with the uni-constant hypothesis for isotropic bodies and instead proved the need for a two-constant theory.

2 Today's Role of Parameter Identification in Model Building

A comparative look on the above historical examples reveals the different conceptions to close the gap between theory and experiment: Newton started his law on gravitation with theory, but he was not able to predict any experiments, which was done only 111 years later by Cavendish. Galilei's investigations on the cantilever beam started with experimental findings, however, a concise theory including material constants was provided only later by others, especially Jacob Bernoulli and Euler. The initial theory of elasticity by Navier, using a hypothesis on the molecular structure, failed to show a satisfying agreement to various experimental data. With the introduction of stresses and strains by Cauchy and the acceptance of a two-constant theory the gap to experiment finally could (almost) be closed.

The above historical examples demonstrate, that experimental data are imperative for realistic model building. Furthermore, the sometimes laborious process on the theoretical part requires *identification* of the correct mathematical model. It can be split into *structure identification*, thus providing the mathematical structure of the model and *parameter identification*, thus providing the constants, which enter it. Concentrating on the latter, the process as accepted nowadays reduces as follows:

$$
\begin{array}{ccc}
 & \text{Data } \bar{d} & \Longrightarrow \quad \begin{array}{c}\text{Comparison}\\ \bar{d} \text{ and } d(\kappa)\end{array} \\
\textbf{Experiment} & & \Downarrow \qquad\qquad\qquad \textbf{Theory}\\
 & \text{Data } d(\kappa) \Longleftarrow \text{Parameters } \kappa \Longleftarrow \begin{array}{c}\text{Mathemati-}\\ \text{cal structure}\end{array}
\end{array}
$$

In this way, today's role of parameter identification is based on the fundamental disciplines of experiment and theory. However, we keep the experimental data $\bar{d}$ and the mathematical structure fixed, and vary only the material parameters κ iteratively until a certain criteria for the simulated data $d(\kappa)$ is satisfied. In case the criteria cannot be satisfied, the model structure has to be revised.

The above mentioned general principle applies to several disciplines in science and engineering, such as dynamics, fracture mechanics, contact mechanics, mechatronics, biomechanics etc. In the following specific attention is paid to parameter identification for constitutive modeling in continuum mechanics. We address various historical topics within this research field and outline also some newer aspects such as inhomogeneous field problems and stochastic modeling.

3 The "Hand Fitting" Method

At the early stages in various physical fields it was observed, that the mathematical structure for input and output variables of the physical process can be illustrated as linear lines in a diagram. The related proportional constants are then nothing else as the sought material parameters, and consequently can be obtained by "hand fitting" in the diagram. In the sequel we give some prominent examples from the early stages in some physical fields.

3.1 One-Parameter Models

We denote by a and b the input- and output quantities of a physical process (the reverse could also be used) as illustrated in Fig. 4. Then in several physical processes they obey the following relations:

Fig. 4 One-parameter model

$$1.\ b \propto a \quad \Longrightarrow \quad 2.\ b = Ca, \qquad (3)$$

that is, there is *one* proportional constant C, which is the material parameter of the state equation (3.2). In Fig.5 three well known examples from mechanics are summarized, relating a) stress σ and strain ε in Hooke's law on elasticity, b) stress σ and strain rate $\dot{\varepsilon}$ in Newton's law on viscosity and c) frictional force R and normal force N in Coulomb's law on friction. Drawing an approximate line through the points of experimental data, represented by the star symbols, reveals the sought material parameter as its slope.

Note, that Hooke himself formulated his law on elasticity in 1678 for springs in the form $F = f \Delta l$, between force F and elongation Δl with stiffness f. The material constant E was introduced more than 130 years later by Navier in 1828, [25], p. 359. Similarly, Newton's law of viscosity originally has been formulated between force and displacement rate.

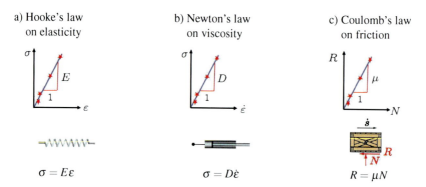

Fig. 5 Three examples of one-parameter models

3.2 One-Parameter Models of Gradient-Type

We denote by a a scalar variable and $\mathbf{b}$ a vector variable, as input- and output quantities of a physical process (the reverse could also be used). Then for several physical processes they obey the following relations:

$$\mathbf{b} \propto \nabla a \implies \mathbf{b} = C \nabla a, \quad (4)$$

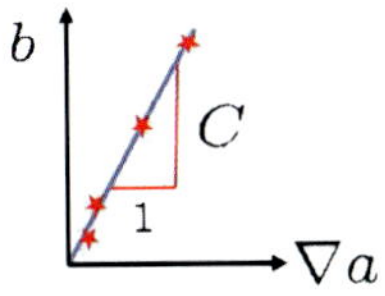

Fig. 6 One-parameter model between $\mathbf{b}$ and gradient of a

where in contrast to Eq. (3) $\mathbf{b}$ is proportional to the spatial gradient of the quantity a, expressed by the material constant C.

A summary of one-parameter models of gradient type is given in Table 1. Here, by a slight misuse of notation we also write Hooke's law in the general structure of Eq. (4), where more precisely b, ∇a and respectively C would be tensors of second and respectively fourth order.

Table 1 Summary of one-parameter models of gradient-type

	a	b	C
Hooke	displacement	stress	elasticity
Fourier	temperature	heat flux	heat conductivity
Darcy	pressure	seepage velocity	hydraulic permeability
Gauss	electric potential	electric displacement	electric permeability
Ohm	electric potential	electric current	electric conductivity
Fick	concentration	diffusion flux	diffusion coefficient

3.3 Power-Laws

We denote by a and b the input- and output quantities of a physical process (the reverse could also be used). Then in some cases they obey the following relations:

$$(\log b - \log C) \propto \log a \implies b = C a^n, \quad (5)$$

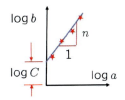

Fig. 7 Power law

where in contrast to Eq. (3) and Eq. (4) the difference $\log b - \log C$ is proportional to the quantity $\log a$, expressed by the material constants C and n.

Two well known examples of power-laws are Norton's law on secondary creep in visco-plasticity and Paris-law on stationary crack growth in fracture mechanics in Fig. 8. Drawing a line in the related diagram renders directly the sought material constants.

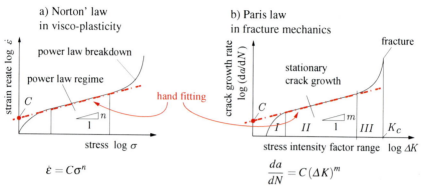

Fig. 8 Examples of power-laws: a) Visco-plasticity, b) Crack growth

3.4 Hand Fitting for "Modulus of Elasticity" in Soil Mechanics

In 1925 Therzaghi published experimental results in [26] within the research field of soil mechanics in order to investigate the compressive strength of clay. In particular he was interested in the elastic properties of clay cubes in dependence of loading.

To this end he performed experimental results on cubes with sizes of 2 cm and 4 cm, respectively. Fig.9 shows the resulting stress-strain diagram, where to be more specific, stress is the pressure (kg/cm^2), and strain is the ratio between the total compression and the reduced height. Furthermore, Therzaghi stated that "The term modulus of elasticity should be confined to the reversible part .... It is equal to the tangent of the angle between the strain axis and the axis of the hysteresis loop". In this way, Therzaghi obtained the modulus simply by hand fitting as indicated in Fig. 9. Furthermore, he established a relation between the intensity of the capillary pressure and the modulus of elasticity.

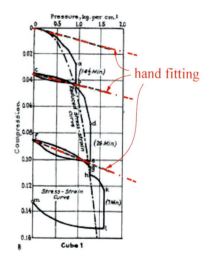

Fig. 9 Strain stress curve for cyclic loading on clay

3.5 Visco-plasticity

The basic rheological elements in Fig. 5, the Hooke element on elasticity and the Newton element on viscosity may be combined to obtain further classical elements of visco-elasticity. In Fig. 10 we distinguish the Maxwell element (J. C. Maxwell, 1831 - 1879), the Kelvin element (Lord Kelvin, 1824 - 1907), the Poynting element, (J. H. Poynting, 1852 - 1914) and the Burgers element (J. M. Burgers, 1895 - 1981).

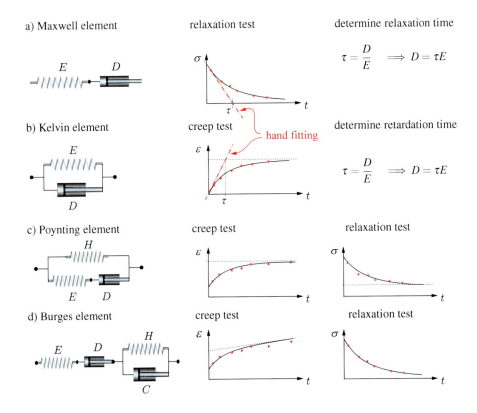

Fig. 10 Four examples of rheological models

Interpretation of the outcome for the combined elements in Fig. 10 can be used to determine some related material parameters by hand-fitting. E.g. drawing a tangent at the initial slope to the stress-time diagram of a relaxation test in Fig. 10.a renders the relaxation time τ graphically. Then with known Young's modulus the damping constant is obtained according to $D = \tau \cdot E$. Analogously, drawing a tangent at the initial slope to the strain-time diagram of a creep test in Fig. 10.b renders the retardation time τ graphically. Again, with known Young's modulus the damping constant is obtained according to $D = \tau \cdot E$. The hand-fitting method applied to the Poynting and the Burgers element (not shown here) becomes more tedious, and thus reveals the limitation of the method.

3.6 Conclusions on the Hand-Fitting Method

The hand-fitting method as a direct method has its attractiveness, when material parameters can directly be related to experimental test curves. For more complex situations it is possible to extend hand fitting in a systematic manner. Basically it involves 1. physical and phenomenological interpretations of the constants in relation to standard uniaxial *ideal* tests, 2. to distinguish between different effects, such as kinematic hardening, recovery and damage separately in the experimental data, 3. certain ad hoc assumptions (e.g. neglecting elastic parts of strains), 4. sequential parameter determination, i.e. data from one test are used to evaluate one or two parameters. These in turn are used as input, along with data from other tests, in order to evaluate more parameters.

Following [21] the hand fitting has the drawbacks, that 1. the ideal test conditions often cannot be realized in the laboratory, e.g. instantaneous change in inelastic strain rate, 2. as a consequence of the sequential parameter determination the results may depend upon the order in which they are determined, 3. the associative ad hoc assumptions may be unrealistic.

4 Optimization

4.1 The Least-Squares Problem

Due to the limitations of the rheological models in the previous section and also due to the improving capabilities of computer resources during the sixties and the seventies of the 20th century, there started an increasing research interest for improved modelling in visco-plasticity. Some of the popular approaches in this new research field, in those times sometimes referred to *advanced modelling*, are done by the following researches: Perzyna (1963) (rate dependency), Bodner/Partom (1972) (rate and temperature dependency), Hart/Miller (1976) (rate and temperature dependency), Chaboche (1977) (isotropic and kinematic hardening), Robinson (1978) (isotropic and kinematic hardening), Johnson Cook (1983) (rate and temperature dependency), Steck (1985) (high temperature dependency), Choi/Krempl (1985) (rate and temperature dependency), Leblond (1989) (Phase Transformation), Voyiadjis (1991) (ratcheting), Ohno/Wang (1993) (ratcheting), McDowell (1995) (ratcheting), etc.

The complexity of these approaches revealed also the limitations of the hand-fitting method. Consequently, identification has been defined as a least–squares problem, in order to minimize the distance between the experimental data $\tilde{d}_i$ to the simulated data $d_i(\kappa)$, with respect to a vector of material parameters κ for $i = 1,...n_d$ data points: It is formulated as

Parameter Identification in Continuum Mechanics

$$\text{Find } \kappa^* \in \mathcal{K} : f(\kappa) := \frac{1}{2} \sum_{i=1}^{n_d} \left(d_i(\kappa) - \tilde{d}_i \right)^2 \longrightarrow \min_{\kappa}. \tag{6}$$

The least-square method has been introduced by Carl Friedrich Gauss in 1795 at the age of eighteen. Eq. (6) constitutes an optimization problem, which requires suitable algorithms for its solution, discussed next.

4.2 The Evolution Strategy

The Evolution Strategy (ES) is an optimization method firstly developed in the early 1960s and refined by Rechenberg and Schwefel [16, 20, 17]. The basic idea of ES is exploiting the ideas of adaptation and evolution. As illustrated in Fig. 11, it allows the interpretation a) of the evolution of mankind, b) of natural selection, according to Darwin's principle of evolution or, c) according to a theme in the Cinderella folk tale, "The good ones go into the pot, the bad ones into your crop".

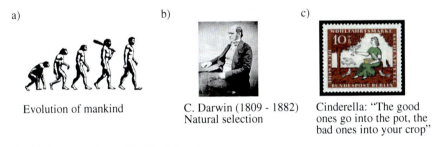

a) Evolution of mankind

b) C. Darwin (1809 - 1882) Natural selection

c) Cinderella: "The good ones go into the pot, the bad ones into your crop"

Fig. 11 Interpretations of the Evolution Strategy

The different versions of the ES are distinguished by the nomenclature $(1+1)$-ES, $(1+\lambda)$-ES, $(1,\lambda)$-ES, $(\mu/\rho,\lambda)$-ES. Here μ denotes the total number of parents, $\rho \leq \mu$ the number of parents, which will be recombined, and λ is the number of offspring. Selection takes place deterministically (i.e., deterministic survivor selection) only among the offspring (komma notation) or among the offspring and parents together (plus notation).

One of the earliest applications of the Evolution Strategy can be found within the german Collaborative Research Centre 319 (Sonderforschungsbereich 319, SFB 319) supported by the german research foundation (Deutsche Forschungsgemeinschaft). Fig. 12.a gives a schematic sketch of the Evolution Strategy and Fig. 12.b shows a comparison between simulation with the Chaboche model and experiment for cyclic loading for an alloy AlMg, [23].

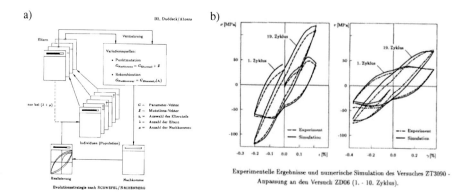

Fig. 12 Evolution Strategy in viscoplastic modeling of Subproject B2 (Duddeck/Ahrens) within the german SFB 319: a) Schematic sketch of the method, b) Comparison between simulation and experiment, taken from [23]

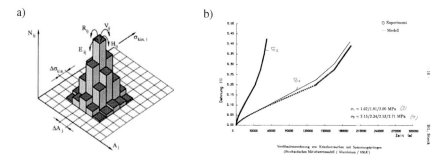

Fig. 13 Evolution Strategy for a stochastic model of Steck: a) Interpretation of the Markov chain, b) Comparison between simulation and experiment, taken from [23]

Fig. 13 shows results obtained within the SFB 319 for a stochastic viscoplastic model accounting for the interaction of plasticity and creep in metals on the basis of Marcov-Chains. Fig. 13.a gives a schematic sketch of the interacting Marcov-process, and Fig. 12.b shows a comparison between simulation and experiment, [23]. A further application of the Evolution Strategy for parameter identification of visco-plastic modelling for concrete has been done in the dissertation of Wedemeier [29] in 1990. Representative results are given in Fig. 14.

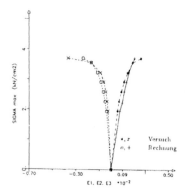

Fig. 14 Evolution strategy for material modeling of concrete, taken from [29]

4.3 Comparative Strategies for Optimization

Although the Evolution Strategy has proved to be very reliable for parameter identification, the long computational time is a main deficiency of the method. This created the need for more efficient optimization methods in parameter identification, however, the debate on the most effective method goes up to the present.

Algorithms for solution of the optimization problem (6) may be grouped into methods, where only function evaluations are needed (zero-order methods) and where additionally its gradients are required (first-order methods). Another classification into deterministic and stochastic methods is possible. Examples of gradient-based deterministic methods are the Gauss-Newton method, the Levenberg-Marquard method or the BFGS method, see e.g. [6, 9]. An example of a gradient free deterministic method is the Simplex method, see e.g. [14]. Examples for stochastic methods are the Monte-Carlo method, the Evolution strategy, the method of Price and the method of Müller, Nollau, see e.g. [20, 18].

An illustrative example on comparison of the iteration behavior for a gradient based method and an evolution strategy for a convex problem

$$f(\kappa) = \sum_{i=1}^{n} \kappa_i^2, \quad -5.12 \leq \kappa_i \leq 5.12, \quad \kappa^* = \mathbf{0}, \quad f(\kappa^*) = 0 \qquad (7)$$

has been presented by [12]. As it becomes obvious in Fig. 15, the gradient-based method approaches the minimum in a "direct" (steepest descent) path, whilst the evolution strategy uses a "zig-zagging" path. Note that repetition of a gradient-based method yields i.g. the *same* path (deterministic method), whilst repetition of the evolution strategy i.g. yields a *different* path (stochastic method).

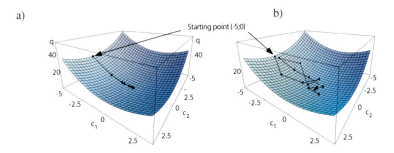

Fig. 15 Comparison of iteration behavior for a) a gradient based method and b) an evolution strategy for a convex problem, taken from [12]

Fig. 16 illustrates a comparison of the iteration behavior for a gradient based method and an evolution strategy for a nonconvex problem

$$f(\kappa) = 10n + \sum_{i=1}^{n} \left(\kappa_i^2 - 10\cos(\omega \kappa_i) \right), \quad -5.12 \leq \kappa_i \leq 5.12, \quad \kappa^* = \mathbf{0}, \quad f(\kappa^*) = 0 \quad (8)$$

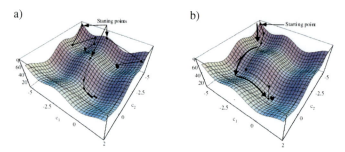

Fig. 16 Comparison of iteration behavior for a) a gradient based method and b) an evolution strategy for a nonconvex problem, taken from [12]

From Fig. 16 we observe: The gradient-based method attains i.g. *one* local minimum for *each* starting value, since it cannot overcome local minima, whereas the evolution strategy attains i.g. *several* local minima for *each* starting value, since it can overcome local minima.

The advantages of both conceptions, gradient-based methods, which efficiently determine local minima and stochastic methods, which can overcome different local minima, have been exploited in a hybrid strategy, see e.g. [15].

5 Stochastic Methods

In general experimental data are uncertain due to disturbances and scattering, see Fig. 17a, and b, and therefore the data should be treated as random variables. Fig. 17.c shows a Maximum-Likelihood adjustment of a straight line with mean value of data. However, also the model can be given a stochastic character by considering the estimates κ as random variables, see e.g. [3].

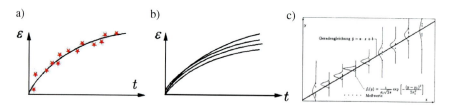

Fig. 17 Uncertainties of data: a) Disturbances, b) Scattering, c) Maximum-Likelihood adjustment of a straight line with mean value of data, taken from [23]

Critical concerns for identification problems are as follows: 1. How can statistical properties be obtained from estimators that are determined with the deterministic approaches? 2. How can estimators for the mean value and the covariance matrix of the random variables $\bar{\mathbf{d}}$ and κ be calculated? 3. How is the uncertainty transferred from the data $\bar{\mathbf{d}}$ to the material parameters κ?

Parameter Identification in Continuum Mechanics

A closed form estimate for parameters and covariances for linear identification problems is given in [3] and [14]. For nonlinear problems an estimate of parameters and covariance can be obtained by the Monte-Carlo method as presented in [3].

In practice, typically, the amount of test data obtained from laboratory experiments is insufficient for a statistical analysis. As a remedy, the performance of experiments may be considered as a stochastic process and each experiment, meaning the experimental data, is a realization of this stochastic process. A stochastic model can be summarized in three steps

1. Nonlinear regression model: We approximate the experimental data with a nonlinear regression function with multiple approach constants.
2. ARMA model: We compute the residuals between the experimental data and the fitted regression functions. The residuals describe measurement errors and other systematic errors.
3. Concatenation: The nonlinear regression model and the ARMA model are combined to one equation.

Fig. 18 summarizes some results of stochastic simulation for an adhesive material: Fig. 18.a shows experimental data and fitted regression function for the strain rate 0.001/s. The creation of 50 artificial data as a result of an AR[1] process is shown in Fig. 18.b. These data can be used for statistical analysis in parameter identification.

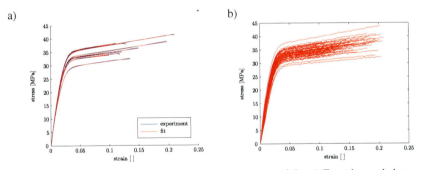

Fig. 18 Results of stochastic simulation for adhesive materials: a) Experimental data and fitted regression function for strain rate 0.001/s, b) 50 artificial data, [19]

6 Inhomogeneities

The assumption of uniformness within the sample for the state variables such as stresses and strains cannot always be guaranteed during the experiment. As summarized in Fig. 19 possible causes for inhomogeneities are barreling, perforated specimens, failure mechanism at height loads, damage and localization.

In the identification process for inhomogeneous problems field quantities are taken into account: To this end an optical method is applied to obtain the experimental data. For example the compact tension specimen, in Fig. 20 with the material

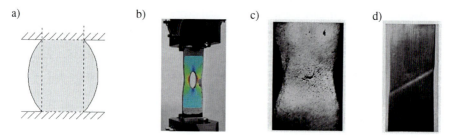

Fig. 19 Inhomogeneities: a) Barreling, b) Perforated strip, c) Necking, b) shear band localization

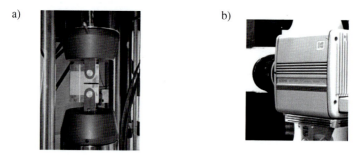

Fig. 20 CT specimen: Specimen with grated surface, CCD-camera from 1996

Baustahl St52 due to german industrial codes is investigated with a so called *grating method*. The underlying photos have been taken with the CCD camera of Fig. 20.b, see [1] and references therein for more information.

The finite element method is used to obtain simulated data. We let $\varphi(\kappa)$ denote the state variable occurring within an appropriate solution space U. It is dependent on the material parameters $\kappa \in \mathscr{K} \subset \mathbb{R}^{n_p}$, where $\mathscr{K}$ defines the parameter space. Then we can formulate

The direct problem : Find $\varphi \in U$, such that $\mathscr{A}(\kappa, \varphi(\kappa)) = 0$. (9)

In particular, the state equation $\mathscr{A}(\kappa, \varphi(\kappa)) = 0$ is the weak form of a finite element formulation. With the observation operator $\mathscr{M}$ mapping the state variables φ to points of the observation space $\mathscr{D}$ we can formulate

The least-squares problem : Find $\kappa \in \mathscr{K}$, such that
$$f(\kappa) = \tfrac{1}{2}\|\mathscr{M}\hat{\varphi}(\kappa) - \mathbf{d}\| \to \min_{\kappa \in \mathscr{K}}, \quad \hat{\varphi} = \arg\{\mathscr{A}(\kappa, \varphi(\kappa)) = 0\}, \quad (10)$$

where **d** are the experimental data obtained within the data space $\mathscr{D}$. Problem (10) requires efficient optimization algorithms, due to the accompanying time-consuming finte-element solution of the related direct problem. For gradient based optimization the gradient of the objective function is required. For this purpose, a

Parameter Identification in Continuum Mechanics

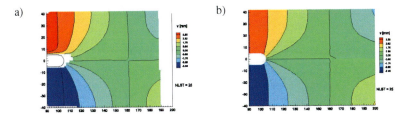

Fig. 21 CT specimen: Simulation and experiment for vertical displacement at the end of loading

recursion formula has been obtained in [10, 11] for history dependent problems. Results on the comparison of simulation and experiment for the CT-specimen in Fig. 20 are given in Fig. 21 for the vertical displacement, thus illustrating the excellent agreement.

More extensive works on the basis of a Lagrangian formulation

$$\mathscr{L}(\kappa,\mathbf{u},\lambda) = f(\kappa,\mathbf{u}) + \lambda \star \mathscr{A}(\kappa,\mathbf{u}) = \mathscr{L}(\mathbf{z}), \quad \mathbf{z} = (\kappa,\mathbf{u},\lambda) \qquad (11)$$

have been done in [28, 7, 4, 30] on strategies for computing goal-oriented a posteriori error measures in parameter identification.

7 Concluding Remarks

The ultimate goal of theory has always been to assimilate the outcome of experiment to the outcome of theory. The achievement of this goal is a process, termed identification, which as seen in three historic examples can be laborious and lengthy. Identification today is split into *structure identification*, thus providing the mathematical structure of the model, and *parameter identification*, thus providing the constants occurring in the model. In this essay, we have revisited some examples of hand-fitting and optimization. Concerning new research challenges, parameter identification will take account uncertainties of data, where e.g. stochastic modeling is a tool to provide experimental data. Furthermore, goal oriented error control for inhomogeneous problems will constitute a future research work.

References

1. Andresen, K., Dannemeyer, S., Friebe, H., Mahnken, R., Ritter, R., Stein, E.: Parameteridentifikation für ein plastisches Stoffgesetz mit FE-Methoden und Rasterverfahren. Der Bauingenieur 71, 21–31 (1996)
2. Bard, Y.: Nonlinear Parameter Estimation. Academic Press, New York (1974)
3. Beck, J.V., Arnold, K.J.: Parameter Estimation in Engineering and Science. Wiley, New York (1977)

4. Becker, R., Vexler, B.: Mesh refinement and numerical sensitivity analysis for parameter calibration of partial differential equations. J. Comput. Phys. 206(1), 95–110 (2005)
5. Cavendish, H. (1798) Experiments to Determine the Density of the Earth. In: MacKenzie, A.S. (ed.) Scientific Memoirs, vol. 9, pp. 59–105. The Laws of Gravitation, American Book Co (1900)
6. Dennis, J.E., Schnabel, R.B.: Numerical Methods for Unconstrained Optimization and Nonlinear Equations. Prentice Hall, New Jersey (1983)
7. Larsson, F., Hansbo, P., Runesson, K.: Strategies for computing goal-oriented a posteriori error measures in non-linear elasticity. Int. J. Numer. Meth. Engng. 55(8), 879–894 (2002)
8. Lemaitre, J., Chaboche, J.L.: Mechanics of solid Materials. Cambridge University Press, Cambridge (1990)
9. Luenberger, D.G.: Linear and nonlinear programming, 2nd edn. Addison-Wesley, Reading (1984)
10. Mahnken, R., Stein, E.: The identification of parameters for visco-plastic models via finite-element-methods and gradient-methods. Modelling and Simulation Material Science and Engineering 2(3A), 597–614 (1994)
11. Mahnken, R., Stein, E.: Parameter identification for finite deformation elasto-plasticity in principal directions. Comp. Meths. Appl. Mech. Eng. 147, 17–39 (1997)
12. Mohr, R.: Modellierung des Hochtemperaturverhaltens metallischer Werkstoffe. Dissertation, Report GKSS 99/E/66, GKSS Research Center, Geesthacht (1999)
13. Neal, B.G.: Structural Theorems and their Applications. Pergamon Press (1964)
14. Press, W.H., Teukolsky, S.A., Vetterling, W.T., Flannery, B.P.: Numerical Recipes in Fortran. Cambridge University Press (1992)
15. Quagliarella, D., Vicini, A.: Coupling genetic algorithms and gradient based optimization techniques. In: Quagliarella, D., et al. (eds.) Genetic Algorithms and Evolution Strategies in Engineering and Computer Science, pp. 289–309. Wiley, Chichester (1997)
16. Rechenberg, I.: (1971) Evolutionsstrategie - Optimierung technischer Systeme nach Prinzipien der biologischen Evolution (PhD thesis). Reprinted by Fromman-Holzboog (1973)
17. Rechenberg, I.: Evolutionsstrategie 1994. Frommann-Holzboog, Stuttgart (1994)
18. Schwan, S.: Identifikation der Parameter inelastischer Werkstoffmodelle: Statistische Analyse und Versuchsplanung. Shaker Verlag, Aachen (2000)
19. Nörenberg, N., Mahnken, R.: A stochastic model for parameter identification. Arch. Appl. Mech. (2012), doi:10.1007/s00419-012-0684-7
20. Schwefel, K.P.: Numerische Optimierung von Computer-Modellen mittels der Evolutionsstrategie. Birkhäuser Verlag, Basel (1977)
21. Senseny, P.E., Brodsky, N.S., De Vries, K.L.: Parameter Evaluation for a Unified Constitutive Model. Transactions of the ASME: J. Eng. Mat. Tech. 115, 157–162 (1993)
22. Steck, E.: A Stochastic Model for the Interaction of Plasticity and Creep in Metals. Nuclear Engineering and Design 114, 285–294 (1989)
23. Steck (ed.): Collaborative Research Centre 319 (SFB 319), Stoffgesetze für das inelastische Verhalten metallischer Werkstoffe. Entwicklung und technische Anwendung, Arbeitsbericht 1991/92/93 (1993)
24. Strathern, P.: Newton und die Schwerkraft. Fischer Taschenbuch Verlag (1998)
25. Szabo, I.: Geschichte der mechanischen Prinzipien, 3. Birkhäuser Verlag, Auflage (1987)
26. Terzaghi, C.: Principles of Soil Mechanics: II-Compressive Strength of Clay (1998); Engineering News-Record 95(20) (1925)
27. Timoshenko, S.P.: History of Strength of Materials (1983)

28. Vexler, B.: Adaptive Finite Element Methods for Parameter Identification Problems. Dissertation, University of Heidelberg (2004)
29. Wedemeier, T.: Beiträge zur Theorie und Numerik von Materialien mit innerer Reibung am Beispiel des Werkstoffes Beton, Dissertation, University of Hannover (1990)
30. Widany, K.U., Mahnken, R.: Adaptivity for parameter identification of incompressible hyperelastic materials using stabilized tetrahedral elements. Comput. Methods Appl. Mech. Engrg. 245-246, 117–131 (2012)
31. http://en.wikipedia.org/wiki

Historical Development of the Knowledge of Shock and Blast Waves

Torsten Döge and Norbert Gebbeken

Abstract. This chapter describes the historical development of the theory of blast and shock waves from the 17th to the 19th century.

Usually, the propagation of blast and shock waves are mathematically described by partial differential equations for conservation of mass, momentum and energy. Additionally, an equation of state for the material (e.g. air) is required. Of course, the material properties must also be known. Because there is a discontinuity at the shock front and the partial differential equations can not be used at the shock front, jump conditions (equations for conservation of mass, momentum and energy) must be set up. With this at hand, a way to find a solution for the equations must be described.

The realization of all this took several centuries. It's development is chronologically described. The theory of blast and shock waves is not trivial. Many problems had to be solved. For example, even the determination of the density of air was a problem to be solved in the 17th century.

1 Introduction

Material properties, an equation of state for the material, and conservation equations for mass, momentum and energy are required for the mathematical description of the propagation of blast and shock waves. This was not set up at once, but took a long way over several centuries. The development of the theory of blast and shock waves is chronologically described in the following sections.

Torsten Döge
Dr. Linse Ingenieure GmbH, Karlstr. 46, 80333 Munich, Germany
e-mail: doege@drlinse.de

Norbert Gebbeken
Institute of Engineering Mechanics and Structural Mechanics, University of the Bundeswehr Munich, Werner-Heisenberg-Weg 39, 85577 Neubiberg, Germany
e-mail: norbert.gebbeken@unibw.de

The content of this chapter was collected when writing the dissertation [11]. Only by reading the original literature it becomes clear, what problems had to be solved, why problems were not solved until a certain time and why some problems were then independently solved by several persons. The authors of this chapter recommend to everyone to obtain and read the original literature. It is very instructive!

2 The 17th Century

MARIN MERSENNE (1588–1648) measured around 1636 the speed of sound (LENIHAN [34, 35]). He applied two methods. One method was to measure the time from the arrival of the flash of a canon, which distance was known, until the arrival of the report of the canon. With this method, he determined the speed of sound to 230 Toisen per second (≈ 448 m/s) [34]. The second method was to measure the time that a sound and its echo took to a wall and back. With this method he determined approximately 316 m/s [35].

GALILEO GALILEI (1564–1642) wrote already in his 1638 published work *Discorsi e dimostrazioni matematiche intorno a due nuove scienze attenenti alla meccanica et i movimenti locali* about experiments on air. Thereby, he described an experiment, with which the density of air can be determined [22, p. 117–118]:

> I took a pretty big Glass Bottle, with a narrow Neck, and tied very close to its narrow Neck a Leathern Cover; to this Cover, and within Side, I put a Valve; thro' this with a Syringe I forced a great Quantity of Air, of which, because it admits of great Condensation, twice or thrice as much may be forced in as the Bottle naturally holds: Then I very carefully weigh'd in a most exact Ballance the Bottle with the compressed Air within it, adjusting their Weight by very fine Sand; then opening the Valve, I let out the Air which was violently contain'd in the Vessel; I put the Bottle again into the Scales, and finding it much lighter than before, I took out of the other Scales so much Sand (keeping it by itself) until the Sand and the Bottle were in *Equilibrio*: Now there can be no Room to doubt but that the Weight of the Sand taken out is = the Weight of the Air which was violently forced into the Bottle, and which afterwards was let out.
>
> But this Experiment assures me of no more than this, *viz.* that the Weight of the Air violently compressed in the Bottle, is equal to the Weight of that reserved Sand: But I have not yet determin'd how much the Air absolutely weighs in respect of Water, or any other heavy Matter; nor can I know this, unless I measure the Quantity of the compressed Air; which may be done by either of the two Ways following.
>
> The former is this: Take such another Bottle, with a Neck exactly of the same Size with that of the former; round which Neck tie very fast another Leather, the other End of which tie also closely over the former Bottle's Neck. Now the Bottom of this second Bottle must be drill'd or bor'd thro', so that thro' the Hole a Wire may be put, wherewith at Pleasure the Valve of the former Bottle may be open'd, to let out the superfluous Air after it hath been weigh'd: But now this second Bottle must be fill'd with Water. All things thus prepar'd, and the Valve open'd by Help of the Wire, the Air issuing out with Impetuosity, and entering the Bottle of Water, shall drive the Water out by the Hole at the Bottom. Now 'tis manifest that the Quantity of Water forc'd out in this manner, is = the Bilk and the Quantity of Air that issued out of the other Bottle; Wherefore keeping that Water, let the Bottle, now lighten'd of the compressed

Air, be again weigh'd, (for I suppose it have been weigh'd before, together with the compressed Air) and then 'tis manifest that the superfluous Sand, reserv'd as before order'd, is exactly equal to the Weight of such a Mass of Air, as is the Mass of Water forc'd out and preserv'd, which if weigh'd, we shall see how many times its Weight shall contain the Weight of the reserv'd Sand; and we may safely affirm, that the Water is so much heavier than the Air, which will not be only ten times, as seems to be *Aristotle*'s Opinion, but nearly 400 times, as this Experiment shews us.

This description shows, which problems had to be solved in the 17th century in order to find facts which are today well known like the density of air.

At the time of GALILEI, it was already known that water in a closed suction tube can rise up to about 10 m. The rise of the water was explained with the *horror vacui*, a force of the nature, which should prevent the creation of a vacuum. EVANGELISTA TORRICELLI (1608–1647) got the idea to repeat the experiment with mercury instead of water. He expected a column of $1/14$ of the length of the water column. The experiment was conducted in 1643 by VIVIANI [37].

MACH described the experiment of TORRICELLI as follows [37]: An about 1 m long, at one end closed and with mercury filled glass tube is closed at the opened end by a finger. This end is brought into mercury and vertically aligned. When the finger is removed, then the mercury column sinks and remains at a height of about 76 cm.

GALILEI measured the weight of air. TORRICELLI was the one, who connected the weight of air with the *horror vacui* and explained the rise of the water and the mercury by the weight pressure of the air [37]. He discovered the air pressure.

BLAISE PASCAL (1623–1662) conducted several experiments on air pressure. He got the idea, that the air pressure depends on the height above sea level. His brother-in-law PERIER conducted such an experiment on the Puy de Dôme in 1648 ([37, p. 105–107], [65, p. 85]). The unit of pressure is named after PASCAL (1 Pa = $1 N/m^2$).

CASPAR SCHOTT (1608–1666) published the book *Mechanica Hydraulico-Pneumatica* in 1657. In this book, he reported about the famous experiments of OTTO VON GUERICKE (1602–1686). ROBERT BOYLE (1627–1691) probably came to know about the experiments of GUERICKE and the invention of the air pump by this book [2, p. 407].

BOYLE together with ROBERT HOOKE (1635–1703) improved the air pump and conducted experiments on air. These experiments were published 1660 [2, p. 407 ff.]. In 1662, BOYLE published another book, in order to comment criticism of FRANCISCUS LINUS [2, p. 652 ff.]. In this book, he published that the pressure is inversely proportional to the volume [2, p. 671]. EDME MARIOTTE (around 1620–1684) published similar results in the year 1676 [53]. Therefore, this relationship is called BOYLE-MARIOTTE law.

WEBSTER pointed out in [66] that this relationship was discovered first by HENRY POWER (1623–1668) and RICHARD TOWNELEY instead by BOYLE. BOYLE came to know of this relationship by letters of POWER and wrote this in [2] on page 669. The description BOYLE-MARIOTTE law is not fully right. But a description like POWER-TOWNELEY-BOYLE-HOOKE-MARIOTTE law would be too bulky.

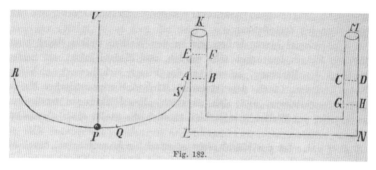

Fig. 1 NEWTON, cycloidal pendulum and U-shaped channel [44, p. 358]

ISAAC NEWTON (1643–1727) published his *Philosophiae Naturalis Principia Mathematica* in the year 1687. In this publication, he also calculated the velocity of waves in fluids. He derived the velocity of waves in fluids from the time of oscillations of a cycloidal pendulum and of water in a U-shaped channel (Figure 1) and formulated it in theorem XXXVIII [43]:

> The speed of pulses propagating in an elastic fluid are in the ratio composed from the direct proportion of the square root of the elastic force or pressure and the inverse proportion of the square root of the density; but only if the same elastic force is supposed for the same proportional condensation.

The derivation of this relationship was a remarkable step at this time. NEWTON was the first who has calculated the speed of sound. Today we would write this theorem in the equation

$$c = \sqrt{\frac{p}{\rho}} \qquad (1)$$

[56, p. 282]. This equation differs from the now known equation for the speed of sound by the factor $\sqrt{\gamma}$ (with speed of sound c, pressure p, density ρ, specific heat ratio γ).

NEWTON calculated the speed of sound in air as follows [43]:

> For since the specific gravities of rain-water and quick-silver are to one another as about 1 to 13 2/3, and when the mercury in the barometer is at the height of 30 inches of our measure, the specific gravities of the air and of rain-water are to one another as about 1 to 870: therefore the specific gravity of air and quick-silver are to each other as 1 to 11890. Therefore when the height of the quick-silver is at 30 inches, a height of uniform air, whose height would be sufficient to compress our air to the density we find it to be of, must be equal to 356700 inches or 29725 feet of our measure. And this is that very height of the medium, which I have called A in the construction of the foregoing proposition. A circle whose radius is 29725 feet is 186768 feet in circumference. And since a pendulum 39 1/5 inches in length compleats one oscillation, composed of its going and return, in two seconds of time, as is commonly known; it follows that a pendulum 29725 feet or 356700 inches in length will perform a like oscillation in 190 3/4 seconds. Therefore in that time a sound will go right onwards 186768 feet, and therefore in one second 979 feet.

NEWTON corrected this calculated speed of sound of 979 feet/second = 298 m/s by additional, confusing factors for particles and vapors in the air and got a velocity of 1142 feet/second = 348 m/s. NEWTON probably introduced these factors because the theoretical result did not match with experiments. If we ignore these additional factors, then we can only admire him for his achievement at this time.

3 The 18th Century

DANIEL BERNOULLI (1700–1782), the son of the mathematician JOHANN BERNOULLI (1667–1748), published his famous *Hydrodynamica* [1] in 1738. He described in the 10th chapter a model of gases, which is composed of many particles with rapid motion. There were other persons before DANIEL BERNOULLI who had tried to describe air by particles. These persons mostly assumed that these particles are embedded in an ether and just have vibrations and rotations. STEPHEN G. BRUSH and CLIFFORD AMBROSE TRUESDELL (1919–2000) considered DANIEL BERNOULLI to be the first who has set up a kinetic theory of gases ([4, p. 20], [63, p. 276]).

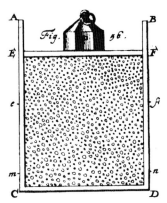

Fig. 2 Model for explanation of kinetic theory of air (DANIEL BERNOULLI [1])

DANIEL BERNOULLI imagined a cylinder which contains air and is compressed with a weight P (Figure 2). He assumed that air consists of many tiny particles with rapid motion. He explained pressure as a result of many hits of the tiny particles against the wall of the cylinder. After some calculations he came to the result that the pressure changes proportional to the square of particle velocity [3, p. 57 ff.].

LEONHARD EULER (1707–1783), a scholar of JOHANN BERNOULLI [20], was one of the most important and most productive mathematicians. The *Verzeichnis der Schriften* LEONHARD EULERs (ENESTRÖM index), which was compiled by GUSTAF ENESTRÖM (1852–1923), contains 866 works of EULER [13]. More than 50 terms, theorems and methods in mathematics and mechanics are named after EULER [25].

EULER repeatedly dealt with sound propagation. Already his first work *Dissertatio physica de sono* (E2, [17, p. 181–196]) in 1727 when he was just 20 years old was on this topic. In the following, just a few but important findings of EULER are presented.

In *Découverte d'un nouveau principle de mécanique* (E177, [14]), which was written in 1750 and published in 1752, EULER gives NEWTON's law the general shape

$$M d^2 x = P dt^2; \quad M d^2 y = Q dt^2; \quad M d^2 z = R dt^2 . \tag{2}$$

Thereby, x, y and z are the coordinates of the point mass M, and P, Q and R are the forces which act on the point mass in x-, y- and z-direction [14, p. 195-196]. There is $2M$ instead M in [14] because the weight M was derived by a gravitation acceleration of $g = 1/2$ [56, p. 22]. EULER commented [14, p. 195]:

> Et c'est cette formule seule, qui renferme tous les principes de la Mecanique.[1]

It is remarkable that these equations were not included in NEWTON's *Principia* and that EULER's method of section was necessary for the derivation of these equations [56, p. 247–248].

EULER assumed the equation of state

$$p = nqr \tag{3}$$

for air in *Sectio prima de statu aequilibrii fluidorum* (1769, E375, [19, p. 1–72], [16, p. 20]). Thereby, p is the pressure, n a constant, q the density and r the heat. This equation of state has great resemblance to the today known equation of state $p = (\gamma - 1)\rho e$ for ideal gases (specific heat ration γ, density ρ, specific internal energy e).

In *Sectio secunda de principiis motus fluidorum* (1770, E396, [19, p. 73–153]), EULER set up the differential equations

$$\frac{du}{dt} = P - \frac{1}{q}\frac{\partial p}{\partial x}$$
$$\frac{dv}{dt} = Q - \frac{1}{q}\frac{\partial p}{\partial y}$$
$$\frac{dw}{dt} = R - \frac{1}{q}\frac{\partial p}{\partial z}, \tag{4}$$

which describe the conservation of momentum. Thereby, the derivatives

$$\frac{du}{dt} = \frac{\partial u}{\partial t} + \frac{\partial u}{\partial x}u + \frac{\partial u}{\partial y}v + \frac{\partial u}{\partial z}w$$
$$\frac{dv}{dt} = \frac{\partial v}{\partial t} + \frac{\partial v}{\partial x}u + \frac{\partial v}{\partial y}v + \frac{\partial v}{\partial z}w$$
$$\frac{dw}{dt} = \frac{\partial w}{\partial t} + \frac{\partial w}{\partial x}u + \frac{\partial w}{\partial y}v + \frac{\partial w}{\partial z}w \tag{5}$$

---

[1] And this is the only formula that contains all principles of mechanics.

which EULER had derived, have to be applied. Additional, he set up the differential equation

$$\frac{\partial q}{\partial t} + \frac{\partial (qu)}{\partial x} + \frac{\partial (qv)}{\partial y} + \frac{\partial (qw)}{\partial z} = 0 \qquad (6)$$

for conservation of mass, which is also called equation of continuity today. In these equations, q is the density and u, v, w are the velocities in the Cartesian coordinates x, y, z. EULER was aware of the significance of these equations [16, p. 140–141]. These equations are now called the EULER equations.

In the year 1775, EULER managed the discovery that the conservation of angular momentum is an own independent principle [56, p. 30]. The conservation equations appear in [15, §29] as

$$S = \int z dM \left(\frac{d\,dy}{dt^2}\right) - \int y dM \left(\frac{d\,dz}{dt^2}\right)$$
$$T = \int x dM \left(\frac{d\,dz}{dt^2}\right) - \int z dM \left(\frac{d\,dx}{dt^2}\right)$$
$$U = \int y dM \left(\frac{d\,dx}{dt^2}\right) - \int x dM \left(\frac{d\,dy}{dt^2}\right) \qquad . \qquad (7)$$

Thereby, there is a z instead of the second x in the third row in [15]. Even many years after EULER, not all physicians had the insight that this is an independent principle [56, p. 19].

By reading EULER's works, it is clearly recognizable how his knowledge developed over the years. He also learned from the achievements of his contemporaries and gained new knowledge from it. He was active well into old age. 34 % of the quantity of his works are from the period when he was older than 68 years and almost completely blind [20].

JOSEPH-LOUIS LAGRANGE (1736–1813) assumed 1760 in *Nouvelles recherches sur la nature et la propagation du son* [32, p. 296 ff.], that in sound waves the relationship

$$\varphi(D) = D^m \qquad (8)$$

between the pressure φ and the density D is valid [21]. Thereby, he estimated the constant m to $1 + \frac{1}{3}$. This constant corresponds to the heat capacity ratio γ and LAGRANGE's estimation comes close to the today's value of 1.4.

4 The 19th Century

The emergence of the steam engine in the 18th century led to the necessity to understand the processes of the steam engine. JOSEPH LOUIS GAY-LUSSAC (1778–1850) published the results of his experiments on the expansion of gases and vapors by heat in 1802 (*Recherches sur la dilatation des gaz et des vapeurs* [23]). He discovered by these experiments that the volume of air, hydrogen, oxygen and nitrogen expands due to heating by almost the same amount of $37.5 \pm 0.02\,\%$. He formulated the now called GAY-LUSSAC's law ([23, p. 174–175], [24, p. 25]):

> 1. Tous les gaz, quelque soient leur densité et la quantité d'eau qu'ils tiennent en dissolution, et toutes les vapeurs, se dilatent également par les mêmes degrés de chaleur.[2]
> 2. Pour les gaz permanens, l'augmentation de volume que chacun d'eux reçoit depuis le degré de la glace fondante jusqu'à celui de l'eau bouillante est égale aux $\frac{80}{213,33}$ du volume primitif pour le thermomètre divisé en 80 parties, ou aux $\frac{100}{266,66}$ du même volume pour le thermomètre centigrade.[3]

As mentioned before, EULER had stated such a relationship already in 1769 (equation (3)).

SIMÉON DENIS POISSON (1781–1840) published his work *Mémoire sur la théorie du son* in 1808 [46] (English translation in [28]). He introduced a factor k in section 22, in order to calculate more accurately the speed of sound. He calculated $k = 0{,}4254$ ([46, p. 362], [36]). $k+1$ corresponds to the today's heat capacity ratio γ.

He investigated the one dimensional motion of air with finite amplitudes in sections 23 and 24. For the equation

$$\frac{d\varphi}{dt} + a\frac{d\varphi}{dx} + \frac{1}{2}\frac{d^2\varphi}{dx^2} = 0 \qquad (9)$$

with the velocity potential φ, the velocity $\frac{d\varphi}{dx}$ and the speed of sound a, he found the solution

$$\frac{d\varphi}{dx} = f\left(x - at - \frac{d\varphi}{dx}t\right) \qquad (10)$$

whereby, f is an arbitrary function.

Another publication of POISSON is *Sur la Chaleur des gaz et des vapeurs* from the year 1823 [48] (English translation by HERAPATH in [47]). There, POISSON derived the equation

$$p' = p\left(\frac{\rho'}{\rho}\right)^k \qquad (11)$$

for adiabatic processes [48, Equ. (5)]. Thereby, $k = \frac{c}{c_v}$ is the ratio of the specific heat capacity c at constant pressure to the specific heat capacity c_v at constant volume. This k corresponds to the today's heat capacity ratio γ. POISSON stated that GAY-LUSSAC had determined the value of k to 1.375.

---

[2] All gases, irrespective of their density and the amount of water they hold in solution, and all vapors, equally expand by the same degree of heat.

[3] For permanent gases, the increase in volume that each of them receives from the degree of melting ice to the boiling water is equal to $\frac{80}{213.33}$ of the original volume for the thermometer divided into 80 parts, or $\frac{100}{266.66}$ of the same volume for the centigrade thermometer.

PIERRE-SIMON DE LAPLACE (1749–1827) succeeded in 1816 to correctly calculate the speed of sound in air [33]. He recognized that a temperature rise is caused by the sound waves. He described the speed of sound:

> La vitesse réelle du son est égale au produit de la vitesse que donne la formule newtoniène, par la racine carrée du rapport de la chaleur spécifique de l'air soumis à la pression constante de l'atmosphère et à diverses températures, à sa chaleur spécifique lorsque son volume reste constant.[4]

This corresponds to the now known equation $c = \sqrt{\gamma p/\rho}$. This is the first correct calculation of the speed of sound, 180 years after MERSENNE's measurement and 129 years after NEWTON's calculation which had not the correct result.

CLAUDE LOUIS MARIE HENRI NAVIER (1785–1836) submitted his work *Mémoire sur les lois du mouvement des fluides* [42] to the French Academy of Sciences on March 18, 1822. He assumed repulsive forces $f(r)$ between molecules, which rapidly decreases with increasing distance r between the molecules. He set up the equations

$$P - \frac{\partial p}{\partial x} = \rho \left(\frac{\partial u}{\partial t} + u \frac{\partial u}{\partial x} + v \frac{\partial u}{\partial y} + w \frac{\partial u}{\partial z} \right) - \varepsilon \left(\frac{\partial^2 u}{\partial x^2} + \frac{\partial^2 u}{\partial y^2} + \frac{\partial^2 u}{\partial z^2} \right)$$
$$Q - \frac{\partial p}{\partial y} = \rho \left(\frac{\partial v}{\partial t} + u \frac{\partial v}{\partial x} + v \frac{\partial v}{\partial y} + w \frac{\partial v}{\partial z} \right) - \varepsilon \left(\frac{\partial^2 v}{\partial x^2} + \frac{\partial^2 v}{\partial y^2} + \frac{\partial^2 v}{\partial z^2} \right)$$
$$R - \frac{\partial p}{\partial z} = \rho \left(\frac{\partial w}{\partial t} + u \frac{\partial w}{\partial x} + v \frac{\partial w}{\partial y} + w \frac{\partial w}{\partial z} \right) - \varepsilon \left(\frac{\partial^2 w}{\partial x^2} + \frac{\partial^2 w}{\partial y^2} + \frac{\partial^2 w}{\partial z^2} \right) \quad (12)$$

which are now called NAVIER-STOKES equations. Thereby, P, Q and R are external forces in the x-, y- and z-directions, e.g. gravitation, and u, v and w are the velocities in the three directions. For more information about the historical development of the NAVIER-STOKES equations see e.g. [56].

JULIUS ROBERT MAYER (1814–1878) was not a physicist, but a physician. He wanted to publish his first paper 1841 in the *Poggendorff-Annalen*. This paper was rejected because it was full of errors and mistakes [41, p. 81]. In 1842, MAYER published the paper *Bemerkungen über die Kräfte der unbelebten Natur* [38], which contains the first time the principle of energy conservation which included heat (an equivalence between potential and kinetic energy was already known at the time of GALILEI). Also in this paper, there are still some inexact physical terms. For example, he calculates the kinetic energy, which he calls living force ("*lebendige Kraft*"), to mc^2 instead of $\frac{mc^2}{2}$.

The term energy was not introduced at this time. The term force was used instead of energy. MAYER was aware of the importance to distinguish between these two terms [40, p. 27]:

---

[4] The real speed of sound is equal to the product of the velocity of NEWTON's formula with the square root of the ratio of the specific heat of air at constant pressure of the atmosphere at various temperatures to the specific heat at constant volume.

Unter den obwaltenden Umständen ist nun nichts übrig, als entweder der Newton'schen todten, oder der Leibnitz'schen lebendigen Kraft die Benennung 'Kraft' zu entziehen, wobei man aber in jedem Falle mit dem herrschenden Sprachgebrauche in Conflict geräth.[5]

Considering that MAYER has been a physician and not a physicist, than his achievement to discover the principle of energy conservation has to be even more appreciated. Not all physicists of his time accepted his principle of energy conservation. Many physicists doubted that kinetic energy can be converted into heat. MAYER had to withstand many hostilities, which affected his health [41, p. 85].

MAYER postulated not only the principle of energy conservation but also calculated the relationship between different types of energy, e.g. between potential energy and heat [38]. He formulated the principle of energy conservation in *Die organische Bewegung in ihrem Zusammenhange mit dem Stoffwechsel* in 1845 for all known energy types [39, p. 32 ff.].

The name of GEORGE GABRIEL STOKES (1819–1903) is nowadays known for the so-called NAVIER-STOKES equations, which he published 1845 in his work *On the Theories of the Internal Friction of Fluids in Motion* (for more information we refer to [56]).

In the year 1848, STOKES published the paper *On a difficulty in the Theory of Sound* [54]. He responded to POISSON's equation (10) of the shape $w = f(z - (a + w)t)$ and showed that the slope of the velocity function $w(z,t)$ is

$$\frac{\frac{dw}{dz}\big|_{t=0}}{1 + \frac{dw}{dz}\big|_{t=0} t} . \qquad (13)$$

STOKES recognized that the curve becomes steeper and steeper when $\frac{dw}{dz}\big|_{t=0}$ is negative and thereby the denominator in (13) decreases (Figure 3).

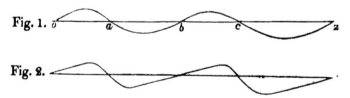

Fig. 3 Figures of STOKES on nonlinear propagation of waves [54]

He also recognized that only values of t can be used, so that the predominator in (13) is not equal to zero. He assumed that a discontinuity might be formed [54]:

---

[5] Under these circumstances, there is nothing left but to withdraw either NEWTON's dead force or LEIBNITZ's living force the term 'force', whereby on gets in conflict with the predominant language use.

Of course, after the instant at which the expression (13) becomes infinite, some motion or other will go on, and we might wish to know what the nature of that motion was. Perhaps the most natural supposition to make for trial is, that a surface of discontinuity is formed, in passing across which there is an abrupt change of density and velocity. The existence of such a surface will presently be shown to be possible, on the two suppositions that the pressure is equal in all directions about the same point, and that it varies as the density. I have however convinced myself, by a train of reasoning which I do not think it worth while to give, inasmuch as the result is merely negative, that even on the supposition of the existence of such a surface of discontinuity, it is not possible to satisfy all the conditions of the problem by means of a single function of the form $f\{z-(a+w)t\}$.

Then, STOKES derived the equations

$$\rho w - \rho' w' = (\rho - \rho') \gamma$$
$$(\rho w - \rho' w') \gamma - (\rho w^2 - \rho' w'^2) = a^2 (\rho - \rho') \quad (14)$$

for conservation of mass and momentum at the discontinuity. Thereby, γ is the velocity of the discontinuity. STOKES wrote further:

> The strange results at which I have arrived appear to be fairly deducible from the two hypotheses already mentioned. It does not follow that the discontinuous motion considered can ever take place in nature, for we have all along been reasoning on an ideal elastic fluid which does not exist in nature. [...] It appears, then, almost certain that the internal friction would effectually prevent the formation of a surface of discontinuity, and even render the motion continuous again if it were for an instant discontinuous.

Apparently, STOKES did not know the principle of energy conservation at that time. The correspondence of STOKES with LORD RAYLEIGH and THOMSON, who advised him on energy conservation, is very interesting [52]. If STOKES would have also set up the equation for energy conservation at the discontinuity, then the so-called RANKINE-HUGONIOT conditions should now named after STOKES. Remarkably, STOKES removed the part with the jump conditions from the reprint of this paper in his collected works [55] in the year 1883 with the remark that the principle of energy conservation is violated or the jump conditions for energy conservation only allow the solution $\rho = \rho'$. The mistake laid in the assumption of an equation of state of the shape $p = p(\rho)$ by disregarding the internal energy. This is remarkable, because EULER formulated already in 1768 the equation of state (3) and because RANKINE's paper [49] from the year 1870 should be known to STOKES.

WILLIAM THOMSON (1824–1907) was ennobled in 1892 and adopted the title 1ST BARON KELVIN OF LARGS. Therefore, he is often described as LORD KELVIN. In 1848, he suggested the introduction of an absolute temperature scale [57] based on CARNOT's investigations (*Réflexions sur la puissance motrice du feu et sur les machines propres à développer cette puissance*, [6]). This suggested temperature scale differs from the later used KELVIN scale. In [57, p. 315], THOMSON still was of the opinion that "the conversion of heat (or caloric) into mechanical effect is probably impossible". Later, he changed this opinion.

He formulated his second suggestion for an absolute temperature scale in 1854 with the words [29, p. 351]:

> If any substance whatever, subjected to a perfectly reversible cycle of operations, takes in a heat only in a locality kept at a uniform temperature, and emits heat only in another locality kept at a uniform temperature, the temperatures of these localities are proportional to the quantities of heat taken in or emitted at them in a complete cycle of the operations.

This can be expressed by the equation

$$\frac{Q_1}{Q_2} = \frac{T_1}{T_2} . \tag{15}$$

Thereby, Q_1 and Q_2 are the heat taken in and emitted in the so-called CARNOT cycle at the absolute temperatures T_1 and T_2 [61, p. 200] (also [58, p. 117], [59, p. 280], [60, p. 123 footnote †]). THOMSON calculated the melting point of water on the new scale to approximately 273.7 [29, p. 352].

RUDOLF JULIUS EMANUEL CLAUSIUS (1822–1888) laid the base for the second law of thermodynamics in 1850 [7, 8], which he described in 1854 by the words [9, p. 504]:

> Die algebraische Summe aller in einem Kreisprocesse vorkommenden Verwandlungen kann nur positiv seyn.[6]

He stated in [9] that the equation $\int \frac{dQ}{T} = 0$ is valid for reversible processes. CLAUSIUS suggested 1865 in [10] the term entropy (*Entropie*), following THOMSON who had suggested the term energy (according to CLAUSIUS). CLAUSIUS set up the inequality

$$\int \frac{dQ}{T} \geq 0 \tag{16}$$

with $\int \frac{dQ}{T} = S - S_0$ in [10]. Thereby, S is the entropy.

The reverend SAMUEL EARNSHAW (1805–1888) submitted 1858 his work *On the mathematical theory of sound*, which was published in 1860 [12]. He developed solutions for the one-dimensional propagation of waves for isothermal, isentropic and arbitrary pressure-density relationships. He recognized that loud sound waves are faster than gentle sound waves:

> I should expect, therefore, that in circumstances where the human voice can be heard at a sufficiently great distance, the *command* to fire a gun, if instantly obeyed, and the *report* of the gun, might be heard at a long distance in an inverse order; i.e. *first* the report of the gun, and *then* the word 'fire.'

---

[6] The algebraic sum of all conversions occurring in a thermodynamic cycle can only be positive.

Thereby, he referred to [45, p. 239], where is written:

The Experiments on the 9th February, 1822, were attended with a singular circumstance, which was – the officers' word of command 'fire,' was several times distinctly heard both by Captain Parry and myself, about one beat of the chronometer *after* the report of the gun; from which it would appear, that the speed of sound depended in some measure on its intensity.

EARNSHAW further described the formation of a discontinuity, which he called bore.

BERNHARD RIEMANN (1826–1866) was an outstanding mathematician, who in spite of his short life impacted the mathematical world significantly. His most important works are on analysis, number theory and differential geometry. But also in the field of blast waves, he managed an important milestone with his work *Ueber die Fortpflanzung ebener Luftwellen von endlicher Schwingungsweite* [50] which was submitted in 1859. In this work, he developed the method of characteristics for the solution of hyperbolic systems of partial differential equations. He introduced r and s, which are called RIEMANN invariants today. He recognized the formation of shock waves and rarefaction waves and explained it by an example, which is now called RIEMANN problem. RIEMANN considered only the conservation equations for mass and momentum in his calculations, but not the conservation equation for energy.

WILLIAM JOHN MACQUORN RANKINE (1820–1872) set up the jump conditions

$$m^2 = \frac{1}{S}\left\{(\gamma+1)\frac{p}{2}+(\gamma-1)\frac{P}{2}\right\}$$

$$a^2 = m^2 S^2 = S\left\{(\gamma+1)\frac{p}{2}+(\gamma-1)\frac{P}{2}\right\}$$

$$u = \frac{p-P}{m} = (p-P)\sqrt{\left\{\frac{S}{(\gamma+1)\frac{p}{2}+(\gamma-1)\frac{P}{2}}\right\}} \qquad (17)$$

for a shock front in his paper *On the Thermodynamic Theory of Waves of Finite Longitudinal Disturbance* [49] which he had submitted in the year 1869. These equations were derived from the conservation equations for mass, momentum and energy. These equations are now called RANKINE-HUGONIOT conditions.

THOMSON wrote to STOKES on March 7, 1870 [52, 67]:

MY DEAR STOKES,
I have read Rankine's paper with great interest. The simple elementary method by which he investigates the condition for sustained uniformity of type is in my opinion very valuable. It ought as soon as it is published to be introduced into every elementary book henceforth written on the subject.

With the knowledge of this letter, it is very surprising that STOKES cut out the part with his jump conditions from his paper from 1848 [54] in his collected works in 1883 [55]. For a more detailed story, please read the paper of SALAS [52].

On October 26, 1885, PIERRE HENRI HUGONIOT (1851–1887) submitted his work to the Academy of Sciences. HUGONIOT died before publication of the work and according to the editor, he was unable to make the modifications and additions to his preliminary draft which he apparently intended to make. HUGONIOT's work was published in two parts. The first part [26] was published in 1887. This part contains the first three chapters with the theory of characteristic curves, the equation of motion for ideal gas and the motion of gases without discontinuities. The second part, which was published in 1889 [27], is particularly interesting because of the discontinuities in the fifth chapter.

The so-called HUGONIOT equation is often mentioned in the shape

$$e_1 - e_0 = \frac{1}{2}(p_1 + p_0)\left(\frac{1}{\rho_0} - \frac{1}{\rho_1}\right) . \tag{18}$$

This equation is not found in this shape in [27], but it is found as

$$\frac{p + p_1}{2} + \frac{p_1 - p}{m - 1}\frac{1}{z_1 - z} + \frac{p_1 z_1 - pz}{m - 1}\frac{1}{z_1 - z} \; [= 0] \tag{19}$$

([27, p. 82], [52]). These both equations can be converted into each other with $e = \frac{p}{\rho(\gamma - 1)}$, $\gamma = m$ and $\frac{1}{\rho} = z + 1$ (with pressure p, specific internal energy e, density ρ, specific heat ratio γ).

5 Conclusion

This chapter described the arduous way in the theory of blast and shock waves from the 17th to the 19th century. It is only a short description. The following further readings are recommended: the books and papers of BRUSH [4, 5], FOX [21], KLEIN [30], KREHL [31], MACH [37], SACHDEV [51], SALAS [52], SCHLOTE [53], SZABÓ [56] and TRUESDELL (introductions in [18, 19], as well as [62, 63, 64]).

References

1. Bernoulli, D.: Hydrodynamica, sive de viribus et motibus fluidorum commentarii. Dulsecker (1738)
2. Boyle, R.: The Philosophical Works of the Honourable Robert Boyle Esq, vol. II (1725), printed for W. and J. Innys; and J. Osborn, and T. Longman, London
3. Brush, S.G.: Kinetic theory. The nature of gases and of heat, vol. 1. Pergamon Press (1965)
4. Brush, S.G.: The kind of motion we call heat. Book 1: Physics and the Atomists. North Holland (1976)
5. Brush, S.G.: The kind of motion we call heat. Book 2: Statistical physics and irreversible processes. North Holland (1976)
6. Carnot, S.: Betrachtungen über die bewegende Kraft des Feuers. Ostwalds Klassiker der exakten Wissenschaften Band 37, 1. edn, Reprint der Bände 37, 180 und 99. Verlag Harri Deutsch (2003)

7. Clausius, R.: Ueber die bewegende Kraft der Wärme und die Gesetze, welche sich daraus für die Wärmelehre selbst ableiten lassen. Annalen der Physik und Chemie 155(3), 368–397 (1850)
8. Clausius, R.: Ueber die bewegende Kraft der Wärme und die Gesetze, welche sich daraus für die Wärmelehre selbst ableiten lassen. Annalen der Physik und Chemie 155(4), 500–524 (1850)
9. Clausius, R.: Ueber eine veränderte Form des zweiten Hauptsatzes der mechanischen Wärmetheorie. Annalen der Physik und Chemie 169(12), 481–506 (1854)
10. Clausius, R.: Ueber verschiedene für die Anwendung bequeme Formen der Hauptgleichungen der mechanischen Wärmetheorie. Annalen der Physik und Chemie 201(7), 353–400 (1865)
11. Döge, T.: Zur Reflexion von Luftstoßwellen an nachgiebigen Materialien und Baustrukturen. PhD thesis, Universität der Bundeswehr München (2012)
12. Earnshaw, S.: On the mathematical theory of sound. Philosophical Transactions of the Royal Society of London 150, 133–148 (1860)
13. Eneström, G.: Verzeichnis der Schriften Leonhard Eulers. Teubner, Leipzig (1913)
14. Euler, L.: Découverte d'un nouveau principle de mécanique. Histoire de l'Académie Royale des Sciences et des Belles-Lettres de Berlin 6, 185–217 (1752)
15. Euler, L.: Nova methodus motum corporum rigidorum degerminandi. Novi Commentarii Academiae Scientiarum Petropolitanae 20, 208–238 (1776)
16. Euler, L.: Die Gesetze des Gleichgewichts und der Bewegung flüssiger Körper. Crusius, Leipzig (1806)
17. Euler, L.: Opera Omnia, Ser. III, Bd. 1. Teubner (1926)
18. Euler, L.: Opera Omnia, Ser. II, Bd. 12. Orell Füssli (1954)
19. Euler, L.: Opera Omnia, Ser. II, Bd. 13. Orell Füssli (1955)
20. Fellmann, E.A.: Leonhard Euler. Birkhäuser (2007)
21. Fox, R.: The Caloric Theory of Gases from Lavoisier to Regnault. Clarendon Press, Oxford (1971)
22. Galilei, G.: Mathematical discourses concerning two new sciences relating to mechanicks and local motion in four dialogues. Printed for J. Hooke, at the Flower-de-Luce, over-against St. Dunstan'S Church in Fleet-street, London (1730)
23. Gay-Lussac, J.L.: Recherches sur la dilatation des gaz et des vapeurs. Annales de Chimie 43, 137–175 (1802)
24. Gay-Lussac, J.L.: Untersuchungen über die Ausdehnung der Gasarten und der Dämpfe durch die Wärme. In: Ausdehnungsgesetz der Gase. Ostwalds Klassiker der exakten Wissenschaften Band 44. Verlag Harri Deutsch (1997)
25. Gottwald, S.: Lexikon bedeutender Mathematiker. Verlag Harri Deutsch (1990)
26. Hugoniot, H.: Mémoire sur la propagation du mouvement dans les corps et spécialement dans les gaz parfaits. Journal de l'École Polytechnique 57, 3–97 (1887)
27. Hugoniot, H.: Mémoire sur la propagation du mouvement dans les corps et spécialement dans les gaz parfaits. Journal de l'École Polytechnique 58, 1–125 (1889)
28. Johnson, J.N., Chéret, R.: Classic papers in shock compression science. Springer (1998)
29. Joule, J.P., Thomson, W.: On the Thermal Effects of Fluids in Motion – Part II. Philosophical Transactions of the Royal Society of London 144, 321–364 (1854)
30. Klein, F.: Vorlesungen über die Entwicklung der Mathematik im 19. Jahrhundert. Chelsea Publishing Company (1967)
31. Krehl, P.O.K.: History of Shock Waves, Explosions and Impact: A Chronological and Biographical Reference. Springer (2009)
32. Lagrange, J.-L.: Nouvelles recherches sur la nature et la propagation du son. In: Serret, J.-A. (ed.) Ouvres de Lagrange, vol. 1, pp. 151–332. Georg Olms Verlag, Hildesheim (1973)

33. Laplace, P.-S.: Sur la Vitesse du Son dans l'air et dans l'eau. Annales des Chimie et de Physique 3, 238–241 (1816)
34. Lenihan, J.M.A.: Mersenne and Gassendi: An early chapter in the history of sound. Acustica 1(2), 96–99 (1951)
35. Lenihan, J.M.A.: The velocity of sound in air. Acustica 2(5), 205–212 (1952)
36. Lipkens, B.: Book review: Classic papers in shock compression science. The Journal of the Acoustical Society of America 111(3), 1143–1144 (2002)
37. Mach, E.: Die Mechanik. Wissenschaftliche Buchgesellschaft, Darmstadt (1976)
38. Mayer, J.R.: Bemerkungen über die Kräfte der unbelebten Natur. Annalen der Chemie und Pharmacie 42(2), 233–240 (1842)
39. Mayer, J.R.: Die organische Bewegung in ihrem Zusammenhange mit dem Stoffwechsel: Ein Beitrag zur Naturkunde. Verlag der C. Drechsler'schen Buchhandlung, Heilbronn (1845)
40. Mayer, J.R.: Bemerkungen über das mechanische Aequivalent der Wärme. Verlag von Johann Ulrich Landherr, Heilbronn (1851)
41. Mayer, J.R.: Die Mechanik der Wärme. Ostwalds Klassiker der exakten Wissenschaften Band 180, Reprint der Bände 37, 180 und 99, 1st edn., Verlag Harri Deutsch (2003)
42. Navier, C.L.M.H.: Mémoire sur les lois du mouvement des fluides. Mémoires de l'Académie des Sciences 6, 389–440 (1823)
43. Newton, I.: The mathematical principles of natural philosophy, vol. II. Printed for Benjamin Motte, London (1729)
44. Newton, I.: Mathematische Principien der Naturlehre. Oppenheim, Berlin (1872)
45. Parry, W.E., Hooker, W.J.: Appendix to Captain Parry's journal of a second voyage. John Murray (1825)
46. Poisson, S.D.: Mémoire sur la théorie du son. Journal de l'École Polytechnique 14(7), 319–392 (1808)
47. Poisson, S.D.: On the Caloric of Gases and Vapours. Philosophical Magazine 62, 328–338 (1823)
48. Poisson, S.D.: Sur la Chaleur des gaz et des vapeurs. Annales des Chimie et de Physique 23, 337–352 (1823)
49. Rankine, W.J.M.: On the Thermodynamic Theory of Waves of Finite Longitudinal Disturbance. Philosophical Transactions of the Royal Society of London 160, 277–288 (1870)
50. Riemann, B.: Ueber die Fortpflanzung ebener Luftwellen von endlicher Schwingungsweite. Abhandlungen der Königlichen Gesellschaft der Wissenschaften zu Göttingen (1860)
51. Sachdev, P.L.: Shock waves and explosions. Monographs and Surveys in Pure and Applied Mathematics, vol. 132. Chapman & Hall/CRC (2004)
52. Salas, M.D.: The curious events leading to the theory of shock waves. Shock Waves 16(6), 477–487 (2007)
53. Schlote, K.-H.: Chronologie der Naturwissenschaften. Verlag Harri Deutsch (2002)
54. Stokes, G.G.: On a difficulty in the theory of sound. Philosophical Magazine 33, 349–356 (1848)
55. Stokes, G.G.: On a difficulty in the theory of sound. Mathematical and Physical Papers 2, 51–55 (1966)
56. Szabó, I.: Geschichte der mechanischen Prinzipien und ihrer wichtigsten Anwendungen. Birkhäuser Verlag (1987)
57. Thomson, W.: On an Absolute Thermometric Scale founded on Carnot's Theory of the Motive Power of Heat, and calculated from Regnault's Observations. Philosophical Magazin 33, 313–317 (1848)

58. Thomson, W.: On the Dynamical Theory of Heat, with numerical results deduced from Mr. Joule's equivalent of a Thermal Unit, and M. Regnault's observations on Steam. Philosophical Magazin 4, 105–117 (1852)
59. Thomson, W.: On the Dynamical Theory of Heat, with numerical results deduced from Mr. Joule's equivalent of a Thermal Unit, and M. Regnault's observations on Steam. Transactions of the Royal Society of Edinburgh 20, 261–288 (1853)
60. Thomson, W.: On the Dynamical Theory of Heat. Transactions of the Royal Society of Edinburgh 21, 123–171 (1857)
61. Thomson, W.: Über die dynamische Theorie der Wärme. Ostwald's Klassiker der exakten Wissenschaften Nr. 193. Verlag von Wilhelm Engelmann, Leipzig (1914)
62. Truesdell, C.A.: Zur Geschichte des Begriffes "innerer Druck". Physikalische Blätter 12 (1956)
63. Truesdell, C.A.: Essays in the history of mechanics. Springer (1968)
64. Truesdell, C.A.: Rückwirkungen der Geschichte der Mechanik auf die moderne Forschung. Humanismus und Technik 13(1), 1–25 (1969)
65. von Guericke, O.: Neue "magdeburgische" Versuche über den leeren Raum (1672). Ostwald's Klassiker der exakten Wissenschaften Nr. 59. Akademische Verlagsgesellschaft mbH, Leipzig (1936)
66. Webster, C.: Richard Towneley and Boyle's Law. Nature 197, 226–228 (1963)
67. Wilson, D.B.: The Correspondence between Sir George Gabriel Stokes and Sir William Thomson, Baron Kelvin of Largs, vol. 2, pp. 1870–1901. Cambridge University Press (1990)

The Historical Development of the Strength of Ships

Eike Lehmann

Abstract. Timber from oaks and soft wood was hundreds of years the only material for building ocean going ships. Sufficient strength of wooden ships based only on the practical experience of their builders. The strength of iron ships has been estimated first by William Fairbairn 1860. His method based on a very modern ultimate load concept. A couple of years later John Macquorn Rankin first published the physically correct formulation of the longitudinal bending moment of the whole ship hull structure including a quasi- static wave effect. On occasion of the sinking of the torpedo boat H.M.S. Cobra 1901 the bending test of a similar ship H.M.S. Wolf in a dry-dock has been prepared. The test results were found not in line with the classical bending theory. After a lot of research work Georg Schnadel found 1929 a proper explanation by taking the post buckling behavior of the thin deck plating under consideration. It needs nearly hundred years to consider the probabilistic nature of the seaway into the design formula of the longitudinal strength of ships. Nowadays almost all structural members of the steel design of ships like the stiffened plates of the shell, bulkheads and decks has be calculated by the classical theory of elasticity, including post buckling effects following the rules of the classification societies. Since the end of the last century the Finite Element Method is used not only for structural details but also for the whole hull structure, including the dynamic loads of the seaway.

Keywords: Strength of wooden ships, longitudinal bending strength of iron ships, torsion of the hull, post buckling, principle of strength of container ships and tanker.

Eike Lehmann
Institute of Ship Structural Design and Analysis,
University of Technology Hamburg-Harburg, Schwarzenberg Str. 95 C, 21073 Hamburg

1 Wooden Ship Strength

The strength of wooden ships never became the subject of intensive theoretical deliberations. The scantlings were chosen according to experience and the availability of appropriate wood. Nevertheless, problems were encountered with the connections between the various components; these difficulties accelerated the introduction of iron and steel as the shipbuilding material of choice, especially owing to the growth in the size of ships in the 19th century – also in view of the fact that England, as the leading shipbuilding nation of the time, began to suffer a chronic shortage of suitable timber.

The latter reason in particular accelerated the change from wood to iron in shipbuilding, although the strength aspect of the timber in relation to iron is not as serious as it may appear. For example, the relationship between the bending strength of good pine timber in relation to the same weight of steel is about 30 and the bending stiffness is in fact far more than 300. Incidentally, this is also the reason why, on the iron sailing ships with wooden decks, a deck beam was only provided at every second frame as support for the deck planks. Such a construction can be seen very well on the museum ship "Rickmer Rickmers" in the Port of Hamburg.

The drawback in using wood for shipbuilding lies rather in the joining technology, which for the case of tension, depending on the joint used, offers no more than 10 to 20 % of the tensile strength of wood as the basic material.

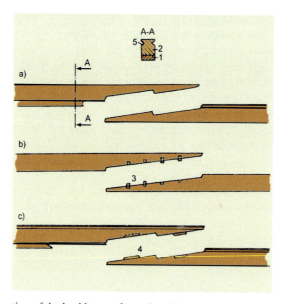

Fig. 1 The connection of the keel beam of wooden ships

The connection of the keel beam of wooden ships was always a weak point, and one which even the diverse joining methods were not able to rectify from a fundamental viewpoint. Keel beam connection, from *"Vorlageblätter für Schiffbauer"* by Gustav David Klawitter, how was the first teacher in naval architecture in Prussia [1].

Over and above this intrinsic drawback, the usual carvel construction exhibited another basic weakness, in that only a limited transfer of shear forces is possible between the planks, and this depends on the quality of the caulking.

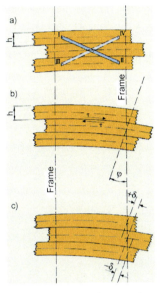

a) German connection; b) English type of connection; c) Danish connection

Fig. 2 Effect of diagonal tie bare braces

Through diagonal braces of steel, the aim was a) to attain an increased shear transfer of the shell planks, which then b) causes the outer shell, as the web of the girder, c) to bend uniformly. If this does not occur and if the caulking was not performed with the necessary care, the individual planks are displaced, which leads to a substantial reduction in the stiffness of the entire hull girder. This effect was occasionally to be seen in earlier times as the "hogs back" of older wooden ships.

Shell structure amidships with diagonal tie bar braces (1) and wooden pressure diagonals (2). The deck beams are supported by iron brackets (3). Experts will recognize that this particular arrangement is only effective for ships subject to loads bending the hull up (hogging). See Figure 3.

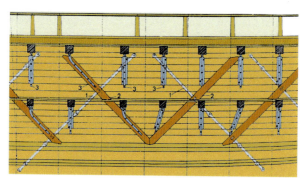

Fig. 3 Shell structure amidships with diagonal tie bar braces

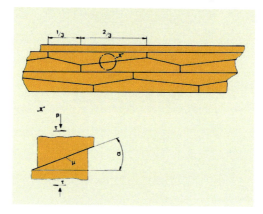

Fig. 4 Function of the anchor stock planks

Another possibility was the use of anchor stock planks, but this method was very time-consuming and usually only applied with naval vessels[2].

Moreover, this was preferentially implemented at the waterline, where the shear stresses are highest.

The thorough scientific analysis of the strength of ships really began with the construction of a very large iron ship. Boasting a length of 211 m and a displacement of 32,000 tons, the "Great Eastern" was built far ahead of its time in 1858/59. The hull had still been constructed purely intuitively, yet with adequate strength.

2 Ultimate Load Estimation of William Fairbairn

However, the desire for a comprehensive scientific treatment of the strength question was evident, because the first paper of scientific note on the strength of ships [3], which was presented by the English industrial entrepreneur William Fairbairn in 1860 before the Institution of Naval Architects in London (founded in the very same year), met with extremely lively interest and heralded the era of strength research in shipbuilding.

The Historical Development of the Strength of Ships

Fig. 5 Fairbairn's determination of load-bearing capacity

His approach to determining the overall strength of ships was simple and, to a certain extent, also quite modern. He believed that an initial appraisal of the necessary strength of a ship was possible if one considered the vessel to be constrained at the height of the midship section and, in a second load case, at each of the ends. He then determined the probability of the ship breaking in such a situation by using the magnitude of the tolerable reaction force W, which results as the supporting force from the strength of the hull. If the supporting forces are smaller than the weight of the ship, then there is no failure; however, if they are greater, the hull will rupture. The reaction force was estimated by Fairbairn as

$$W = \frac{a \cdot d \cdot c}{l}$$

where "l" is the length of the sagging ship, "d" the effective moulded depth, "a" the load-bearing cross-sections and "c" the breaking strength of the riveting. In view of the difficulty in assessing the strength of wooden decks of ships otherwise made of iron, this is a very sensible initial approach, which can easily be confirmed by a calculation of the moments. This is therefore a very modern ultimate strength calculation, if one views $a \cdot d$ to be the section modulus effectively leading to fracture. It should be noted that microstrain measurements were not yet possible at the time. The only way was to determine the breaking load of individual components by using a tensile testing machine. The breaking strength of the riveting of those times varied between 60% and 80% of the breaking strength of the basic material [4], as Lloyd's Register had determined empirically for various riveted joints. These values have remained more or less unchanged to the present day and hence demonstrate the weakness of the riveting technique in relation to today's welding technology, which for the welded connection realizes the same strength as the parent metal.

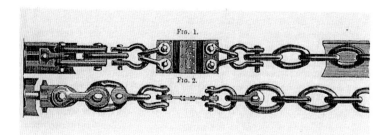

Fig. 6 Lloyd's Register: breaking tests of riveted sheet, 1858

At the same session at which Fairbairn had submitted his estimates, a certain John Grantham [5] also warned that the ship must be viewed as an elastic beam and that the distribution of buoyancy and weight had to be considered properly. The correct moment of inertia for determining the bending stress was also calculated by Robert Mansell [6] for the first time in the same year, so that the Bernoulli-Euler beam theory – already generally known in the British scientific community of the time through the work of William John Macquorn Rankine, Manual of Applied Mechanics [7] of 1858 – could now also be applied by naval architects.

In addition to that, Rankine himself made detailed statements on the overall strength of ships in his "Shipbuilding – Theoretical and Practical" published in 1866 jointly with Isaac Watt, Fredrick K. Barnes and Robert Napier [8]. Here Rankine [9] recommends assuming for merchant vessels the maximum moment for the case "ship on wave crest" as

$$M_{WVH} = D \cdot L/(37 \div 59)$$

and in still water as

$$M_{SW} = D \cdot L/(40 \div 59)$$

One therefore obtains

$$M_{TH} = D \cdot L/(19 \div 30)$$

It took more than 10 years – until 1871 – before the Director of Naval Construction of the Royal Navy, Edward James Reed (1830–1906) [10], published precise calculations on the buoyancy and weight distribution of ships of the Royal Navy, which were the prerequisite for integration of both the shear force and the bending moments and their distribution over the ship's length [11, 12].The precise determination of the distribution of the buoyancy and the weights of ship and cargo had already been the subject of scientific discussion since 1746 by

Fig. 7 William John Macquorn Rankine (1820–1872) Edward James Reed (1830-1906)

Pierre Bouguer (1698–1758), but this was only conducted in its modern form by Edward James Reed [13] – first of all in still water, and then for the shear forces and moments in a seaway [14]. For the ironclad "HMS Minotaur" (9800 tons and 400 ft in length), Reed calculated a max. still water moment of 44,545 ft-ts, such that this results from

$$M_{SW} = D \cdot L / 88$$

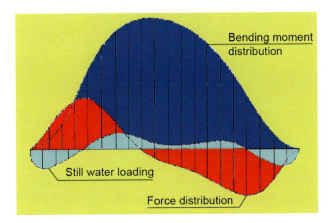

Fig. 8 Still water loading, shear force and bending moment distribution of the British ironclad "Minotaur", Reed 1872

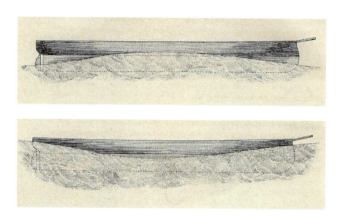

Fig. 9 Ship in wave crest (hogging) and wave trough (sagging), Reed 1873

Of course, the still water calculation does not suffice to assess the stress induced in a seaway, especially as this was usually much smaller for the warships of the time than for the conventional steamships of the merchant navy. For this reason, Reed carried out such a calculation for a seaway case as early as 1873.

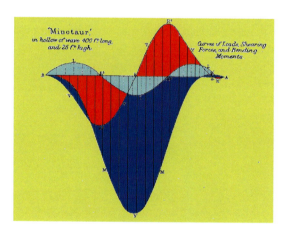

Fig. 10 Ship in wave trough (sagging): wave-induced loading (L L L), shear force distribution (V V V) and bending moment distribution (M M M) of the British ironclad "Minotaur", Reed 1873

Rankine's assumption that a sine wave of the same length as the ship leads to the greatest stress was rapidly accepted. What wave height should be taken was more of an arbitrary choice, however. Reed made his calculations with a wave height of $\lambda/16$, which yielded 25 ft for a ship length of 400 ft.

For the case "ship on wave crest", he obtained a total moment of

$$M_{TH} = D \cdot L / 28$$

and for the case "ship in wave trough", a moment of

$$M_{TS} = -D \cdot L/53.$$

If the still water moment is calculated as

$$M_{SW} = D \cdot L/88$$

one obtains for the case "ship on wave crest"

$$M_{WVH} = D \cdot L/41$$

and for the case "ship in wave trough"

$$M_{WVS} = -D \cdot L/33.$$

That is, Reed had already identified the fact that the stress on a ship caused by the waves is much greater for the load case "ship in wave trough" than when the vessel is on the wave crest. The verifying calculation of various service-proven ships performed by Willliam John, the Chief Surveyor of Lloyd's Register (the leading classification society at the time), then led to the agreement in 1874 that a normal mercantile vessel would have to bear, in the most unfavourable case "ship on wave crest", a max. total bending moment of

$$M_{TH} = D \cdot L/35.$$

Without damage [15]. It was possible to make this definition because the cargo ships coming into question here exhibited a "hogging" still water bending moment in the empty load case, hence accounting for the worst possible situation.

The insights into the bending moment in a seaway were then expanded at the end of the 19th century through the work of W.E. Smith (1883), who first described the nonlinear pressure distribution in a wave [16] (now referred to as the Smith effect) as well as of T.C. Read in 1890, who estimated the dynamic effect of a wave on the bending moment for the first time as lying in the order of 40 % [17] and of Alexander Kryloff in 1896 [18], who examined the influence on the bending moment of squatting and pitching in a seaway. The effect of the hull shape on the longitudinal strength was already described in detail by F.H. Alexander in 1905 [19]. For this, he used the shape influence of the waterline coefficient, whereas today, the somewhat more precise block coefficient c_B of the hull is applied. Despite the fact that these theoretical insights into the dimensioning of ships were already available at the end of the 19th century, the scantlings of merchant vessels were still chosen without the benefit of such direct strength analyses and solely according to the experience gained with previous ships. On the basis of "scanting numerals" that were more or less plausible, plate thicknesses and section sizes were compiled as tables and corresponding instructions in the construction rules of the classification societies. If a definition proved to be inappropriate in practice, these tables were revised accordingly. The system was of practical benefit, with the result that it proved possible to build reliable ships without any in-depth knowledge of strength and statics. The ships were reliable because the ship sizes changed only very slowly over many decades, with the exception of the large, fast transatlantic steamers. It is therefore not surprising that this design practice was generally considered adequate right up until the 1960s.

Even Georg Schnadel, who made unparalleled contributions to hull strength and, as the Chairman of the Executive Board of Germanischer Lloyd, was in the best position to foster and assert the dimensioning of ships with the methods of the technical beam theory, was not able to achieve this goal. It took several years after he left Germanischer Lloyd in 1959 that the direct calculation of longitudinal strength began in 1964 to be incorporated into the construction rules.

Today's usual subdivision of the computation of the bending moment into one for still water and one considering the additional loading in a seaway has a very practical but also a formal reason. The practical aspect is that there is a particular safety-related interest in examining precisely how far a ship has been loaded in the port, i.e. in still water, so that it can sail safety over the ocean. This is augmented by a methodical reason. The still water stress is a deterministic task and must be checked by direct means before leaving a harbour. For the wave-induced stress, which is of a probabilistic nature, this is much more difficult and, moreover, open to interpretation, because the all-important question of what wave heights must be assumed for the deterministic computation of the bending moments remains a matter of the assumption and assessment of the seaway (cf. wave coefficient). In order to investigate the question of the stress acting on a ship in a natural sea state using a real ship, an ocean trial was therefore undertaken under the leadership of Georg Schnadel, with the involvement of Otto Lienau, Fritz Horn, Georg Weinblum, Georg Weiss et al. on the motor vessel "San Francisco", which took place in the autumn of 1934 from Hamburg to the US West Coast and back. This trial voyage, which united Germany's scientific experts for shipbuilding for the first time in a large-scale research campaign on hull strength, was met with an unusually high level of interest [20, 21, 22, 23].

Besides benefitting from careful preparation, the success of the trial was based on the use of the various measurement facilities of the German Research Institute for Aviation, with which Schnadel maintained a close cooperation in Berlin.

The main results were as follows:

- Although the design wave should have a length equal to that of the ship, the wave heights and steepnesses proved to be much greater than previously assumed.
- The theoretical wave heights calculated from the longitudinal strength measurements were only half as high as the observed wave heights.

- However, the well-known Smith effect did not fully explain this discrepancy.
- Furthermore, it was shown that, for the case "ship on wave crest" (hogging), the bending moments are smaller than for the case "ship in wave trough" (sagging). The significance of this is that higher waves must be assumed for the case "ship in wave trough" than for the wave crest. The latter finding confirmed Edward Reed's deliberations from 1873 and is still considered today in the construction rules of the classification societies for calculating the vertical wave bending moment.

Fig. 11 Cargo motor vessel "San Francisco", built in 1928 by Deutsche Werft for HAPAG (Length Lpp 136.80 m, beam 18 m, moulded depth 9.06 m, displacement 13,070 t) [24]

Today's dimensioning practice considers, by international agreement, a minimum section modulus and inertial moment of the main section. Since the admissible maximum stress is likewise defined and the vertical wave bending moment is also defined with this design wave by means of a formula, the design equation for the usual merchant vessels offers only a few possibilities for optimization.

$$M_T = M_{SW} + M_{WV} \leq W_{min} \cdot \sigma_P \cdot 10^3$$

Here σ_P is the permissible stress of 175 N/mm² when using normal shipbuilding steel, M_{SW}, M_{WV} are the still water and vertical wave bending moments, and W_{min} the minimum section modulus of the main frame.

$$W_{min} = L^2 \cdot B \cdot c_0 \cdot (c_B + 0{,}7) \cdot 10^{-6} \ [m^3]$$

Today, the vertical wave bending moment is uniformly calculated as

$$M_{WV} = L^2 \cdot B \cdot c_0 \cdot c_1 \ [kNm]$$

where unimportant factors have been neglected. The ship's length and breadth B are inserted in metres. The inconspicuous coefficient c_0 accounts for the special character of the natural wave action and is defined as

$$c_0 = 10{,}75 - \left(\frac{300-L}{100}\right)^{\frac{1}{3}}$$

This "wave coefficient" is the result of the seaway calculations commonly applied since the end of the sixties, based on long-term statistics of observed waves that accurately reflect the natural sea state. The wave environment to be found in the North Atlantic is used. The expected value of the maximum bending moment occurring once every 10^8 load cycles, with a service lifetime of a ship lasting some 30 years, was set more or less arbitrarily. With the aid of the magnitude of the quadratic transfer function of the bending moment, the seaway spectrum is transformed into the spectrum of the bending moments, the mean value of which leads to the above wave coefficient.

The fact that the bending moment in the case "ship on wave crest" is smaller than for "ship in wave trough" is considered by the factor c_1, which for the former case (hogging) is set to $c_{1H} = 0{,}19 \cdot c_B$ and for the latter (sagging) to $c_{1S} = -0{,}11 \cdot (c_B + 0{,}7)$.

The idiosyncratic determination of a minimum section modulus is based upon the notion that one fundamentally assumes a still water bending moment amounting to at least approx. 60 % of the vertical wave bending moment for the case "ship in wave trough" (sagging).

Since not only the permissible stress but also the minimum section modulus is laid down internationally as mandatory values, one can really only attempt nowadays to adjust the still water bending moment by a judicious arrangement of the fuel and ballast tanks such that the design equation is fulfilled with an equal sign. If this is not the case, the section modulus W must be increased above W_{min}, which naturally involves more steel and increased costs.

A storm of protest is occasionally raised against this "minimum" rule, i.e. the minimum section modulus. Why is the naval architect not allowed to design ships with a still water bending moment that is almost zero?

Must we really maintain a longitudinal strength that permits a safe passage to America across the stormy North Atlantic when we only intend to sail through the comparatively calm Baltic?

Can we not simply avoid heavy weather with the aid of advance information by satellite? Can we not, in such cases, use a part of this spare seakeeping ability for an "overdraft" on the permissible still water bending moment, which would offer a substantial economic benefit?

By voicing these and similar arguments, the shipowners seek to increase the profitability of their ships, which is a legitimate idea, seeing that it would indeed be feasible. Nevertheless, the classification societies agree that, in the interest of uniform standards for the ships engaged in international trade and from the experience that the operational region of ships can certainly change during their life cycles, this provision has remained in place.

3   None Linear Structural Analysis. The Sinking of the Torpedo Boat HMS "Cobra"

The actual strength issues requiring thorough scientific treatment arose towards the end of the 19th century on naval vessels. The sinking of the torpedo boat HMS "Cobra" of the Royal Navy in 1901 provided the inducement for the first measurement on a real ship.

Fig. 12 HMS "Cobra"

The "Cobra" had been built as a test ship by Armstrong Withworth, Newcastle on Tyne, in 1899 to demonstrate to the British Admiralty the superiority of the propulsion of fast torpedo cruisers with a steam turbine, then a new invention by Charles Parson. After it had been shown that this 68 m vessel had reached 36.63 knots – constituting the gigantic Froude number of 0.73, the Royal Navy bought the ship for testing purposes. In 1901, however, the ship broke into two parts in a strong North Sea swell, fracturing abaft the last funnel and sinking with the loss of a large number of naval crewmen and Parson's test personnel.

It was then decided in 1903/04 to perform a large-scale test with a structurally similar vessel, HMS "Wolf", in a Portsmouth dry dock; the ship was to be loaded solely amidships and then, as a second load case, solely at the bow and the stern – just as Fairbairn had already examined theoretically in his first appraisal 40 years earlier. Under the ship's own weight, this yields the simple load cases of a cantilever beam with tensile loading in the deck and compression in the bottom and, in the second case, a beam resting on two supports with compressive stresses in the deck and tension in the bottom.

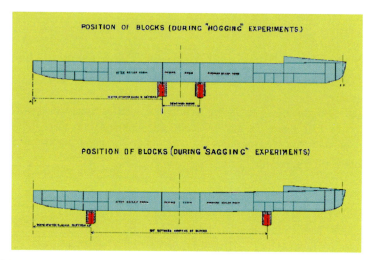

Fig. 13 Large-scale test with the torpedo boat HMS "Wolf" in 1902

The measurements were performed with great care by Sir John Biles and his staff [25]. However, the verifying calculation with the aid of the technical beam theory showed that correspondence with the measurements was only found when one uses an elastic modulus that is considerably smaller than that of the base material and, further, is assumed to be dependent on the magnitude of the stress as well as on the load case.

Naturally, the experts refused to accept this. First of all, various authors tried to calculate the substitute moment of inertia that applied for the beam theory, taking into the account the openings in the deck, but this turned out to be only a minor effect.

Moreover, considering the effect known today as "effective breadth" or also as "shear lag" did not provide a satisfactory explanation. The search for an acceptable solution to the "Wolf" problem led to the situation, however, that the naval architects attempted to at least approach the problems of calculating the hull structure in a fundamental manner by using the tools offered by the theory of elasticity. For example, the effective breadths for a wide range of cases were calculated by Georg Schnadel in 1926 with the aid of the Airy differential equation. It is easily determined that Schnadel's calculations correspond well to computations using the method of finite elements [26]. This is all the more surprising, because shear stresses (not considered in the Schnadel solution) do not satisfy the equilibrium of forces in the outer shell, which was certainly noticed at the time in the discussion by Hans Reissner [27]. In fact, the teamwork between the naval architects and the mechanics specialists in Berlin turned out to be extremely useful. Besides Hans Reissner, Moritz Weber, who made a name for himself – not only in shipbuilding but also for the entire field of technical testing – with the so-called π theorem for formulating similitude laws, also deserves mention [28, 29].

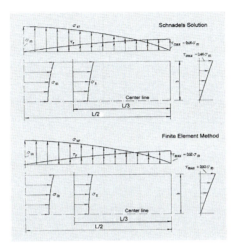

Fig. 14 Effective breadth using the Airy differential equation, calculated according to Schnadel and by the Finite Element Method

It was only after more than 25 years, in 1931, that Georg Schnadel presented a satisfactory solution by considering the post-buckling effect for the outer shell and main deck of HMS "Wolf", which was very thin at 3.11 to 4.97 mm [30].

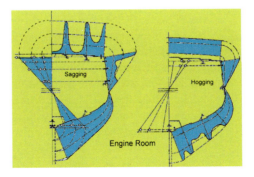

Fig. 15 Stress distribution: left with compression in the deck (sagging); right with pressure in the bottom (hogging) of the torpedo boat HMS "Wolf", as calculated by Georg Schnadel

The calculation of the post-buckling behaviour of individual plates had already been performed by Maurice Levy [31] in 1899, and by Hans Reissner, Georg Hartley Bryan and Stephen P. Timoshenko in the 1920s. Schnadel then carried out this calculation for the continuous plate fields of hull structures with the aid of the method of determining the minimum of the strain energy according to Timoshenko [32]. It is therefore of great interest to compare the various solutions with the results offered by modern numerical methods.

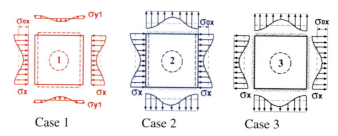

Case 1 Case 2 Case 3

Fig. 16 Different boundary conditions of clamped plates

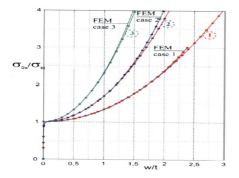

Fig. 17 Post buckling solutions of Levy and Schnadel in comparison with FE calculations

The Schnadel solution is obtained when one assumes that the edges are immoveable. This assumption was arrived at by Schnadel by taking a chessboard-like buckling mode for a stiffened plate field. He assumed that, for a chessboard buckling mode, the edges of each plate field will not be displaced and will remain straight. However, it is then unclear as to how stresses within the individual plate fields can arise through an external load acting on the plate field. If we have arrived at an "FE solution" nonetheless then only through a physically inadmissible trick by first proceeding as for solution 2, by fixing the buckling surface and then returning the edge to its original position.

Of course, Schnadel assumed an elastic structural behaviour. With today's hull structures, plasticisation usually takes place at an early load stage, so that the classic cases best described by the *von Kármán* equations do not, in general, cover the real situation in shipbuilding correctly.

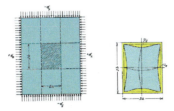

Fig. 18 Schnadel's model for calculating the post-buckling behaviour of thin plate structures on ships

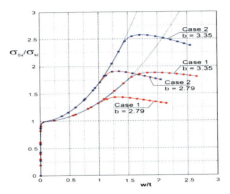

Fig. 19 The post-buckling behaviour of the square plate with consideration of an elasto-plastic material behaviour $b = \frac{2a}{t}\sqrt{\frac{\sigma_y}{E}}$

4 Torsion of the Ship Hull

Another problem that initially resisted a purely theoretical treatment was the torsion of the hull. The first to address this question empirically was Otto Lienau, who in 1928 investigated simple ship-like hollow bodies with large hatchways [33]. In 1927, Danzig University of Technology was the first institution of higher learning in Germany to set up a strength testing laboratory for investigating ship structures, albeit within a modest scope.

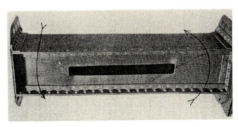

Fig. 20 First test body for studying the torsional stresses in an idealized ship hull with a hatchway

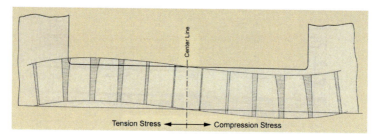

Fig. 21 Normal stresses in the deck caused by a torsional moment, according to Lienau

Besides the already known shear stresses, the twisting of the test body resulted in considerable normal stresses, to which Lienau remarked: *The entire process taking place in the deck plate adjacent to the hatchway may possibly be described theoretically as a bending process at the deck plate, which, beginning from the middle of the hatch, experiences on the one side a concave and on the other side a convex shape change with, however, only tensile or compressive stresses arising over the corresponding plate width. To what extent these problems can be expressed accurately by theoretical means is still a matter for research.*

Today, we know – thanks to the warping torsion theory of Vlasov [34] – that Lienau was describing the warping stresses, which now play a significant role in dimensioning the structural members of container ships. It is of interest that Stephen Timoshenko, in continuing the doctoral thesis of Ludwig Prandtl as a student of his in Göttingen, had already concerned himself in 1905 with the nature of the normal stresses in T-beams under torsional load, and had also derived the applicable differential equation [35]. Evidently, this work was not known to naval architects, because neither Lienau nor Schnadel mentioned it. A numerical verifying calculation of the Danzig tests using the method of finite elements confirmed Lienau's supposition.

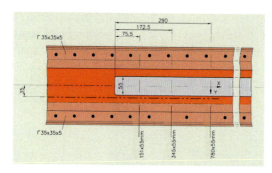

Fig. 22 Test models with three hatchways of different length, as used by Lienau

Fig. 23 Cross-section of Lienau's test models

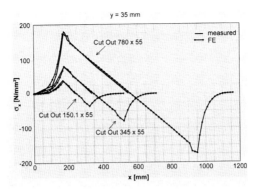

Fig. 24 Measurements of the warping stresses correspond with the FE computations surprisingly well

What the measurements of Lienau do not cover, however, are the notch stresses in the hatchway corners. When container ships were to be built in the 1960s, the question of torsional strength came into the focus of attention. Besides accounting for the warping stresses in the longitudinal strength, it was also necessary to determine the stresses in the hatchway corners in a reliable manner. The method for calculating the warping stresses was based on the differential equation already established by Timoshenko:

$$E \cdot J_\omega(x) \cdot \varphi^{III} + G \cdot J_d(x) \cdot \varphi^I = T(x)$$

or

$$\varphi^{III} + \lambda(x) \cdot \varphi = T(x) / E \cdot J_\omega(x)$$

where:

$$\lambda(x) = G \cdot J_d(x) / E \cdot J_\omega(x).$$

This differential equation only becomes an ordinary differential equation when λ is a constant. Since the shear centre changes in height due to the shape of the ship and hence both the torsional moment of inertia J_d and the warping moment of inertia J_ω are dependent on x, one arrives at a relatively simple solution if the same curve is assumed for both, so that the decay factor λ becomes a constant. Even a section-by-section treatment leads to a jump in the stress resultants, which is magnified by the very localized warping restraint of the transverse box girders. Moreover, even "smearing" the deck area with a notional deck thickness only leads to the situation that the actual critical stresses in the hatchway corners are not included at all. All in all, this modelling approach using the theory of warping torsion does not suffice for very large container ships, so that only a complete FE model is viewed as the necessary basis for dimensioning.

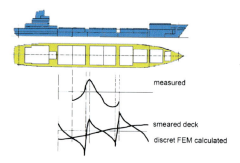

Fig. 25 Warping stresses of a container ship: measured, with smeared deck thickness as a closed body proposed by de Wilde and with discrete consideration of the transverse box-girder [37]

In view of this uncertainty, comprehensive measurements were conducted in Japan and also in Germany on ship-like models, together with diverse shipboard measurements. In Hamburg, a model made of aluminium was examined on the large strength testing unit of the Institute for Shipbuilding of Hamburg University by the Institute for Structural Design and Strength at the Technical University of Hanover [38]. Aluminium was chosen in order to measure strains and displacements adequately. We verified these measurements 35 years later using the finite element method and found an astounding degree of correlation.

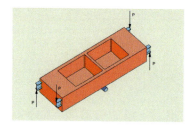

Fig. 26 Ship-like torsion model with large hatchway, with the dimensions of length approx. 10.0 m, breadth 2.40 m and height 1.80 m

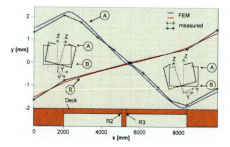

Fig. 27 Measured and calculated angle of twist

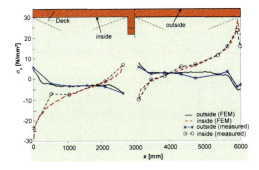

Fig. 28 Measured and calculated warping stress

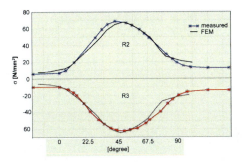

Fig. 29 Good correlation, even for the stresses in the hatch corner radii

The intensive fundamental research into the torsional strength of container ships worldwide, and also to a special degree at Germanischer Lloyd and the universities in Germany, is a major reason why we have not suffered any major structural failure for this sensitive ship type up to the present day. However, Vlasov's theory is no longer applied today; it has been supplanted by numerical computation using the Finite Element Method. Since these models are extremely large, precise modelling of the especially critical hatchway corners is replaced by notch factors.

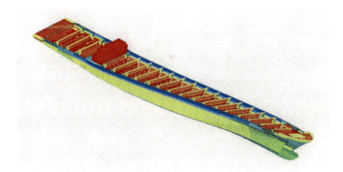

Fig. 30 FE computer model by Germanischer Lloyd of a container ship, here with a torsion load case. Scantling up to 6 mm red, up to 12 mm yellow and up to 60 mm blue painted.

5 The Disaster of the Big Tankers

An exception to this happy state of affairs must also be mentioned. With the construction of the large crude oil tankers at the end of the sixties of the last century, serious structural failure occurred worldwide – also to a major extent on ships built by German shipyards. The cause lay in the design practice, commonplace up to that time, of using design tables for the dimensioning of the scantlings proposed by the classification societies. The few direct calculations with the aid of the classic slender beam theory, when applied to deep-web frame structures, were so inappropriate that the results did not even exhibit any similarity with the measurements that had been performed. This was the hour of the Finite Element Method in shipbuilding in Germany. With the plane stress elements, it was possible to compute these transversal deep-web frames, which are to be viewed rather as an assembly of membrane plates, in a more realistic manner than as an arrangement of slender beams.

Fig. 31 Jumbo tanker "Esso Scotia", built by the "AG Weser" shipyard in 1969

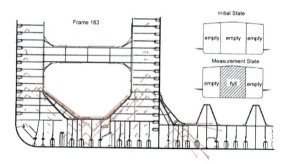

Fig. 32 Measurement of the strengthened web frame of the tanker "Esso Scotia", 1969

However, the FEM mesh was much too coarse, because of the lack of computing power, so that sufficiently precise results could not be expected. Despite this drawback, the results surpassed those of the beam theory by far.

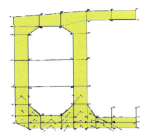

Fig. 33 Computation of the stresses on the tanker "Esso Scotia" from 1969. Excerpt from a 3D FEM model: for reasons of computing power, the chosen mesh was far too coarse.

From the measurements on the tanker "Esso Norway" a sister ship of the "Esso Scotia", it became apparent that the thickness of the webs with 11 mm was so low that buckling effects during the watertight tank test, specifically through the shear stress, were involved in the failure.

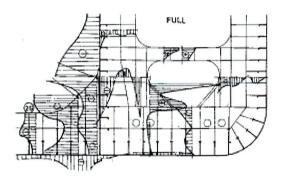

Fig. 34 Measurement of the membrane stresses in a transverse of the "Esso Norway"

The classic buckling analyses, which are based on orthogonal plate fields with simplified stress distributions, were not suitable for providing adequate proofs. The Finite Element Method – or at least suitable software to render these proofs – was not yet at the required state of development in Germany. Moreover, the existing capacity of one staff member precluded any development of own software. It was; however, clear that sooner or later the method would be advanced, and so efforts were focused on investigating which major nonlinear problems in ship design would be of significance for future ships.

Here, the issues involved collision resistance within the scope of safety surveys of nuclear-powered merchant vessels. On this topic, the Society for the Use of Nuclear Energy in Shipbuilding and Shipping (GKSS) carried out a series of collision tests, which met with great interest [39]. Similar trials were also performed in Japan [40] and Italy [41]. The energy dissipated in the event of collision was estimated very coarsely by heuristic means through the deformed steel volume [42]. The first numerical method was developed by Karl-August

Reckling [43], who subdivided the various groups of structural components by differentiating between three typical responses when absorbing the collision energy: component groups which exhibit a concertina effect, i.e. decks, bulkheads and generally plate structures under longitudinal loading; components torn open in the longitudinal direction; and components under membrane-like loading after tearing. For these types of structural components, Reckling then attempted to estimate the energy absorbed. This idea was pursued in the years that followed by Tomasz Wierzbicki. The method requires only simple inputs and yields the dissipated energy. By now, it has been superseded by direct FE calculations, e.g. using DYNA 3D.

Naturally, in the course of developing mechanics-based treatments of ship strength, a large number of methods were taken over, specifically from civil engineering, and adapted to the needs of shipbuilding. For example, comprehensive analyses of transverse strength with multi-level frames and grillage calculations using familiar methods were already carried out before the First World War. For the actual assessment of structural strength, however, this was only done for certain special cases up until the 1960s.

Nonetheless, a particular problem deserves mention one that had concerned naval architects for many decades: the strength of transverse bulkheads and tank walls. With the Merchant Shipping Act of 1854 [44], all British ships over 100 tons deadweight had been required to have watertight transverse bulkheads, with the aim of achieving a substantial improvement in safety at sea.

Initially, it was unclear as to what the most suitable spacing and technical implementation could be. It took some 60 years until certain procedures and ideas found acceptance through the work of the British Bulkhead Committee in 1915 [45]. For the ship owners, this signified an increase in the cost of procurement. The fact that an appreciable extra weight had to be transported over the oceans generated further complaints. This weight represented a reduction in the cargo carried capacity and hence financial losses; and all that just for the relatively rare case of a serious casualty.

In principle, there was an agreement that, for the case of a casualty in which the full water pressure acts on the bulkheads through ingress of water, they may be allowed to become permanently deformed, provided there is no leakage. Presenting such a theoretical proof for a riveted design was something one was not confident of attempting in the time before the First World War, since springing a leak was determined by failure of the riveting, which would take place long before the ultimate strength of the structure was reached. The corresponding tests on board existing ships could be ruled out for various reasons. For this reason, various types of test tanks were built [46, 47]. The designs were chosen so that the load-bearing capability of the stiffeners and their end attachments could be investigated to a large degree. The stiffener spacing was chosen to be so close that no permanent deformation of the plating would arise.

This meant that the immense ultimate strength margin offered by the plating through the membrane effect was neglected for large degrees of sagging. On the basis of these trials, standardized scantlings were therefore laid down as a statutory measure in the form of the corresponding tables.

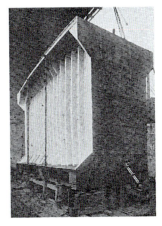

Fig. 35 Test tank with a height of 20 ft

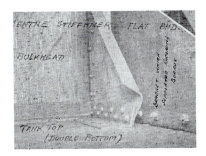

Fig. 36 Plastic failure at the end attachment of the bulkhead stiffeners

At the Institute of Ship Design and Statics of the University of Hanover and the Institute of Shipbuilding University of Hamburg, comparable pressure tanks were examined both empirically and theoretically 60 years later, this time as welded structures. The finding was that the plating can be made much thinner than is usual even today, since the load-bearing capability provided by the membrane effect must be assessed as extremely high; furthermore, greater distances between the stiffeners are thus admissible [48].

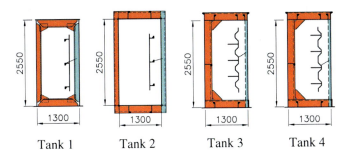

Fig. 37 Test tanks

Tank 5 has been calculated entirely without stiffeners. The comparison of the tank with interior stiffeners to that without stiffeners shows, first of all, the dominating influence of the stiffeners. After failure, the stiffened tank exhibits the same behaviour as the unstiffened tank (Tank 1/ Tank5). A similar but not quite as pronounced effect is seen in the comparison of external stiffeners with the unstiffened tank (Tank 2/Tank 5). Tanks No. 3 and No. 4 are stiffened by so-called bulkhead corrugations. Here one can see very well that the membrane effect for corrugated tanks can also occur only in one direction and thus with a much lesser influence, with the result that the load-carrying behaviour differs from that of the tank without stiffeners.

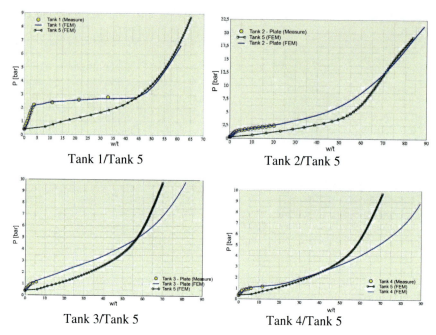

Fig. 38 Load/ deflexion graphs

This examination also explains why ships have sunk with the empty or partial filled fuel oil tanks frequently intact and there is no spillage, despite very high overloading – as was the case with the German ironclad "Blücher", which went down in 180 m of water in the Oslo-Fjord in the Second World War without any spillage.

More recent research aimed at determining the actual ultimate strength of a hull structure. Trailblazing work was done by John Caldwell 1965 [49], who was the first to determine an upper limit to the ultimate strength of ships by calculating the fully plastic bending moment of the hull girder cross-section. Later authors then refined this method by considering local buckling effects and the failure of girders and frames. Here two essential factors determine the quality of the calculations. First of all, the elastic-plastic reserve of a hull girder cross-section is relatively low, as must be expected for thin-walled girders subject to bending; it ranges between shape factor 1.1 and 1.4. Moreover, it must be considered that the loading necessary for reaching the ultimate strength is also physically possible, which in the static case occasionally, arises through improper loading or unloading. In other cases, a limit load can lead to a situation in which the ship has already capsized or has already sunk.

Although the design load had been defined with consideration for certain stochastic aspects, it is a little surprising that ships, quite in contrast to offshore installations, were calculated primarily by deterministic methods, in spite of the fact that the randomness of the wave loads also has a predominant influence on ships. The cause is that ships have to be built within the scope of the very sophisticated construction rules of classification societies, which in part exhibit the stringent nature of textbooks, in order to be granted an international sailing permit. To a very great extent, these construction rules are of a deterministic nature. More recent deliberations that deviate from this stringent approach are under discussion nowadays as "goal-based standards".

6   Summary

The designers of ships were relatively late in recognizing and applying the possibilities of theoretical stress mechanics as an effective tool. Initially, there were the difficulties associated with definition of the loading and the complex geometry of the hull girder, before viable strength analyses could be performed on the basis of the classic beam and plate theory. Added to that, the more or less heuristic dimensioning rules laid down at a very early stage by the classification societies did not allow a thorough application of the strength theories in ship design. Until the advent of numerical methods in the 1960s, theoretical methods have really been limited to isolated cases and the explaining of certain phenomena.

Today, however, one has the impression that the use of the numerical methods of solid mechanics and hydromechanics are being applied extremely intensively in shipbuilding and offshore technology. To sum up, cultivating an appreciation for the science of mechanics is always worthwhile and, conversely, those in the world of mechanics would do well to take note of the needs and concerns of the ship designers.

References

[1] Klawitter, D.G.: Vorlege-Blätter für Schiff- Bauer. Königlich technische Deputation für Gewerbe, Berlin (1835)
[2] van Hüllen, A.: Schiffbau. Verlag von Lipsius & Tischer, Kiel (1888)
[3] Fairbairn, W.: The Strength of Iron Ships. Trans. Inst. of Nav. Arch. I (1860)
[4] Lloyd's Experiments upon Plates and Modes of Riveting applicable to the Construction of Iron Ships. Discussion contribution to [3]
[5] Grantham, J.: The Strength of Iron Ships. Trans. Inst. of Nav. Arch. I (1860)
[6] Mansell, R.: On the Comparative Strength of Iron Ships. Proceedings of the Scottish Shipbuilders' Association (1860)
[7] Rankine, W.J.M.: Manual of Applied Mechanics (1858)
[8] Rankine, W.J.M., Watt, I., Barnes, F.K., Napier, R.: Shipbuilding – Theoretical and Practical. William Mackenzie, London (1866)
[9] The index WVH means wave bending moment, vertical hogging; SW still water bending moment; and TH total bending moment, hogging
[10] Walker, F.M.: Ships & Shipbuilders. Seaforth Publishing (2010)

[11] Murray, J.M.: Development of Basis of Longitudinal Strength Standards for Merchant Ships. Trans. of Royal Inst. of Nav. Arch. 108 (1966)
[12] Reed, E.J.: The Strains of Ships in Still Water. Naval Science I, 351 (1872)
[13] Id.: The Distribution of Weight and Buoyancy in Ships. Naval Science I (1872)
[14] Id.: The Strains of Ships at Sea. Naval Science II, 12 (1873)
[15] John, W.: On the Strength of Iron Ships. Trans. of Royal Inst. of Nav. Arch. XV (1874)
[16] Smith, W.E.: Hogging and Sagging Strains in a Seaway as Influenced by Wave Structure. Trans. of Inst. of Nav. Arch. XXIV (1883)
[17] Read, T.C.: On the Variation of the Stresses on Vessels at Sea due to Wave Motion. Trans. of Inst. of Nav. Arch. XXXI (1890)
[18] Kryloff, A.: A New Theory of the Pitching Motion of Ships on Waves, and of the Stresses Produced by this Motion. Nav. Arch. XXXVII (1896)
[19] Alexander, F.H.: The influence of the proportions and form of Ships upon their longitudinal bending Moments among Waves. Trans. of Inst. of Nav. Arch. XLVII (1905)
[20] Schnadel, G.: Die Beanspruchung des Schiffes im Seegang. Dehnungs- und Durchbiegungsmessungen an Bord des MS „San Francisco" der Hamburg- Amerika Linie. J. STG 37 (1936)
[21] Horn, F.: Hochseemessfahrt, Schwingungs- und Beschleunigungsmessungen. J. STG 37 (1936)
[22] Weinblum, G., Block, W.: Stereophotografische Wellenaufnahmen. J. STG 37 (1936)
[23] Lienau, O.: Messungen über das Arbeiten des Schiffsboden und der Deckbeplattung während der Hochseemessfahrt 1934. J. STG 36 (1937)
[24] Claviez, W.: 50 Jahre Deutsche Werft 1918-1968, Hamburg (1968)
[25] Biles, J.H.: The Strength of Ships, with Special Reference to Experiments and Calculations made upon H.M.S. "Wolf". Trans. Inst. of Nav. Arch. XVII (1905)
[26] Lehmann, E.: 100 Jahre Schiffbautechnische Festigkeitsforschung. J. STG 98 (2004)
[27] Discussion contribution by Hans Reissner on Schnadel. G.: Die Spannungsverteilung in den Flanschen dünnwandiger Kastenträger. J. STG 27 (1926)
[28] Weber, M.: Grundlagen der Ähnlichkeitsmechanik. J. STG 20 (1919)
[29] Weber, M.: Allgemeines Ähnlichkeitsprinzip der Physik und sein Zusammenhang mit der Dimensionslehre und dem Modelwissenschaft. J. STG 31 (1930)
[30] Schnadel, G.: Elastizitätstheorie und Versuch. J. STG 32 (1931)
[31] Levy, M.: Sur l'équilibre élastique d'une plaque rectangulaire. l'Académie des Sciences de Paris 127 (1899)
[32] Schnadel, G.: Über die Knickung von Platten. J. STG 30 (1929)
[33] Lienau, O.: Versuchseinrichtungen und Ergebnisse des Instituts für Schiffsfestigkeit der Technischen Hochschule Danzig. J. STG 29 (1928)
[34] Wlassov, W.S.: Dünnwandige elastische Stäbe. VEB Verlag für Bauwesen, Berlin (1964)
[35] Timoshenko, S.P.: Erinnerungen, p. 98. Ernst & Sohn, Berlin (2006)
[36] Schnadel, G.: Torsionsversuche. J. STG 34 (1933)
[37] Schultz, H.-G.: Festigkeitsprobleme im Großschiffbau. J. STG 63 (1969)
[38] Nießen, E.: Statische Messungen an einem Aluminium-Modell eines Container-Schiffes, Bericht Nr. 4/1968. Forschungszentrum des Deutschen Schiffbaus, Hamburg (1968)
[39] Woisin, G.: Kollisionsprobleme bei Atomschiffen, p. 999. Hansa (1964)

[40] N.N.: Die Widerstandsfähigkeit verschiedener Bordwandkonstruktionen bei Zusammenstößen, p. 2174. Hansa (1962)
[41] Spinelli, F.: Schutz von Kernreaktoren auf Schiffen gegen Kollisionen, p. 148. S+H (1964)
[42] Minorsky, V.U.: Eine Studie über Schiffskollisionen mit Bezug auf schiffbauliche Schutzmaßnahmen für Kernenergie-Antriebsanlagen, p. S. 163. S+H (1960)
[43] Reckling, A.: Beitrag der Elasto- und Plastomechanik zur Untersuchung von Schiffskollisionen. J. STG 70 (1976)
[44] Welch, J.J.: The Subdivision of Ships. Trans. of Inst. of Nav. Arch. LVII (1915)
[45] Denny, W.: Subdivision of Merchant Vessels: Reports of the Bulkhead Committee, 1912-1915. Trans. of Inst. of Nav. Arch. LVIII (1916)
[46] Height of the tank 17 ft $\approx$ 5.17 m, 19.5 ft $\approx$ 5.49 m and 20 ft $\approx$ 6.1 m
[47] Foster King, J.: Strength of Watertight Bulkheads. Trans. of Inst. of Nav. Arch. LVIII (1916)
[48] Ref. [26]
[49] Caldwell, J.B.: Ultimate Longitudinal Strength. Trans. R.I.N.A. 107 (1965)

Part III
Theories, Engineering Solutions and Applications in Fluid Dynamics

The Development of Fluid Mechanics from Archimedes to Stokes and Reynolds[*]

Oskar Mahrenholtz

Abstract. Short presentation of the fundamentals and the development of fluid mechanics from ancient times until the end of 19th century.

Keywords: Fluid Mechanics.

1 Introduction

The evolution of life is inseparably connected to water. But water was not only the cradle of mankind, it served – besides wind – as a natural source of energy. Our ancestors have used water for multiple purposes, such as irrigation and water supply, to name only some. With great inventiveness water related mechanisms have been built, heuristically, without knowledge of basic principles. These were developed step by step with an increasing speed in the Renaissance era. In this 'tide of fortune' of Mechanics and Mathematics basic principles came into existence. This framework was brought to perfection in the 19th century. Nowadays, we experience an unforeseen progress in the world of Computational Sciences, and in particular of Fluid Dynamics. It is impossible to present the development of several millennia in detail on a few pages. A (subjective) choice had to be done. The expert may miss something, but it is expected that the main stream of development comes to light. The author still appreciates the discussions with Lothar Gaul, Stuttgart, on this and other historical topics, 30 years ago [1].

Oskar Mahrenholtz
Institute of Mechanics and Ocean Engineering, Technische Universität Hamburg-Harburg, 21071 Hamburg, Germany

[*] Dedicated to *Heinz Ismar*, Saarbrücken, for his remarkable contributions to material physics and engineering application on the occasion of his 75th birthday.

2 From Ancient Times to Medieval Times

The Old Testament reports on two remarkable events where water plays an important role: about the Noachian flood and about the Exodus (The 'Children Israel's' leave Egypt.)

The Noachian flood (Noah saved one of each species in his Ark) could have been caused by a sea earthquake in the Mediterranean, generating a Tsunami. During the Exodus the 'Children Israel' had to cross the Red Sea at a straits near Suez. The Old Testament describes, that there was a strong wind when Moses had divided the (reed-) Sea with his stick. This phenomenon can be modeled nowadays, but in ancient times it was a miracle.

Real was the art of hydraulic constructions. The mother of Egypt, the Nile river area, was for three millennia witnessing developments and inventions like embankment dams, irrigation channels and water pumps. The Romans became famous for their aqueducts (Fig. 1): Water supply for large cities, and that started more than two thousand years ago [1].

Fig. 1 Roman aqueduct, Pont du Gard, Nimes, last quarter of 1st century B.C. From Propyläen, Kunstgeschichte

But there was no systematic penetration of the subject. Archimedes' (287 – 282 B.C.) famous investigations about buoyancy of swimming and dipped bodies in water including their stability had no followers. Congenial, but more application oriented, was Heron of Alexandria (about 150 – 100 B.C.). He invented Heron's

Ball or aeolipile: Expanding steam brings by repulsion the ball into rotation. The aeolipile is the forefather of the steam engine [2].

With the decline of the Roman Empire the fertile soil for the development of science got lost. But there were heirs of the knowledge of the ancient world: The Arabs. They became not only mediators but also contributors. Prominent – for our topic – was Ibn-Razzaz al Ğazarī (al Dschazarī) who designed various water driven machines. He presented around 1205 his famous opus on mechanical-technical artwork (*book about the knowledge of ingenious mechanical devices*; about 550 pages and 150 drawings). He described the design and construction of various useful and entertaining devices such as water meters (clepsydrae), piston pumps, safety locks and jukeboxes [3] [1]. Fig. 2 shows as an example a hydro powered bucket chain.

The widely published book (translation from Arabic into Turkish and Persian) is the most important source of the high level of Arabian technics in the Middle Ages, far advanced compared to the contemporary European technics [3]. Al Ğazarī's designs were influenced by Heron of Alexandria, Phylon of Byzanz and the Arabian engineer Banū Mūsā. They contain plenty of technological knowledge, in particular from hydro techniques. The described equipment has partially been reconstructed, and has been proven as functional [3]. But while the Europeans aimed for technical applications the Arabians were in principle more satisfied by the pure functioning of their devices.

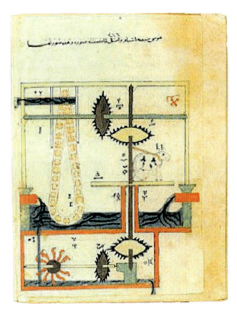

Fig. 2 Al Dschazarī's hydro powered bucket chain, from a hand writing (about 1205). (Bodleian Library, Oxford/Ms. Graves 27).

3 Hydrostatics

In the final stage of the Renaissance Age, the rebirth of Greek and Roman culture (Cinquecento), the works of Archimedes were enlightened by a man from Flanders, Simon Stevin, born 1548 in Brügge. He was working at the fiscal authority of the free port of his hometown, he visited Prussia, Poland and Scandinavia, and he studied – after professional success – since 1583 at the University of Leyden. He published only three years later his two "opera magna": *De Beghinselen des Waterwichts* and *De Beghinselen der Weeghconst* [1].

To demonstrate his influence a small excursion to mathematics and mechanics may be useful. Stevin was the first to represent rational numbers by decimal fraction. In 1585 he published a little booklet for daily use, *De Thiende* (The Tenth), in Duytsch (Dutch). He became the teacher of his countrymen in elementary calculations. He was convinced that Dutch was the only language appropriate for science. An English translation (*disme*) has inspired the founding fathers of the United States of America to introduce decimal currency. A tenth of a Dollar is a *dime*, a hundredth a *cent*.

Also in 1585, Stevin published *L'arithmétique*, a consistent method for the solution of quadratic equations, and in an appendix 1594 *a numerical method for the solution of algebraic equations of higher order* by successive approximation [4].

Fig. 3 Simon Stevin (1548 – 1620). Fortress Builder, Mathematician, Engineer [1] [4]

De Beghinseleen der Weeghconst is Stevin's most important work. *Weeghconst* means nowadays *statics*. Following Archimedes Stevin brings heavy bodies into equilibrium. Fig. 4 demonstrates an example leading to the parallelogram of forces [4].

Fig. 4 Simon Stevin, *De Beghinselen der Weeghconst*, Leyden, 1583. Practice of Weighing.

Stevin disproved by experiment Aristotle's opinion that heavy bodies fall faster than light ones (1586), before Galileo did so [4].

Stevin commented on Archimedes (with his own words; free translation): "I do not know what Archimedes had in mind when he left us his 'book of things that are supported in water', and where he portrayed nature in a wonderful manner. However, I know, and I admit it freely, that it was him, who has led me to bring this topic in such a form that we have given it. I also admit that we had better preconditions as Archimedes, namely the language that is Duytsch while his was only Greek. Because one should know that the amenity of the language is not only beneficial to learn arts but also for the research of scholars." [5]

István Szabó has dealt with this development in a likewise challenging and critical manner in his monograph *Geschichte der mechanischen Prinzipien*. He identifies Stevin as the probably first scholar who has combined theory and practice in the most effective manner and therefore he is entitled to be seen as a real "engineer" [5].

Stevin has intensively studied the pressure forces in a resting fluid: *hydrostatics,* as the fluid was water. The term hydrostatics stems from Stevin. He has his thoughts – and probably experiments – applied to a multitude of containers filled with water. He came – long before Pascal – to the conclusion that the bottom pressure depends only and only on the height of the water column and not on the shape of the container (*Hydrostatic Paradox*). Stevin came close to the pressure forces on the side walls of a container. Fig. 5 illustrates some of the problems he had tackled.

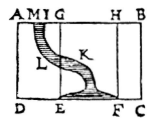 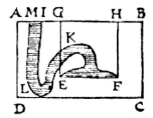

Fig. 5 Simon Stevin, *De Beghinselen des Waterwichts*, Leyden, 1583. Corrolarium II: Pressure forces on a container bottom.

There was an interesting event, a "meeting" of mechanics and fluid mechanics, and Stevin was firm in both subjects. He constructed a sailing carriage on wheels, for 25 people, Fig. 6. The order came from Maurice of Nassau, Prince of Orange, governor of the Netherlands. Stevin was the tutor of the 19 years younger Prince. The carriage made it onshore in about two hours from Scheveningen to Petten (85 km), with a speed of about 40 km/h, unbelievable at that time [4].

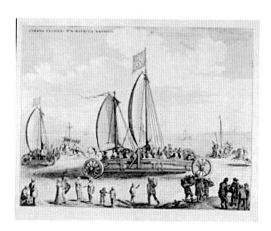

Fig. 6 Stevin's sailing carriage, by Joan Blaeu, Stedeboec (city book, 1649). [4]

Gravity leads to pressure forces in water. This is daily experience. That also air produces such pressure forces was shown by Evangelista Torricelli (1608 – 1647). He became Galileo's assistant (fall 1641), and later his successor as chief mathematician at the court of Florence and Professor at the local Academy. Unfortunately, they had only three months together (Galileo, born 1564, died in January 1642). Six years later, Torricelli himself passed away, rather young, at the age of 39. Torricelli invented the barometer (the air pressure meter). He started from Galileo's observation that water in a vertical tube follows only about 10 m a (sucking) piston. He arranged for an experiment with mercury (which is 13,6 times more dense than water). The upper end of the tube was closed, the tube itself was filled with mercury while the lower end was dunked into a dish filled with mercury. The result: The mercury column was sinking from top towards bottom until it had reached a level of 1½ cubit above the mercury level of the dish. Between the mercury column and the closure head of the tube was emptiness, hence no pressure (*Torricelli's emptiness*. Descartes, his peer (1596 – 1650), made the sarcastic comment, emptiness is only in the head of Torricelli; Descartes was wrong.). Torricelli gave the correct explanation that the mercury column of 1½ cubit height was in equilibrium with the air column of the atmosphere. He explained the small variability of the mercury column as variability of the air pressure [6]. (1 Tuscan cubit = 0,54 m; 1½ cubit = 810 mm. Mean value of air pressure is about 760 mm mercury = 760 Torr(icelli)). He concluded (correctly) that wind results from air pressure and temperature differences.

Fig. 7 Evangelista Torricelli (1608 – 1647) [7]

Torricelli's experiments of the flow of water out of a vessel under gravity, again triggered by Galileo, rendered the same result as Galileo's famous law of free fall: The rate of outflow is proportional to the square root of the water downgrade. Torricelli was seen as a second Galileo [6].

Fig. 8 Otto v. Guericke (1602 – 1686) [8]

The effect of air pressure was dramatically shown (in 1654, 1657) by Otto v. Guericke (1602 – 1686), then Mayor of Magdeburg, a town nearly completely destroyed and massacred by count Tilly's Catholic troops in 1631 during the 30-years religious war in Germany (1618 – 1648). This has to be mentioned to admire that despite such miserable circumstances scientific progress was flourishing.

Guericke designed and constructed the "Magdeburg half spheres", he evacuated the combined hollow spheres, and let eight horses act on each half sphere, Fig. 9. The horses were not able to separate the half spheres. These were compressed by (external) air pressure. (With a diameter of the sphere of 42 cm the separation force is approximately 14 kN – or in old figures 1,4 tons.).

Fig. 9 Experiment with "Magdeburg half spheres", 8 horses (on each side), 1657. Ottonis de Guericke, *Experimenta nova Magdeburgica de Vacuo Spatio*, Amsterdam 1672.

Guericke's first public demonstration was at the Reichstag in Regensburg 1654, the second one 1657 at the Court of Friedrich Wilhelm, elector of Brandenburg (later Prussia). It was not at all easy to achieve this epoch making result. Guericke started from idea of air as 'a bodylike something' that is compressible, expands by heat and contracts by coldness. Air has weight, presses on itself and on everything. For his investigations he needed a really functioning air pump, and he had to work hard to come up with what may be seen as his greatest invention, Fig. 10. He started his evacuation experiments with a barrel, but failed. Spheres of copper sheet followed. The first was 'compressed like cloth squeezed between fingers'. It was geometrically inaccurate. The experiment was eventually succssful with a more carefully produced sphere. [8]

Fig. 10 Guericke's 'Magdeburg half spheres', 1657, with the transportable air pump (3. Version, 1663) [8]

At the end of the 17th century the influence of pressure was in general known, in particular the influence on walls. The clear formulation of pressure as an *internal pressure* valid for any (ideal) fluid was formulated much later by Leonhard Euler (1707 – 1783) middle of 18th century. The internal pressure does not need a wall to act on, it acts on itself.

4 Classical Hydrodynamics

Seen from today it is only a small step from hydrostatics, the model of a fluid at rest, to hydrodynamics, the model of a fluid in motion. Indeed, fluid particles form a continuum of unknown shape, and instead of a single element an entire flow field has to be determined. If one comes to the conclusion that the motion of a fluid element follows the same laws than that of a rigid body, Newton's laws can be applied [1]. But this step was subjected to Euler, who did it around the middle of the 18th century.

Before that, in the first half of the 18th century, it was very troublesome to bring single phenomena into a coincident model. The main problem was to solve the question of conversion of pressure (difference) into velocity. To mention are here the attempts of (Sir) Isaac Newton (1642 – 1726) himself, Johann I. Bernoulli (1667 – 1748) and his son Daniel (1700 – 1782), Jean le Rond d'Alembert (1717 – 1783), and – initially – also Euler. As interesting as these attempts are in detail here is no space to follow their pursuit.

But some notes may be permitted. Newton became famous already in younger years for his magnum opus *Philosophiae Naturalis Principia Mathematica*, 1687, the bible of mechanics. It consists of three volumes. The first book – the well-known one – is mainly on gravitation and dynamics with the famous laws of motion (of rigid bodies): 'Newtonian mechanics'. The second book is on motion in viscous fluids (Newton disproves Descartes' eddy theory of planetary motion), and the third book deals with the application of the results of the first two books on the motion of celestial bodies. Newton did not make use of his (new) method of *fluxions*, 1671 (his method corresponds to Leibniz' (1646 – 1716) *infinitesimal calculus*, 1675). Instead of he worked strictly geometrical what makes the books hard to read.

Our interest is in particular with the second (the fluidics) book. Clifford Truesdell (1919 – 2000), an American mathematician, natural philosopher and historian of sciences with an inclination to polemics noted that 'Newton's *principia* are a masterpiece. In the first book the physical background was almost known, and everything was perfect, but in the second book (on fluidics) nearly everything is new but almost wrong. Every chain of thoughts goes with a new hypothesis.'(Cited after [5])

Truesdell's comment on Newton's fluidics book is characteristic for the non deductive procedure until Euler. The fundamental result of the numerous investigations can be shown at a curved, narrowing duct (Fig. 11).

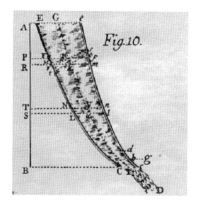

Fig. 11 Curved, narrowing duct. From Johann Bernoulli's *Hydraulica* [5]

The volume flow dV/dt (time t) requires for water (as a practically incompressible fluid)

$$A \cdot v = \text{constant}$$

(A is the cross section of the duct, v the mean velocity of the fluid flow in the duct). This so-called *continuity equation* is a plausible expression.

It was much more difficult to find the relation between pressure p in the fluid and velocity v. A stone pit of models, leaning on wall forces, led eventually to success. Daniel Bernoulli (Fig. 13a)), has treated – like Torricelli – the outflow of water under gravity. Some examples are shown in Fig. 12a). They stem from his book *Hydrodynamica,* published 1738. The manuscript was older, from Daniel's time at St. Petersburg. He came to Basel University as Professor of Medicine, and he mentions on the title page that he is Johann's son (Fig. 12b). That was before his bitter struggle with his father Johann Bernoulli about the priority of the *Bernoulli equation*. Matter-of-factly, Daniel's approach states more the conservation of energy of a filament of flow (of an inviscid (ideal) fluid).

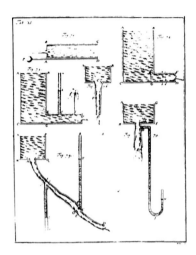 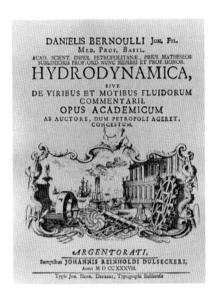

Fig. 12 a) Outflow from receptacles under gravity. From Daniel Bernoulli's *Hydrodynamica* [9]

Fig. 12 b) Daniel Bernoulli *Hydrodynamica sive de viribus et motibus fluidorum commentarii*, 1738 [9]

Johann Bernoulli (Fig. 13b)) has followed the flow shown in Fig.11. It is obvious that in a narrowing duct the velocity (v) increases (see *continuity equation* above), that the pressure (p) decreases, and that the effect should depend on the density (ρ) of the fluid. He arrived at the famous *Bernoulli equation*

$$p + \tfrac{1}{2} \rho\, v^2 = \text{constant}$$

(horizontal duct; no influence of gravity). The equation can be found in his work *Hydraulica,* printed in 1742, but backdated to gain priority over his son Daniel. The dispute between son and father has not been settled during Johann Bernoulli's lifetime. Euler has plaid a debatable role as he wrote a very supportive foreword for the *Hydraulica* (Ad Auctorem) where he admired Johann Bernoulli. [1] [5]

The ingenious mathematician, physicist and engineer Leonhard Euler has consequently applied the infinitesimal *calculus* on Newton's axioms. He made extensively use of the method of sections (*naturam secare* – Descartes). Seen from today his procedure was like this: He makes a snapshot of the drifty fluid. He denotes the velocity **v** of the element dm in Cartesian co-ordinates **v** = [u,v, w]. Now, Euler makes an important step. He does not, as the Bernoullis

Fig. 13 a) Daniel Bernoulli (1700 – 1782) *Hydrodynamica sive de viribus et motibus fluidorum commentarii,* 1738

Fig. 13 b) Johann Bernoulli (1667 – 1748), father of Daniel. *Hydraulica, nunc primum detecta at que demonstrata directe ex fundamentis pure mechanicis,* printed in 1742, but backdated. Priority dispute with son Daniel

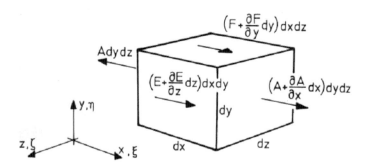

Fig. 14 Euler´s procedure at the volume element dV = (dxdydz), cut out of the fluid [1]. The fixed reference frame is x, y, z. The forces shown on the surfaces of the cuboid are those in x-direction. As the ideal fluid has no shear forces the in plane components (E, F) disappear. The only remaining force is A = - p (dydz), the pressure force in direction x

did, start from the wall, but from the interior of the fluid. He denotes the pressure p as *interior pressure*. This pressure is discrete. Its introduction by Euler is a highlight of 18th century mechanics [9].

From Fig. 14 follows, with Newton's law of momentum (*lex secunda*),

$dF_x = [p - (p + (\partial p/\partial x) dx)](dydz) + \rho P (dxdydz) = dm\, du/dt = \rho (dxdydz)\, du/dt,$
$- \partial p/\partial x + \rho P = \rho\, du/dt = \rho [\partial u/\partial t + (\partial u/\partial x) u + (\partial v/\partial y) v + (\partial u/\partial z) w].$

This is the equation of motion in direction x, with P as the impressed volume force. The equations for directions y,z arise from cyclic permutation. Then, Euler is treating the conservation of mass, dm´ = dm during the time increment dt = t´- t: time t, dm = ρ dV = ρ (dxdydz); time t´ = (t + dt), dm´ = ρ´ dV´ = ρ´ (dx´dy´dz´). The control cuboid changes its volume from dV to dV´,

$dV' = dV [1+ (\partial u/\partial x + \partial v/\partial y + \partial w/\partial z) dt].$

The density changes from ρ to ρ´. With Euler's differentiation rule follows

$\rho' = \rho + [\partial \rho/\partial t + (\partial \rho/\partial x) u + (\partial \rho/\partial y) v + (\partial \rho/\partial z) w]\, dt.$

Hence, ρ´ dV´ = [ρ + (∂ρ/∂t + ∂ρ/∂x u + ∂ρ/∂y v + ∂ρ/∂z w) dt] dV [1+ (∂u/∂x + ∂v/∂y + ∂w/∂z) dt] = ρ dV + ∂ρ/∂t + ∂(ρu)∂x + ∂(ρv)/∂y + ∂(ρw)/∂z,

and the *equation of conservation of mass* reads

$\partial \rho/\partial t + \partial(\rho u)/\partial x + \partial(\rho v)/\partial y + \partial(\rho w)/\partial z = 0.$

Fig. 15 Leonard Euler (1707 – 1783), ingenious mathematician, physicist and engineer Pastel picture by Emanuel Handmann, Museum of Arts, Basel, 1753

Euler himself stated [5]: 'These three [partial differential] equations [for the velocity field [u,v,w], combined with the one found relation between density and velocity, contain the whole theory of motion of fluid bodies. Our whole business would consist of determining the quantities p,ρ,u,v,w as such functions of x,y,z and t which satisfy those equations'.

The Bernoulli equation can be derived from Euler's equations. The solution of these partial differential equations depends on the boundary conditions (ideal fluid rests vertical to the wall) and on the initial state of the flow. These field equations were much ahead of their time. Their disadvantage: They fail for viscous flow, when viscous forces dominate over inertia forces, and they are nonlinear. Therefore, they had a bad reputation in pipe flow. It seems, Euler had all the tools to implement shear forces into his equations, but he did not do so.

5 Tube Hydraulics

'Dear friend, all theory is grey, And green the golden tree of life' says Mephisto to the student, and so said the *hydraulic specialists*. *Hydraulic* originates from *hýdor* (water) and *aulós* (tube, pipe). Machinery, operated with and by water, in particular pipes and pipe systems, belong to *hydraulics*. Even if the fluid is oil the term oil hydraulics is customary. Hydraulics does not need field equations, inertia forces in the fluid are negligible, dominant parameter is the viscosity of the fluid. The central device is the tube or pipe.

Hydraulics has mainly been developed during the 19th century, parallel to classical fluid mechanics, here experience, there theory. If one checks literature of hydraulics one comes immediately across Poiseuille, pressure drop, turbulence. Indeed, the laminar tube flow, the Poiseuille flow, is a basic module of tube hydraulics.

The French physician and physiologist Jean Louis Marie Poiseuille (1797 – 1869) has intensively studied the physiology of blood circulation, the flow of venous blood, 1832, and of capillary blood, 1839 [10]. His model was a constant volume flow of a viscous fluid passing through a cylindrical tube of circular cross section. The fluid particles move parallel to the tube wall, and the sheet close to the wall sticks to the wall. There are no inertia forces. Fig. 16 shows the procedure

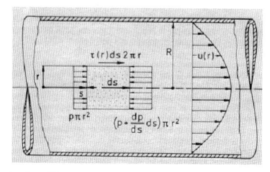

Fig. 16 Laminar tube flow [1]

and the result. Because there is no radial particle motion the pressure over the cross section must be constant. Equilibrium at an inner cylinder of radius r and length ds, cut out of the fluid, requires τ (r) $(2\pi r\, ds) = (dp/ds)\, (ds\, \pi r^2)$, hence $dp/ds = \tau$ (r) $(2/r)$ or τ (r) = const r. An assumed parabolic velocity profile, $u(r) = u(0)\, (1 - r^2/R^2)$, corresponds to $du/dr = u(0)\, (-2r/R^2) = -$ const r. This profile requires a shear stress τ proportional to shear rate (du/dr). Hidden in the parabolic profile is Newton's shear law (Chapter 6). The Poiseuille model yields a surprising pressure drop Δp along the section Δl of the tube axis $\Delta p/\Delta l \sim (1/R^4)$ for the same volume flow dV/dt. Gotthilf Heinrich Ludwig Hagen (1797 – 1884), Geheimer Oberbaurath at Berlin, found the same relationship in 1839, experimentally, by very precise measurements, Fig. 17. [11]. Therefore in Germany: Law of Hagen-Poiseuille.

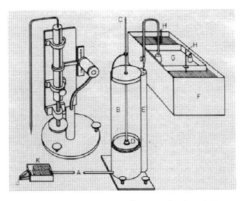

Fig. 17 Gotthilf Hagen's eperimental set-ups for precie (out-) low measurements (1839) [11]

Laminar flow does not stay stable. With increasing velocity radial components develop, and at a critical velocity (which depends on the quality of the laminar inlet flow and on the surface roughness of the tube) the flow gets turbulent. This was shown in a famous experiment by Osborne Reynolds (1842 – 1912) in 1883 (Fig. 18). This observation has triggered countless follower experiments and investigations. The question of instability and turbulence is to some extent still open.

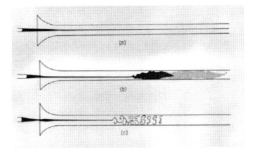

Fig. 18 Tube flow, experiment by Osborne Reynolds, 1883 [12] a) Laminar flow b) Transition laminar – turbulent c) Turbulent flow

Shear stress (τ), proportional to shear rate (du/dt), is – in tube flow – only important near the tube wall where high velocity gradients occur. It is here, where nearly the whole dissipation of energy and production of turbulence takes place. Reynolds' thoughts about flow transition (laminar - turbulent) were facilitated by the so-called Navier-Stokes equations (Chapter 6). These equations include viscosity, they close the gap between fluid mechanics and hydraulics.

6   Navier-Stokes Equations (N-S eqs.)

Friedrich II. (the Great) king of Prussia, wanted to crest the garden of his (new) castle *Sanssouci* in Potsdam by a fountain. The fountain never worked during his lifetime. It was one hundred years later, after 1840, that the fountain did its duty. It was then a success of hydraulic experts. It is said that Euler has warned king Friedrich against the debacle. Euler chaired at that time the mathematical class of the Brandenburg Society of Sciences. But his equations could not cope with viscous tube flow, and the king might have made sarcastic comments. Matter-of-factly, Euler left Berlin after 25 years in 1766 for St. Petersburg, where he worked at the Academy for the rest of his life.

In the second book of his *Principia* Newton describes a thought experiment, derived from observation and experiment, shown in Fig. 19. Between two parallel plates is a viscous fluid. The near wall fluid sticks to the plates. The upper plate moves with the velocity U in direction x. The force needed for motion is F. The coated area is A. Then, τ = F/A is the shear stress. Newton states F ~ (U/a) A, hence τ ~ (U/a), where a is the distance between the two plates. He assumes further a linear distribution u(y) = U y/a of the fluid velocity over the height y. The shear velocity follows as du/dy = dγ/dt = U/a; γ is the angle of shear. Then it counts τ ~ (U/a) = dγ/dt. Newton states τ = μ (dγ/dt). The factor μ of proportionality is called *dynamic viscosity*. As a physical quantity it defines and describes a *Newtonian fluid*.

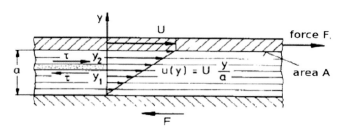

Fig. 19 Velocity distribution in a fluid between a pulled (velocity U) and a resting plate [1]. This experiment cannot be carried out in reality as the fluid would flow out at the margin

To arrive at the equations of viscous flow shear stresses have to be implemented into the Euler model. It should have been easy for Euler to do so with Newton's shear stress in mind. But it has taken seven more decades until shear went into the continuum, and it was first with elastic bodies. To name are here Augustin Louis Cauchy (1789 – 1857), 1821, Claude Louis Marie Henry Navier (1785 – 1836) with his 1822 exposé *Mémoire sur les lois du mouvement des fluides*. Navier developed the correct structure of the equations of motion of incompressible viscous fluids [1].

Remarkable is the work of Barré de Saint-Venant (1797 – 1886). He presented his paper 1834 to the *Académie des Sciences;* 1843 appeared a note in *Compte Rendu*. Saint-Venant treates the relative velocity of the fluid particles in the sense of Newton as decisive for the additional stresses that have to be superimposed on the pressure field. He refers to the material law of the elastic body of Cauchy. He replaces the displacements ξ,η,ζ in x,y,z – direction by the velocities u,v,w. Shear becomes shear velocity as illustrated in Fig. 20 for the x,z – plane. In generalization of Newton's shear law one arrives at the equation of motion (law of momentum) in vector notation

$$\rho \, \partial v/\partial t + \rho(v \, \nabla)v = - \nabla p + \mu \, \Delta v + (\lambda + \mu)\nabla(\nabla \cdot v) + f,$$

(density ρ; pressure p; particle velocity **v**; density of volume force **f**, dynamical viscosity μ; Lamé constant λ).

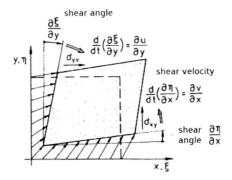

Fig. 20 Illustration of shear and shear velocity in the xy-plane [1]

These are the Navier-Stokes equations. The law of conservation of mass has to be added. These partial differential equations are nonlinear (like the Euler equations). They are the most numerically treated equations, and special Navier-Stokes solvers have been developed. The literature concerned with the N-S eqs. is highly visible. It is regrettable that the contribution of Saint-Venant was not appreciated.

Finally, the contributions of (Sir) George Gabriel Stokes (1819 – 1903) should be mentioned. He became famous by his contributions to the behaviour of viscous fluids. His fundamental contribution *On the Theory of the Internal Friction of Fluids in Motion* appeared 1845, 14 years after Poisson's similar approach (Siméon Denis Poisson (17812 – 1840), brilliant French mathematician) that was unknown to Stokes. He solved as a limiting case of the N-S eqs. the problem of creeping motion of a sphere in a viscous fluid: $F_R = 6 \pi r \mu v$. From force (weight) F_R and velocity v the dynamic viscosity can be determined (*Stokes flow*). The N-S eqs. are based on the Boltzmann Axiom (no internal moments), and on the assumption of a linear shear law (Newton). This is no disadvantage for engineering application.

References

[1] Mahrenholtz, O., Gaul, L.: Die Entwicklung der Mechanik: Die Rolle der Strömungsmechanik. Z. Universität Hannover 1(2), 6–29 (1982)
[2] Schaefer, H.W.: Aiolu pylai (transliterated from ancient Greek). In: Paulys Realencyclopädie der Classischen Altertumswissenschaft (RE). Band I,1, Sp. 1042, Stuttgart (1893)
[3] Hill, D.R. (Hrsg.): The Book of Knowledge of Ingenious Mechanical Devices by Ibn al-Razzaz al-Jazari. Reidel, Dordrecht (1974)
[4] Devreese, J.T., Van den Berghe, G.: Magic is no magic. The wonderful World of Simon Stevin 1548-1620. WITpress, Southampton (2007)
[5] Szabó, I.: Geschichte der mechanischen Prinzipien und ihrer wichtigsten Anwendungen. Birkhäuser, Basel (1977)
[6] Matschoß, C. (Hrsg): Männer der Technik. VDI, Berlin (1925)
[7] O'Connor, J.O., Robertson, E.F.: Evangelista Torricelli. University of St. Andrews: MacTutor History of Mathematics archive (2002)
[8] München, D.M. (Hrsg.): Große Forscher und Erfinder. Leben und Werk. Schätze im Deutschen Museum. VDI, 2. Aufl., Düsseldorf (1978)
[9] Truesdell, C.: Essays in the History of Mechanics. Springer, Berlin (1968)
[10] Schiller, L.: Drei Klassiker der Strömungslehre: Gotthilf Hagen, Jean L Poiseuille, Eduard Hagenbach. Akademische Verlagsgesellschaft, Leipzig (1933)
[11] Hagen, G.: Poggendorfs Annalen der Physik und Chemie, vol. 46, pp. 423–442 (1842)
[12] Fung, Y.C.: First Course in Continuum Mechanics. Prentice Hall, Englewood Cliffs (1969)

The Millennium-Problem of Fluid Mechanics – The Solution of the Navier-Stokes Equations

Egon Krause

Abstract. The paper sets out from the inclusion of the analysis of existence and regularity of solutions of the Navier-Stokes equations in the seven unsolved problems of mathematics by the Clay Mathematics Institute in the USA in 2000. It then mentions d' Alembert's early paradoxical attempts to determine the aerodynamic drag and the insolvability of the Euler equations at the time when they were first published. After the discussion of the derivation of the Navier-Stokes equations early results as for example Helmholtz's vorticity theorems and Reynolds' approach to turbulent flows are introduced, followed by Prandtl's revolutionary boundary-layer theory and lifting-line theory. The next section sketches the rapid development of modern computing machines, enabling the introduction of numerical methods into fluid mechanics. Arrangement of computational grids and solution techniques are briefly discussed. The results of a recent international workshop on drag prediction and an example showing the use of numerical methods in aerodynamic design are used to demonstrate the state of the art. The summary concludes with a look on future problems.

Keywords: Navier Stokes equations, boundary-layer theory, numerical solutions.

1 D' Alembert's Paradoxon

It was in the year 2000 that the Clay Mathematics Institute in Cambridge (USA) selected the analysis of existence and regularity of solutions of the Navier-Stokes equations for three-dimensional incompressible flows, the millennium-problem of fluid mechanics, as one of the seven unsolved problems of mathematics [1]. One million US dollars were offered as prize money for the solution. Although more than ten years have passed in the meantime successful solution approaches did not

---

Egon Krause
Aerodynamisches Institut,
RWTH Aachen University, Wüllnerstr. 5a,
52062 Aachen, Germany

become known as yet, a disappointing result, especially, if one realizes that a large body of literature pertaining to the subject exists. The understanding of the mathematical nature of the Navier-Stokes equations is still rather limited, as Charles L. Fefferman remarks in the official description of the problem [2].

But disappointments have always accompanied research in fluid mechanics. For example the former Secretary General of the Académie Francaise and member of the Académie des Sciences and also of the Preußische Akademie der Wissenschaften, Jean-Baptiste le Rond d' Alembert, writes 1752 in [3] in the translation of [4]:

"I do not see then, I admit, how one can explain the resistance of fluids by theory in a satisfactory manner. It seems to me on the contrary that this theory, dealt with profound attention, gives, at least in most cases, resistance absolutely zero; a singular paradox which I leave to geometricians to explain."

The last part of the statement, the reference to the geometricians remains vague and indeed discouraging. It is not easy to see, how the geometricians could possibly take care of the paradoxical result, unless one wants to imply that a very large amount of numerical work is required for an accurate description of complex aerodynamic configurations. But most likely it was not this problem which d´ Alembert meant, when he described his non-satisfying result of his carefully worked-out theory, yielding no resistance at all. Almost another century was necessary to find a suitable answer.

Fig. 1 Jean-Baptiste le Rond d´ Alembert, (1718 - 1783), Secretary General of the Académie Francaise, member of the Académie des Sciences and of the Prussian Academy of Sciences, exchanged letters with Catharine II. the Great and retired on a pension of Frederick II the Great of Prussia. The above remark can be found in J. le R. d´ Alembert: "Essai d´ une nouvelle théorie de la résistance des fluides", Paris, 1752, Opuscules mathematiques, Paris, 1768, V, 132 – 138.

D' Alembert's Paradox nowadays can easily be verified by assuming that the flow is inviscid and can be described by a potential. For simplicity the flow around a circular cylinder is considered. The solution of the potential equation yields the velocity v on the surface of the cylinder to $v = 2v_\infty sin\varphi$, where v_∞ is the free-stream velocity and φ is the polar angle. Bernoulli's equation [5] gives the pressure on the cylinder to $p = p_\infty + \rho(v^2_\infty - v^2)/2$. Insertion of the velocity v into Bernoulli's equation gives the dimensionless pressure coefficient Π to

$$\Pi = 2(p - p_\infty)/(\rho v_\infty^2) = 1 - 4\sin^2\varphi \qquad (1)$$

Equ. (1) shows that Π neither depends on p_∞ nor on v_∞. There are two stagnations points, the forward one at $\varphi = \pi$ and the rearward at $\varphi = 0$, and two extremes, one at $\varphi = \pi/2$, and the other at $\varphi = 3\pi/2$. Because of the symmetry of the pressure distribution the resulting force with which the flow acts on the cylinder is zero. This result holds for every arbitrary body in a steady inviscid irrotational flow, very much contradicting experience.

Also Leonhard Euler offers certain skepticism only a short time later, after his successful formulation of the equations of motion for inviscid fluid flow in 1755, later named after him, he remarks in [6]:

"I hope to reach the goal with some luck, so that the remaining difficulties are only of analytical but not of mechanical nature."

Euler's remark is best be understood if one remembers, that Johann Bernoulli was the first who applied the fundamental laws of mechanics to describe one-dimensional fluid motion, published in his Hydraulica in 1742 [5].

Fig. 2 Leonhard Euler, (1707 - 1783), Professor at the University in Saint Petersburg, was appointed 1741 by Frederick II. to the Royal Prussian Academy of Sciences in Berlin; returned to Saint Petersburg in 1766. Catharine II. the Great strongly supported him, even when he lost his sight completely in 1771. His equations of motion for inviscid flows were first published in Principe's géneraux du mouvement des fluides, Memoires de l´ Acad. des Sciences de Berlin *11*, 274 – 315, also in Opera Omnia, II 12. 54 – 91, 1755.

It was only thirteen years later that Euler had expanded Johann Bernoulli's new approach to describe fluid motion to general incompressible, unsteady three-dimensional flow, a gigantic step forward at that time. But also Euler's skepticism was justified: About 200 years had to pass, until finally solutions of the Euler equations could be constructed for the description of flows about aerodynamic configurations. The non-linearities appearing in them prohibited direct applications for a long time.

2 Equations of Motion for Viscous Flows

Almost another seventy years went by, until the complete equations of motions for viscous flows, today called the Navier-Stokes equations, could be formulated. They were first published in 1823 by Claude-Louis-Marie-Henry Navier in [7]. The equations derived by him describe the conservation of mass and momentum for an infinitesimally small volume element of incompressible fluid in motion.

Fig. 3 Claude-Louis-Marie-Henry Navier, (1785 – 1836), Professor of mechanics at the École des Ponts et Chaussées and later Professor of calculus and mechanics at the École Polytechnic, brought the theory of elasticity into a usable form; he is considered to be the founder of modern structural mechanics, his main contribution is the derivation of the Navier-Stokes equations, first published in Mémoire sur les lois du mouvement des fluides, Mémoires de l´Academie des Sciences 6, 389 – 416, 1823

The naming of the equations is not clear, if it is remembered that Adhémer Barré de Saint-Venant already in 1843 published the equations in the form given in [8], two years before the publication of Sir Georg Gabriel Stokes [9]. With the usual notation, v denoting the velocity, p the pressure, ρ the density, μ the viscosity, and f the volume force, the equations read as follows:

$$\nabla \cdot \vec{V} = 0 \qquad (2)$$

$$\rho\left(\frac{\partial \vec{V}}{\partial t} + (\vec{V}\cdot\nabla)\vec{V}\right) = -\nabla p + \mu\nabla^2\vec{V} + \vec{f} \qquad (3)$$

Fig. 4 Adhémar Jean Claude Barré de Saint-Venant; (1797 -1886), Professor of mathematics at the École des Ponts et Chaussées in Paris, member of the Académie des Sciences; introduced the vector calculus in France; in 1843 he published a correct derivation of the Navier-Stokes equations, two years prior to Stokes, in his Mémoire sur le équations générales de l´équilibre et du mouvement des corps solides élastiques et des fluides in Journal de École Polytechnique 13, pp. 1-174, Cahier XX, 1831.

De Saint-Venant was also first to recognize in the derivation that the viscosity coefficient could replace the shear modulus and serve as a multiplicative factor of the velocity gradients. But still the above equations were not named after him. Perhaps the reason is, that Stokes provided solutions for two problems of the equations of motion simplified for very slow motion - that is the second term of the left-hand side could be left out - in [10] in 1851.

Fig. 5 Sir George Gabriel Stokes (1819 - 1903), Lucasian Professor of mathematics at Cambridge University, member, secretary, and president of the Royal Society, worked in pure mathematics, mathematical and experimental physics, his theoretical works were mainly in hydrodynamics. His derivation of the equations of motion was first published under the title "On the Theories of the Internal Friction of Fluids in Motion", Transactions of the Cambridge Philosophical Society, 8, 287 – 305, 1845.

3 First Exact Solutions

In the so-called Stokes´ first problem Stokes analyzed a flow situation, which is generated, when a plane wall, bounding a fluid, is suddenly accelerated. As the pressure gradient is zero for laminar flow, that is to say the fluid moves in parallel layers, the momentum equations reduce to a partial differential equation describing the conduction of heat in a half plane. A further reduction to an ordinary differential equation is then possible, and a solution is readily available in the form of the complementary error function. For the solution boundary and initial conditions have to be specified. In [10] Stokes postulated that the fluid during its motion has to adhere to the wall and therefore in its vicinity move with the velocity of the wall. This condition of adherence of a moving viscous fluid to a rigid wall is of general importance as it manifests one of the major differences between the flow of an inviscid and a viscous fluid. It serves as a boundary condition for the solution of the Navier-Stokes equations and was later on termed as the Stokes no-slip condition. The solution is briefly established in the following.

Under the conditions mentioned above only the following two terms of the momentum equation Eq. (3) are retained:

$$\frac{\partial u}{\partial t} = v \frac{\partial^2 u}{\partial y^2} \tag{4}$$

In Eq. (4), u represents the velocity component in the x-direction, chosen along the plate, t the time, and v the kinematic viscosity coefficient. The initial and boundary conditions to be employed for the solution of Eq. (4) are

$$t \leq 0: u = 0 \quad \text{for } y \geq 0;$$

$$t > 0: u = U_0 \quad \text{for } y = 0; \qquad u = 0 \quad \text{for } y = \infty. \tag{5}$$

The reduction of the partial differential equation Eq. (4) to an ordinary differential equation is facilitated by the introduction of a similarity variable η together with an assumption for the dependence of u on η:

$$\eta = y/2(vt)^{1/2} \quad \text{and} \quad u = U_0 f(\eta), \tag{6}$$

yielding the ordinary differential equation for $f(\eta)$ and its solution

$$f'' + 2\eta f' = 0 \quad \text{and} \quad u = U_0 \operatorname{erfc} \eta. \tag{7}$$

In Eq. (7) the dashes indicate differentiation with respect to η and erfc η is the complementary error function, which exists in tabulated form.

In dealing with his second problem Stokes provided a solution describing the slow motion of a fluid near an infinite flat plate harmonically oscillating parallel to itself. As in the first problem for laminar flow the Navier-Stokes equations can be reduced to the heat-conduction equation, and the solution is again dictated by the no-slip condition to be obeyed at the solid wall. When the steady state case is considered, for which initial conditions do not have to be prescribed, the velocity profile has the form of a damped harmonic oscillation. However, the solution of the first problem can also be employed to describe the flow for the case that the wall begins to oscillate and the fluid is at rest.

These two problems could be solved with the solution techniques available at that time. Two noteworthy results had already been obtained before, but without the use of the Navier-Stokes equations: In 1839 Gotthilf Heinrich Ludwig Hagen in [11], and about the same time Jean Louis Marie Poiseuille in [12] published a simple relation describing the flow of water in a pipe of radius R. The results were obtained from experimental studies, for which the flow was assumed to be laminar, and the pipe to have a constant circular cross-section with a diameter much smaller than its length. The frictional forces were balanced by a pressure difference along the axis of the pipe, and the pressure was assumed to be constant in the radial direction for each cross-section. The radial velocity profile of the flow $u(r)$ was found to have a parabolic shape, with the maximum at the axis and decreasing to zero velocity at the wall of the pipe.

$$u(r) = -\frac{1}{4\mu}\frac{dp}{dx}(R^2 - r^2) \tag{8}$$

The momentum balance described in [11] and [12] yielded an expression for the volume rate of flow in terms of the constant pressure gradient along the axis.

$$Q = \frac{\pi R^4}{8\mu}\left(-\frac{dp}{dx}\right) \tag{9}$$

This result can also be considered as an exact solution of the Navier-Stokes equations reduced to the conditions given above, but specified only for a certain range of velocities, viscosity coefficients, and the length of the pipe, expressed by a similarity parameter, the Reynolds number in the form $Re = ud/v$.

Another early exact solution of the simplified Navier-Stokes equations was formulated by Maurice Marie Alfred Couette in 1890 [13]. In his study Couette considered the steady laminar flow in a two-dimensional channel, formed by two parallel flat plates, at distance h apart from each other. In the simplest case the fluid can be assumed to be driven by a constant pressure gradient in the direction of the flow, yielding a parabolic velocity profile as in the case of the Hagen-Poiseuille flow. The solution can be extended to the case that one of the parallel flat plates is moving with velocity U and the other is held still:

$$u = \frac{y}{h}U - \frac{h^2}{2\mu}\frac{dp}{dx}\frac{y}{h}\left(1-\frac{y}{h}\right) \qquad (10)$$

Then the velocity profile changes its shape, and back flow may set in, depending of the magnitude of the pressure gradient and the velocity of the moving wall. The Couette flow later on furnished the basis for the lubrication theory.

An exact solution of the Navier-Stokes equations could also be found for the laminar incompressible flow in the gap between two concentric cylinders, moving with different speeds of rotation. According to [14] the essential parts of the solution were already given by Stokes in [9]. An extension of the solution to the case of unsteady flow motion was provided by Carl Wilhelm Oseen in 1910 [15]. Oseen considered the flow around a single cylinder rotating in a fluid of infinite extent. This situation corresponds to the process of decay of a straight vortex in the course of time. The corresponding distribution of the tangential velocity component $w(r,t)$ is given by

$$w(r,t) = \frac{\Gamma_0}{2\pi r}\left[1 - \exp(-r^2/4\nu t)\right], \qquad (11)$$

where Γ_0 is the initial strength of the vortex.

In the following years it was also possible to find exact solutions of the although reduced Navier-Stokes equations for laminar incompressible plane and axially symmetric flow near a stagnation point. Karl Hiemenz in 1911 was able to reduce the momentum equations for two-dimensional flow to two non-linear ordinary differential equations and solve them in his doctoral thesis, published in [16]. The solution for the axially symmetric flow could be provided in a similar fashion. The reduction of the momentum equations again led to two non-linear ordinary differential equations which were solved for the first time only in 1936 [17].

In the last example demonstrating an exact solution of the Navier-Stokes equations the laminar flow in convergent and divergent channels with straight walls is considered. The reduction of the momentum equations to a non-linear ordinary differential equation could also successfully be applied in the solution of this problem. The Navier-Stokes equations were written in polar co-ordinates and the solution could be constructed with an ansatz for the radial velocity component of the form $v \sim F(\varphi)/r$, whereby the continuity equation Eq. (3) is satisfied. The momentum equation Eq. (3) reduces to an ordinary differential equations for $F(\varphi)$

$$2FF' + 4F' + F''' = 0, \tag{12}$$

which after integration gives

$$F^2 + 4F + F'' + K = 0. \tag{13}$$

In Eq. (13) the constant K denotes the pressure gradient at the walls. The function F can be expressed explicitly as an elliptic function of φ. The solution was found independently from each other by Georg Barker Jeffery in 1915 [18] and one year later by Georg Hamel [19].

Although these solutions and others not summarized here constituted a great success, the methods used did not lend themselves to a more rigorous approach in solutions of more complex problems and cope with them successfully. In addition to the results available, flow observations and experiments gave evidence that new ways of flow investigations in theory and experiment were necessary to gain more insight into the flow structures observed. According to E. A. Mueller the solution techniques available in the nineteenth century were not sufficiently adaptable to the ever changing problems of the technical sciences [20]. Time had still to wait for new more suitable solution techniques.

4 New Findings

Since the integration of the complete Navier-Stokes equations was not yet possible, further success concerning their solution could not be reported. However, already in 1858, 35 years after their derivation, Hermann von Helmholtz, at that time professor for physiology in Heidelberg, was able to derive a new relation from the equations of motion, which later became known as the vorticity transport equation. In his derivation he used the Euler equations – that is the equations given above without the last but one term. By introducing the definition of the notion of vorticity in [21] he obtained a vector equation with which the motion of vortices in inviscid fluid flows could be described.

The vorticity characterizes the tendency of a fluid particle to rotate about its own axis. Its mathematical representation is the vector Ω given by the curl of the velocity vector. By taking the curl of the Euler equations von Helmholtz arrived at the following equation:

$$\frac{\partial \vec{\Omega}}{\partial t} + \left(\vec{V} \cdot \nabla\right)\vec{\Omega} = \left(\vec{\Omega} \cdot \nabla\right)\vec{V} + \nu \nabla^2 \vec{\Omega} \tag{14}$$

The last term in Eq. (14) was not contained in the derivation of [21], as viscous forces were not yet considered.

Fig. 6 Hermann von Helmholtz, (1821 - 1894), Professor of pathology and physiology in Königsberg, Bonn, and Heidelberg; universal scholar, was ennobled 1883 and elected first president of the newly founded Physikalisch-Technische Reichsanstalt in 1888. His work on the vorticity transport equation appeared under the title "Über Integrale der hydrodynamischen Gleichungen, welche den Wirbelbewegungen entsprechen", Celles J. 55, 25, 1858.

Helmholtz was also able to derive three theorems, which characterize the behavior of vortex filaments in inviscid flows. As formulated in his original paper [21], they read in the translation:

1. *A water particle, which does not rotate from the beginning on, cannot begin to rotate at a later time.*
2. *Water particles which belong to a vortex filament at an arbitrary time will always belong to that same filament, even when the particles are in motion.*
3. *The vortex filaments must therefore form closed loops in the fluid or can end only at its boundaries.*

The vortex theorems played an important role in the developments that followed. The realization, that a vortex ring cannot be cut is a direct consequence of Helmholtz´ third theorem. It also became the starting point in the development of the theory of lift. Needless to say that the vorticity transport equation could not be solved either because of the non-linearities of the terms describing the convective acceleration.

Another fundamental cornerstone of the nineteenth century was laid five years later in 1883 by Osborne Reynolds, professor of mechanics in Manchester [22]. He showed in an experiment that under certain conditions – today characterized by the dimensionless similarity parameter, mentioned earlier and called Reynolds number – that an originally plain laminated pipe flow would generate pressure and velocity fluctuations, later on termed laminar-turbulent transition, eventually turning into fully turbulent flow.

Reynolds also introduced the concept of turbulent fluctuations into the Navier-Stokes equations and postulated that after time-averaging the equations could be used for describing turbulent flows. Since information is lost in the averaging process new unknown terms appeared, which were called apparent or Reynolds stresses.

Fig. 7 Osborne Reynolds (1842 - 1912), Professor of civil and mechanical engineering at Owens College in Manchester, 1877 Fellow of the Royal Society; worked in fluid mechanics, electrical engineering, magnetism, and astrophysics; most important similarity parameter in fluid dynamics is named after Reynolds. His 1883 experiments were published in "An Experimental Investigation of the Circumstances whether the Motion of Water Shall Be Direct or Sinuous and of the Law of Resistance in Parallel Channels", Philosophical Transactions of the Royal Society of London, series A, *174*, 1883, 935 -982.

To arrive at the apparent stresses, according to Reynolds' hypothesis, the instantaneous flow quantities are split into a time-averaged value and a fluctuating part, for example the velocity component u

$$u(x,y,z,t) = \bar{u}(x,y,z,) + u'(x,y,z,t), \quad (15)$$

and the time-averaged value of u is given by

$$\bar{u} = \frac{1}{T}\int_0^T u\, dt. \quad (16)$$

Introducing Eqs. (15) and (16) and similar relations for the other velocity component into the Navier-Stokes equations there result the three time-averaged momentum equations. When written in Einstein notation, for steady flow they read

$$\rho \bar{u}_j \frac{\partial \bar{u}_i}{\partial x_j} = \frac{\partial}{\partial x_j}\left[-\bar{p}\delta_{ij} + \mu\left(\frac{\partial \bar{u}_i}{\partial x_j} + \frac{\partial \bar{u}_j}{\partial x_i}\right) - \rho\overline{u'_i u'_j}\right] + \rho \bar{f}_i. \quad (17)$$

The terms on the left-hand side of Eq. (17) are the convective terms of the time-averaged flow, on the right the first two terms represent the pressure gradient of the mean flow and its viscous stresses, respectively. The third term describes the Reynolds´ stresses, resulting from the fluctuations of the flow velocity, and the last is the time-averaged volume force. From the time of Reynolds' experiment on the construction of suitable closure relations of Eq. (17) is regarded as the central problem of turbulent flow research, a problem of great technical importance.

5 New Theories

In the year 1904 Ludwig Prandtl presented a paper at the III. International Mathematics Congress in which he showed with an order-of-magnitude analysis that in fluids with small viscosity the frictional forces come into play only in the vicinity of solid boundaries [23].

Fig. 8 Ludwig Prandtl, (1875 - 1953), Professor at the Universities of Hanover and Göttingen, Director of the Institute for Technical Physics, built the first closed-circuit wind tunnel in 1909; President of the Aerodynamic Research Laboratory at Göttingen; developed the concept of the boundary-layer theory, first published in "Über Flüssigkeitsbewegung bei sehr kleiner Reibung", Verhandl. III. Intern. Math. Kongr. Heidelberg, 484 – 491, 1904; his fundamental work on wing theory was published in "Tragflügeltheorie I. u. II." Mitt., Nachr. Ges. Wiss. Göttingen, Math.-Phys. Kl., 451 - 477, 1918; 107 - 137, 1919.

In his derivation, which soon became known as boundary-layer theory, Prandtl was able to simplify the Navier-Stokes equations to the boundary-layer equations.

By comparing the order of magnitude of the inertia forces and the friction forces per unit volume Prandtl could show that the thickness of the layer δ, in which the viscous forces cannot be neglected, is proportional to the inverse of the square root of the Reynolds number, when referenced to a characteristic body length l:

$$\delta/l = O\left(1/\sqrt{\mu/(\rho u l)}\right) = O\left(1/\sqrt{\mathrm{Re}_l}\right) \tag{18}$$

According to Eq. (18) the thickness δ of the "viscous layer", the boundary layer, for large Reynolds numbers is very small in comparison to the characteristic body length. The nondimensionalized velocity component in the direction normal to the main flow should therefore also be of $O(Re_l)^{-1/2}$, and moreover, the dimensionless pressure gradient in the normal direction turns out to be of $O(Re_l)^{-1}$. These findings enabled the reduction of the Navier-Stokes equations to the boundary-layer equations. For steady, incompressible two-dimensional incompressible flow they are

$$\frac{\partial u}{\partial x} + \frac{\partial v}{\partial y} = 0 \tag{19}$$

$$\rho u \frac{\partial u}{\partial x} + \rho v \frac{\partial u}{\partial y} = -\frac{\partial p}{\partial x} + \mu \frac{\partial^2 u}{\partial x^2} \tag{20}$$

The first solution of Eqs. (19) and (20) was constructed with the aid of the similarity technique by Prandtl's student Heinrich Blasius in 1908 [24]. By introducing dimensionless co-ordinates ξ and η, and the stream function ψ in a dimensionless form $f(\xi,\eta)$

$$\xi = x/l, \quad \eta = y\sqrt{\frac{u_\infty}{vx}}, \quad f(\xi,\eta) = \frac{1}{\sqrt{vxu_\infty}} \psi(x,y) \tag{21}$$

Eq. (19) could be satisfied, and Eq. (20) for the case of the flow around a flat plate could be transformed into the following equation:

$$2f''' + ff'' = 0 \tag{22}$$

Blasius obtained the solution to Eq. (22) with a power series solution, from which also the drag coefficient c_D of a flat plate with length L at zero incidence could be computed by integration the local skin friction coefficient c_f:

$$c_D = \frac{1}{L}\int_0^L c_f dx = 1.328/\sqrt{\text{Re}_L} \tag{23}$$

Since the solution required a large effort an approximate integration technique was introduced by Theodore von Kármán in 1921 [25], providing a better applicability to technical flow problems. Therein the displacement of the inviscid flow by the boundary layer was expressed by the displacement thickness δ_1 and the loss of momentum due to the friction forces by the momentum thickness δ_2, defined as follows:

$$\delta_1 = \int_0^\infty (1 - u/u_e)dy \qquad \delta_2 = \int_0^\infty (u/u_e)(1 - u/u_e)dy \tag{24}$$

Insertion of δ_1 and δ_2 into the momentum equation Eq. (20) and integration yielded the von Kármán integral relation

$$\frac{d\delta_2}{dx} + \frac{1}{u_e}\frac{du_e}{dx}(2\delta_2 + \delta_1) + \frac{\tau(y=0)}{\rho u_e^2} = 0. \tag{25}$$

The subscript e indicates the local velocity component u external to the boundary layer. If δ_1 and δ_2 are known, the wall shear stress $\tau(y = 0)$ can be determined. As shown in [25], values of the wall shear stress and the drag coefficient obtained with Eq. (25) agree very satisfactorily with those obtained with Eq. (23).

Fig. 9 Theodore von Kármán (1881 – 1963), Professor of mechanics and aerodynamics and director of the Institute of Aerodynamics at the TH Aachen 1913 - 1934; director of the Guggenheim Aeronautical Laboratory at the California Institute of Technology 1930; founded with others the Jet Propulsion Laboratory in 1944, AGARD in 1951, the International Council of the Aeronautical Sciences 1956, the International Academy of Astronautics in 1960. His Integral relation was first published in ZAMM 1, 233 - 252, 1921: "Über laminare und turbulente Reibung"; also Abh. H. 1, S. 1, 1921.

With the aid of the solution to Prandtl's boundary-layer equations the drag of bodies moving in fluids and gases could now be determined, although with some limitations of the theory proposed. In some sense the paradoxical result of d' Alembert was resolved.

Prandtl proposed another pacesetting theory in 1918, when he was able to formulate the lifting-line theory, often referred to as Lanchester-Prandtl wing theory [26]. With the aid of Helmholtz vortex filaments, shaped into the form of horseshoes, it was possible to show that at the wing, the location of the bounded part of the horseshoe vortex system, because of the finite span, a downwash velocity w_i was induced by the vortex elements leaving the wing. If it is assumed that the circulation distribution $\Gamma(y)$ on the wing is known, with y the co-ordinate in the spanwise direction, the downwash velocity w_i can be computed from the lifting-line theory published in 1918:

$$w_i = \frac{1}{4\pi} \int_{-s}^{s} \frac{d\Gamma}{dy'} \frac{dy'}{y - y'} \tag{26}$$

The integration is to be extended over the entire span, $-s \leq y \leq s$, and $y - y'$ is the distance between the point considered and the location of the vortex line leaving the wing. The downwash velocity reduces the angle of attack of the wing by the amount $\alpha_i = w_i/V$, where V is the undisturbed free stream velocity. According to the Kutta-Joukowsky theorem the resultant aerodynamic force dR acting on a wing element stands normal to the resultant oncoming flow direction. The lift component acting on a wing element with width dy in the spanwise direction then is $dL = dR\cos\alpha_i \approx dR$, and the drag component, which is parallel to the undisturbed flow direction $dD_i = dR\sin\alpha_i \approx dR\alpha_i$, is the induced drag acting on the element. There results for the entire wing

$$D_i = \int_{-s}^{s} \Gamma(y) w_i(y) dy \tag{27}$$

Eq. (27) can be used to determine the induced drag for special circulation distributions prescribed. The well-known value for the elliptic circulation distribution shows that the induced drag coefficient is inversely proportional to the square of the span divided the planform of the wing, the aspect ratio, hinting at the usually large span of modern aeroplanes. This is another result of theoretical approaches to describe the motion of flows around bodies and their forces acting on them: From then on aerodynamicists were enabled to determine lift and drag of wings of finite span. The lifting-line theory became the fundament of all the later developments of wing theories that followed.

Prandtl also foresaw the development of modern computing. One of his former students, later professor of mathematics at Freiburg University, Henry Görtler reports in [27], that before and during World War II Prandtl developed a mechanical computing machine, he had dreamed of to solve the initial-two-point boundary-value problem of the boundary-layer theory with. The machine never worked, and only a design drawing dating back to the year 1941, just a short time before the advent of the electronic computing machines, was saved.

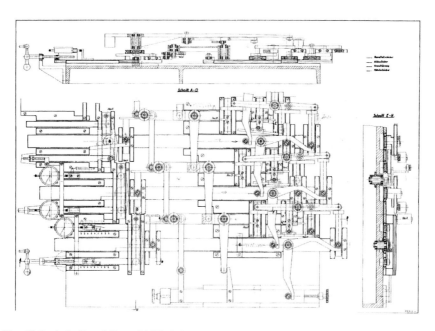

Fig. 10 Ludwig Prandtl's unfulfilled dream: A mechanical computing machine, he thought he could solve the initial–two-point boundary-value problem of his boundary-layer theory with. The design draft shown dates back to 1941, only a few years before the advent of the electronic computing machines. The machine was never completed, as reported in [27], only the draft was saved.

6 The Ascent of Numerical Solutions

Even in the twenties solutions of the complete Navier-Stokes equations could not be availed for the description of viscous flows. If flows with large Reynolds numbers were to be described, the flow field would be divided into an inviscid part and the boundary layer. The inviscid flow was then computed with the potential-flow theory and the flow in the boundary layer with the new integral method for the solution of Prandtl's boundary-layer equations, derived by Theodore von Kármán [25].

Since the wind-tunnel technique was substantially advanced by Prandtl and his students in Göttingen from the turn of the century on, extensive measurements of pressure distributions were used for flow diagnostics. This scenario prevailed until the sixties. Astonishingly enough this situation was confirmed by one of the fathers of informatics, John von Neumann, after whom the architecture of modern computing machines is named and whom we owe the scientification of modern computing. Von Neumann strongly shaped and pushed forward the unforeseen development of modern computers and computational techniques. He developed the first computer of the Institute for Advanced Studies in Princeton. In 1963 von Neumann remarked in [28]:

"Thus, wind tunnels are, for example, used at present, at least in large part, as computing devices of the so-called analogy type (.....) to integrate the non-linear partial differential equations of fluid dynamics."

This remark offered by one of the leading mathematicians of his time came as a surprise, in as much as von Neumann emphasized that wind tunnels were used as computing machines and not so much as instruments to determine flow characteristics with by measuring, for example, certain quantities, like pressure or velocities.

Fig. 11 John von Neumann, (1903 - 1958), after appointments as university lecturer in Göttingen, Berlin, and Hamburg, followed an invitation to Princeton University in 1930; joined the newly founded Institute for Advanced Studies in 1933. Known as one of the fathers of informatics, the architecture of modern computing machines is named after him. The above remark may be found in an article entitled: "On the Principles of Large Scale Computing Machines", Collected Works, Pergamon Press, Oxford, 1 – 34, 1963, by J. H. Goldstine, and J. von Neumann.

The following fifty years witnessed a development of the information and communication technology that not only affected all branches of science and technology but also our entire life. For example in 1991 the US Congress promulgated the High Performance Computing and Communication Act, thereby proposing high-speed communications systems in order to enhance science and education in the USA. In [29] it is reported that the Japanese government in 2006 declared the supercomputer technology as one of the key technologies of national importance:

"The Japanese government selected the supercomputing technology as one of the key technologies of national importance and launched the Next Generation Supercomputer project in 2006. ...The system with 10 Petaflops class perperformance is planned to be completed in 2012. ... One of the goals of this project is to develop ... the grand challenge applications."

The aim of this project is to increase the computational performance of presently available machines so that solutions of the "Grand Challenges" can be provided. The Grand Challenges were formulated in the 1980s in the USA as those fundamental problems in science and technology with many possibilities of applications, but with the expectation that their solutions can only be obtained in the near future with the aid of supercomputers. Turbulent flow research also belongs to these problems. The Nobel laureate Richard Feynman remarked already in the 1960s: "Turbulence is the most important unsolved problem in classical physics".

From the mid 1960s on the rapid development of the computer technology enabled the construction of numerical solutions of the equations of motion in fluid dynamics. This requires the choice of a suitable mathematical model for the problem to be analyzed, as either the complete Navier-Stokes equations, the Euler equations, or Prandtl's boundary-layer equations. The non-linear partial differential equations have to be discretized and replaced by algebraic equations, which can be solved on supercomputers. A computational grid has to be designed for the discretisation, which must be arranged in such a way, that the local flow structures can properly be resolved. Because of limited storage capacity this is not always possible. The last step of the numerical solution requires the presentation of the results.

In the discretisation of the equations of motion a certain condition has to be met, which guarantees that the domain of dependence of the differential equations is always included by the domain of dependence of the discretisation scheme chosen, or else the convergence of the numerical solution cannot always be ensured. It surprises one today that this condition, the domain-of-dependence condition, was already derived in 1928 by Richard Courant, Kurt Otto Friedrich, and Hans Lewy in Göttingen, published in [30], in a time, when numerical solutions of the Navier-Stokes equations were not in sight.

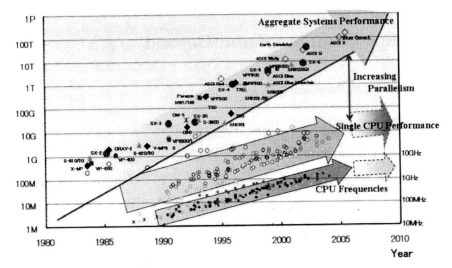

Fig. 12 Increase of the performance of supercomputers during the past 25 years. Data were published by T. Watanabe and M. Nomura in their article: "Petaflops Computers and Beyond", Notes on Numerical Fluid Mechanics and Multidisciplinary Design, 100, 481 – 490, 2009.

In addition to the Courant-Friedrichs-Lewy condition the discretisation of the equations of motion has to satisfy additional conditions to ensure convergence. Because of the non-linearity of Eqs. (2) and (3) a rigid approach is not possible, and the construction of the solution has to rely on results obtained for linear partial differential equations with constant coefficients. In particular Lax's equivalence theorem, strictly valid only for linear partial differential equations [31], serves as an auxiliary tool and states the conditions to be observed:

Given a properly posed initial-value problem and a finite-difference approximotion to it that satisfies the consistency condition, stability is the necessary and sufficient condition for convergence.

If it is assumed that the equivalence theorem holds at least locally for the non-linear problems of fluid mechanics, finite-difference approximations have to be shown to be consistent with the Eqs. (2) and (3), which must be casted into stable finite-difference schemes. This will briefly be demonstrated for the reduced momentum equation of Stokes' first problem.

Following [31] a two-level difference scheme is used to discretize Eq. (4). If n denotes the time level and i the location of the co-ordinate y, the finite-difference approximation to Eq. (4) is

$$\frac{u_i^{n+1} - u_i^n}{\Delta t} = \nu \frac{u_{i+1}^n - 2u_i^n + u_{i-1}^n}{(\Delta y)^2}. \tag{28}$$

The scheme given by Eq. (28) is consistent with Eq. (4), if the truncation error defined as

$$E(u) = \frac{u_i^{n+1} - u_i^n}{\Delta t} - v\frac{(\delta^2 u)_i^n}{(\Delta y)^2}, \quad (29)$$

where $u(y,t)$ is the exact solution, approaches zero when the grid steps are refined. With a Taylor's series expansion about the point (y,t) the truncation error can be shown to be

$$E(u) = O(\Delta t) + O[(\Delta y)^2]. \quad (30)$$

Since $E(u)$ does approach zero when the grid steps are refined, the scheme given by Eq. (28) is consistent with Eq. (4) and the consistency condition is satisfied.

The numerical stability of the Eq. (28) can be shown with the aid of the von Neumann condition [31]. When the difference equation Eq. (28) is expressed through a Fourier series, it can be demonstrated that the growth factor ξ is given by

$$\xi = \xi(m) = 1 - \frac{2v\Delta t}{(\Delta y)^2}(1 - \cos m\Delta y). \quad (31)$$

Since the von Neumann condition requires for numerical stability that ξ does not exceed unity, there results the von Neumann stability condition to

$$\frac{2v\Delta t}{(\Delta y)^2} \le 1. \quad (32)$$

This simple example may demonstrate how the partial differential equations given above can be approximated by algebraic equations that can be solved with numerical techniques.

There are several ways in which the numerical solutions of the Navier-Stokes equations can be constructed. For example in the direct numerical simulation (DNS) the Navier-Stokes equations are discretized and solved on a grid, which guarantees that the local flow structures are adequately resolved. This method is suited for the simulation of flows characterized by relatively small Reynolds numbers. Another method of solution is the large-eddy-simulation technique (LES). In this approach the large vortex structures are directly computed, and the smaller vortex structures are described with modeled approximations. The large-eddy-simulation technique is used when flows with moderate Reynolds numbers are to be described. The third technique mentioned here consists in the solution of the Reynolds-averaged Navier-Stokes equations (RANS), as introduced by Reynolds in 1883. Prior to their solution these equations have to be closed with closure relations for the description of the Reynolds' stress tensor.

As for the numerical solutions there also exist several possibilities for the discretisation of the Navier-Stokes equations. If straight-forward finite differences are used, the derivatives in the differential equations are substituted by difference approximations, and the resulting difference equations are solved with explicit or implicit algorithms. In the finite-volume technique the integral forms of the conservation equations are discretized. A third method is the finite-element method: The flow field is subdivided into finitely large elements. In a second step shape functions are defined for the elements, which when inserted in the conservation equations supply a system of algebraic equations that can be solved on computers.

The computational grids are generated by discrete decomposition of the flow field into area or volume elements. The grids can be structured, following certain prescribed regularities, as for example Cartesian volume elements. Also unstructured grids are used: They are generated by prescribing the coordinates of the vertices of the elements. The unstructured grids offer good adaptation possibilities for the resolution of local flow structures.

The following Fig. 13 shows an example for the generation of a structured hybrid Cartesian grid for the computation of the flow around an aerodynamic shape. The grid is composed of rectangular cells, based on an octree-data structure, with an imbedded grid with triangular – prismatic cells for the computation of the flow in the boundary layer as described in [32]. It is clear from Fig. 13 that a large effort is necessary for the grid generation. Perhaps it was this difficulty d´Alembert was hinting at in [3], that could not be overcome in the eighteenth century with computing machines not available.

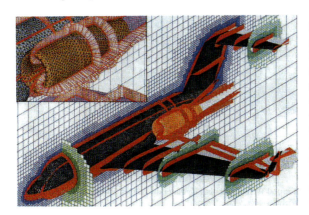

Fig. 13 Example for a generation of a structured hybrid Cartesian grid for computation of the flow around an aerodynamic shape. The grid consists of rectangular cells of different size. A second grid with triangular prismatic cells is imbedded near the surface for computation of the boundary layer [32].

Many technical flows are characterized by extraordinary large Reynolds numbers, as for example, flows about airplanes $Re: O(10^8)$. A direct simulation of such flows on supercomputers presently available is not possible, since according to [33] a computer performance of about 10^{23} *flops* would be required for the solution. The large-eddy simulations today allow investigations of modeled flow problems.

The Reynolds-number restriction was already discovered in 1963 by Jacob E. Fromm. He remarked in [34], that in his simulation of the flow about a rectangular block the solution would fail at a Reynolds number of $Re = 6 \times 10^3$:

"It is believed, however, that the Re = 6000 case is about as far as the calcolational method can be extended in its current form, since here the instability discussed earlier tends to confuse the display patterns."

The instability of Fromm's numerical solution of the Navier-Stokes equations can also be recognized in a comparison with experimental flow visualization provided by A. M. Lippich in 1958, shown in Fig. 14.

Although actual flow conditions are still difficult to simulate on supercomputers, today flow computations, for example to determine lift and drag of airplane configurations can be carried out at much larger Reynolds numbers, see [36], than those chosen in [34] and [35]. Nevertheless it remains a laborious task for several reasons. One, of course, is the difficulty of correctly predicting the Reynolds stresses, with often little or no reliable information available. Turbulent flow research still has to depend on judicious assumptions. Another reason is the approximate nature of the numerical solutions due to their dependence on grid spacing and orders of approximation of the discretisation procedure chosen.

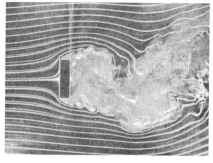

Fig. 14 Fromm´s numerical visualization of the flow around a rectangular block, obtained from the Navier-Stokes equations in [34]   Lippich´s experimental flow visualization of the flow around a rectangular block obtain ed by smoke injection described in [35]

For these and other reasons the American Institute of Aeronautics and Astronautics organizes „Drag Prediction Workshops" in order to determine the accuracy of predictions of numerical solutions of the conservation equations, so far restricted to aerodynamic problems. The fourth workshop took place in 2009. Test computations were carried out for the transonic flow about the "NASA Common Research Transonic Wing-Body-Tail Model" for Reynolds numbers of the order $Re = 2 \times 10^7$, about one order of magnitude smaller than those of flight conditions, but four orders of magnitude larger than the Reynolds number in Fromm's 1963 computation.

The workshop was organized as an international meeting: Nineteen groups from various countries participated with 29 solutions. Researchers from the USA and Europe provided eleven solutions each and seven were submitted from Asia and Russia. Industry presented results of seven solutions, research establishments and venders nine each, and academia four. The presentations showed that grid generation still occupies a large portion of the work, with structured grids used in nine solutions, but unstructured in twenty.

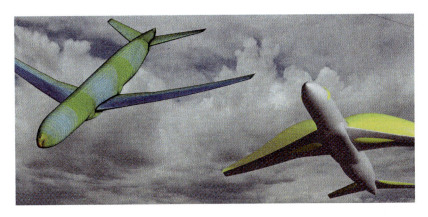

Fig. 15 The NASA Common Transonic Research Model, consisting of a wing-body-tail configuration, used in the AIAA "Drag Prediction Workshops" for comparison calculations for a Reynolds number $Re = 2 \times 10^7$, see, for example [36]

An example for the results obtained is pictured in Fig. 15, taken from [36]. Shown is the Lilienthal polar diagram, which gives the dimensionless lift coefficient as a function of the dimensionless drag coefficient. The computed data are compared with experimental results.

Fig. 16 contains data for several test conditions. For example, the influence of the grid spacing on the accuracy of the solution was investigated by varying the number of grid cells between 3.5 to 105 million cells. A second test case was focused on the influence of the Reynolds number on the drag coefficient. In these studies the Reynolds number was varied between $Re = 5 \times 10^6$ and $Re = 2 \times 10^7$.

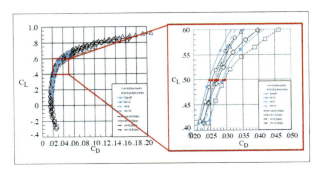

Fig. 16 Comparison of numerical and experimental results of the 4th AIAA Drag Prediction Workshop of 2009. Shown is the lift coefficient as a function of the drag coefficient in Lilienthal's polar diagram for several test conditions [36].

It is reported that the computed results matched the experimental data best at the higher Reynolds number. The workshop also included the investigation of other problems, for example, studies of the influence of the stabilizer on the drag and lift coefficients [36].

Altogether the results of the fourth workshop were summarized in [37]. In the general conclusion it is manifested, that a number of numerical solutions of the Navier-Stokes equations is available, which for all test cases yield good agreement. The authors state in [37], that

> ..."there is a set of CFD codes whose members ... agree relatively well with each other ... over all the test cases. ...it is comprised of flow solvers that are based on all types of grids. Hence several structured, unstructured, and hybrid mesh solvers have matured sufficiently to be useful CFD tools for accurate drag prediction."

Fig. 17 Present and future applications of numerical methods in industrial design and optimization of aerodynamic configurations. Solutions of Reynolds-averaged Navier-Stokes equations are preferably used as described in an article by K. Becker, J. Vassberg: "Numerical Aerodynamics in Transport Aircraft Design", in Notes on Numerical Fluid Mechanics and Multidisciplinary Design, Vol. 100, 2009.

In the past fifty years numerical solutions of the Navier-Stokes equations have matured to such a degree that the accurate determination of the drag for the flow conditions mentioned is possible. According to [38] numerical methods are already intensively being used in analyses of aerodynamic characteristics of airplane configurations in the design and optimization phase. These investigations mainly prefer solutions of the Reynolds-averaged Navier-Stokes equations. Fig. 17 published in [38] shows the many applications of numerical methods in industrial applications.

7 Concluding Remarks

The enormous progress in the development of modern computing machines and numerical solutions of the Navier-stokes equations during the past fifty years makes it possible today to determine lift and drag of aerodynamic configurations for flows up to moderate Reynolds numbers. Counting from d´Alembert´s first attempts in 1752 it took more than 200 years until the drag of a body moving in a fluid could be computed with presently available numerical simulation techniques, although with some severe restrictions: Transition of laminar to turbulent flow and fully turbulent flows still belong to the most important unsolved problems of fluid mechanics. They pose problems in the construction of numerical solutions for flow simulation, as approximations have to be introduced.

Flows with high Reynolds numbers still remain inaccessible for direct simulation. The validity of results of numerical flow computations can therefore only be checked by comparing with experimental results, because of the approximations that have to be introduced in the solutions. Presently available computational speeds and storage capacities, although large, pose another problem: They restrict the temporal and spatial resolution of the flow motion.

Nevertheless, the new computational methods in fluid mechanics in the mean time have become an indispensible tool for fluid flow research and development. Many applications demonstrate the usefulness of modern simulation techniques, how incomplete they still may be. Some of the most difficult fundamental problems still wait for their solution. For these reasons the millennium problem formulated by the Clay Mathematics Institute, the analysis of existence and regularity of solutions of the Navier-Stokes equations for the description of three-dimensional flows remains of outmost importance.

Acknowledgement. Names, titles, and citations of references [3], [5] – [10] can be found in István Szabó: Geschichte der mechanischen Prinzipien, Birkhäuser Verlag, Basel und Stuttgart, 1976. Wikipedia is acknowledged for the pictures shown in Figs. 1 – 8, and 10. Extended version of original text published in Journal of Computational Technology, Russian Academy of Sciences, Tom 17, N0 1, pp. 3-16, Copyright provided by Editor in Chief Prof. Yu. I. Shokin.

References

[1] Jaffe, A.M.: The Millenium Grand Challenge in Mathematics. Notes of the AMS, 652–660 (June/July 2006)
[2] Fefferman, C.L.: Existence and Smoothness of the Navier-Stokes equation. The Millennium Problems, Official Statement of the Problem, Clay Mathematics Institute, Cambridge, MA, USA (2000)
[3] de Alembert, J. le R.: Essai d' une nouvelle théorie de la résistance des fluides. Opuscules Mathematiques V, 132–138 (1752, 1768)
[4] von Kármán, T.: Aerodynamics, Selected Topics in the Light of Their Historical Development. Cornell University Press, Ithaca (1954)

[5] Bernoulli, J.: Hydraulica. Opera Omnia (1742)
[6] Euler, L.: Principes géneraux du mouvement des fluides. Memoires de l' Acad. des Sciences de Berlin 11, 274–315, also in Opera Omnia II 12, 54–91 (1755)
[7] Navier, C.L.M.H.: Mémoire sur les lois du mouvement des fluides. Mémoires de l' Academie des Sciences 6, 389–416 (1823)
[8] de Saint-Venant, A.J.C.B.: Mémoire sur le équations générales de l'équilibre et du mouvement des corps solides élastiques et des fluides. Journal de École Polytechnique 13, 1–174, also Mémoire sur la dynamique des fluides. Comptes Rendus 17, 1240–1243 (1843)
[9] Stokes, G.G.: On the Theories of the Internal Friction of Fluids in Motion. Transactions of the Cambridge Philosophical Society 8, 287–305 (1845)
[10] Stokes, G.G.: On the effect of the internal friction of fluids on the motion of pendulums. Cambr. Phil. Trans. IX, 8 (1851)
[11] Hagen, G.: Über die Bewegung des Wassers in engen zylindrischen Röhren. Pogg. Ann. 46, 423 (1839)
[12] Poiseuille, J.: Récherches expérimentelles sur le mouvement des liquids dans les tubes de très petits diamètres. Comptes Rendus 11, 961 and 1041 (1840)
[13] Couette, M.M.A.C.: Etudes sur le frottement de liquides. Ann. Chim. Phys. 6, Ser. 21, 433–510 (1890)
[14] Lamb, H.: Hydrodynamics, 6th edn., ch. XI, pp. 587–588. Dover Publications, New York (1932)
[15] Oseen, C.W.: Ark. F. Astron. Och. Fys. (1911), Neuere Methoden und Ergebnisse in der Hydromechanik. Akademische Verlagsgesellschaft m. b. H., Leipzig (1927)
[16] Hiemenz, K.: Die Grenzschicht an einem in den Flüssigkeitsstrom eingetauchten geraden Kreiszylinder. (Thesis Göttingen 1911). Dingl. Polytech. J. 326, 321 (1911)
[17] Homann, F.: Der Einfluß großer Zähigkeit bei der Strömung um den Zylinder und um die Kugel. ZAMM 16, 153–164 (1936)
[18] Jeffery, G.B.: Steady motions of a viscous fluid. Phil. Mag. 29, 455 (1915)
[19] Hamel, G.: Spiralförmige Bewegung zäher Flüssigkeiten, Jahresber. d. Dt. Mathematiker-Vereinigung 34 (1916); see also Tollmien, W.: Handbuch der Experimental-Physik IV(Pt. 1), 257
[20] Mueller, E.A.: Theododor von Kármán und die angewandte Mathematik. Max-Planck-Institut für Stroemungsforschung, Bericht 9 (1981)
[21] von Helmholtz, H.: Über Integrale der hydrodynamischen Gleichungen, welche den Wirbelbewegungen entsprechen. Celles J. 55, 25 (1858)
[22] Reynolds, O.: An Experimental Investigation of the Circumstances whether the Motion of Water Shall Be Direct or Sinuous and of the Law of Resistance in Parallel Channels. Philosophical Transactions of the Royal Society of London, Series A 174, 935–982 (1883)
[23] Prandtl, L.: Über Flüssigkeitsbewegung bei sehr kleiner Reibung. Verhandl. III, pp. 484–491. Intern. Math. Kongr., Heidelberg (1904)
[24] Blasius, H.: Grenzschichten in Flüssigkeiten mit kleiner Reibung. Z. Math. u. Physik 56(1) (1908)
[25] von Kármán, T.: Über laminare und turbulente Reibung. ZAMM 1, 233–252 (1921)
[26] Prandtl, L.: Tragflügeltheorie I. u. II. Mitt., Nachr. Ges. Wiss. Göttingen. Math.-Phys. Kl., 451–477 (1918); 107–137 (1919)
[27] Görtler, H,: Ludwig Prandtl – Perönlichkeit und Wirken. Z. Flugwissenschaft 23, Heft 5, 153–162 (1975)

28. Goldstine, J.H.H., von Neumann, J.: On the Principles of Large Scale Computing Machines, Collected Works, pp. 1–34. Pergamon Press, Oxford (1963)
29. Watanabe, T., Nomura, M.: Petaflops Computers and Beyond. In: Hirschel, E.H., Krause, E. (eds.) Notes on Numerical Fluid Mechanics and Multidisciplinary Design, vol. 100, pp. 481–490 (2009)
30. Courant, R., Friedrichs, K., Lewy, H.: Über die partiellen Differentialgleichun-gen der mathematischen Physik. Mathematische Annalen 100, 32–74 (1928)
31. Richtmyer, R.D., Morton, K.W.: Difference Methods for Initial-Value Problems, 2nd edn., p. 45. Interscience Publishers, New York (1967)
32. Deister, F., Rocher, D., Hirschel, E.H., Monnoyer, F.: Self-Organizing Hybrid Cartesian Grid Generation and Solutions for Arbitrary Geometries. In: Hirschel, E.H. (ed.) Notes on Numerical Fluid Mechanics and Multidisciplinary Design. Numerical Flow Simulation II, vol. 75, pp. 19–33 (2001)
33. St. Pope, B.: Turbulent flows. Cambridge University Press, Cambridge (2000)
34. Fromm, J.E.: A Method for Computing Non Steady Incompressible, Viscous Flows, Rep. No. LA-2910, Los Alamos Scientific Laboratory (1963)
35. Lippisch, A.M.: Experimental Flow visualization. Aeronaut. Eng. Rev. 17(2), 24 (1958)
36. Feldhaus, U.: Simulation mit CFD. Luft- und Raumfahrt Heft 1, 28–29 (2011)
37. Vassberg, J.C., et al.: Summary of the Fourth AIAA CFD Drag Prediction Workshop, AIAA Paper 2010 - 4547 (June 2010)
38. Becker, K., Vassberg, J.: Numerical Aerodynamics in Transport Aircraft Design. In: Hirschel, E.H., Krause, E. (eds.) Notes on Numerical Fluid Mechanics and Multidisciplinary Design, vol. 100, pp. 209–220 (2009)

On Non-uniqueness Issues Associated with Fröhlich's Solution for Boussinesq's Concentrated Force Problem for an Isotropic Elastic Halfspace

A. Patrick S. Selvadurai[*]

Abstract. This paper examines O.K. Fröhlich's solution to Boussinesq's problem for the action of a concentrated normal force on the surface of an isotropic elastic halfspace. In this endeavour, Fröhlich introduced the concept of a *"Concentration Factor"* that would allow for variations in the stress diffusion within the halfspace region. This note draws attention to the possible limitations of the solution as it pertains to the evaluation of the resulting displacement field in the halfspace region due to violation of the conventional Beltrami-Michell compatibility criteria applicable to continua.

Keywords: Fröhlich's concentration factor, Boussinesq's solution, compatibility equations, diffusion of load in a halfspace, uniqueness of solution.

1   Introduction

The solution to the classical elasticity problem dealing with the action of a concentrated normal force on an isotropic halfspace region was first presented by Boussinesq (1885) and represents a classical result that is widely used in geomechanics and applied mechanics (Davis and Selvadurai, 1996; Selvadurai, 2007). Boussinesq's approach for solving the problem of the loading of a halfspace by a concentrated normal force (Fig.1) takes into account all the equations governing the classical theory of elasticity and the relevant boundary

---

A.P.S. Selvadurai
Department of Civil Engineering and Applied Mechanics
McGill University, Montréal, QC, Canada H3A 0C3

[*] *William Scott Professor* and *James McGill Professor*.

conditions and regularity conditions. The method of solution relies on results of potential theory and yields an *exact closed form solution* for the displacements and stresses within the halfspace region. An alternative approach to obtaining a solution to Boussinesq's problem was presented by Selvadurai (2001) and the classical result by Mindlin (1936) represents the generalized result from which both Boussinesq's solution for the problem of the normal loading of the surface of a halfspace region by a concentrated force and the solution of Kelvin (1848) (see also Love (1927)) for the interior loading of an isotropic elastic halfspace by a concentrated force can be recovered as special cases. These fundamental solutions have been extended to include both anisotropy and heterogeneity of the elastic medium (Ting, 1996; Hetnarski and Ignaczak, 2004).

The notion of a *"concentration factor"* (n) for examining the load transfer to the interior of an isotropic elastic halfspace region was considered by Griffith (1929) and introduced and documented in the book by Fröhlich (1934). The possibility of varying the load transfer pattern within the halfspace region by altering a single coefficient was a major attraction of the theory and for the specific value of the *"concentration factor"*, $n = 3$, Boussinesq's solution is recovered. The justification for the *"concentration factor"* was, however, not provided by Fröhlich (1934). Ohde (1939) and Borowicka (1942) indicate that in addition to the limit of $n = 3$, the case $n = 4$ corresponds to an elastic halfspace region where the linear elastic shear modulus varies linearly with depth. The result of Borowicka (1942) involves an infinite series solution in terms of the Poisson number. The complete solution to the problem of the surface loading of an isotropic elastic halfspace with a linear variation of the linear elastic shear modulus was first given by Gibson (1967) (see also Selvadurai (2007), Selvadurai and Katebi (2013) and Katebi and Selvadurai (2013)). Ohde (1939) arrives at a relationship between the *"concentration factor"* n and Poisson's ratio v in the form $n = (1 + v^{-1})$, giving the result $n = 3$, which is required for the solution to reduce to Boussinesq's classical result for the special case of an incompressible elastic material. The interest in the application of Fröhlich's concentration factor is largely because of the potential of the result to explain the deviations in the stress transfer at depth due to the effects of soil compaction. Measurements of stress distribution within soil masses during the application of surface loads has been documented by several investigators and a recent review with applications to mechanics of soil tillage is given by Keller et al. (2013). It should be noted that the measurement of stresses within soil masses using embedded contact pressure cells is a difficult exercise, largely due to the fact that the cell-action factor that is needed to correctly interpret the stress state will be governed by a variety of responses including the constitutive relationship for the soil itself. Extensive reviews of contact stress measurement and the development of techniques for the interpretation of results from soil pressure cells are given by Hvorslev (1976), Selvadurai (1979) and Selvadurai et al. (2007). Results derived from embedded pressure cells are considerably more difficult to interpret than data from pressure cells that are located at a rigid boundary. For this reason, the experimental results themselves can be prone to incorrect interpretation. In spite of this limitation, Fröhlich's result is extensively used in current approaches to examine compaction-induced alterations of the soil fabric and its load transfer capabilities. The theoretical developments of Fröhlich therefore merit further discussion so that its basis can be examined in the context of theoretical geomechanics.

2 Boussinesq's Solution for the Concentrated Force Problem

Boussinesq's problem that deals with the action of a concentrated normal force P_B on the surface of a semi-infinite isotropic elastic medium (Fig.1) can be solved using a variety of approaches. These are discussed in several articles (e.g. Timoshenko and Goodier, 1970; Little, 1973; Davis and Selvadurai, 1996; Selvadurai, 2000, 2007). For example, the formal integral expressions (see e.g. Sneddon and Berry, 1958; Gurtin, 1972) for the non-zero displacements components $u_r(r,z)$ and $u_z(r,z)$, referred to the cylindrical polar coordinate system take the forms

$$u_r(r,z) = -\frac{P_B}{4\pi\mu} \int_0^\infty [(1-2\nu) - \xi z] \exp(-\xi z) J_1(\xi r) d\xi \qquad (1)$$

$$u_z(r,z) = \frac{P_B}{4\pi\mu} \int_0^\infty [2(1-\nu) + \xi z] \exp(-\xi z) J_0(\xi r) d\xi \qquad (2)$$

where J_0 and J_1 are, respectively, the zeroth-order and first order Bessel functions of the first kind, μ is the linear elastic shear modulus and ν is Poisson's ratio. Taking into consideration the direction of application of the Boussinesq force, these results can also be expressed in spherical polar coordinates (Selvadurai, 2001) in the forms

$$2\mu u_R = \frac{P_B}{2\pi R}[4(1-\nu)\cos\Theta - (1-2\nu)] \qquad (3)$$

$$2\mu u_\Theta = \frac{P_B \sin\Theta}{2\pi R}\left(-(3-4\nu) + \frac{(1-2\nu)}{(1+\cos\Theta)}\right) \qquad (4)$$

where $r = R\sin\Theta, z = R\cos\Theta$ with $R(=\sqrt{r^2+z^2}) \in (0,\infty)$ and $\Theta \in (0, \pi/2)$. By appeal to the constitutitve equations governing classical elasticity, the state of stress in the halfspace region can be determined uniquely from these results. As is evident, the displacements (1) to (4) satisfy the regularity conditions necessary and sufficient to ensure uniqueness of the solution, Similarly, the stress state derived from (1) to (4) also satisfies the equations of equlibrium, the regularlity conditions and the traction boundary conditions at the surface $z = 0$. Furthermore, the singularities in the stress state are integrable and contribute to a traction resultant identical to the normal force applied at the surface of the halfspace.

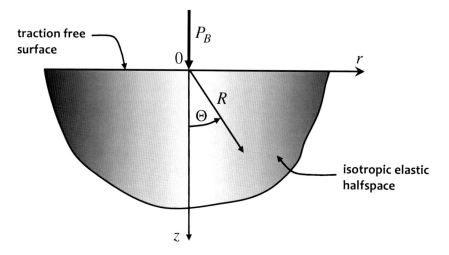

Fig. 1 Boussinesq's problem for an isotropic elastic halfspace

3 Fröhlich's Solution for the Concentrated Force Problem

A solution to the problem of the loading of an isotropic incompressible elastic halfspace by a concentrated normal force was presented by Fröhlich's (1934). This work is an empirical development, which adjusts the form of the vertical stress $\sigma_{zz}(r,z)$ due to Boussinesq's solution by introducing a "concentration factor n". It allows the alteration of the decay pattern to suit experimental observations. In the ensuing, the sign convention assumes tensile stresses to be positive and the concentration factor v used by Fröhlich is replaced by n to avoid confusion with Poisson's ratio v. It should be remarked at the outset that this semi-empirical modification is proposed *only* for incompressible ($v = 1/2$) elastic materials. The state of stress in the halfspace region obtained by Fröhlich takes the form

$$\sigma_{zz}(r,z) = -\frac{nP_B z^n}{2\pi R^{n+2}} \tag{5}$$

$$\sigma_{rr}(r,z) = -\frac{nP_B r^2 z^{n-2}}{2\pi R^{n+2}}; \quad \sigma_{\theta\theta} = 0; \quad \sigma_{rz} = -\frac{nP_B rz^{n-1}}{2\pi R^{n+2}} \tag{6}$$

where $R = \sqrt{r^2 + z^2}$. The mathematical basis for introducing this concept is lacking; there appears to be no formal linear solution of the equations governing the classical theory of elasticity that will yield a solution to Boussinesq's problem for the action of a concentrated force P_B normal to the surface of a an elastic halfspace in the above forms. It can, however, be verified that, in the absence of body forces, the stress state defined by (5) and (6) also satisfies the non-trivial axisymmetric equations of equilibrium expressed in cylindrical polar coordinates:

$$\frac{\partial \sigma_{rr}}{\partial r} + \frac{\partial \sigma_{rz}}{\partial z} + \frac{\sigma_{rr} - \sigma_{\theta\theta}}{r} = 0$$
$$\frac{\partial \sigma_{rz}}{\partial r} + \frac{\partial \sigma_{zz}}{\partial z} + \frac{\sigma_{rz}}{r} = 0 \tag{7}$$

It can also be verified that on any plane $z = \text{const.}$,

$$\int_0^\infty \int_0^{2\pi} \sigma_{zz}(r,z) \, r \, dr \, d\theta + P_B = 0 \tag{8}$$

indicating that vertical equilibrium is satisfied at any plane $z = \text{const.}$, within the halfspace region. Similarly it can be shown that on any cylindrical surface $r > 0$,

$$2\pi r \int_0^\infty \sigma_{rz}(r,z) \, dz + P_B = 0 \tag{9}$$

It should be noted that the results (8) and (9) are valid irrespective of the concentration factor n. Also, the classical Boussinesq's solution for $\sigma_{zz}(r,z)$ in an incompressible elastic medium is recovered when $n = 3$. Typical results for the distribution of the axial stress $|\sigma_{zz}(r,z)|$ for various values of n are shown in Figure 2. Since the interpretation of the concentration factor-based analysis in terms of classical elasticity is valid only for $n = 3$, for any other choice of n the solution should deviate from classical elastic behaviour, which satisfies only the equations of equilibrium and may not, *in general*, satisfy other equations applicable to classical elasticity. The objective of this note is to examine whether all governing equations of compatibility applicable to strains (Timoshenko and Goodier, 1970; Davis and Selvadurai, 1996; Selvadurai, 2000) are satisfied to provide validity to the continuum concept. The Beltrami-Michell equations of compatibility applicable to a continuum region take the form

$$\nabla \times \varepsilon \times \nabla = 0 \qquad (10)$$

where ε is the linearized strain tensor referred to the appropriate coordinate system, ∇ denotes the gradient operator and $\times$ denotes the cross product.

Considering a cylindrical polar coordinate system (r, θ, z) and a state of axial symmetry characterized by the displacement field $\{u_r(r,z), 0, u_z(r,z)\}$, it can be shown that the non-zero components of the linearized strain tensor

$$\varepsilon = \begin{pmatrix} \varepsilon_{rr} & 0 & \varepsilon_{rz} \\ 0 & \varepsilon_{\theta\theta} & 0 \\ \varepsilon_{rz} & 0 & \varepsilon_{zz} \end{pmatrix} = \begin{pmatrix} \dfrac{\partial u_r}{\partial r} & 0 & \dfrac{1}{2}\left(\dfrac{\partial u_r}{\partial z} + \dfrac{\partial u_z}{\partial r}\right) \\ 0 & \dfrac{u_r}{r} & 0 \\ \dfrac{1}{2}\left(\dfrac{\partial u_r}{\partial z} + \dfrac{\partial u_z}{\partial r}\right) & 0 & \dfrac{\partial u_z}{\partial z} \end{pmatrix} \qquad (11)$$

should satisfy the four compatibility equations (10).

Considering the stress state given by (5) and (6) and Hooke's law applicable to an *incompressible* isotropic elastic material, it can be shown that

$$\varepsilon_{rr} = -\frac{nP_B}{4\pi ER^{n+2}}(2r^2 z^{n-2} - z^n) \quad ; \quad \varepsilon_{\theta\theta} = \frac{nP_B}{4\pi ER^{n+2}}(z^n + r^2 z^{n-2})$$

$$\varepsilon_{zz} = -\frac{nP_B}{4\pi ER^{n+2}}(2z^n - r^2 z^{n-2}) \quad ; \quad \varepsilon_{rz} = -\frac{nP_B}{4\pi ER^{n+2}}(3r\, z^{n-1}) \qquad (12)$$

It can be seen that the strain field (12) satisfies the incompressibility constraint

$$\text{tr}\,\varepsilon = \varepsilon_{rr} + \varepsilon_{\theta\theta} + \varepsilon_{zz} \equiv 0 \qquad (13)$$

for any choice of n. It should also be noted that the incompressibility condition is obtained through a constitutive constraint rather than a kinematic constraint on the solution. Substituting (12) in (10) it can be shown that the compatibility equations will be satisfied *if and only if* $n \equiv 3$, and the compatibility equation is violated for all other values of n.

It would appear that Fröhlich's modification to Boussinesq's result to account for either *concentration* or *diffusion* of the load transmission pattern is a *statically admissible solution* but an incomplete result that violates the kinematics of deformation of a continuum. In hindsight, it is clear that *if* Fröhlich's solution for Boussinesq's concentrated force problem (which *completely satisfies* all the governing equations of classical elasticity, including equilibrium, compatibility and the constitutive equations of linear elasticity) is recovered *only* when $n \equiv 3$, then from consideration of Kirchhoff's uniqueness theorem in classical elasticity (Gurtin; 1972; Davis and Selvadurai, 1996; Selvadurai, 2000) the solution must violate one or more of the governing equations when either $n < 3$ or $n > 3$. We have shown that the equations of equilibrium are satisfied for all choices of n, consequently, the compatibility equations must be violated for all choices of $n \neq 3$. The implications of violation of the compatibility conditions presents itself in a non-uniqueness of the integration of the strain-displacement equations when determining the displacement components $u_r(r,z)$ and $u_z(r,z)$. For example, the displacement component $u_r(r,z)$ can be directly determined using the results for $\varepsilon_{\theta\theta}$ in (12) in the strain-displacement relationships (11).i.e.

$$u_r(r,z) = \frac{nP_B r \, z^{n-2}}{4\pi ER^n} \tag{14}$$

We can also obtain an expression for $u_r(r,z)$ by integrating the result in (12) for ε_{rr}, which gives

$$u_r(r,z) = \frac{P_B r \, z^{n-2}}{4\pi ER^n}\left(3 + \frac{(n-3)R^n}{z^n} \, {}_2F_1[\frac{1}{2},\frac{n}{2},\frac{3}{2};-\frac{r^2}{z^2}]\right) + G(z,n) \tag{15}$$

where ${}_2F_1[a,b,c,d]$ is the hypergeometric function (Abramowitz and Stegun, 1964)) and (15) is indeterminate to within $G(z,n)$, an arbitrary function of z. This function can be evaluated through an integration of the expressions for ε_{zz} and substituting the resulting expression for $u_z(r,z)$ and the expression (14) for $u_r(r,z)$ into the expression for ε_{rz}. For the present discussion it is sufficient to consider the expressions (14) and (15); it is clear that the two expressions are distinctly different for $n \neq 3$ but when $n = 3$, both expressions reduce to

$$u_r(r,z) = \frac{3P_B \, rz}{4\pi E(r^2 + z^2)^{3/2}} \tag{16}$$

which is identical to Boussinesq's solution for the radial displacement in an incompressible elastic halfspace due to the action of the concentrated normal force. From (15), the arbitrary function $G(z,n)$ cannot produce a solution for

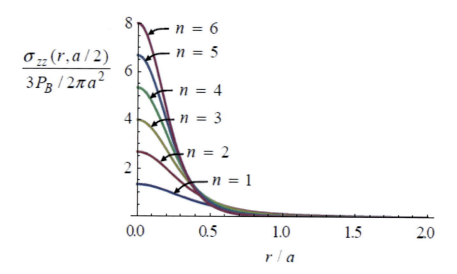

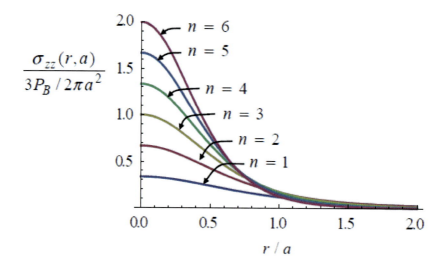

Fig. 2 Influence of the concenration factor n on the distribution of axial compressive stress $\sigma_{zz}(r,z)$

$n \neq 3$ to completely eliminate the second term within the brackets in (15) to yield the result (16). Similar results are encountered in the evaluation of the axial displacement $u_z(r,z)$.

The detailed evaluation of the arbitrary functions $G(z,n)$ is possible but the results are not central to the basic theme of the paper. It is sufficient to note that there is non-uniqueness in the evaluation of the displacement components from the integration of the strain-displacement relations and it would appear that this stems from the violation of the equations of compatibility, which are *necessary and sufficient* for the purposes of integration of the relevant equations.

4   Concluding Remarks

The concentration factor was introduced by Fröhlich with the intention of providing an approach to account for the departure of observed results from predictions made using the results based on the classical theory of elasticity. In particular, the exact analytical solution for Boussinesq's problem for an isotropic elastic halfspace region is used to calibrate the "concentration factor", n, that can alter the shape of either the *spreading* or the *concentration* of the load at depth. At the outset it is clear that from Kirchhoff's uniqueness theorem in classical elasticity, if Fröhlich's result converges to Boussinesq's result for $n \equiv 3$, the solution will not satisfy all the governing equations of elasticity when $n \neq 3$. The results presented in the paper indicate that Fröhlich's solution satisfies the equations of equilibrium, the boundary conditions and the stress-strain relations applicable to incompressible elastic materials but violates the equations of compatibility applicable to continua. This manifests in the form of a non-uniqueness in the displacement field obtained through the integration of the strain-displacement relations. Alternatively, if the kinematical relationships are not satisfied by Fröhlich's result, then the halfspace region must exhibit traits of a discontinuum similar to that of a particulate medium. It would also imply that the stress state associated with Fröhlich's solution is one that could be described by appeal to particulate mechanics similar to discrete element techniques where the particle shape and re-orientation can influence the stress transfer process.

Acknowledgements. The work described in this paper was supported in part by a Discovery Grant awarded by the Natural Sciences and Engineering Research Council of Canada. The interest of Dr. Thomas Keller of Agroscope, Zurich and Dr. Mathieu Lamandé of the Swedish University of Agricultural Sciences, Uppsala, Sweden is gratefully acknowledged. The author is grateful to Professor Dr. habil. Michael Hanss, Institut für Technische und Numerische Mechanik, Universität Stuttgart, Germany for providing a copy of Fröhlich's book and to Professor Helmut Schweiger, Computational Geomechanics Group, Technische Universität Graz, Austria for historical information. The author is also grateful to Ms. Nikki Tummon of The Schulich Library of Science and Engineering at McGill University for obtaining some of the older references.

References

Abramowitz, M., Stegun, I.A.: Handbook of Mathematical Functions with Formulas, Graphs and Mathematical Tables. Dover, New York (1964)

Borowicka, H.: Die druckausbreitung im halbraum bei linear zunehmendem elastizitätsmodul. Ingenieur-Archive 14, 75–82 (1942)

Boussinesq, J.: Application des Potentiels á l'Étude de l'Équilibre et du Mouvement des Solides Élastiques. Gauthier-Villars, Paris (1885)

Davis, R.O., Selvadurai, A.P.S.: Elasticity and Geomechanics. Cambridge University Press, Cambridge (1996)

Fröhlich, O.K.: Druckverteilung im Baugrunde. Mit Besonderer Berücksichtigung der Plastischen Erscheinungen. Julius Springer, Wien (1934)

Gibson, R.E.: Some results concerning displacements and stresses in a non-homogeneous elastic halfspace. Géotechnique 17, 58–67 (1967)

Griffith, J.H.: The pressure under substructures. Engineering Contracting 68, 113–119 (1929)

Gurtin, M.E.: The linear theory of elasticity. In: Flugge, S. (ed.) Mechanics of Solids II. Encyclopaedia of Physics, vol. VIa(2), pp. 1–295. Springer, Berlin (1972)

Hetnarski, R.B., Ignaczak, J.: Mathematical Theory of Elasticity. Taylor and Francis, New York (2004)

Hvorslev, M.J.: The Changeable Interaction between Soils and Pressure Cells: Tests and Reviews at the Waterways Experiment Station. Technical Report No. 5-76-7, U.S. Army Corps of Engineers, Waterways Experiment Station, Vicksburg, MI (1976)

Katebi, A., Selvadurai, A.P.S.: Undrained behaviour of a non-homogeneous elastic medium: the influence of variations in the elastic shear modulus with depth. Géotechnique (accepted, 2013)

Keller, T., Lamandé, M., Peth, S., Berli, M., Delenne, J.-Y., Baumgarten, W., Rabbel, W., Radjai, F., Rajchenbach, J., Selvadurai, A.P.S., Or, D.: An interdisciplinary approach towards improved understanding of soil deformation during compaction. Soil and Tillage Research 128, 61–80 (2013)

Kelvin, L. (Thompson, W.): On the equations of equilibrium of an elastic solid. The Cambridge and Dublin Mathematical Journal 3, 87–89 (1848)

Little, R.W.: Elasticity. Prentice-Hall, Upper Saddle River (1973)

Love, A.E.H.: A Treatise on the Mathematical Theory of Elasticity. Cambridge University Press, Cambridge (1927)

Mindlin, R.D.: Force at a point in the interior of a semi-infinite solid. Physics 7, 195–202 (1936)

Ohde, J.: Zur Theorie der Druckverteilung im Baugrund. Bauingenieur, 451 (1939)

Selvadurai, A.P.S.: Elastic Analysis of Soil-Foundation Interaction. Developments in Geotechnical Engineering, vol. 17. Elsevier Sci. Publ., The Netherlands (1979)

Selvadurai, A.P.S.: Partial Differential Equations in Mechanics II. The Biharmonic Equation, Poisson's Equation. Springer, Berlin (2000)

Selvadurai, A.P.S.: On Boussinesq's problem. International Journal of Engineering Science 39, 317–322 (2001)

Selvadurai, A.P.S.: The analytical method in geomechanics. Applied Mechanics Reviews 60, 87–106 (2007)

Selvadurai, A.P.S., Katebi, A.: Mindlin's problem for an incompressible elastic halfspace with an exponential variation in the linear elastic shear modulus. International Journal of Engineering Science 65, 9–21 (2013)

Selvadurai, A.P.S., Labanieh, S., Boulon, M.J.: On the mechanics of contact between a flexible transducer diaphragm located at a rigid boundary and an elastic material. International Journal for Numerical and Analytical Methods in Geomechanics 31, 933–952 (2007)

Sneddon, I.N., Berry, D.S.: The classical theory of elasticity. In: Flugge, S. (ed.) Handbuch der Physik, vol. VI, pp. 1–123. Springer, Berlin (1958)

Timoshenko, S.P., Goodier, J.N.: Theory of Elasticity. McGraw-Hill, New York (1970)

Ting, T.C.T.: Anisotropic Elasticity. Oxford University Press, Oxford (1996)

Essential Contributions of Austria to Fluid Dynamics Prior to the End of World War II

Helmut Sockel

Abstract. Important contributions to fluid dynamics of persons borne in the Habsburg empire prior to the end of World War II are discussed. In 1827 the forester J. Ressel obtained a privilege, as patents were called at that time, for ship propulsion with a screw. His important invention was the positioning of the screw between stem and rudder. V. Kaplan worked in the field of water turbines. At first he attempted to improve the efficiency of Francis turbines, but his main contribution is the invention of the Kaplan turbine. F. L. Schneider is well known for the invention of the Voith-Schneider cycloidal propeller providing ships with outstanding maneuverability. R. Knoller was the first to theorize that an airfoil may experience a thrust in a flow with periodically changing direction. He also designed and built a wind tunnel with a free jet under atmospheric conditions and he designed many airplanes. R. Katzmayr performed experiments in Knoller´s wind tunnel that verified Knoller´s theory. He also initiated experiments with special airfoil models in three different aerodynamic research stations in Europe (Vienna, Göttingen, St.Cyr) to establish the reliability of wind tunnel testing

Keywords: Marine screw propeller, Kaplan turbine, Voith-Schneider propeller, free jet wind tunnel.

1   Introduction

At the beginning of the 19th century the Habsburg empire was a multi-cultural and a multi-lingual state comprising many territories, such as Austria, Hungary, Bohemia and Moravia (the current Czech Republic), Slovakia, Slovenia, Istria, parts of Italy (such as Trieste and Venice) and other territories. These parts of the Habsburg empire were centrally administrated by ministries located in Vienna and the

---

Helmut Sockel
Institute for Fluid Mechanics and Heat Transfer TU Wien
e-mail: `helmut.sockel@tuwien.ac.at`

army (with German as commend language) served as another unifying organization. As a consequence, persons born in one part of the empire often found themselves serving in quite different parts throughout their careers. In this paper I highlight the life and the work of five major contributors to fluid dynamics who lived and worked in the Habsburg empire or in one of its succession states after the empire´s collapse at the end of World War I in 1918.

2 Josef Ressel

2.1 Ressel a Forester

Josef Ressel [1-5], Fig.1, was born on 29 June 1796 as son of a tax collector in Pardubice (Bohemia). He attended the Czech primary school in Chrudim (Bohemia) and the secondary school (gymnasium) in Linz, Austria, from 1806 to 1809. The following two years he served in an artillery regiment in Ceske Budejovice. This regiment was well known by its requirement for instruction in geometry, algebra, trigonometry and engineering drawing. From 1812 to 1814 he studied at Vienna University. He enrolled in courses like Chemistry, Natural Sciences and General Technology, pharmacy etc. But he was also interested in lectures in mechanics, hydraulics and civil architecture. Since his parents had problems to pay the high tuition fees at the university he applied for a scholarship at the Academy of Forestry in Mariabrunn. He finished this training with excellent success. In 1817 he started his professional career as a district forestry official in Pleterje (Slowenia). His whole professional life was dedicated to forestry and so it is very astonishing that he had enough time for numerous inventions. Between 1820 to1825 he had to move back and forth between Ljubljana and Trieste. Large forests around Trieste supplied the imperial shipyards with wood. In 1821 Ressel married and had three children, one of them died when it was a child. In 1825 Ressel was promoted to forestry master in Trieste. After his wife´s death 1826 he married again four years later and had seven more children. Due to a reorganization of the forestry administration Ressel was reassigned to a much lower paid position in 1832. In 1835 he was appointed provisional head forester in Motovan. He changed to the field of marine forestry in 1838, but his pay was reduced to fifty percent from 1845 to 1848. He continued his marine work until the end of his life. Ressel died from typhus in Ljubljana on 8 October 1857.

2.2 Ressel´s Inventions

Ressel applied for several "privileges", as patents were called at that time. For instance he had privileges concerning a press-roll-machine, a press for wine and oil, rolling- and ball-bearings, a hydraulic steam engine etc. Of direct relevance in his job, he made suggestions for the reforestation of karst areas and for river drainage. Moreover he was the first person who introduced new methods concerning sustainable forest management. But his most important invention is the marine screw propeller.

Fig. 1 Josef Ressel 1796-1857 Vienna Technical Museum

2.3 The Marine Screw Propeller

The oldest known screw propeller, Fig.2, for the propulsion of a ship was proposed by Daniel Bernoulli 1752. Though he obtained the price offered by the French Academy of Sciences for the best project for impelling vessels without the aid of wind, no practical application is known. In 1768 Paucton proposed a screw

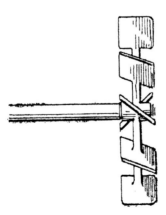

Fig. 2 Screw by Daniel Bernoulli 1752 [5]

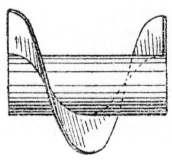

Fig. 3 Screw by M.Paucton 1768 [5]

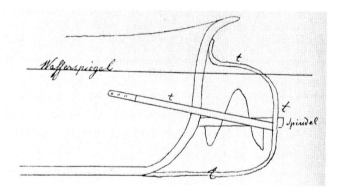

Fig. 4 Screw by Josef Ressel 1827 [2]

propeller based on Archimedes' screw, Fig.3. Two screws were positioned at the side of the ship or only one at the front part. But it is unknown whether this propulsion was a success. Based on Paucton's idea many similar propulsion systems were proposed as by W. Lyttleton 1794.

In his application for a privilege for a ship propulsion screw Ressel mentions in 1826: "I will place the screw propeller at the bow of the first steamer....".He documented this with a drawing in the document, Fig.4. There is some discussion about another drawing showing a screw of Archimedes type at the rear of the ship, Fig.5. In this document we find a confirmation of 1858, that it is an autographic drawing of Ressel. The date 1812 on it is in opposition to the drawing added to the application for the privilege. Therefore it is assumed that this drawing should be dated about 1828-29.

On 27 February 1827 Ressel received a privilege for a screw "without an end" for the propulsion of ships. At that time paddle steamers were already in service. The Englishman Morgan owned a privilege for regular paddle steamer cruises between Venice and Trieste. Morgan did not want a competition for his business. Due to his limited finances Ressel had to look for a sponsor who would support the demonstration of his screw. He could convince the whole sale trader Fontana.

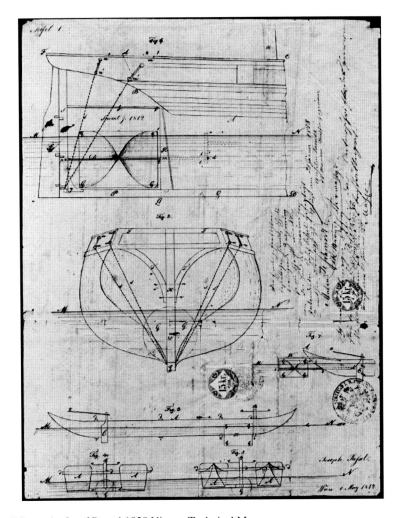

Fig. 5 Screw by Josef Ressel 1828 Vienna Technical Museum

Ressel wanted a very broad application of his privilege and therefore he tried to establish contacts in France and Britain with the help of an agent who was authorized to show the drawings of Ressel´s screw. But this effort proved to be counterproductive because the Frenchman Malar received a patent in 1828 in which only Ressel´s name was cited. The English merchant Cummerow received a patent in 1829 which even contained an obvious copy, Fig.6, of Ressel´s drawing, Fig.5.

Ressel was able to persuade the shipbuilder Zanoni to build a ship to demonstrate his screw. For the test a screw with two blades with a length of a half pitch was built in 1829 by Hermann in Trieste. Morgan went to court to block Ressel and Fontana, but Ressel and Fontana were allowed to undertake occasional cruises between Venice and Trieste and governments support was promised for a project within the Habsburg monarchy. For this reason Ressel could not use an advanced

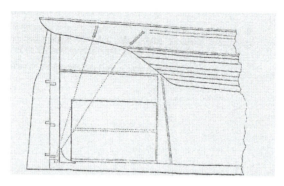

Fig. 6 Screw by Ch.Cummerow 1829 [1]

Fig. 7 Model of Civetta Vienna Technical Museum

English steam engine. Instead, he had to choose an Austrian engine of inferior quality. Moreover Ressel wanted a 30 PS engine, but Fontana ordered a 6 PS engine which delivered insufficient power for a ship like Civetta with an overall length of 24.25m, Fig.7.

It is reported that the discovery of many defects during the assembly of the engine delayed the official tests in Trieste by several months. Although the exact date of the tests is not known, they probably occurred end of 1829. The many invited spectators witnessed a disaster because few minutes after the start the steam pipe started to leak and the experiment had to be stopped.

Further tests of the Civetta were forbidden and Ressel´s privilege for his screw ended in 1830, because Fontana refused to pay the fees and Ressel had no money. However he continued to occupy himself with propellers for ship propulsion. There exists a sketch from the year 1854 which shows a pivoted propeller with a cardan shaft to enable ship maneuvering, Fig.8.

In 1836 the American Francis Pettit Smith obtained a patent on a one-bladed screw propeller with a length of one and a half pitch. His promoter enabled him to use a very good steam engine. He performed his experiments with the three-master "Archimedes". The screw broke and the effect was that the ship ran faster. This inspired Smith to shorten the length of the screw and to increase the number of blades. Due to the success of the experiments with this modified screw, the ship "Great Britain" was fitted with a propeller in 1843. It was the first ship with propeller propulsion which crossed the Atlantic Ocean in 1845. This was the start of the triumph of the marine screw propeller.

Ressel´s important invention is the positioning of the screw between stem and rudder. This is an arrangement still in use today.

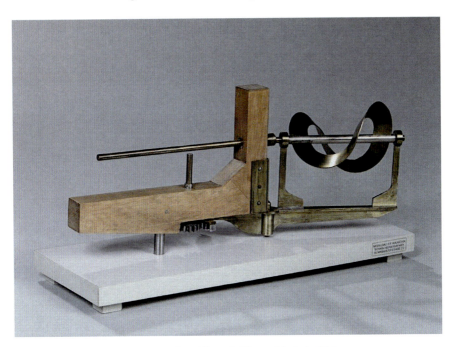

Fig. 8 Model of steering screw by Josef Ressel Vienna Technical Museum

3   Viktor Kaplan

3.1   Kaplan´s Early Life

Viktor Kaplan [6-10], Fig.9, was born 27 November 1875 as a son of a railway official who lived at that time in the railway station building in Muerzzuschlag

(Austria). He attended the primary schools in Neuberg at the Muerz and Vienna (Austria) and successfully completed the secondary school in Vienna in 1895. Then Kaplan studied mechanical engineering at the Vienna College of Technology from 1895 to 1900. He did his military service at the imperial naval training station in Pola, Istria (Croatia).

In 1901 Kaplan started his professional career in the mechanical engineering industry where he was concerned with Diesel engines. To improve their efficiency he proposed to inject high compressed air or liquified gases in the cylinder immediately after the explosion. But he could never prove the correctness of his theoretical considerations.

Fig. 9 Viktor Kaplan 1875-1934 [10]

3.2 Kaplan and the Francis Turbine

In 1903 Kaplan was employed as a design engineer at the Brno College of Technology (Slovakia). Here he was mainly concerned with Francis turbines, Fig.11. After passing a guide vane apparatus in a plane normal to the axis of the turbine the flow enters the rotating blade row. In the rotating blade row system the flow is turned in the axial direction and behind this there is a draft tube. In 1908 Kaplan published a book with the title: "Construction of efficient Francis turbines" [7].

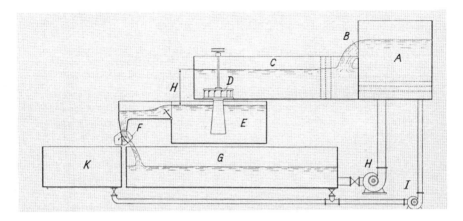

Fig. 10 Hydraulic laboratory for turbine experiments in Brno [9]

He presented this book to Vienna College of Technology for the award of the degree of Doctor of Technical Sciences. He was awarded the degree in 1909 and he received the right to lecture on hydraulic engines at the Brno College of Technology. Also in the same year he got married.

The characteristic quantity for hydraulic machines is the so called specific speed n_s

$$n_s = n \frac{\sqrt{N}}{H^{5/4}}$$

N is the power, H is the flow head and n the number of rotations. The range of the specific speed for a high efficiency Francis turbines is about 300. The number of rotations n has to be compatible with the generator speed or a set of gears has to be used causing a reduction of the efficiency. For a low head the consequence is a low power N. For a given head two or more turbines are necessary. Therefore Kaplan tried to increase the specific speed by changing the shape of the blades of a Francis turbine, thereby producing a fast runner with an optimum specific speed at about 450.

Then he started to establish a laboratory for hydraulic engines in Brno, Fig 10. It shows water flowing from a tank A passing an overflow used as measuring system in the lower tank C. From there it flows through the model turbine with a draft tube and then it is collected in tank E. Pump H or I delivers the water from tank G back to tank A. Tank K is for the calibration of the overflow B. The draft tube is made of glass for the observation of the flow behind the runner for improving the efficiency of the turbine and of the draft tube.

Kaplan published his experimental results in papers concerning the influence of the three-dimensionality of the flow and of the effect of friction on the efficiency of a turbine. Moreover, he realized that a further increase of specific speed with a turbine of Francis type is not possible, because this causes an increase of the number and length of the blades resulting in an increase of friction effects and a decrease of the efficiency, Fig.11.

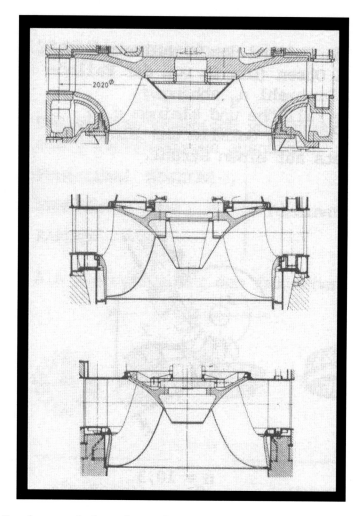

Fig. 11 Francis runners for increasing numbers of specific speed [6]

3.3 The Kaplan Turbine

In 1912 Kaplan applied for a patent for a turbine with guide vanes with a flow mainly in radial direction and a runner with a flow mainly in axial direction, Fig.12. In 1913 other patents followed, for instance for adjustable blades of the runner, Fig.13, 14. In the course of his life Kaplan obtained 33 patents. In 1913 the emperor appointed him professor at Brno College of Technology.

In 1913 he presented first results of experiments in the hydraulic laboratory which showed especially the high efficiencies for high values of specific speed.

Essential Contributions of Austria to Fluid Dynamics Prior 365

Experts doubted the validity of these results. Kaplan contacted representatives of well-known companies in many countries (Austria, Hungaria, France, Norway, Sweden, USA). Although his negotiations with some companies were successful (Norway) Voith and Escher Wyss were able to delay the issuance of Kaplan´s turbine patent for two years until 1916. In 1919 the first commercial Kaplan turbine started running in Velm (Austria), Fig.15. It demonstrated that the use of adjustable blades allowed the maintenance of nearly constant efficiency for different values of the specific speed, Fig.16.

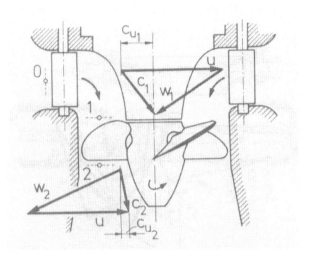

Fig. 12 Kaplan turbine [6]

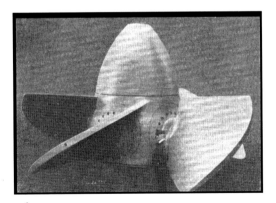

Fig. 13 Kaplan runner with blades open [9]

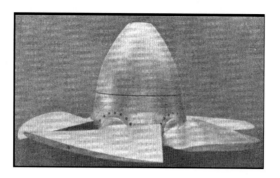

Fig. 14 Kaplan runner with blades closed [9]

Fig. 15 First commercial Kaplan turbine in Velm 1919 [9]

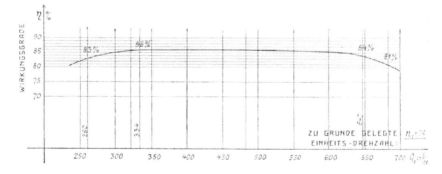

Fig. 16 Efficiency of the turbine in Velm [9]

Fig. 17 Kaplan turbine in Lilla Edit 11200 PS 1925 [9]

In 1920 first problems occurred during the operation of Kaplan turbines. Its efficiency was much smaller than that of the model turbine. Kaplan immediately started the investigation of this problem and traced the cause to cavitation in the operational turbine which did not occur in the model turbine due to its low head. He fell seriously ill in 1923 from a disease whose precise nature remained undiagnosed. Therefore it took him until 1924 to publish a paper on cavitation in high-speed water turbines.

In 1925 the triumph of the Kaplan turbine started in Lilla Edet (Sweden) with the largest water turbine at that time with a power of 11 200 PS and a runner with 5.8m diameter, Fig.17. In 1931 the book "Theory and construction of fast running turbines" was published by Kaplan and Lechner [9]. Because of his deteriorating health he applied for his retirement in 1931, but his application was accepted only in 1934 because the Brno College of Technology hoped for his eventual recovery. He spent the last years of his life in Unterach am Attersee (Austria), where he died on 23 August 1934 caused by a stroke.

Fig. 18 Ernst Leo Schneider 1894-1975 [11]

4 Ernst Leo Schneider

4.1 First Inventions

Ernst Leo Schneider [11], Fig.18, was born in Gaya (Bohemia) on 18 June 1894. He went to school in Gmunden and Linz (Austria). After his military service he started to study mechanical engineering at Vienna College of Technology, but he never graduated.

In 1923, while still a student, he came up with the idea of a screw propeller whose blades were shaped like bird wings. He presented the idea to Maschinenfabrik J. M. Voith. But tests carried out with a single propeller resulted in an inadequate efficiency curve.

Another idea for a ducted propeller even reached the testing stage. Schneider had no financial resources and he could not find an investor. This idea was patented in 1934 by Ludwig Kort and in practical application is nowadays known as Kort nozzle.

After this period of failure Schneider befriended with Voith engineer Ludwig Kober with whom he continued experiments on a purely private basis.

4.2 The Blade-Wheel Propeller

It is reported, that Schneider was inspired by the Wels propulsion system to his first propeller for propulsion. This was based on the use of a swinging blade on a boat, Fig.19.

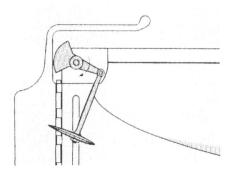

Fig. 19 Experimental boat with thrust surface [11]

Then Schneider proposed a rotating plate wheel with vertical axis. The blades are arranged in a circle on the blade wheel and are parallel to the axis of rotation. Schneider developed his propeller for the best angular position for the blades for zero thrust, Fig.20. For this case the blades have to be angled in such a way that as they rotate they are aligned exactly in the direction of the resultant inflow. The surprising result is that this occurs, if all the normals of the plates pass through one point N within the blade circle, Fig.21.

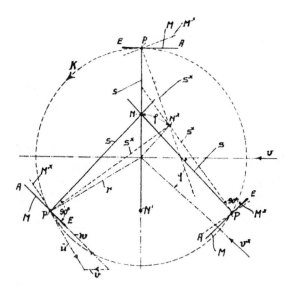

Fig. 20 Sketch of the blade wheel in the patent of Schneider 1927 [11]

If the propeller is to produce thrust, the blades must assume an angle relative to the resultant inflow. Now all the normals of the blades pass through one point N´ too, Fig.22. The absolute path of the blade through the water is a cycloid. Therefore the blade-wheel propeller is also referred to as a cycloidal propeller. In Fig.23. there is shown the state of zero thrust, in Fig.24 there is shown the passage of a blade generating thrust in the direction of the movement of the propeller. By shifting this point N´ it is possible to generate thrust components in any direction.

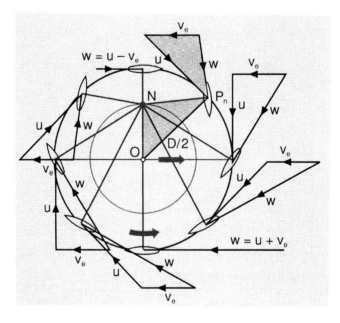

Fig. 21 Velocity triangles at the VPS blades in 8 positions during one revolution, zero thrust [11]

Originally Schneider had been thinking of using the propeller for a power turbine. After a successful experiment with a high-speed rotor model he drew up a patent application on December 7, 1925 with financial backing from Ludwig Kober. In February 1926 Schneider presented his invention to Voith in St.Pölten (Austria).

Essential Contributions of Austria to Fluid Dynamics Prior 371

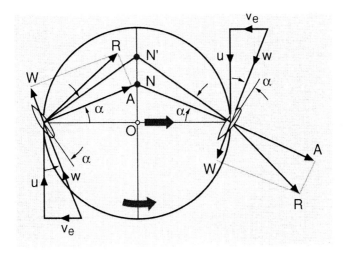

Fig. 22 Velocity triangles at the VPS blades for thrust [11]

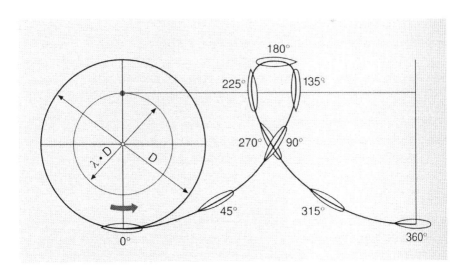

Fig. 23 Cycloidal path of a blade through the water, zero thrust [11]

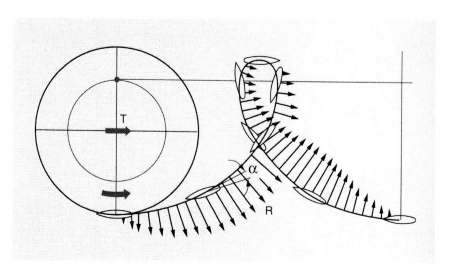

Fig. 24 Path of the blade through the water when generating thrust [11]

4.3 From the Invention to the Commercially Viable Product

In May 1926 a cooperation agreement between Voith and Ernst Schneider was signed in which Schneider surrendered the sole and exclusive rights to the exploitation of his invention to Voith. Any further proposals and improvements made by Schneider relating to his propeller were to be communicated exclusively to Voith. Schneider benefited from the sales. Voith paid the patent fees and expenses.

In June 1926 Schneider applied for a German patent with the title "Blade wheel" which was granted in December 1927. Dr. Keithner from Voith applied modern hydrofoil theory for the optimization of the wheel and Josef Erhart from Voith developed the kinematic aspect. A large number of trials on models in different places, Vienna, St.Pölten, Hamburg and Heidenheim started in July 1926. In the course of time, the number, width and length of the blades were varied and various blade outlines were studied. In January 1928 Ernst Schneider filed a patent covering the ring kinematics.

In 1928 it turned out as a result of tests by Voith, that the new turbine was less efficient than comparable existing turbines. So Schneider`s invention was implemented as a ship`s propeller, just as it had been intended by the inventor.

After the tests 1926/7 the experimental boat `Torqueo`, Fig.25, was launched fitted with a Voith-Schneider propeller VSP, as Schneider`s invention was then called. The results exceeded all expectations thanks to its outstanding maneuverability. In 1930 Voith signed a contract with Gutehoffnungshütte (GHH), hoping that the shipyards of this company could solve the marine engineering problems and to benefit from the sales network of GHH. In 1931 three ships on Lake Constance entered in service, Fig.26. The reason for choosing ships with VSP was that screw driven ships had difficulties when entering or leaving narrow harbors.

Fig. 25 Experimental vessel Torquea 1927 [11]

Fig. 26 Lake Constance, ship with VPS passing the narrow entrance to Lindau Harbor 1934 [11]

In 1931 Voith started the investigation of medium- and high-power propulsion systems for sea going ships with a hydraulic blade actuating gear. But in that case the efficiency of the propeller is important and not the maneuverability. In the same year

Voith detected two similar patents in France and USA. An agreement was reached to make clear distinction between aeronautical and water-based applications.

In 1939 the sea-going passenger ship "Helgoland" fitted with a VSP made its maiden voyage in the North Sea resort service. Many minesweepers were fitted with two VSPs each. After the Second World War the US Navy discovered a mine sweeper that had escaped the destruction. Since no suitable ship was available for transport to USA the ship had to cross the Atlantic under its own power. It had to pass two storms, the second one was a hurricane. This was a further proof for the excellent maneuverability of a ship fitted with VSP.

On June 1, 1975, Ernst Schneider died in Vienna.

5 Richard Knoller

5.1 Knoller's Work in Machine Industry

Richard Knoller [12-25], Fig.27, was born on 25 April 1869 in Vienna. He studied mechanical engineering at the Vienna College of Technology and graduated with distinction in 1892. From 1893 to 95 he made study trips to machine factories in France, England and Scotland. After his return he first worked in a steam engines manufacturing company and then applied to become an assistant at the Chair for Mechanical Engineering at the Vienna College of Technology. In 1899 he founded a car manufacturing company together with a gentleman named Goebel in Vienna. In 1904 he and a gentleman named Friedmann jointly developed a steam driven car with a four-wheel brake system. From 1904 to 1907 he was head of car production at Weyer Richmond in Paris.

Fig. 27 Richard Knoller 1869-1926 Internationale Flugausstellung Wien 1912

5.2 First Contacts with Aviation

During his stay in Paris he came in contact with practical aviation problems which inspired him in 1909 to publish a paper on fluid mechanic drag and propeller theory and another on aerodynamic drag [16]. He showed in his second paper that an airfoil may experience a thrust when moving in a wavy flow causing the angle of attack to change periodically. He was the first to show this effect. In 1912 Betz unaware of Knoller's paper came to the same conclusion in his paper on the "theory of gliding flight" [17]. In 1922 Katzmayr experimentally demonstrated in wind tunnel tests the validity of the Knoller-Betz effect. Its complete computational analysis was first given in 1936 by Garrick [22]. More recent analysis using modern computational fluid dynamics were given by K. D. Jones et al.1998 [23].

5.3 Knoller, an Aviation and Aerodynamics Expert

30 November 1909 Knoller was appointed associate professor for aircraft construction and automotive engineering and head of the division with the same name at Vienna College of Technology. He immediately started the planning of an aerodynamics laboratory. The prototype for his construction was the free-jet tunnel by Eiffel in Paris where a jet crosses a test chamber. Facing severe financial and space problems he decided on a vertical tunnel in the courtyard of the main building of the Vienna College of Technology, Fig.28. It is the first free-jet tunnel under atmospheric conditions, Fig.29. The air flows from a settling chamber at the centre top of the building through an eight angled nozzle with guide vanes and then enters the section as a free jet with a cross section of 2.4 m^2, Fig.30,. From there the air flows through an octagonal diffuser into the base room. In the corners of the base room there are 4 fans pumping the air back to the settling chamber. With a power of 36 PS it was possible to reach a flow velocity of 90 km/h in the test section. Knoller designed balances for the measurement of drag and lift, as a function of the angle of attack. The financial support came from the Krupp iron and steel company. The tunnel became operational in 1913.

This special type of wind tunnel was introduced by Prandtl in Göttingen after his visit to Vienna and, thereafter, was adopted by many other research institutes. However, since these institutions had no space limitations, all were of horizontal type and are now well-known as Göttinger type tunnels.

Knoller performed many model experiments in the tunnel for the Austrian airplane industry. He designed several airplanes for the emperial air force, Fig.31. In one of his planes he demonstrated a drag reduction by replacing the wires between the wings by struts, Fig.32. In 1917 Knoller designed a propeller test rig with a diameter up to 4m, Fig.33. It served as a special wind tunnel such that the flow was produced by the propeller could be adjusted by a nozzle ahead of the propeller.

Fig. 28 Wind tunnel building by Knoller 1913 [12]

Since Knoller was overstressed, the construction of the tunnel was supervised by Theodore von Karman. He was since 1913 professor for mechanics and aerodynamics for flight at the Aachen College of Technology (Germany). Karman was born in Budapest (Hungary) 1881 and so he got the call to the imperial air force 1914 and he served until the end of the First World War in 1918. At that time the rig was not yet finished and only few experiments had been performed.

In 1919 Knoller was appointed full professor at the Vienna College of Technology. The collapse of the Austrian aircraft industry after the First World War forced Knoller to turn to theoretical problems and motivated him to publish the paper "The development of theoretical aerodynamics". At the 150th anniversary of the Vienna College of Technology in 1965 Prof. Kirste called Knoller the inventor of the science of aeronautics in Austria [24].

Prof. Richard Knoller died 4 March 1926.

Essential Contributions of Austria to Fluid Dynamics Prior 377

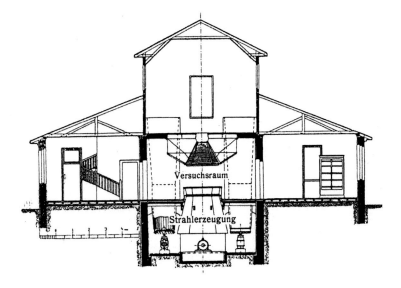

Fig. 29 Cross-section of Knoller´s wind tunnel building [15]

Fig. 30 Test section of the wind tunnel [13]

Fig. 31 Knoller BI 35 [15]

Fig. 32 Knoller CII [15]

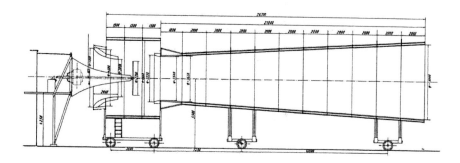

Fig. 33 Propeller test rig in Fischamend [15]

6 Richard Katzmayr

6.1 Katzmayr´s Life

Richard Katzmayr [13-25], no picture available, was born in Vienna 3 November 1884. He finished his mechanical engineering studies at Vienna College of Technology in 1909. After a brief employment at the Institute of Hydraulic Engines at the Vienna College of Technology he joined AEG Union as a metallurgical engineer. But already in 1910 he was employed as mechanical engineer at the printing plant of the Austrian government. From 1923 to 45 he served as public relations officer of the government´s administration of buildings.

In 1914 Katzmayr was suspended from his duties to assist Knoller in the conduct of wind tunnel experiments. From 1914 to 18 he served in the imperial air force, but he remained assigned to Knoller´s laboratory which was administrated by the airforce due to the importance of aerodynamic experiments for imperial air force projects. Katzmayr´s main duties were to support Knoller in the laboratory and to give lectures in place of Knoller. In 1928, after Knoller´s death, Katzmayr was appointed associate lecturer for aviation. He remained in this position until the end of Second World War, with a short brake from 1937 to 1938. In 1934 he received the honorary associate professor title.

6.2 Katzmayr´s Contributions to Fluid Dynamics

He performed many wind tunnel experiments in the aerodynamics laboratory of Vienna Institute of Technology. As already mentioned in 5.2, he was first in conducting wind tunnel experiments which proved Knoller´s theoretical considerations that an airfoil experiences a thrust when exposed to a waving flow with periodically changing angle of attack. This effect therefore is often referred to as Katzmayr effect.

Katzmayr also made important contributions to the understanding of the effect of boundary layer blowing on the airfoil aerodynamics. He called wings equipped with nozzles to eject air in the flow direction nozzle wings. The ejection of mass in the direction of the flow in the boundary layer of an airfoil increases the momentum of the flow near the surface. The effect is, that the point of separation is shifted downstream, what results in higher lift values. Fig.34 shows a model made of wood used by Katzmayr for his experiments. The cross-sections of two different wing positions are also shown. The results of the measurement are given as a polar diagram in Fig. 35 where the lift coefficient is plotted on the ordinate and the drag coefficient on the abscissa. The lower curves show the results without ejection, the upper curves show results for two different ejection speeds. Katzmayr mentions in his paper describing these experiments, that after finishing it, he learned about similar experiments in Germany and Switzerland.

Katzmayr invited the aerodynamic laboratories in Göttingen and St.Cyr to conduct wind tunnel experiments with special airfoils for the purpose of exploring the reliability of wind tunnel tests. The main differences in the polar diagrams and moment coefficients, Fig.36, were caused by different turbulence intensities of the flow in the tunnels. In another series of tests the change of center of pressure position of a half wing with changing angle of attack was investigated.

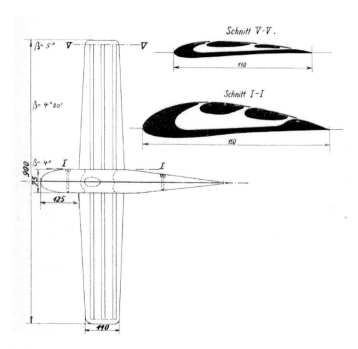

Fig. 34 Airplane model of wood for blowing experiments [13]

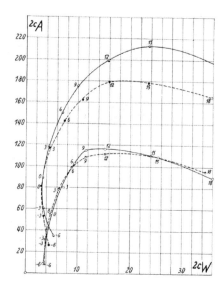

Fig. 35 The influence of blowing on the polar curves of a profile [13]

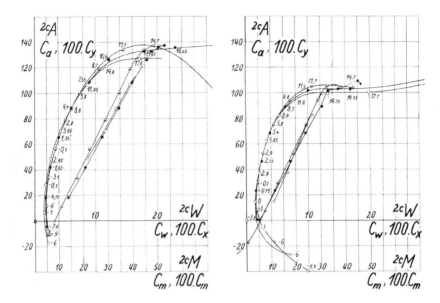

Fig. 36 Comparisons of polar diagrams from experiments in different wind tunnels [13]

All the results are published in 1928 as "Reports of the Aerodynamic Research Center Vienna" [13]. These reports are similar to the "Reports of the Aerodynamic Research Center in Göttingen", published in the same time period. In the Viennese report an extensive description of the wind tunnel in Vienna is presented.

At the occasion of the 25th anniversary of the aerodynamic laboratory in Vienna Katzmayr edited an anniversary volume titled "Contributions to Aeronautics" including papers of, R.Katzmayr, L.Kirste W.B. Klemperer, F.Magyar, R.v. Mises, among others [14].

Katzmayr died on the 12 April.1945.

Acknowledgments. I am very indebted to the editor E.Stein and Th.Ditzinger from Springer who helped me to solve problems when preparing this paper. I would like to express my warm thanks to Vienna Technical Museum and R.Keimel for the provision with excellent original pictures. My special thanks to Voith Turbo Schneider Propulsion providing me with the book "The fascination of the Voith-Schneider Propeller, History and Engineering" and the permission to publish figures of the book. I would like to thank E.Käppeli for the permission to publish figures of his book. My special thanks also to M. Platzer for his extensive work in improving my English.

References

[1] Bourne, J.: A treatise of the screw propeller, 458 p. Longman, London (1867)
[2] Josef Ressel Denkschrift (ed.): Comite für die Centenarfeier Josef Ressel, Wien, 274 p., Anhang 117 p. (1893)
[3] Wess, A.: Blätter für Technikgeschichte, vol. 19, pp. 1–19. Springer, Heft (1957)

[4] Murko, V.: Blätter für Technikgeschichte, vol. 34, pp. 1–64. Springer, Heft (1973)
[5] Geissler, R.: Der Schraubenpropeller, 87 p. Springer (1918)
[6] Käppeli, E.: Strömungslehre und Strömungsmaschinen, 2. Auflage, 358 p. Selbstverlag (1983)
[7] Kaplan, V.: Bau rationeller Francisturbinen, 346 p. Oldenbourg (1908)
[8] Kaplan, V.: Wasserkraftjahrbuch, pp. 421-435. Pflaum (1924)
[9] Kaplan, V., Lechner, A.: Theorie und Bau von Turbinenschnelläufern, 301 p. Oldenbourg (1931)
[10] Lechner, A.: Viktor Kaplan, 59 p. Springer (1936)
[11] Jürgens, B., Fork, W.: The Fascination of the Voith-Schneider Propeller; history and engineering, 208 p. Koehler (2002)
[12] Doblhoff, W.: Flugtechnik u. Motorluftschiffahrt V. Heft 7, 105–115 und Heft 8, 125–129 (1914)
[13] Katzmayer, R.: Berichte der Aerodynamischen Versuchsanstalt in Wien, Band I, 76 p. Schmidt & Co. (1928)
[14] Katzmayer, R.: Beiträge zur Flugtechnik, 43 p. Springer (1937)
[15] Keimel, R.: Luftfahrzeugbau in Österreich, Enzyklopädie, 408 p. Technisches Museum Wien (2003)
[16] Knoller, R.: Flug-und Motortechnik III/21, 1–7 and III/22, 1–6 (1909)
[17] Betz, A.: Flugtechnik und Motorluftschiffahrt III/21, 269–272 (1912)
[18] Katzmayr, R.: Flugtechnik und Motorluftschiffahrt XI/13, 192–194 (1920)
[19] Katzmayr, R.: Flugtechnik und Motorluftschiffahrt XIII/6, 80–82 (1922)
[20] Katzmayr, R.: Flugtechnik und Motorluftschiffahrt XIII/7, 95–101 (1922)
[21] Katzmayr, R.: Zeitschrift des Österr. Ing.-und Architektenvereins, Heft 17/18, 97–100 (1923)
[22] Garrick, I.E.: NACA Report 567 (1936)
[23] Jones, K.D., Dohring, C.M., Platzer, M.F.: AIAA J. 36(3), 1240-1246 (1998)
[24] Kirste, L.: 150 Jahre Technische Hochschule Wien, Ed. A. Sequenz, pp. 396–400. Technische Hochschule Wien (1965)
[25] Oswatitsch, K.: 150 Jahre Technische Hochschule Wien, Ed. A. Sequenz, pp. 336–339. Technische Hochschule Wien (1965)

Part IV
Numerical Methods in Solid Mechanics from Engineering Intuition and Variational Calculus

From Newton's Principia via Lord Rayleigh's Theory of Sound to Finite Elements

Lothar Gaul

1 Introduction

The contribution summarizes the route from synthetic mechanics based on Newton's and Euler's axioms via analytical mechanics based on Rayleigh-Ritz method to Finite Elements.

2 Newton's and Euler's Axioms

Is it correct, what Ernst MACH (1838 – 1916), famous experimental physicist and philosopher, states in his book "Die Mechanik in ihrer Entwicklung (The Development of Mechanics)", which appeared in 1888, with respect to Isaac Newton's (1643 – 1727) "Philosophiae naturalis principia mathematica"?

Ernst Mach states that Newton's laws of mechanics are sufficient; even the motion of fluids can be determined by them as Mach implies.

Indeed, the first book of the principia describes the "motion of bodies" and contains the axioms or laws of motion "Axiomata sive Leges Motus".

It has been clearly worked out by Clifford Truesdell and Istvan Szabó that Newton by no means gives "classical" mechanics its present form, nor were his principles clear and definite enough to do so. But Newton displays in the highest every ability of a great theorist

- to organize and recast known but separated laws and phenomena,
- to create new concepts.

Lothar Gaul
Universität Stuttgart, Institute of Applied and Experimental Mechanics,
Pfaffenwaldring 9, 70550 Stuttgart
e-mail: gaul@iam.uni-stuttgart.de

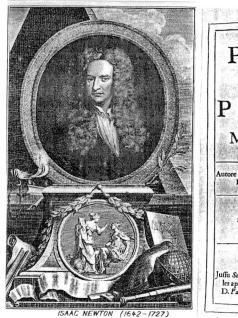

Fig. 1 Isaac Newton

PHILOSOPHIÆ
NATURALIS
PRINCIPIA
MATHEMATICA

Autore *J S.* NEWTON, *Trin. Coll. Cantab. Soc.* Matheſeos Profeſſore *Lucaſiano*, & Societatis Regalis Sodali.

IMPRIMATUR.
S. PEPYS, *Reg. Soc.* PRÆSES.
Julii 5. 1686.

LONDINI,
Juſſu *Societatis Regiæ* ac Typis *Joſephi Streater*. Proſtant Venales apud *Sam. Smith* ad inſignia Principis *Walliæ* in Cœmiterio D. *Pauli*, alioſq; nonnullos Bibliopolas. *Anno* MDCLXXXVII.

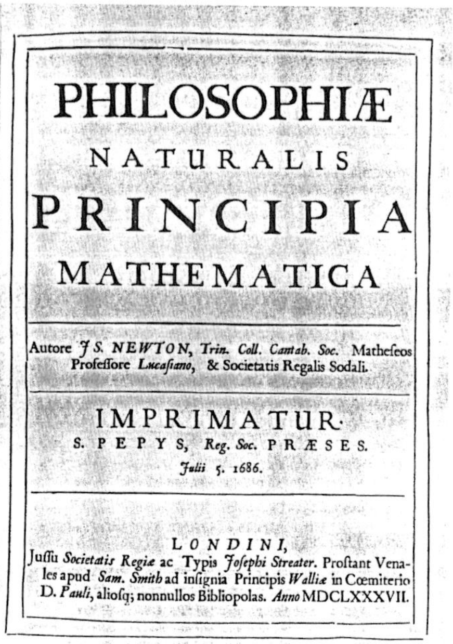

Fig. 2 Front page of the first edition of the"Principia"

The first of the axioms or laws of motion "Axiomata sive Leges Motus" has already been stated by Galilei.

Lex I: *Corpus omne perseverare in statu suo quiescendi vel movendi uniformiter in directum, nisi quantenus a viribus impressis cogitur statum illum mutare.*

Every body remains at rest or in uniform motion along a straight line, if it is not forced to change this state by impressed forces.

The second law was new at that time.

Lex II: *Mutationem motus proportionalem esse vi motrici impressae, et fieri secundum lineam rectam qua vis illa imprimitur.*

The change of momentum is proportional to the impressed moving force and acts in the direction of that straight line, along which the force acts.

Lex III is the reaction principle: *Actioni contrariam semper et aequalem esse reactionem: sive corporum duorum actiones in se mutuo semper esse aequales et in partes contrarias dirigi.* Each action is associated with an equal reaction, or the actions of two bodies on each other are always equal and of opposite direction.

Are, according to Mach, the laws of mechanics sufficiently stated by Newton?

The answer is: no! it is necessary to incorporate findings later obtained to give Newton's formulations the well-known meaning.

Especially in the second law the content is not clear. The momentum (motus) as well as the moving force (vis motrix) are not clearly defined.

Fig. 3 Leonhard Euler

In his "Principia" Newton uses the geometrical approach applied already by Huygens. Compared with the elegant differential calculus by Leibniz, his approach is based on the less efficient "Fluxion" calculus.

The introduction of a radius vector $\vec{r}$ and the coordinate description are due to Euler (1707 – 1783) 50 years later.

> **224**   D E   M O T V
>
> ea nullum nafcitur momentum pro hoc axe; pro axe autem I B nafcetur momentum $= z\,d\,M\left(\frac{d\,d\,x}{d\,t^2}\right)$ et pro axe I C momentum $= y\,d\,M\left(\frac{d\,d\,x}{d\,t^2}\right)$. Simili modo ex vi acceleratrice fecundum directionem I B, quae eft $d\,M\left(\frac{d\,d\,y}{d\,t^2}\right)$ nafcitur momentum pro axe I A $= z\,d\,M\left(\frac{d\,d\,y}{d\,t^2}\right)$, at pro axe I C momentum $= x\,d\,M\left(\frac{d\,d\,y}{d\,t^2}\right)$. Denique ex vi acceleratrice fecundum I C, quae eft $d\,M\left(\frac{d\,d\,z}{d\,t^2}\right)$ nafcitur momentum pro axe I A $= y\,d\,M\left(\frac{d\,d\,z}{d\,t^2}\right)$ et pro axe I B momentum $= x\,d\,M\left(\frac{d\,d\,z}{d\,t^2}\right)$. Hinc igitur pro quolibet axe habemus bina momenta elementaria, quae in partes contrarias vergunt; vnde pro axe I A fumma omnium momentorum elementarium erit
>
> $+ \int z\,d\,M\left(\frac{d\,d\,y}{d\,t^2}\right) - \int y\,d\,M\left(\frac{d\,d\,z}{d\,t^2}\right) = i\,S.$
>
> Eodem modo pro axe I B obtinebimus hanc aequationem:
>
> $\int x\,d\,M\left(\frac{d\,d\,z}{d\,t^2}\right) - \int z\,d\,M\left(\frac{d\,d\,x}{d\,t^2}\right) = i\,T.$
>
> Tertia vero aequatio erit pro axe I C
>
> $\int y\,d\,M\left(\frac{d\,d\,x}{d\,t^2}\right) - \int x\,d\,M\left(\frac{d\,d\,y}{d\,t^2}\right) = i\,U.$
>
> §. 29. Hac igitur ratione fex nacti fumus aequationes, quas hic coniunctim confpectui exponamus
>
> I. $\int d\,M\left(\frac{d\,d\,x}{d\,t^2}\right) = i\,P$ IV. $\int z\,d\,M\left(\frac{d\,d\,y}{d\,t^2}\right) - \int y\,d\,M\left(\frac{d\,d\,z}{d\,t^2}\right) = i\,S$
>
> II. $\int d\,M\left(\frac{d\,d\,y}{d\,t^2}\right) = i\,Q$ V. $\int x\,d\,M\left(\frac{d\,d\,z}{d\,t^2}\right) - \int z\,d\,M\left(\frac{d\,d\,x}{d\,t^2}\right) = i\,T$
>
> III. $\int d\,M\left(\frac{d\,d\,z}{d\,t^2}\right) = i\,R$ VI. $\int y\,d\,M\left(\frac{d\,d\,x}{d\,t^2}\right) - \int x\,d\,M\left(\frac{d\,d\,y}{d\,t^2}\right) = i\,U.$

$$\int_B dm\,\ddot{\vec{r}} = \vec{F} \qquad \vec{L} = \frac{d}{dt}\int_B \vec{r} \times \dot{\vec{r}}\,dm = \vec{M}$$

NEWTON – EULERS LAWS OF MECHANICS

L. EULER *Novi Commentarii Academiae Scientarium Petropolitanae*, Vol. 20, 1775

Fig. 4 Novi Commentarii Academiae Scientarium Petropolitanae

The conclusion is:

We don't find Newtons's Mechanics, as it is thought today, in the "Principia". Truesdell states in his "Essays in the History of Mechanics":

"It is not the function of the historian to guess what NEWTON might have done or could have done, nor is what MACH could do with NEWTON's principles relevant; the cold fact is, the equations are not in NEWTON's book!

The first definitive formulation of the principle of linear momentum is due to Leonhard Euler (1707 – 1783). His first results on a general principle of dynamics are formulated 1750 and published 1752 in his "Discovery of a new principle of mechanics", "Decouverte d'un nouveau principe de la méchanique". 1750 he writes the principle of linear momentum in components as it appears in his later work of 1775 in the Novi Commentarii Academiae Scientiarum Petropolitane and mentions the mass M may be either finite or infinitesimal (un corps infiniment petit) in his axiom

$$2M\,ddx = P dt^2 \qquad 2M\,ddy = Q dt^2 \qquad 2M\,ddz = R dt^2$$

The reason why these equations deviate from those of the 20th century is due to the different system of units used by Euler. He used only two basic units "length" and "force". The mass is described in units of force, so that the acceleration is measured dimensionless. Euler's equations lead to our notation by substituting

e. g. $2M\,ddx = P dt^2 \Rightarrow 2Mg\,ddx = P dt^2 (\sqrt{2g})^2 \Rightarrow M d^2 x = P dt^2$

Later, Euler calls the equations "the first principles of mechanics". It required the accumulated mechanical experience to much of the rest of his life before he saw that a second principle was needed. In his paper written in 1775, the old Euler laid down as "fundamental, general and independent laws of mechanics, for all kinds of motions of all kinds of bodies", the principles of linear momentum and moment of momentum of each element of mass.

3   Rayleigh-Ritz Method

It is worth-while to note that the basic ideas of both methods already appear in a fundamental paper of Daniel Bernoulli (1700-1782).

He treats the impact of a free-free beam. This can be considered as a first wave theory of impact.

In his work "Examen physico-mechanicum de motu mixto qui laminis elasticis a percussione simul imprimatur" he considers two problems of motion of a body initiated by impact. The second deals with a free-free beam, moving after an impact in the middle. Bernoulli asks for this motion consisting of the translation of the center of gravity and the superimposed vibration. The impact leads to the red coloured deflection. Bernoulli considers the velocities of the beam elements

Fig. 5 Daniel Bernoulli

to be proportional to the deflections (this is true for time-harmonic motion $f(x)\sin(\omega t + \alpha)$). The squared relative displacement $h - w(x)$ integrated with respect to the beam length $2l$ is thus the doubled kinetic energy, his "Vis viva" or living force.

Bernoulli requires the true shape $w(x)$ to minimize the energy L.

He reduced the variational problem to an ordinary extremum problem by using already a one parameter approximation $\widetilde{w}(x) = a\varphi(x)$, which is later referred to as Ritz-Ansatz. Walter Ritz published his approach in 1909.

DANIEL BERNOULLIS CONTRIBUTION (1770) TO WAVE THEORY OF IMPACT

EXAMEN PHYSICO-MECHANICUM DE MOTU MIXTO QUI LAMINIS ELASTICIS A PERCUSSIONE SIMUL IMPRIMITUR, NOVI COMMENTARII ACADEMIAE SCIENTIARUM IMPERIALIS PETROPOLITANAE, XV, pp. 361 (1770)

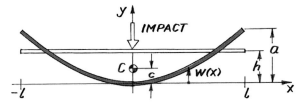

'Vis viva' with velocity $v(x)$ proportional to deflection $h-w(x)$

$$L = \int_{-l}^{l} [h-w(x)]^2 dx$$

BERNOULLI requires the true shape $w(x)$ to minimize L.
His 'Ansatz' $\tilde{w}(x) = a\varphi(x)$, later referred to W. RITZ (1909) leads by $dL(a)/da = 0$ to a quotient

$$h = \frac{\int_{-l}^{l} \tilde{w}^2(x) dx}{\int_{-l}^{l} \tilde{w}(x) dx}$$

similar to J.W. STRUTTS, LORD RAYLEIGH (1877).
BERNOULLI approximates the 'Curva elastica'
$w(x) = \alpha e^{x/f} + \beta e^{-x/f} + \gamma \sin(\frac{x}{f}+\varepsilon)$ by the parabola $\tilde{w}(x) = a(\frac{x}{l})^2$
to obtain $h/a = 3/5$, $c = a/3$.

Kinetic Energy 'Vis viva' is partitioned by

$$L = L(\text{Translation}) + L(\text{Vibration}) = \int_{-l}^{l}(h-c)^2 dx + \int_{-l}^{l}[c-\tilde{w}(x)]^2 dx$$

$$\frac{L(\text{Vibration})}{L(\text{Translation})} = \frac{5}{9}$$

Fig. 6 Examen physico-mechanicum de motu mixto

The obtained ordinary extremum problem is solved by the vanishing first derivative of L with respect to the parameter a, $\frac{dL(a)}{da} = 0$. The unknown distance h of the deflected beam measured from the undeformed state is obtained by a quotient similar to the one given by John William Strutt, the III Baron Rayleigh in 1877. The difference with Lord Rayleigh's quotient for the lowest squared eigenfrequency, relating the maximum stored energy of deformation to the maximum reduced kinetic energy is due to the fact that Bernoulli considers a minimum principle for the kinetic energy, only.

Fig. 7 John William Strutt, III Baron Rayleigh

Daniel Bernoulli then continues in the sense of the Rayleigh-Ritz approach and approximates the "Curva elastica" by the parabola $\tilde{w}(x) = a(\frac{x}{l})^2$.

Whereas in a lot of papers appearing later than Bernoulli's work, the kinetic energy of vibration has been neglected, it is admirable that Daniel Bernoulli already succeeded to relate the energy of beam vibration to the translational energy associated with the motion of the centre of gravity. He showed that the vibrational portion of energy is almost one half of the translational energy.

One of the first applications of Rayleigh's quotient is part of the first volume of his Theory of Sound. Lord Rayleigh, who followed James Clerk Maxwell on the chair of experimental physics in Cambridge, started writing the manuscript during a boat expedition on the river Nile. He spent the winter of 1872 in warm Egypt after a severe sickness (... At these times it was difficult to persuade him to land, even to see the most enchanting temple, stated his wife).

> 156   Another problem worth notice occurs when the load at the free end is great in comparison with the mass of the rod. In this case we may assume as the type of vibration, a condition of uniform extension along the length of the rod.
>
> If ξ be the displacement of the load M, the kinetic energy is
>
> $$T = \tfrac{1}{2} M \dot{\xi}^2 + \tfrac{1}{2} \dot{\xi}^2 \int_0^l \rho \omega \frac{x^2}{l^2} dx = \tfrac{1}{2} \dot{\xi}^2 (M + \tfrac{1}{3} \rho \omega l) \dots \dots (1).$$
>
> The tension corresponding to the displacement ξ is $q\omega\,\xi/l$, and thus the potential energy of the displacement is
>
> $$V = \frac{q\omega \xi^2}{2l} \dots \dots \dots \dots \dots \dots (2).$$
>
> The equation of motion is
>
> $$(M + \tfrac{1}{3}\rho\omega l)\,\ddot{\xi} + \frac{q\omega}{l}\xi = 0,$$
>
> and if $\xi \propto \cos pt$
>
> $$p^2 = \frac{q\omega}{l} \div (M + \tfrac{1}{3}\rho\omega l) \dots \dots \dots \dots (3).$$
>
> The correction due to the inertia of the rod is thus equivalent to the addition to M of one-third of the mass of the rod.

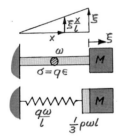

The THEORY OF SOUND
John William Strutt,
III Baron RAYLEIGH, Vol. I
first Ed. 1877, sec. Ed. 1894

RAYLEIGHS QUOTIENT
$$p^2 = V / T^*$$

Fig. 8 The Theory of Sound, Vol. I

In chapter 156 he considers the harmonic vibration of a concentrated mass m fixed at the end if a flexible rod with much smaller distributed mass $q\omega l$; ω denotes the cross section, q is Young's modulus. An approximation of kinetic energy is obtained in eq (1) by assuming a linear deflection distribution $\xi \frac{x}{l}$ along the rod. Dividing the maximum potential energy V by the kinetic energy T^* reduced of the time dependency by assuming harmonic motion

$\xi \alpha \cos(pt)$ with circular frequency p in the sense of Rayleigh's quotient leads to a correction due to the inertia of the rod, which is equivalent to the addition to M of one-third of the mass of the rod. The route paved from Rayleigh-Ritz to the Finite Element Method has been outlined by O. Mahrenholtz and L. Gaul: Mechanics in the 19th Century, 1978 (in German) and in the Den Hartog Lecture of Professor Leonard Meirovitch at the 17th ASME Biennial Conference on Mechanical Vibration and Noise, September 13, 1999.

4   From Rayleigh-Ritz to the Finite Element Method

The author worked on the link between the Rayleigh Principle and the Rayleigh-Ritz method with the mathematical foundations of the Finite Element Method with Professor Leonard Meirovitch , Virginia Polytechnic Institute Blacksburg, during his visit as Alexander von Humboldt Senior Scientist 1992 at the University of the Federal Armed Forces Hamburg(Helmut—Schmidt – University).

An outline of the results is given in abbreviated form below. The full content has been published by Professor Meirovitch in the cited Den Hartog Lecture 1999 [8].

Foreward

Although conceived [10] as a new method for the stress analysis of complex distributed-parameter structures, the finite element method is demonstrated to be a variant of the venerable Rayleigh-Ritz method, albeit the most important one.

Historical Notes

- The interest lies in differential eigenvalue problems for conservative systems that do not admit analytical solutions, so that one must be content with an approximate solution.
- The commonly accepted approach is to formulate the eigenvalue problem as a variational problem.
- The "direct method" of solving the variational problem consists of constructing a solution in the form of a minimizing sequence in terms of a finite number n of parameters and passing to the limit as $n \to \infty$.
- In vibrations, the direct method par excellence is the (classical) Rayleigh-Ritz method, whereby the minimizing sequence has the form of a linear combination of known admissible functions (shape functions), with the coefficients of the admissible functions playing the role of the parameters to be determined.
- The finite element method represents the most important variant of the Rayleigh-Ritz method, the main difference being in the nature of

- the admissible functions. This difference has enormous implications, which give the finite element method great versatility and computational efficiency.
- The variational approach to the eigenvalue problem is based on *Rayleigh's principle*, expounded by Lord Rayleigh in his *Theory of Sound* (1st edition 1877)[3], as well as in earlier papers. It states that: The lowest eigenvalue λ_1 is the minimum value Rayleigh's quotient $R(w(P))$ can take as the trial function $w(P)$ varies over the entire function space.
- The preferred form of Rayleigh's quotient is the energy form, in which the numerator is a measure of the potential energy and the denominator a measure of the kinetic energy.
- The implication of Rayleigh's principle is that, if a trial function $w(P)$ differing from the lowest eigenfunction $w_1(P)$ by a small quantity of first order, $w(P) = w_1(P) + O(\varepsilon(P))$, is inserted into Rayleigh's quotient, then the resulting $\lambda = R(w(P))$ differs from the lowest eigenvalue λ_1 by small quantity of second order, $\lambda = \lambda_1 + O(\varepsilon^2)$.
- The natural thing to do is to devise a trial function $w(P)$ that lowers $\lambda = R(w)$ in the secure knowledge that any lowering represents an improvement, as $R(w)$ cannot drop below λ_1.
- The Rayleigh-Ritz method envisions a *minimizing sequence*
$$w^{(n)}(P) = \sum_{i=1}^{n} a_i \phi_i(P) \quad (n = 2,3,...)$$
where $\phi_1(P), \phi_2(P),...$ are known trial functions such that Rayleigh's quotient is defined for every one of them and a_i are coefficients to be determined by rendering Rayleigh's quotient stationary. The method works very well provided suitable trial function can be found.
- To this day, there is some controversy as to who developed the method. The majority opinion is that both Lord Rayleigh and Walter Ritz conceived of the method independently, an opinion disputed vigorously by some.
- The best exposition of the method was provided by Ritz in two companion papers (1908 and 1909), and for many years it was known as the Ritz method. The modern exposition of the method is very close to that of Ritz.
- Rayleigh disputed the originality of Ritz's method, citing his own papers dating as far back as 1870, but his exposition bears little resemblance to the modern version. Moreover, there is some ambiguity as to Rayleigh's objectives. This doubt is reinforced by

- Rayleigh's own statement that he did not expect Ritz's success in the case of higher modes of vibrations.
- Perhaps the best way of settling the controversy is by deferring to a paper by Courant (1943), who stated:"...envisioned by two physicists, Lord Rayleigh and Walter Ritz; they independently conceived the idea... Rayleigh, in his classical work – *Theory of Sound* – and in other publications, was the first to use such procedure. But only the spectacular success of Walter Ritz and its tragic circumstances caught the general interest. In two publications of 1908 and 1909, Ritz, conscious of his imminent death from consumption, gave a masterly account of his history."
- The Rayleigh-Ritz method is based on a rigorous mathematical foundation. Its main disadvantage is that its usefulness is confined to structures with simple geometry.
- Motivated by the need to analyze structures with complex geometry in the aircraft industry, Turner, Clough, Martin and Topp (1956) developed a numerical technique that sometime later came to be known as the finite element method [10].
- The finite element method, originally developed for static stress analysis, required the solution of large sets of algebraic equations. The timing of the finite element method was perfect, as about the same time digital computers capable of solving such sets of equations were being developed. As a result, the finite element became an instant success. Since then, the method has acquired a life of its own, expanding well beyond the original scope.
- The method soon attracted the attention of mathematicians, who demonstrated that the finite element method represents a variant of the classical Rayleigh-Ritz method (Strang and Fix, 1973).
- In the classical Rayleigh-Ritz method, the trial functions are defined over the entire domain and tend to be complex. Accuracy is improved by increasing the number of trial functions, which is the general idea of a minimizing sequence.
- In the finite element method, the trial functions are defined over small finite elements and they are low-degree polynomials. As a result, the method lends itself to easy computer coding and can accommodate complex structures with irregular boundaries.
- Two Pertinent facts:
  – Without detracting in any way from the original finite element method (Turner et al., 1956), it should be mentioned that in 1943 R. Courant used a variational approach in conjunction with linear elements defined over small triangular subdomains to produce an approximate solution to St.Venant's torsion problem, thus preceding the finite element method by 13 years. The paper received little attention at that time, because the solution of large

sets of algebraic equations was impractical before the digital computer.
- In 1860, W.C.Hurty developed an extension of the Rayleigh-Ritz method to flexible multibody systems. The method, known as component-mode synthesis, was seen as a competitor to the finite element method, but in fact it complements it.

The Finite Element Method

The finite element method can be regarded as a Rayleigh-Ritz method, thus benefiting from the fine theory of the latter. It differs from the classical Rayleigh-Ritz method in that

- The admissible functions, called interpolation functions, are defined over small finite elements and are zero everywhere else, so that they are nearly orthogonal, resulting in banded mass and stiffness matrices.
- The finite elements, being small, can accommodate complex systems with wide parameter variations and irregular boundaries.
- The interpolation functions tend to be polynomials of lowest degree admissible.
- The coefficients $a_i^{(n)}$ represent actual displacements at the "nodal" points, and not abstract quantities as in the classical Rayleigh-Ritz method.
- Both the formulation and numerical solution lend themselves to efficient computer programming.

The above advantages have combined to make the finite element method the most successful development in analysis in the last five decades.

One disadvantage is that convergence can be slow, but this disadvantage is fading fast in importance as the capability of the computers increases.

Closure

- Rayleigh contributed the analytical developments on which the Rayleigh-Ritz method is based, but seems to have stopped short of developing the method itself. His efforts seem to have been concerned with improving the estimate of the lowest frequency, rather than with estimating higher frequencies.
- Ritz seems to have developed the method as it is known today.
- The Rayleigh-Ritz method is based on a fine mathematical foundation.
- Its strength is that it can yield accurate results with a small number of degrees of freedom.

- Its weakness is that the distributed systems for which suitable functions can be found are limited to one-dimensional systems and two-dimensional systems with simple geometry.
- The finite element method overcomes virtually all the difficulties encountered by the classical Rayleigh-Ritz method.
- The fact that the interpolation functions are defined over small elements gives it enormous versatility, manifested in the variety of complex problems it can handle and in the computational efficiency with which it can produce solutions. In fact, its application to many engineering areas far exceeds the scope of the original structural analysis.
- For vibrations, regarding the finite element method as a Rayleigh-Ritz method permits it to enjoy the mathematical theory developed for the latter, without any penalties. This, in turn, permits a deeper understanding of the finite element method.

5 Summary

In vibrations, a direct approach aiming at approximate solutions is the Rayleigh-Ritz method, whereby the minimizing sequence has the form of a linear combination of known admissible functions (shape functions), with the coefficients of admissible functions playing the role of the parameters to be determined.

For this class of problems, the finite element method represents the most important variant of the Rayleigh-Ritz method, the main difference being in the nature of the admissible functions. This difference has enormous implications, which give the finite element method great versatility and computational efficiency.

References

[1] Newton, I.: PHILOSOPHIÆ Naturalis Principia Mathematica. Imprimatur, London (Juli 5, 1686)
[2] Eulero, L.: Novi Commentarii Academiae Petropolitanae, vol. 20 (1775)
[3] Strutt, J.W., Rayleigh III, B.: The Theory of Sound, 1st edn., vol. 1 (1877)
[4] Ritz, W.: Journal für Mathematik 135(H.1), 1–61 (1909)
[5] Truesdell, C.: Essays in the History of Mechanics, p. 91, 92, 116, 260. Springer, Berlin (1968)
[6] Szabó, I.: Geschichte der mechanischen Prinzipien, p. 4, 7, 12, 19, 30. Birkhäuser, Basel (1979)
[7] Mikhailov, G.K., Schmidt, G., Sedov, L.I.: ZAMM 64, 73–82 (1984)
[8] Meirovitch, L.: Proceedings 17th ASME Conf. on Mechanical Vibration and Noise (1999)
[9] Mahrenholtz, O., Gaul, L.: Zeitschr. d. Universität Hannover 5, Heft 2, pp. 28 – 48 (1978)
[10] Turner, M.J., Clough, R.W., Martin, H.C., Topp, L.J.: Stiffness and Deflection Analysis of Complex Structures. Journal of Aeronautical Sciences 23, 805–823 (1956)

History of the Finite Element Method – Mathematics Meets Mechanics – Part I: Engineering Developments

Erwin Stein

Abstract. The birth of variational calculus and the principle of virtual work goes back to the 17th and 18th century, and the first draft of a discrete variational method with "elementwise" triangular shape functions was given by Leibniz (1697). First analytical studies were made by Schellbach (1851) and then, already with numerical results, by Rayleigh (1877). The mathematician Ritz (1909) marks the first discrete (direct) variational method for the linear elastic Kirchhoff plate, and the engineer Galerkin (1915) published his seminal article on FEM for linear elastic continua, postulating the orthogonality of the residua of equilibrium with respect to the test functions, but both, Ritz and Galerkin, used test and trial functions within the whole domain as supports.

Courant (1943) was the first to introduce triangular and rectangular "finite elements" for the 2D-St.-Venant torsion problem of a prismatic bar (Poisson equ.), and Clough and his team (1956) published the first modern 2D-FEM for arrowed aircraft wings. Also Wilson, Melosh and Taylor with their important schools, e.g. Bathe and Simo, promoted in Berkeley the new discipline of "Computational Mechanics". Argyris (since 1959 in Stuttgart) and Zienkiewicz (since 1965 in Swansea), together with Irons e.g., developed primal FEM in a systematic way: hierarchical classes of FEs with different topologies and ansatz techniques for solid and fluid mechanics.

Mixed finite elements, advocated for non-robust problems, are usually based on dual mixed variational functionals (yielding saddle point problems) by Hellinger (1914), Prange (1916) and Reissner (1950). Important elements came, e.g., from Cruceix, Raviart (1973), Raviart, Thomas (1977) and Brezzi, Douglas, Marini (1987). For numerical stability of saddle point problems, the Brezzi-Babuška global (infsup) stability conditions have to be fulfilled.

Erwin Stein
Institute of Mechanics and Computational Mechanics,
Leibniz Universität Hannover,
Appelstr. 9A, 30167 Hannover, Germany
e-mail: stein@ibnm.uni-hannover.de

Rather new extensions and variants of FEM are XFEM by Belytschko (1996) and SFM (Singular Function Method) by Fix, Strang (1973) and Grisvard (1992), as well as Isogeometric Analysis by Hughes (2005).

This chapter and also chapter 23 are restricted to linear elastic solid and structural mechanics. The topics are designed not only in a descriptive way with verbal comments on the origins, main features and the importance of publications on FEM, but moreover, essential steps of continuous and discrete variational theory, finite element algorithms and error estimators as a basis for adaptivity are communicated as concentrated as possible. This seems to be advantageous for conceiving the subjects.

1 A Short Historical Overview of Energy-Based Numerical Methods in Solid and Structural Mechanics

1.1 Eminent Scientists in the Development of FEM

Fig.1 shows main traces of developments by eminent scientists over the last 300 years, who fundamentally contributed to discrete energy-based variational methods in solid and structural mechanics, governed by BVPs of elliptic partial differential equations (PDEs).

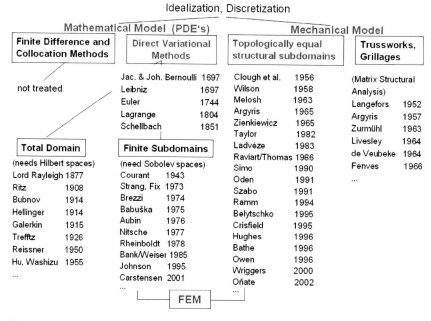

Fig. 1 Eminent scientists in numerical and structural analysis of elasto-mechanics since the 17th century

The discovery of variational calculus by Jacob Bernoulli (1697) and the first proposal for a direct (discrete) variational approximation by Leibniz, both published in Leibniz (1697), [1], as well as other important developments of conservation principles in mechanics in the 17$^\text{th}$ century were outlined in Stein (2011), [2], Stein, Wiechmann (2003), [3]. Following the ideas of Jacob and Johann Bernoulli and especially of Leibniz, Schellbach (1851), [4], published the algebraic equations for the brachistochrone and other optimization problems, in total 12, by analytical calculation of the integrals for equidistant support points and triangular test functions between three adjacent points.

The first discretized mechanical formulation of eigenvalue problems for linear elastic solids via an energy principle was given by Strutt, the 3$^\text{rd}$ Lord Rayleigh (1877), Fig. 2, in his famous book *Theory of Sound*, [5], especially by introducing the Rayleigh-quotient for the eigenvalues with an upper bound property.

1.2 The First Analysis of Direct Variational Calculus with Shape Functions in the Whole Domain and Application to Clamped Plates in Bending by Ritz in 1909

The first direct variational approach for statically loaded clamped plates, modeled by the 4$^\text{th}$ order Kirchhoff 2D-PDE was presented by Ritz, Fig. 2, using products of trigonometrical functions in the whole rectangular domain of a plate.

Ritz, at last professor of mathematics at the Universität Göttingen, published his famous article, [6], translated: On a new method for the solution of certain problems in mathematical physics, in 1909.

We cite from the introduction with translation from German into English:

"The boundary value problems in mathematical physics usually require the representation of finite, continuous functions in prescribed finite domains. Only exeptionally, the expansion in power series is possible, and even more seldom, this is numerically usable in the total domain...
Thus, there is a request for an approximated representation of the integrals in the total prescribed domain by a polynomial of given degree n... in a way that for growing n the accuracy is growing unlimited, finally resulting in polynomial expansions of the integrals ... Also Fourier series can be useful. In general, one can use most of those functions $\psi_1, \psi_2, \ldots, \psi_n$ which can be chosen according to qualitative observation."

In general, Ritz uses power series of linear independent and relative complete functions, e.g. $\psi_i(x,y)$ for 2D problems with unknown so-called Ritz-parameters $a_i, i = 1 \ldots n$. The stationary value for the approximated energy functional J_h of the total potential energy is got by differentiating this functional with respect to the Ritz-parameters as $\partial J_h/\partial a_i = 0, i = 1, 2, \ldots, n$.

Fig. 2 Lord Rayleigh, John William Strutt, the 3rd Baron Rayleigh (1842-1919) (left), Walter Ritz (1878-1909) (middle), Boris Grigoryevich Galerkin (1871-1945) (right)

Ritz especially treats BVPs of the Kirchhoff biharmonic elastic plate equation for rectangular clamped domains, using biharmonic shape functions in the full domain. He proved that the solutions yield a minimum of J for an infinite number of shape functions and that consistency and convergence hold.

1.3 The Principle Idea of Direct Variation by the Orthogonality of the Equilibrium Residuum and the Test Functions by Galerkin in 1915

Galerkin, Fig. 2, was the first to publish a key idea of primal FEM: calculating the unknown coefficients of the trial functions by the orthogonality condition of the residuum of equilibrium with respect to the test functions in the whole domain of a system, which can be seen as a variant of the principle of virtual work, Galerkin (1915), [7]. This reads in operator notation (Galerkin used Cartesian coordinates for 2D problems) for the 2D-BVP of the Lamé PDEs for static linear elasticity

$$\mathcal{L}[\mathbf{u}] = \mathbf{f} \text{ in } \mathbf{\Omega} \in \mathbb{R}^2, \tag{1}$$

with the kinematic (Dirichlet) boundary conditions

$$\mathcal{B}_D[\mathbf{u}] = \mathbf{0} \text{ on } \partial\mathbf{\Omega}. \tag{2}$$

Introducing equal trial and test functions

$$\tilde{\mathbf{u}} = \sum_{j=1}^{n} \alpha_j \mathbf{u}_j(x,y), \tag{3}$$

with the necessary conditions for kinematic admissibility

$$\mathcal{B}_d[\tilde{u}] = \mathbf{0} \text{ on } \partial\Omega, \tag{4}$$

the "Galerkin orthogonality" of the residuum of equilibrium with respect to the test functions reads

$$\int_\Omega \underbrace{(\mathcal{L}[\tilde{u}] - f)}_{\mathcal{R}[\tilde{u}]} \cdot u_j \, dA = \mathbf{0}, \ j = 1, 2, \ldots, n. \tag{5}$$

Mathematically the Galerkin approximation is an inner product projection, not a $\mathcal{L}^2$ projection, of the whole analytical solution space $\mathcal{V}(\Omega)$ of the BVP into the reduced discretized function space $\tilde{\mathcal{V}} \equiv \mathcal{V}_h \subset \mathcal{V}$.

The next important step, namely the orthogonality of the test functions with respect to the error, i.e. with $f = \mathcal{L}[\tilde{u}]$, yielding

$$\int_\Omega (\mathcal{L}[\tilde{u}] - \mathcal{L}[u]) \cdot u_j \, dA = \mathbf{0}, \tag{6}$$

was not given by Galerkin. With the discretization error $e := \tilde{u} - u$, and the linear operator $\mathcal{L}$ we get the important orthogonality of the discretization error with respect to the test functions

$$\int_\Omega \mathcal{L}\underbrace{[\tilde{u} - u]}_{e} \cdot \tilde{u} \, dA = 0. \tag{7}$$

1.4 The Introduction of Finite Elements in Finite Subdomains by Courant in 1943

The second important idea for FEM is the introduction of finite subdomains Ω_e – the finite elements – as supports of the trial and test functions which was first given by Courant (1943), [8], Fig. 3, for the 2D-Poisson PDE, modeling St. Venant torsion of a prismatic bar and choosing linear triangles and quadrangles as FEs.

The involved matrix calculus, especially for beam and rod systems, had been developed by Langefors (1952), [9], Argyris (1954/55), [10], Livesley (1964), [11], Zurmühl (1963), [12], Fenves (1966), [13], et al., not to forget the first integrated view of structural analysis and automatic digital computing by Zuse since 1936, Zuse (1962), [14].

However, the structural approximation of plates, shells and 3D structures by 1D beams, i.e. with different topological structural elements compared with the real structure, makes the backward analysis from the discrete beam system to the real 2D or 3D structure very complicated and not unique concerning the real states of stresses; for this an error analysis is nearly impossible.

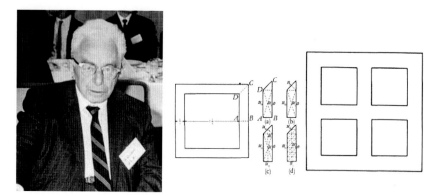

Fig. 3 The mathematician Richard Courant (1888-1972) who was the first to introduce finite subdomains and related shape functions as finite elements (left); finite elements for the St. Vernant torsion problem of a prismatic bar (right)

1.5 Remarks on Trefftz's Method from 1926

A remarkable counterpart to Ritz's direct variational calculus is Trefftz's method, Trefftz (1926), [15]. Herein, the homogeneous PDEs are fulfilled a priori in the whole domain, either by analytical solutions for the displacements or by stress functions, fulfilling the bipotential Beltrami PDEs. Opposite to Ritz's method, mixed boundary conditions are fulfilled approximately by the discrete variational method.

For 2nd order elliptic self-adjoint and well-posed BVPs this yields symmetric positive definite system matrices for admissible test and trial functions at the boundaries, and convergence is assured for complete test and trial polynomials at the boundary – in Trefftz's original work considering the whole boundary – as in Ritz's method.

In case of the 4th order bipotential Kirchhoff plate equation or of corresponding shell bending theories, Trefftz's method is only consistent for pure displacement boundary conditions (BCs), i.e. for clamped edges. In case of so-called Navier BCs and of free edges (with pure Neumann conditions), an additional least-squares term for the integral of the errors of the kinematical BCs, multiplied with a penalty factor, has to be inserted for achieving unambiguous and thus converging results of Trefftz's approximations.

These problems were treated in the author's doctor thesis from 1964, in the doctor thesis of Weidner from 1967, furthermore by Jirousek, Guex (1986), [16], Peters, Stein, Wagner (1994), [17], as well as by Piltner, Zielinski and others who developed generalized Trefftz's methods for discrete boundary element techniques.

By generalizing Trefftz's method to finite (discrete) boundary elements there arose a competition with the important discrete boundary integral equation method (BIEM or simply BEM) with the advantageous kernel functions derived from integral transformations based on the Signorini identity.

However, there are recent improvements of Trefftz's methods by chosing hybrid mixed variational formulations and special test and trial functions, e.g. applying a domain decomposition method for efficient solving of the algebraic equations.

In this article Trefftz's method is not presented in detail because then also BIEM had to be outlined with comparisons what would have increased the size considerably.

1.6 The Method by Hrennikoff in 1940

Hrennikoff published a special version of a discrete element method, mainly with triangles and quadrangles for plane stress and plate bending analysis, which are constructed with rods (for plane stress) and beams (for plate bending), Hrennikoff (1940), [18]. For a fixed Poisson's ratio $\nu = 1/3$ one can calculate the cross-section areas or inertia moments, respectively, for the condition of $\mathcal{C}^1$-strains and $\mathcal{C}^0$-displacements in triangular elements. The method can be extended to quadrangles under restricted conditions. There were no applications in a wider frame, as it is also an application of not topologically equal structural subdomains.

1.7 The First Engineering Version of the Finite Element Method by the Berkeley Group around Clough in 1956

Different from the direct structural approximation – e.g. approximating plates in bending by beam elements – the finite element method is based on isotopological subdomains of the real structure as the supports for the test and trial functions. The variational approximation is (i) based mathematically either on the weak form of equilibrium (represented by the bilinear and linear forms) or on the minimum principle for the related discrete variational functional, see Stein, Part II, Mathematical foundation of primal FEM, error analysis and adaptivity (following chapter), and (ii) based mechanically either on the discrete principle of virtual work or the discrete Dirichlet principle of minimum of the total potential energy, e.g. of a statically loaded linear elastic system. The strain energy depends on trial displacements $\mathbf{u}_h \epsilon H_0^1(\Omega_e), \Omega_e \epsilon \Omega$ and related test functions $v_h \epsilon H_0^1(\Omega_e)$, for calculating displacements, strains and stresses of 3D and 2D solids and structures (especially shells and plates), as well as 1D beam and rod systems.

The first engineering version of the finite element method (FEM), based on the principle of virtual work, for 2D stress analysis of thin-walled metal sheets was developed by the Berkeley team around Clough, Fig. 4, with application to stress analysis of arrowed aircraft wings in collaboration with the Boeing Company. Fig. 5 shows the plane triangular element with bilinear displacment approach and below its stiffness matrix, Turner, Clough, Martin, Topp (1956), [19]. Also Melosh contributed an engineering derivation of linear elastic FEM, Melosh (1963), [20]. Wilson, Fig. 4, made essential contributions on multistory buildings, on the use of incompatible displacement modes for stabilizing FEM, for non-linear elastodynamics improving Newmark's step-by-step integration method by the Wilson-θ method, as well as the solution of large eigenvalue problems, and last but not least the design of the programs SAP 4 and NONSAP by Wilson, Farhoomand, Bathe, (1973), [21], as well as FEAP by Taylor in Berkeley, see Sect. 6.

Fig. 4 Ray W. Clough (*1920), who developed the first engineering version of the finite element method with challenging applications to the stress analysis of aircraft wings and Edward L. Wilson (*1931), one of the pioneers of the finite element method

Strong impacts for the rapid development of FEM came from Argyris, [10], Argyris, Scharpf (1969), [22], (in Stuttgart since 1959) and Zienkiewicz (in Swansea since 1965), Fig. 6, see Zienkiewicz, Cheung (1967), [23], Zienkiewicz, Taylor (2000), [24]. Their main goals were: first, the systematic derivation and comparative applications of classes of primal finite elements in solid, structural and fluid computational mechanics, including various material equations, and secondly the development of well-structured computer programs with the seperation of all important steps, like ASKA by Argyris and Schrem.

The new sight and important change of structural analysis in the 1960s can be seen in Fig. 1, namely in the transition from the right column "Trussworks, Grillage" to the neighbored left column "Topologically equal structural subdomains".

Three important pioneers of FEM can be seen in Fig. 7 during the first European Conference on Computational Mechanics (ECCM) 1999 in Munich. It is a remarkable picture with scarcity value.

2 Five Steps for Primal FEM Calculations

The algorithms for FEM require the following five steps where each of them needed huge research efforts since the 1960s, see the important contributions by Bathe and Wilson (1976), [25], Hughes (1987), [26], as well as Szabó and I. Babuška (1991), [27].

History of the Finite Element Method, Part I

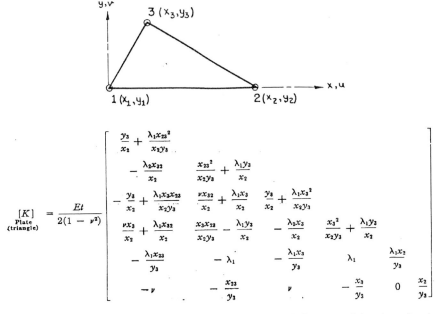

Fig. 5 Triangular plane stress element with 6 nodal degrees of freedom (top); symmetric stiffness matrix of order 8 x 8 for the plane stress triangle (bottom)

Fig. 6 John H. Argyris (1913-2004) and O.C. Zienkiewicz (1921-2009), two emininent pioneers of the finite element method

2.1 Topologies, Geometries, Generation and Refinement of Meshes

Triangular and quadrangular finite subdomains (elements) are suitable for 2D problems as well as tetrahedral and hexahedral elements for 3D problems. The element boundaries have to be convex, and the meshes of those elements must not have intersections or gaps. Furthermore, the distortion of elements

Fig. 7 John H. Argyris, Ray W. Clough and Olgierd C. Zienkiewicz at the 1st ECCM Conference in Munich 1999, photo by Prof. E. Ramm

has to be limited, expressed by angles of adjacent sides or surfaces, and the ratio of element diagonals should be near to one, because distortions can cause rank deficiencies; this can be controlled by element patch tests, [24], vol. 1.

Several mesh-generation methods have been developed since the 1960s based on Delaunay triangulations, Delauney (1934), [28], according to the Bowser-Watson algorithm, see George (1991), [29], Löhner, Parikh (1988), [30], further hierarchical spatial decompositions by constructing trees, especially quadtree and octree techniques by Shephard and Georges (1991), [31], advancing front method, beginning at the boundary, first creating points and then point connections. Special problems are caused by inclusions, especially if they are not convex.

Local mesh refinements of 2D elements

Local mesh refinements of triangular elements are realized by:

(a) Standard refinements without irregular nodes with new nodes at midside points

(b) Halfening of the longest side by Rivara (1984), [32], Fig. 8

Rectangular and quadrangular elements can be locally refined by:

(c) avoiding irregular nodes by introducing intermediate elements, or

(d) using hanging nodes, i.e. midside nodes, which are correctly eliminated in the next refinement step by interpolation with the shape functions of the element, Fig. 9, [24], vol. 1.

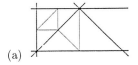 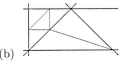

(a)   (b)

Fig. 8 (a) Standard refinements of triangles without irregular nodes (b) Halvening of longest side (Rivara)

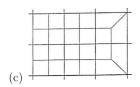 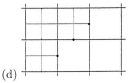

(c)   (d)

Fig. 9 Local refinement techniques for quadrilateral elements; (c) Transition elements for avoiding irregular nodes, non-regular elements are resolved in the next refinement step; (d) Preserving regular elements by eliminating irregular nodes via hanging node technique

Local mesh refinements of 3D elements

Tetrahedral elements can be regularly refined similar to triangles by:

(a) choosing new nodes at mid-surface points, or
(b) selecting the mid-point of the largest surface and using already existing element nodes, see Niekamp, Stein (2002), [33]

Node-regular refinements of hexahedral elements cause significant problems. Using new element points at the third-points of the edges, a node-regular division is possible, but with the big number of 27 child elements, [33]. In order to restrict the number of child elements to 8, two types of transition elements can be used:

(a) with additional 3 degenerated hexahedrons, called heptahedrons, or
(b) trihedrons as degenerated hexahedrons with 5 nodes and 3 hyperbolic surfaces in the inner of the parent domain. In case of mesh adaptivity with regular refinements only, 80 new hexahedral elements are generated, which is by far too much for efficient local refinements.

Therefore, *refinements* with non-regular hexahedral elements by introducing *hanging nodes* at mid-surface points are more efficient due to much better locality, and they are easier to treat in the algorithms and in the codes.

2.2 Interpolation Polynomials as Test and Trial Functions

The ansatz functions must be relatively complete, and they have to fulfill the partition of unity condition (PUM) by Melenk (1996), [34], i.e. the sum of the shape functions of an element has to be one. Of course, rigid body modes must be included.

For hierarchical elements C^0 Lagrange polynomials, the Pascal triangle and tetrahedron provide a framework for defining a form of natural hierarchy: the number and position of the nodes of successively higher-degrees elements are nested in successive levels of the Pascal pattern.

Area coordinates are often used for triangular elements and volume coordinates for tetrahedrons. Tensor products of Lagrange and Legendre polynomials (as linear combinations of Lagrange polynomials with orthogonality properties) are commonly used as nodal basis functions for quadrangles and hexahedrons in case of 2nd order PDEs. Interpolations for the h-version of finite element spaces are treated by Apel (2004), [35], Brenner and Carstensen (2004), [36], as well as by Szabó, Düster and Rank (2004), [37].

Hermitian polynomials are Lagrangian polynomials with C^1 continuity which are adequate for PDEs of 4th order, especially the Kirchhoff plate equation, see also [6]. Herein nodal displacements and nodal derivatives are unknown parameters.

A hierarchical basis of Legendre polynomials $L_n(r)$ is defined as

$$L_0(r) := \frac{1}{2}(1-r) \quad L_1(r) := \frac{1}{2}(1+r), \tag{8}$$

and for $n \geq 2$ it holds

$$L_n(r) := \int_{-1}^{r} P_{n-1}(x) \, \mathrm{d}x = \int_{-1}^{r} \frac{1}{2n-1} \frac{\mathrm{d}}{\mathrm{d}x}(P_n - P_{n-2}) \, \mathrm{d}x \tag{9}$$

and finally

$$L_n(r) := [\frac{1}{2n-1}(P_n - P_{n-2})]_{-1}^{r} = \frac{1}{2n-1}[P_n(r) - P_{n-2}(r)]. \tag{10}$$

If n is the maximal order of the actual approximation space then the hierarchically expanded test space is derived by the orthogonality condition

$$\int_{-1}^{+1} [L_n(x) \cdot L_k(x)] \, \mathrm{d}x = 0 \text{ for } k < n \text{ and } k \neq n-2. \tag{11}$$

Remark: The extension of 1D shape functions to 2D and 3D problems, using tensor products of the 1D functions, is possible for Cartesian, polar, cylindrical and spherical coordinates.

Nodal based linear Lagrangian shape functions, hierarchical Lagrangian shape functions and hierarchical basis functions for p-adaptivity are shown in Fig.10-13, see Borouchaki et al. (1997), [38], as well as [24], [36], [37].

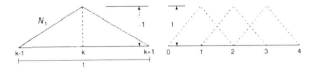

Fig. 10 1D nodal basis of linear Lagrangian shape functions

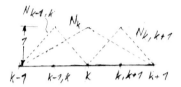

Fig. 11 1D hierarchical nodal basis of linear Lagrangian shape functions

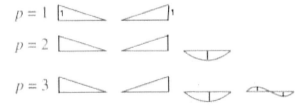

Fig. 12 Lagrangian shape functions of a 1D element for $p = 1; 2; 3$

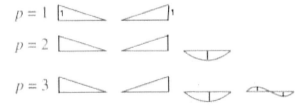

Fig. 13 1D hierarchical basis functions of the p-version for $p = 1; 2; 3$

Isoparametric quadrilateral and hexahedral elements with tensor products of 1D shape functions for the same approximation of the geometry and FE displacements are constructed by bijective mappings with respect to the unit square and unit cube via the Jacobian matrix, Fig. 14, see Irons and Zienkiewicz (1968), [39].

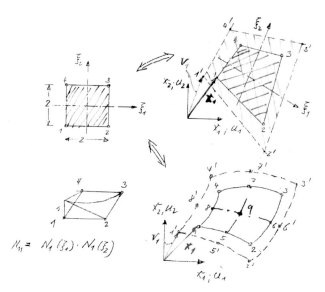

Fig. 14 Isoparametric finite element shape functions with bilinear and biquadratic polynomials

2.3 Integration of Shape Functions and Their Derivatives in Element Domains by Using Gaussian Integration Points

Super-convergent integration orders are achieved by using non-equidistant Gaussian integration points, Gauß (posthum 1863-1874), [40]. The accuracy of numerical integration has to be adapted to the polynomial order of approximation.

2.4 Element Stiffness Matrices, Load Vectors and Assembling to the Total System

It is useful for the further subjects to shortly outline the algorithmic basis of primal FEM. Matrix notation is used for 2D problems. Different from a tensorial notation, this has the advantage of similar and easily understandable algorithmic structures, e.g. concerning symmetries, for different types of elliptic PDEs. In the sequel

Eight Steps for Establishing and Solving Linear FE Problems
are shortly described, see, e.g. Buck, Scharpf (1973), [41], Stein, Wunderlich (1973), [42], Bathe (1982), [43], especially the *Encyclopedia of Computaional Mechanics* from 2004, Stein, de Borst, Hughes (2004), [44], Stein, Rüter (2004), [45], as well as [24], [26]:

(a) **The displacement vector** $u(x)$ of a point $x \in \Omega \in \mathbb{R}^2$ is

$$u^T(x) = \{u_1(x)\, u_2(x)\}^T;\ x^T = \{x_1\, x_2\};\ x = x_i\, e_i \qquad (12)$$

(Eucledian vector with Cartesian basis e_i), and the geometric boundary conditions are $u - \bar{u} = 0$ at Γ_D; $\Gamma_D \cup \Gamma_N = \Gamma = \partial\Omega$.

(b) **The components of the linear 2D strain tensor** are defined by the symmetric parts of the displacement gradient (kinematic relations) $\varepsilon := \nabla_{sym} \otimes u(x)$. In matrix notation they are condensed in the column vector $\varepsilon(x)^T := \{\varepsilon_{11}(x)\, \varepsilon_{22}(x)\, 2\varepsilon_{21}(x)\}$, with the kinematic relations and the differential operator matrix

$$\varepsilon(x) := D\, u(x);\quad D = \begin{bmatrix} \partial_1 \\ \partial_2 \\ \partial_2\ \partial_1 \end{bmatrix};\ \partial_i = \frac{\partial}{\partial x_i}. \qquad (13)$$

(c) **The components of the 2D stress tensor** are in matrix notation $\sigma^T := \{\sigma_{11}\, \sigma_{22}\, \sigma_{12}\}$, and the local equilibrium conditions are, in tensor notation $\mathrm{div}\,\sigma + \rho b = 0$ in $\Omega \subset \mathbb{R}^2$, and in matrix notation $D^T\sigma + \rho b = 0$ in Ω_e and $\sigma_{12} = \sigma_{21}$ for 2D problems.

The tractions $t(x, n)$ at arbitrary cuts and at element interfaces with the normal vector $n = n_i e_i$; $n_i = \cos n, e_i$ at point x are – according to the Cauchy theorem $t(x, n) = \sigma(x) \cdot n$ – in matrix notation

$$t(x, n) = \begin{Bmatrix} t_1 \\ t_2 \end{Bmatrix} = \underbrace{\begin{bmatrix} \cos n, e_1 & 0 & \cos n, e_2 \\ 0 & \cos x, e_2 & \cos n, e_1 \end{bmatrix}}_{\mathcal{N}^T} \underbrace{\begin{Bmatrix} \sigma_{11} \\ \sigma_{22} \\ \sigma_{12} \end{Bmatrix}}_{\sigma}. \qquad (14)$$

The BCs at Neumann boundaries are $\bar{t} - t(x, n) = 0$ at Γ_N, with $\mathcal{N}^T\sigma = t(x, n) = \bar{t}$ at $\Gamma_{N,e}$; $\Gamma_{D,e} \cup \Gamma_{N,e} = \Gamma_e \qquad \forall \cup \Gamma_e = \Gamma = \partial\Omega$, with prescribed tractions at Neumann boundaries.

One recognizes the same index structure of the matrix of direction cosinus, $\mathcal{N}^T$ and of the differential operator matrix D^T in the field equilibrium conditions.

Remark: 2D strain states are not possible in 2D stress states in case of non-vanishing Poisson's ratio $\nu = (E - 2\mu)/2\mu$, with ν Poisson's ratio, E Young's modulus and μ the shear modulus.

(d) **The linear isotopic elasticity law** reads in matrix notation $\sigma = \mathbb{C}\varepsilon$, with the matrix of elastic constants from the elasticity tensor, here for 2D stress states

$$\mathbb{C} = \frac{E}{1-\nu^2}\begin{bmatrix} 1 & \nu & 0 \\ \nu & 1 & 0 \\ 0 & 0 & \frac{1}{2}(1-\nu) \end{bmatrix}, \tag{15}$$

with $\mathbb{C} = \mathbb{C}^T$ and $\varepsilon^T \mathbb{C}\varepsilon \begin{cases} > 0 \text{ for } \varepsilon \neq \mathbf{0} \\ = 0 \text{ for } \varepsilon = \mathbf{0} \end{cases}$.

(e) **The principle of virtual work** for the finite element system is presented for mixed boundary conditions, i.e. the test and trial functions have to fulfill the Dirichlet boundary conditions $\bar{\mathbf{u}} - \mathbf{u} = \mathbf{0}$ at $\Gamma_{D,e}$.

The approximation of $\mathbf{u}(\mathbf{x})$ in Ω_e by admissible trial functions (tensor products of polynomials) is $\mathbf{u}_{h,e} = \sum_{i=1}^n \boldsymbol{\phi}_{u,i}(\mathbf{x})\hat{\mathbf{u}}_{e,i}$, with the displacement shape functions $\phi_{u,i}(\mathbf{x})$, as presented in (ii). The trial functions read in matrix notation

$$\underbrace{\begin{Bmatrix} u_{1,h,e} \\ u_{2,h,e} \end{Bmatrix}}_{\mathbf{u}_{h,e}} = \underbrace{\begin{bmatrix} \underbrace{\phi_{u_1;1}(\mathbf{x})\ \phi_{u_1;2}(\mathbf{x})\ldots\phi_{u_1;n}(\mathbf{x})}_{\mathbf{N}_{u_1}(\mathbf{x})} & \\ & \underbrace{\phi_{u_2;i}(\mathbf{x}), i=1\ldots n}_{\mathbf{N}_{u_2}(\mathbf{x})} \end{bmatrix}}_{\mathbf{N}_u(\mathbf{x})} \underbrace{\begin{Bmatrix} \hat{\mathbf{u}}_{1,e} \\ \hat{\mathbf{u}}_{2,e} \end{Bmatrix}}_{\hat{\mathbf{u}}_e}, \tag{16}$$

in total $\mathbf{u}_{h,e}(\mathbf{x}) = \mathbf{N}_u(\mathbf{x})\hat{\mathbf{u}}_e$.

The virtual displacements or test functions are chosen equally to the trial functions as $\delta\mathbf{u}_{h,e} \equiv \mathbf{v}_{h,e} = \mathbf{N}_u(\mathbf{x})\hat{\mathbf{v}}_e$, in order to realize symmetry and maximum rank of the stiffness matrices.

The test space for primal FEM is based on $\mathcal{C}^1$-continuous displacements in the element domains and $\mathcal{C}^0$-continuous displacements at element interfaces. Thus, the test space reads

$$\mathcal{V}_h = \{\mathbf{v}_{h,e} \in H_0^1(\Omega), \mathbf{v}_{h,e} = \mathbf{0} \text{ at } \Gamma_{D,e},\ \underbrace{[\![\mathbf{v}_{h,e}]\!]}_{\mathbf{v}_{h,e}^+ - \mathbf{v}_{h,e}^-} = \mathbf{0} \text{ at } \Gamma_e\}, \mathcal{V}_h \subset \mathcal{V}. \tag{17}$$

The discretized strains are $\boldsymbol{\epsilon}_{h,e}(\mathbf{u}_{h,e}) = \mathbf{D}(\mathbf{x})\mathbf{u}_{h,e} = \mathbf{D}\mathbf{N}_u(\mathbf{x})\hat{\mathbf{u}}_e = \mathbf{B}(\mathbf{x})\hat{\mathbf{u}}_e$ and the virtual strains $\delta\boldsymbol{\epsilon}_{h,e} = \mathbf{D}(\mathbf{x})\mathbf{v}_{h,e} = \mathbf{B}(\mathbf{x})\hat{\mathbf{v}}_e$. The elasticity law is pre-fulfilled in primal FEM, as $\boldsymbol{\sigma}_{h,e} = \mathbb{C}_e\boldsymbol{\epsilon}_{h,e}(\mathbf{u}_{h,e}) = \mathbb{C}_e\mathbf{N}_u\hat{\mathbf{u}}_e$.

Then the kinematic conditions $\boldsymbol{\epsilon}_{h,e}(\mathbf{u}_{h,e}) = \mathbb{C}^{-1}\boldsymbol{\sigma}_{h,e}(\mathbf{u}_{h,e})$ are fulfilled a priori.

Now the principle of virtual work for 2D discretized systems in static equilibrium reads

$$\delta\mathcal{A}_h = \bigcup_e \{\int_{\Omega_e} \mathbf{v}_h^T [\mathbf{D}^T \boldsymbol{\sigma}(\mathbf{u}_h) + \rho\mathbf{b}]\,\mathrm{d}\Omega + \int_{\Gamma_{N,e}} [\bar{\mathbf{t}} - \mathbf{v}_h^T \mathbf{t}_h(\mathbf{u}_h)]\,\mathrm{d}\Gamma\} \stackrel{!}{=} 0. \tag{18}$$

We keep in mind: the test functions $v_h \in \Omega_e$ have to fulfill the kinematic conditions within the element domains, at element interfaces and at Neumann element boundaries. As output we get the approximations in the weak sense of the equilibrium conditions within all element domains, at all element interfaces and at all Neumann element boundaries.

(f) Partial integration of the first term in the first integral and applying the divergence theorem yield after introducing the shape functions

$$\delta \mathcal{A}_h = \bigcup_e \{\hat{v}_e^T \int_{\Gamma_e} \underbrace{N_u^T \underbrace{[\mathcal{N}^T \sigma(u_h)]}_{[\![t_e(u_h)]\!]}}_{a_e(u_h, v_h)} d\Gamma$$

$$\underbrace{- \hat{v}_e^T \underbrace{(\int_{\Omega_e} \underbrace{(N_u^T D^T)}_{B^T} \mathbb{C}_e \underbrace{DN_u}_{B} d\Omega) \hat{u}_e}_{k_e = k_e^T; \det k_e = 0} + \underbrace{\hat{v}_e^T (\int_{\Omega_e} N_u^T \rho b d\Omega + \int_{\Gamma_{N,e}} N_u^T \bar{t} d\Gamma)}_{\hat{\bar{p}}_e}}_{l_e(u_h)}$$

$$- \hat{v}_e^T \int_{\Gamma_e} N_u^T [\![\mathcal{N}^T \sigma(u_h)]\!] \, d\Gamma \} \overset{!}{=} 0.$$

(19)

The first and the last term in the above equation cancel each other, and the result is

$$\delta \mathcal{A}_h = \bigcup_e \{-\hat{v}_e^T \underbrace{(\int_{\Omega_e} B^T(x) \mathbb{C} B(x) \, d\Omega)}_{k_e = k_e^T; \det k_e = 0} \hat{u}_e + \hat{v}_e^T \underbrace{[\int_{\Omega_e} N^T(x) \rho b \, d\Omega + \int_{\Gamma_{N,e}} \bar{t} \, d\Gamma]}_{\hat{\bar{p}}_e} \} \overset{!}{=} 0.$$

(20)

On element level, we see the equality of fictitious nodal forces $k_e \hat{u}_e$ and nodal forces due to the given loads $\hat{\bar{p}}_e$, where k_e is the element stiffness matrix; its rank decay is equal to the number of admitted rigid body modes, caused by the differential operator matrix D in $B = DN_u$.

(g) Assembling of the elements to the system by Boolean matrices a_e, according to the unknown global reduced nodal displacement vector $\hat{U}$, where *global* means the geometric assembling of the elements at the nodes and *reduced* means the elimination of nodal displacements at Dirichlet boundaries $\Gamma_{D,e}$, at least avoiding rigid body displacements without linear dependencies, results in

$$\hat{u}_e = a_e \hat{U}; \quad \delta \hat{u}_e = a_e \delta \hat{U}; \quad \hat{\bar{p}}_e = a_e \hat{\bar{P}}. \quad (21)$$

Of course, these large rectangular incidence matrices a_e, which are mainly occupied with zeros, are not directly used but realized by index lists attached as column vectors to the element stiffness matrices (stored as upper triangular matrices), from which they are inserted into the global reduced stiffness matrix.

(h) Kinematic assembling yields the global linear equation system

$$\underbrace{\delta \hat{U}^{\mathrm{T}}}_{\neq 0}\{\bigcup_e \underbrace{a_e^{\mathrm{T}} k_e a_e}_{K=K^{\mathrm{T}};\ \det K \neq 0}]\hat{U}\} = \delta \hat{U}^{\mathrm{T}} \{\underbrace{\bigcup_e a_e^{\mathrm{T}} \hat{\bar{p}}_e}_{\hat{\bar{P}}}\}, \quad (22)$$

and thus the linear algebraic equation system for solving the global nodal displacement vector $\hat{U}$ reads

$$K\hat{U} = \hat{\bar{P}} \rightsquigarrow \hat{U} = K^{-1}\hat{\bar{P}}. \quad (23)$$

2.5 Direct and Iterative Solvers for Algebraic Equation Systems

For medium-size dimensions of the equation systems (today about 1 million degrees of freedom) direct Gauß-Cholesky solvers are used, and for large dimensions different types of iterative solvers are applied, such as conjugated gradients, multigrid and domain decomposition, also in connection with parallel processing, see e.g. Braess (2001), [46].

2.6 Computation of Nodal Displacements and Stresses

With the global nodal displacement vector $\hat{U}$ the nodal displacements $u_{h,e}$ of element e follow as

$$u_{h,e}(x) = N(x)a_e\hat{U}, \quad (24)$$

as well as the stresses

$$\sigma_{h,e}(x) = \mathbb{C}B(x)a_e\hat{U}. \quad (25)$$

3 Complementary or Dual Finite Element Methods with Discretized Stresses As Well As the Hybrid Stress Method

3.1 Survey

In the development of structural analysis first the so-called "force method" was developed in the 19th and 20th century, based on the early energy theorems for linear elastic beam systems by Menabrea (1870), [47], Betti (1872), [48], and Castigliano (1879), [49]. Primal and dual energy principles based on the principles of virtual work and of complementary virtual work (virtual forces or stresses) were systematically elaborated in the 20th century by the already cited articles [9], [10], [11], [12] and [13], together with the upcoming digital computers, strongly developed and promoted by Konrad Zuse, [14], since the 1940s. This encouraged the development of more efficient solution algorithms for complex problems in structural analysis. The finite element displacement method (primal FEM) for rods and beams can be easily interpreted as a generalization of the classical displacement method, by Ostenfeld (1920), [50]; herein all adjacing beams in a node are rotated together, whereas in primal FEM the connections of the beams with a node are seperated and only reconnected by kinematic assembling the finite elements to the global system.

In the complementary alternative of primal FEM, namely the finite element force (or stress) method (complementary FEM) can be seen as a generalization of the classical force method for calculating statically overdeterminded beam systems, e.g. frames and trusses; therein the statically overdetermined system has to be weakened by dissolving virtually kinematic continuities, e.g. introducing hinges in 1D continuous beams at the supports and at the same time introducing bending moments as related statically overdetermined unknowns. The introduction of linear independent statically determined systems is rather easy for trained engineers, but it is complicated in case of automatic generation for arbitrary systems, realized in a computer program. For this purpose, one needs topology matrices for the recognition of statically determined systems, e.g. in Klöppel, Reuschling (1966), [51] and Möller, Wagemann (1966), [52]. Many attempts have been made from the 1950s to the 1970s to automize the traditional *force method* towards a rather general finite element stress method, assigned predominantly to de Veubeke (1964), [53], and Robinson (1966), [54].

3.2 *Dual or Stress Finite Element Method in Matrix Notation*

Complementary to primal FEM, the discrete principle of virtual stresses directly approximates the stresses on element level, requiring trial and test stresses which are in equilibrium within element domains, at element interfaces and at Neumann element boundaries. Thus, the input into the principle of virtual stresses is: the trial stresses have to fulfill the equilibrium conditions in all element domains, at element interfaces and at Neumann element boundaries. The output of the principle is: the approximated fulfillment of the kinematic equations in the weak sense

$$\int_{\Omega_e} \delta\boldsymbol{\epsilon}^{\mathrm{T}} (\underbrace{\boldsymbol{\nabla}_{sym} \otimes \boldsymbol{u}(\boldsymbol{\sigma}_h)}_{\boldsymbol{\epsilon}_{h,kin}} - \underbrace{\mathbb{C}^{-1}\boldsymbol{\sigma}_h}_{\boldsymbol{\epsilon}_{h,phys}}) \, \mathrm{d}\Omega = 0, \qquad (26)$$

the fulfillment of the kinematic conditions at all element interfaces

$$\int_{\Gamma_e} \delta t^{\mathrm{T}}(\boldsymbol{\sigma}_h)[\![\boldsymbol{u}_h]\!] \, \mathrm{d}\Gamma = 0 \tag{27}$$

and at all Dirichlet boundaries

$$\int_{\Gamma_{D,e}} \delta t^{\mathrm{T}}(\bar{\boldsymbol{u}} - \boldsymbol{u}) \, \mathrm{d}\Gamma = 0. \tag{28}$$

This shows that elements which fulfill these conditions are very rare and the application is practically restricted to BVPs with pure Neumann BCs, i.e. for self-equilibrated states of stresses.

The discretized trial stresses are

$$\boldsymbol{\sigma}_{h,e} = \boldsymbol{N}_\sigma(\boldsymbol{x})\hat{\boldsymbol{s}}_e = \begin{Bmatrix} \sigma_{11,h,e} \\ \sigma_{22,h,e} \\ \sigma_{12,h,e} \end{Bmatrix} = \begin{bmatrix} \boldsymbol{N}_{\sigma_{11}}^{\mathrm{T}} & & \\ & \boldsymbol{N}_{\sigma_{22}}^{\mathrm{T}} & \\ & & \boldsymbol{N}_{\sigma_{12}}^{\mathrm{T}} \end{bmatrix} \begin{Bmatrix} \hat{s}_{11,e} \\ \hat{s}_{22,3} \\ \hat{s}_{12,e} \end{Bmatrix} \tag{29}$$

with the equilibrium conditions

$$\boldsymbol{D}^{\mathrm{T}}\boldsymbol{\sigma}_{h,e} + \rho \boldsymbol{b} = \boldsymbol{0} \text{ in } \Omega_e; \text{ and } \boldsymbol{\mathcal{N}}^{\mathrm{T}}[\![\underbrace{\boldsymbol{\sigma}_{h,e}}_{\boldsymbol{t}_e^+ - \boldsymbol{t}_e^-}]\!] = \boldsymbol{0} \text{ at } \Gamma_e; \, \bar{\boldsymbol{t}} - \boldsymbol{\mathcal{N}}^{\mathrm{T}}\boldsymbol{\sigma}_{h,e} = \boldsymbol{0} \text{ at } \Gamma_{N,e}. \tag{30}$$

Correspondingly, the test stresses are

$$\delta\boldsymbol{\sigma}_{h,e} := \boldsymbol{\tau}_{h,e} = \boldsymbol{N}_\sigma(\boldsymbol{x})\delta\hat{\boldsymbol{s}}_e. \tag{31}$$

The test space for the stresses is

$$\boldsymbol{\sigma}_{h,e} = \{\boldsymbol{\sigma}_{h,e} \in \boldsymbol{\mathcal{V}}_{\sigma,h} \subset H_{div}, \, [\![\boldsymbol{t}_h]\!] = \boldsymbol{0} \text{ at } \Gamma_e, \, \bar{\boldsymbol{t}} - \boldsymbol{t} = \boldsymbol{0} \text{ at } \Gamma_{N,e}\}. \tag{32}$$

The principle of complementary virtual work (weak form of the kinematic equations) of a 2D elastic system with pure Dirichlet boundary conditions reads in discretized form, compare the principle of virtual work (18) to (21),

$$\delta \overset{*}{\mathcal{A}}_h = \cup_e \{ \int_{\Omega_e} \delta\boldsymbol{\sigma}_h^{\mathrm{T}}(\mathbb{C}^{-1}\boldsymbol{\sigma}_h - \boldsymbol{D}\boldsymbol{u}) \, \mathrm{d}\Omega \\ + \int_{\Gamma_e} \delta t^{\mathrm{T}}(\delta\boldsymbol{\sigma}_h) \underbrace{[\![\boldsymbol{u}_h(\boldsymbol{\sigma}_h)]\!]}_{\boldsymbol{u}_h^+ - \boldsymbol{u}_h^-} \, \mathrm{d}\Gamma + \int_{\Gamma_{D,e}} \delta t^{\mathrm{T}}(\bar{\boldsymbol{u}} - \boldsymbol{u}_h) \, \mathrm{d}\Gamma \} = 0. \tag{33}$$

Applying partial integration and the divergence theorem to the second term in the first integral yields

$$\delta\overset{*}{\mathcal{A}}_h = \cup_e\{\underbrace{\int_{\Omega_e} \delta\boldsymbol{\sigma}_h^\mathrm{T}(\boldsymbol{x})\mathbb{C}^{-1}\boldsymbol{\sigma}_h(\boldsymbol{x})\,\mathrm{d}\Omega}_{\overset{*}{a}_e(\boldsymbol{\sigma}_h,\delta\boldsymbol{\sigma}_h)} - \underbrace{\int_{\Gamma_{D,e}} \delta\boldsymbol{t}_h^\mathrm{T}(\boldsymbol{x})\bar{\boldsymbol{u}}\,\mathrm{d}\Gamma\}}_{\overset{*}{l}_e(\delta\boldsymbol{\sigma}_h)} \overset{!}{=} 0, \qquad (34)$$

and

$$\delta\overset{*}{\mathcal{A}}_h = \cup_e\{\delta\hat{\boldsymbol{s}}_e^\mathrm{T} \underbrace{\int_{\Omega_e} \boldsymbol{N}_\sigma^\mathrm{T}\mathbb{C}^{-1}\boldsymbol{N}_\sigma\,\mathrm{d}\Omega}_{\boldsymbol{f}_e = \boldsymbol{f}_e^\mathrm{T};\ \det \boldsymbol{f}_e \neq 0}\,\hat{\boldsymbol{s}}_e - \delta\hat{\boldsymbol{s}}_e^\mathrm{T} \underbrace{\int_{\Gamma_{D,e}} \boldsymbol{N}_\sigma^\mathrm{T}\bar{\boldsymbol{u}}\,\mathrm{d}\Gamma\}}_{\hat{\boldsymbol{q}}_e} \overset{!}{=} 0. \qquad (35)$$

The element flexibility matrix $\boldsymbol{f}_e$ has full rank, because $\boldsymbol{N}_\sigma$ consists of linear independent shape functions. In comparison to this, the element stiffness matrix $\boldsymbol{k}_e$ in the displacement method is singular because the differentiation operator $\boldsymbol{D}$ applied to the matrix $\boldsymbol{N}_u$ of linear independent displacement shape functions yields the rank decay of $\boldsymbol{k}_e$, which is equal to the admitted rigid body kinematic freedoms of the system.

We then get the weak kinematic conditions as

$$\delta\overset{*}{\mathcal{A}}_h = \cup_e\{\delta\hat{\boldsymbol{s}}_e^\mathrm{T}(\boldsymbol{f}_e\hat{\boldsymbol{s}}_e - \hat{\boldsymbol{q}}_e)\} = 0. \qquad (36)$$

Formal assembling including boundary conditions via Boolean matrices $\boldsymbol{b}_e$ and the related unknown global stress vector $\hat{\boldsymbol{S}}$, as

$$\hat{\boldsymbol{s}}_e = \boldsymbol{b}_e\hat{\boldsymbol{S}} \text{ and } \delta\hat{\boldsymbol{s}}_e = \boldsymbol{b}_e\delta\hat{\boldsymbol{S}}, \qquad (37)$$

yields the global algebraic equation system

$$\delta\overset{*}{\mathcal{A}}_h = \underbrace{\delta\hat{\boldsymbol{S}}}_{\neq 0}{}^\mathrm{T}\{\underbrace{[\cup_e(\boldsymbol{b}_e^T\boldsymbol{f}_e\boldsymbol{b}_e)]}_{\boldsymbol{F}=\boldsymbol{F}^T;\ \det \boldsymbol{F} \neq 0}\,\hat{\boldsymbol{S}} - \underbrace{\cup_e(\boldsymbol{b}_e^T\hat{\boldsymbol{q}}_e)}_{\hat{\boldsymbol{Q}}}\} = 0, \qquad (38)$$

with the global flexibility matrix $\boldsymbol{F}$ and the global given nodal displacement vector $\hat{\boldsymbol{Q}}$. The nodal stress vector follows from the system of linear algebraic equations

$$\boldsymbol{F}\hat{\boldsymbol{S}}_e = \hat{\boldsymbol{Q}} \rightsquigarrow \hat{\boldsymbol{S}} = \boldsymbol{F}^{-1}\hat{\boldsymbol{Q}}. \qquad (39)$$

Only a plane stress triangular element with quintic polynomials, i.e. $n = 3 \times 21 = 63$ parameters and the reductions for fulfilling the equilibrium conditions within the element and at the element boundaries, i.e. $\Delta n = 3 \times 6 + 2 \times 6 = 30$, in total $n_{red} = 63 - 30 = 33$ parameters, is reasonable, yielding $3 \times 3 = 9$ parameters at the corner points and $2 \times 3 \times 4 = 24$ parameters in 4 nodes in each of the 3 sides, in total $9 + 24 = 33$ nodal degrees of freedom.

3.3 Hybrid Stress Method

Dual FEM with pure stress discretizations can only be developed for special cases, usually for pure Neumann boundary, i.e. self-equilibrated systems, see Sect. 3.2. Therefore, Pian proposed in 1964 a hybrid stress method by weakening the equilibrium conditions at element interfaces and Neumann boundaries by adding them in the Lagrangian fashion to the complementary total potential energy functional, multiplied with Langrangian parameters which result – as expected – in the boundary displacements as additional unknowns, Pian (1964), [55], Pian, Tong (1964), [56].

The function spaces for stresses in element domains and displacements at element interfaces are

$$\mathcal{T}_h = \{\boldsymbol{\tau}_h = \boldsymbol{\tau}_h^T \in H_{\text{div}}(\Omega_e), \text{ div } \boldsymbol{\tau}_h + \rho \boldsymbol{b} = \boldsymbol{0} \text{ in } \Omega_e\}; \mathcal{T}_h \subset \mathcal{T}, \tag{40a}$$

$$\mathcal{V}_h = \{\boldsymbol{v}_h \in H^1(\Gamma_e)\}; \mathcal{V}_h \subset \mathcal{V}. \tag{40b}$$

The variation of the complementary energy functional is

$$\bigcup_e \left\{ \underbrace{\{\delta \hat{\boldsymbol{s}}_e^T \; \delta \hat{\boldsymbol{u}}_e^T\}}_{\neq 0} \underbrace{\left(\begin{bmatrix} \boldsymbol{f}_e & -\boldsymbol{l}_e \\ -\boldsymbol{l}_e^T & 0 \end{bmatrix} \begin{Bmatrix} \hat{\boldsymbol{s}}_e^T \\ \hat{\boldsymbol{u}}_e^T \end{Bmatrix} - \begin{Bmatrix} \hat{\boldsymbol{u}}_{D,e} \\ \hat{\boldsymbol{t}}_{N,e} \end{Bmatrix} \right)}_{\stackrel{!}{=}0} \right\} \stackrel{!}{=} 0, \tag{41}$$

with the nodal stress vector $\hat{\boldsymbol{s}}_e$, the nodal displacement vector $\hat{\boldsymbol{u}}_e$ and further the element flexibility matrix $\boldsymbol{F}_e$ (like for the pure stress method) and the Lagrangian matrix $\boldsymbol{l}_e$ with shape functions at the interfaces only. They are defined as

$$\boldsymbol{f}_e = \int_{\Omega_e} \boldsymbol{N}_\sigma^T(\boldsymbol{x}) \mathbb{C}^{-1} \boldsymbol{N}_\sigma \mathrm{d}\Omega \; ; \; \boldsymbol{l}_e = \int_{\Gamma_e} [\boldsymbol{N}_\sigma^T(\boldsymbol{x}) \mathcal{N}(\boldsymbol{x})] \boldsymbol{N}_u(\boldsymbol{x}) \mathrm{d}\Gamma. \tag{42}$$

The matrix of shape functions for the stresses is

$$\boldsymbol{N}_\sigma(\boldsymbol{x}) = \begin{bmatrix} \boldsymbol{\Phi}_{\sigma_{11}}^T(\boldsymbol{x}) & & \\ & \boldsymbol{\Phi}_{22}^T(\boldsymbol{x}) & \\ & & \boldsymbol{\Phi}_{21}^T(\boldsymbol{x}) \end{bmatrix} ; \boldsymbol{\Phi}_{\sigma j} = \{\Phi_1 \; \Phi_2 \; \Phi_k \ldots \Phi_n\}. \tag{43}$$

The matrix $\mathcal{N}$ of direction cosines at element interfaces, equ. (14), has the same index structure as the differential operator matrix, $\boldsymbol{D}$, equ. (13).

Due to the regularity of $\boldsymbol{f}_e$, the unknown nodal stresses can be eliminated on element level. The first equation yields $\hat{\boldsymbol{s}}_e = \boldsymbol{f}_e^{-1} \boldsymbol{l}_e \hat{\boldsymbol{u}}_e + \boldsymbol{f}_e^{-1} \hat{\boldsymbol{u}}_{D,e}$, and inserted in the second one results in

$$\underbrace{\boldsymbol{l}_e^{\mathrm{T}} \boldsymbol{f}_e^{-1} \boldsymbol{l}_e}_{\overset{*}{\boldsymbol{k}}_e = \overset{*}{\boldsymbol{k}}_e^{\mathrm{T}}} \hat{\boldsymbol{u}}_e + \underbrace{\boldsymbol{l}_e^{\mathrm{T}} \boldsymbol{f}_e^{-1} \hat{\boldsymbol{u}}_{D,e} + \hat{\boldsymbol{t}}_{N,e}}_{\hat{\boldsymbol{t}}^*} = \boldsymbol{0}. \quad (44)$$

$\overset{*}{\boldsymbol{k}}_e$ can be called a hybrid element stiffness matrix. Assembling to the global hybrid stiffness matrix is performed for the unknown nodal displacements as in primal FEM, eqs. (21) and (22).

A Hybrid Rectangular Plane Stress Element

The ansatz of trial stress functions is $\sigma_{x,h} = \beta_1 + \beta_2 y$; $\sigma_{y,h} = \beta_3 + \beta_4 x$; $\tau_{xy,h} = \beta_5$, which fulfill the homogeneous equilibrium conditions in Ω_e. Then the boundary tractions are got via the direction cosines at element boundaries. The trial interface displacements are linear polynomials according to the Lagrangian parameters.

The Lagrangian coupling matrix $\boldsymbol{l}_e$ with integrals over the element boundaries is $\boldsymbol{l}_e = \int_{\Gamma_e} (\boldsymbol{N}_\sigma^{\mathrm{T}}(x,y) \boldsymbol{\mathcal{N}}) \boldsymbol{N}_u(x,y) \, \mathrm{d}s$.

The element flexibility matrix $\boldsymbol{f}_e = \int_{\Omega_e} \boldsymbol{N}_\sigma^{\mathrm{T}} \mathbb{C}^{-1} \boldsymbol{N}_\sigma \, \mathrm{d}\Omega$ follows as in the pure stress method.

Various articles on hybrid finite elements for plates and shells were published from the 1960th to the 1980th, but these developments were not continued because of complicated shape functions for general element geometries with the necessity to fulfill the equilibrium conditions in the element domains.

A better access is Trefftz's method, [15], especially the use of Beltrami stress functions, [17], or other variants of equilibrated stresses.

4 Mixed Finite Elements

4.1 General Aspects

Dual mixed and hybrid dual mixed FEMs are required, or at least mathematically advocated, in case of non-robust primal problems for which a global interpolation constant C in the a priori discretization error in the energy norm, $|||\boldsymbol{e}||| = Ch^p$; $\boldsymbol{e} := \boldsymbol{u} - \boldsymbol{u}^h$, h a characteristic element length and p the polynomial degree, does not exist, see Part II in the following chapter.

This mixed discretization method is usually derived from the Hellinger-Prange-Reissner two-field variational functional, Hellinger (1914), [57], Prange (1916), [58], Reissner (1950), [59], and yields a saddle point problem for trial and test stresses $\boldsymbol{\tau}_h \in H(\mathrm{div}, \Omega) := \{\boldsymbol{\tau}_h \in L_2(\Omega); \mathrm{div}\, \boldsymbol{\tau}_h \in L_2(\Omega)\}$, as well as trial and test displacements $\boldsymbol{u}_h \epsilon H_0^1(\Omega_e)$, e.g. with the same polynomial orders for stresses and displacements. The advantage of direct stress approximation (in difference to primal FEM where the stresses are derivatives of the approximated displacements multiplied with the elasticity constants) is affected by the only approximated fulfillment of the kinematic equations $\varepsilon_h^{phys} \stackrel{!}{=} \varepsilon_h^{geom}$; $\varepsilon_h^{phys} := \mathcal{C}^{-1}\boldsymbol{\sigma}_h$; $\varepsilon_h^{geom} := \boldsymbol{\nabla}_{sym} \otimes \boldsymbol{u}_h$.

Another general access to mixed methods comes from the Legendre transformation, Legendre (1811/17), [60], as introduced in the Euler-Lagrange equations of motion, extremizing the Lagrangian functional $\int_{t_0}^{t_1} \mathcal{L}(\boldsymbol{x},\dot{\boldsymbol{x}},t) \, \mathrm{d}t$ by the stationarity condition $\partial \mathcal{L}/\partial \boldsymbol{x} - \mathrm{d}/\mathrm{d}t(\partial \mathcal{L}/\partial \dot{\boldsymbol{x}}) = 0$. Introducing the Legendre transformation $\boldsymbol{p} := \partial \mathcal{L}/\partial \dot{\boldsymbol{x}}$ yields $\boldsymbol{p} = -\partial H/\partial \dot{\boldsymbol{x}}$ with $\partial H/\partial \boldsymbol{p} = \dot{\boldsymbol{x}}$ where $H = \dot{\boldsymbol{x}} \cdot \boldsymbol{p} - \mathcal{L}$.

Furthermore, the Prager-Synge hypercircle theorem is a basis for mixed functionals and their a priori and a posteriori error analysis, Prager, Synge (1949), [61], Synge (1957), [62].

Important examples are nearly incompressible elastic materials and thin-walled plates, shells and beams with bending and transverse shear stresses, using the geometrical hypothesis of plane rotated cross-sections, i.e. without the normality rule with respect to the tangents of the mid-surface, yielding the Timoshenko beam theory and the Reissner-Mindlin plate and shell theories.

With engineering intuition, using bilinear shape functions for the plate problem with lateral displacements w and rotation angles ϕ_x and ϕ_y of the cross-section traces (although ϕ_x, ϕ_y are in essence derivatives of w) are used. For this type of problems shear-locking has to be regarded as a numerical instability problem, resulting in the global infsup-condition by Babuška (1973), [63], and Brezzi (1974), [64], a global stability condition for the related saddle point problem, for which elementwise conditions are necessary but not sufficient, Fig. 15.

Fig. 15 Franco Brezzi (*1945), an eminent mathematician, cooperating with engineers, who fundamentally contributed to the mathematics of mixed finite elements and to generalized FEM, especially by deriving the infsup conditions for numerical stability for two- and three-field variational functionals

Reduced integration of the shear terms in the element stiffness matrices for avoiding shear locking and getting their correct rank was developed with several variants, [24]. Also element patch tests were developed for checking numerical stability of distorted elements for well-posed structural problems. They are mathematically necessary but not sufficient because the infsup-condition holds for the whole static system.

Some important classes of dual-mixed and hybrid dual-mixed finite elements were developed. Most of these elements allow the elimination of stresses on element level such that the assembling process of elements at element nodes to the system is purely kinematic and thus performed like for a primal displacement method; this is of importance for implementing these elements into existing finite element program systems.

4.2 The Hellinger-Prange-Reissner Two-Field Functional and Its Discretization by a Dual Mixed Finite Element Method

The (one-field) displacement FEM has the significant drawback that the approximation of the stress field usually has jumps on element interfaces, since it is obtained from derivatives of $\mathcal{C}^0$-continuous displacements at element interfaces.

Moreover, low order displacement elements may provide robustness problems with non-existing global interpolation constants, showing up as numerical instabilities, e.g. as locking phenomena in case of nearly incompressible elastic materials or in case of Timoshenko beam theory and Reissner-Mindlin plate and shell theories with approximated constant transverse shear deformations over the thickness. It can be shown that in general robustness problems require mixed stress and displacement approximations resulting in saddle point problems with global infsup-conditions for achieving numerical stability.

It has to be mentioned that many engineering attempts have been made for overcoming the global infsup-conditions by applying reduced integration as well as by element-wise counting and balancing the nodal stresses and displacements, also using various patch tests. But these techniques can not fully replace the infsup-condition for the mixed functional of the total system in general, Zienkiewicz et al. (1986), [65].

As already shown in the previous Sect. 3 for the dual and hybrid dual FEM, a mixed FEM based on primal FEM is not advantageous for mixed generalizations because the element stiffness matrices have reduced rank and thus can not be inverted and eliminated on element level. The elimination of the stresses before assembling yields the global algebraic equation system with only unknown nodal displacements. Thus, this type of elements can be easily implemented into a program system for primal FEM.

The dual mixed FEM can also be understood as a generalization of the hybrid stress method or as a method sui generis, in order to approximate both, stresses and displacements in the element domains. This can be achieved either by a Lagangian extension of the complementary energy functional or via the Legendre transformation of the complementary energy. The elements of the Lagrangian coupling matrices are integrals over the finite element domains, i.e. not only on element interfaces as for the hybrid stress method. From mechanical point of view, the equilibrium conditions for trial and test stresses σ and τ are a priori fulfilled in the total domain, whereas the kinematic conditions $\epsilon_{kin}(v) = \epsilon_{phys}(\tau)$; $\epsilon(\tau) = \mathbb{C}^{-1}\tau$ are only approximately fulfilled in the weak sense of variational calculus, see [57], [58], [59], and Prager (1968), [66].

The complementary volume-specific strain-energy functional $\overset{*}{u}(\sigma)$, with the functional of the system $\overset{*}{\mathcal{U}}(\sigma) = \int_\Omega \overset{*}{u}(\sigma)\,d\Omega$, is defined as the Legendre transformation, again in matrix notation

$$\overset{*}{u}(\sigma) := \sigma^T\epsilon_{phys} - u(\epsilon_{geom}) \text{ with } \epsilon_{geom} := Du; \; \overset{*}{u} = \frac{1}{2}\sigma^T\mathbb{C}^{-1}\sigma, \quad (45)$$

for which the total differential

$$du(\epsilon_{geom}) = d\sigma^T \frac{\partial u}{\partial \epsilon_{geom}} = d\sigma^T \mathbb{C}\epsilon \quad (46)$$

exists, and thus the path independence of the integral holds. Then also $\overset{*}{u}(\sigma)$ has the total differential

$$d\overset{*}{u} = d\sigma^T \epsilon_{phys}; \; \epsilon_{phys} = \frac{\partial \overset{*}{u}}{\partial \sigma} = \mathbb{C}^{-1}\sigma, \quad (47)$$

and recalling (45), the specific stress energy functional takes the form

$$\overset{*}{u}(\sigma) = \frac{1}{2}\sigma^T\mathbb{C}^{-1}\sigma; \; \mathbb{C}^{-1} =: \overset{*}{\mathbb{C}}. \quad (48)$$

The Hellinger-Reissner functional for the test stresses $\tau(x)$ and the test displacements $v(x)$ is defined for a continuous system with mixed boundary conditions as

$$\mathcal{F}_{HR}(\tau, v) := \int_\Omega [\tau^T\epsilon(v) - \frac{1}{2}\tau^T\mathbb{C}^{-1}\tau]\,d\Omega + \Pi_{ext}(v);$$

$$\Pi_{ext}(v) = -\int_\Omega v^T\rho b\,d\Omega - \int_{\Gamma_N} v^T\bar{t}\,d\Gamma. \quad (49)$$

Different from the previous Sect. 3 we chose the shorter notation $\tau \equiv \delta\sigma$ for the test stresses (virtual stresses) and $v \equiv \delta u$ for the test displacements (virtual displacements) in the domain Ω. They have to fulfill the following conditions:

$\boldsymbol{\tau} \in \mathcal{T}$; $\boldsymbol{\tau}(\boldsymbol{x}) = \boldsymbol{\tau}^{\mathrm{T}}(\boldsymbol{x})$ is square-integrable in Ω, $\boldsymbol{\tau} = \boldsymbol{0}$ at Γ_N; $\boldsymbol{v} \in \mathcal{V}$, $\boldsymbol{v}$ is $\mathcal{C}^1$-continous in Ω and zero at Dirichlet boundaries, $\boldsymbol{v} = \boldsymbol{0}$ at Γ_D.

The stresses $\boldsymbol{\sigma}(\boldsymbol{x})$ and displacements $\boldsymbol{u}(\boldsymbol{x})$ are determined for the condition that the functional (49) becomes stationary for the saddle point problem

$$\mathcal{F}_{HR}(\boldsymbol{\sigma}, \boldsymbol{u}) = \inf_{\boldsymbol{\tau} \in \mathcal{T}} \sup_{\boldsymbol{v} \in \mathcal{V}} \mathcal{F}_{HR}(\boldsymbol{\tau}, \boldsymbol{v}). \tag{50}$$

As the necessary condition the first variation has to be zero, as

$$\delta \mathcal{F}_{HR} = \delta_{\boldsymbol{\tau}} \mathcal{F}_{HR} + \delta_{\boldsymbol{v}} \mathcal{F}_{HR} \stackrel{!}{=} 0, \; \delta \mathcal{F}_{HR} := \boldsymbol{\tau}^{\mathrm{T}} \left. \frac{\partial \mathcal{F}_{HR}}{\partial \boldsymbol{\tau}} \right|_{\boldsymbol{\tau} = \boldsymbol{\sigma}} + \boldsymbol{v}^{\mathrm{T}} \left. \frac{\partial \mathcal{F}_{HR}}{\partial \boldsymbol{v}} \right|_{\boldsymbol{v} = \boldsymbol{u}} \stackrel{!}{=} 0, \tag{51}$$

yielding

$$\delta_{\boldsymbol{\tau}} \mathcal{F}_{HR} = \int_{\Omega} \boldsymbol{\tau}^{\mathrm{T}} [\underbrace{\boldsymbol{Du}}_{\boldsymbol{\epsilon}_{geom}} - \underbrace{\mathbb{C}^{-1} \boldsymbol{\sigma}}_{\boldsymbol{\epsilon}_{phys}}] \, \mathrm{d}\Omega \stackrel{!}{=} 0 \; \forall \boldsymbol{\tau} \in \mathcal{T}, \tag{52a}$$

$$\delta_{\boldsymbol{v}} \mathcal{F}_{HR} = \int_{\Gamma_N} \boldsymbol{v}^{\mathrm{T}} (\underbrace{\boldsymbol{N}^{\mathrm{T}} \boldsymbol{\sigma}}_{\boldsymbol{t}}) \, \mathrm{d}\Gamma - \int_{\Omega} (\boldsymbol{v}^{\mathrm{T}} \boldsymbol{D}^{\mathrm{T}}) \boldsymbol{\sigma} \, \mathrm{d}\Omega - \int_{\Omega} \boldsymbol{v}^{\mathrm{T}} \rho \boldsymbol{b} \, \mathrm{d}\Omega - \int_{\Gamma_N} \boldsymbol{v}^{\mathrm{T}} \bar{\boldsymbol{t}} \, \mathrm{d}\Gamma$$

$$= -\int_{\Omega} \boldsymbol{v}^{\mathrm{T}} [\boldsymbol{D}^{\mathrm{T}} \boldsymbol{\sigma} + \rho \boldsymbol{b}] \, \mathrm{d}\Omega + \int_{\Gamma_N} \boldsymbol{v}^{\mathrm{T}} [\boldsymbol{t} - \bar{\boldsymbol{t}}] \, \mathrm{d}\Gamma \stackrel{!}{=} 0 \; \forall \boldsymbol{v} \in \mathcal{V}, \tag{52b}$$

where in (52b) partial integration and the divergence theorem were applied to the first term $\int_{\Omega} \boldsymbol{v}^{\mathrm{T}} (\boldsymbol{D}^{\mathrm{T}} \boldsymbol{\sigma}) \, \mathrm{d}\Omega$.

The square brackets in (52a, b) are the weak conditions for the kinematic equations, i.e. $\boldsymbol{\epsilon}_{phys} = \boldsymbol{\epsilon}_{geom}$, but in difference to the dual functional $\overset{*}{\mathcal{F}}$ and the hybrid dual functional $\overset{*}{\mathcal{F}}_{hyb}$, the test stresses do not need to fulfill the equilibrium conditions in the domain; they are fulfilled weakly as can be seen from (52b). The same holds for the equilibrium condition at Neumann boundary. Thus, the directly approximated fulfillment of the equilibrium conditions by the Hellinger-Reissner functional is obtained together with the fact that the kinematic conditions are also approximately fulfilled only. As a 1D example, using linear shape functions for displacements and stresses, the dual mixed FEM (Hellinger-Reissner) yields linear displacements and linear stresses but constant geometrical and linear physical strains; but using linear displacements for primal FEM we get linear displacements and constant stresses, combined with constant geometrical and physical strains.

Algebraic Structure of Linear Mixed Finite Element Methods
The discretized dual mixed formulation, e.g. by Oden, Carey (1983), [67], reads: find $(\boldsymbol{u}_h, \boldsymbol{p}_h) \in H_h \times Q_h$, such that

$$a(\boldsymbol{u}_h,\boldsymbol{v}_h)+b(\boldsymbol{v}_h,\boldsymbol{p}_h) = f(\boldsymbol{v}_h) \; \forall \boldsymbol{v}_h \in \mathcal{V}_h$$
$$b(\boldsymbol{u}_h,\boldsymbol{q}_h) = g(\boldsymbol{q}_h) \; \forall \boldsymbol{q}_h \in \mathcal{Q}_h, \tag{53}$$

and the related discrete Babuška-Brezzi condition, Babuška (1973), [68], Brezzi (1974), [64], for numerical stability of the saddle point problem is

$$\exists \beta_h > 0, \text{ such that sup } \frac{|b(\boldsymbol{v}_h,\boldsymbol{q}_h)|}{||\boldsymbol{v}_h||_H} \geq \beta_h ||[\boldsymbol{q}_h]||_{\mathcal{Z}_h} \; \forall \boldsymbol{q}_h \in \mathcal{Q}_h; \mathcal{Z}_h = \mathcal{Q}_h/\ker\mathcal{B}_h. \tag{54}$$

The resulting global linear discretized algebraic equation system of a mixed finite element method has the form

$$\begin{bmatrix} \boldsymbol{0} & \boldsymbol{L} \\ \boldsymbol{L}^{\mathrm{T}} & \boldsymbol{F} \end{bmatrix} \begin{Bmatrix} \hat{\boldsymbol{U}} \\ \hat{\boldsymbol{S}} \end{Bmatrix} = \begin{Bmatrix} \hat{\boldsymbol{P}} \\ \hat{\boldsymbol{Q}} \end{Bmatrix}, \tag{55}$$

with $\boldsymbol{F} = \boldsymbol{F}^{\mathrm{T}}$, $\det \boldsymbol{F} \neq 0$ the global flexibility matrix, $\boldsymbol{L}$ the rectangular row-regular matrix according to the Lagrangian multipliers, $\hat{\boldsymbol{S}}$ the global unknown nodal stress values which can be eliminated, $\hat{\boldsymbol{U}}$ the element-interface displacements, $\hat{\boldsymbol{P}}$ the global energy-equivalent nodal vector of physical loads and $\hat{\boldsymbol{Q}}$ the load term at the boundary.

For stable solvability, it is not sufficient that the matrix in (55) is non-singular. Moreover, the continuous dependence of the solutions upon the data requires the inequality, see [64], and Auricchio, Brezzi, Lovadina (2004), [69],

$$\|\hat{\boldsymbol{S}}\| + \|\hat{\boldsymbol{U}}\| \leq C(\|\hat{\boldsymbol{0}}\| + \|\hat{\boldsymbol{P}}\|). \tag{56}$$

The stability analysis ends up in the global infsup condition (71) in subsection 4.3.

4.3 The Hellinger-Reissner Dual Mixed FEM for Linear Elastic Boundary Value Problems

As in the previous subsections we proceed from the continuous problem to a discretized finite element system. Then, from (49) we arrive at

$$\mathcal{F}_{HR,h}(\boldsymbol{\tau}_h,\boldsymbol{v}_h) = \bigcup_e \{ \int_{\Omega_e} \boldsymbol{\tau}_h^{\mathrm{T}}(\boldsymbol{D}\boldsymbol{v}_h) \, d\Omega - \frac{1}{2} \int_{\Omega_e} \boldsymbol{\tau}_h^{\mathrm{T}} \overset{*}{\mathbb{C}} \boldsymbol{\tau}_h \, d\Omega - \int_{\Omega_e} \boldsymbol{v}_h^{\mathrm{T}} \rho \boldsymbol{b} \, d\Omega$$
$$- \int_{\Gamma_{N,e}} \boldsymbol{v}_h^{\mathrm{T}} \bar{\boldsymbol{t}} \, d\Gamma \} \to \underset{\boldsymbol{\tau}_h,\boldsymbol{v}_h}{\text{stat.}}, \tag{57}$$

with the test space for the stresses

$$\mathcal{T}_h = \{\boldsymbol{\tau}_h = \boldsymbol{\tau}_h^{\mathrm{T}} \in H_{\mathrm{div}}(\Omega_e)\}; \ \mathcal{T}_h \subset \mathcal{T}, \tag{58}$$

and for the displacements

$$\mathcal{V}_h = \{\boldsymbol{v}_h \in H^1(\Omega_e), \boldsymbol{v}_h = \boldsymbol{0} \text{ at } \Gamma_{D,e}, [\![\boldsymbol{v}_h = \boldsymbol{0}]\!] \text{ at } \Gamma_e\}; \ \mathcal{V}_h \subset \mathcal{V}. \tag{59}$$

The first variation results in, compare (52),

$$\delta_{\boldsymbol{\tau}_h}\mathcal{F}_{HR} = \bigcup_e \left[\underbrace{\int_{\Omega_e} \boldsymbol{\tau}_h^{\mathrm{T}}(\boldsymbol{D}\boldsymbol{u}_h) \, \mathrm{d}\Omega}_{-b_e(\boldsymbol{\tau}_h,\boldsymbol{u}_h)} - \underbrace{\int_{\Omega_e} \boldsymbol{\tau}_h^{\mathrm{T}} \overset{*}{\mathbb{C}} \boldsymbol{\sigma}_h \, \mathrm{d}\Omega}_{-a_e(\boldsymbol{\tau}_h,\boldsymbol{\sigma}_h)} \right] \overset{!}{=} 0, \tag{60a}$$

$$\delta_{\boldsymbol{v}_h}\mathcal{F}_{HR} = \bigcup_e \left[\underbrace{\int_{\Omega_e} \boldsymbol{v}_h^{\mathrm{T}}(\boldsymbol{D}^{\mathrm{T}}\boldsymbol{\sigma}) \, \mathrm{d}\Omega}_{-b_e^{\mathrm{T}}(\boldsymbol{v}_h,\boldsymbol{\sigma}_h)} - \underbrace{\int_{\Omega_e} \boldsymbol{v}_h^{\mathrm{T}} \rho \boldsymbol{b} \, \mathrm{d}\Omega - \int_{\Gamma_{N,e}} \boldsymbol{v}_h^{\mathrm{T}} \bar{\boldsymbol{t}} \, \mathrm{d}\Gamma}_{F_e(\boldsymbol{v}_h)} \right] \overset{!}{=} 0, \tag{60b}$$

$$\bigcup_e \{a_e(\boldsymbol{\tau}_h, \boldsymbol{\sigma}_h) + b_e(\boldsymbol{\tau}_h, \boldsymbol{u}_h)\} = 0 \ \forall \boldsymbol{\tau}_h \in \mathcal{T}_h \subset \mathcal{T}, \tag{61a}$$

$$\bigcup_e \{b_e^{\mathrm{T}}(\boldsymbol{v}_h, \boldsymbol{\sigma}_h) + F_e(\boldsymbol{v}_h)\} = 0 \ \forall \boldsymbol{v}_h \in \mathcal{V}_h \subset \mathcal{V}. \tag{61b}$$

Discrete test and trial functions are introduced as tensor products of Lagrangian or Legendre polynomials with unknown nodal values, as

$$\boldsymbol{\tau}_h(\boldsymbol{x}) = \boldsymbol{N}_\sigma(\boldsymbol{x})\delta\hat{\boldsymbol{s}}_e, \ \boldsymbol{\sigma}_h(\boldsymbol{x}) = \boldsymbol{N}_\sigma(\boldsymbol{x})\hat{\boldsymbol{s}}_e \text{ in } \Omega_e, \tag{62a}$$

$$\boldsymbol{v}_h(\boldsymbol{x}) = \boldsymbol{N}_u(\boldsymbol{x})\delta\hat{\boldsymbol{u}}_e, \ \boldsymbol{u}_h(\boldsymbol{x}) = \boldsymbol{N}_u(\boldsymbol{x})\hat{\boldsymbol{u}}_e \text{ in } \Omega_e, \tag{62b}$$

with the shape functions

$$\boldsymbol{N}_\sigma(\boldsymbol{x}) = \begin{bmatrix} \boldsymbol{\phi}_{\sigma_1 1}^{\mathrm{T}}(\boldsymbol{x}) & & & \\ & \boldsymbol{\phi}_{\sigma_2 2}^{\mathrm{T}}(\boldsymbol{x}) & & \\ & & \ddots & \\ & & & \boldsymbol{\phi}_{\sigma_3 1}^{\mathrm{T}}(\boldsymbol{x}) \end{bmatrix}, \tag{63}$$

and

$$\boldsymbol{N}_u(\boldsymbol{x}) = \begin{bmatrix} \boldsymbol{\phi}_{u_1}^{\mathrm{T}}(\boldsymbol{x}) & & \\ & \boldsymbol{\phi}_{u_2}^{\mathrm{T}}(\boldsymbol{x}) & \\ & & \boldsymbol{\phi}_{u_3}^{\mathrm{T}}(\boldsymbol{x}) \end{bmatrix}. \tag{64}$$

We then get the system of algebraic equations, multiplying (60a) with -1

$$\bigcup_e \left\{ \begin{Bmatrix} \delta \hat{s}_e^T & \delta \hat{u}_e^T \\ \neq 0 & \neq 0 \end{Bmatrix} \left(\begin{bmatrix} \underbrace{\int_{\Omega_e} N_\sigma^T \overset{*}{C} N_\sigma \, d\Omega}_{f_e = f_e^T;\ \det f_e \neq 0} & \underbrace{-\int_{\Omega_e} N_\sigma^T D N_u \, d\Omega}_{l_e} \\ \underbrace{-\int_{\Omega_e} N_u^T D^T N_\sigma \, d\Omega}_{l_e^T} & 0 \end{bmatrix} \begin{Bmatrix} \hat{s}_e \\ \hat{u}_e \end{Bmatrix} \right. \right.$$

$$\left. \left. - \begin{Bmatrix} 0 \\ \underbrace{\int_{\Omega_e} N_u^T \rho b \, d\Omega + \int_{\Gamma_{N,e}} N_u^T \bar{t} \, d\Gamma}_{\hat{\bar{p}}_e} \end{Bmatrix} \right) \right\} = 0, \tag{65}$$

in which the terms within the large round brackets have to become zero. Using the above notations for the matrices and vectors we arrive at the global mixed FE system

$$\bigcup_e \left\{ \begin{bmatrix} f_e & l_e \\ l_e^T & 0 \end{bmatrix} \begin{Bmatrix} \hat{s}_e \\ \hat{u}_e \end{Bmatrix} - \begin{Bmatrix} 0 \\ \hat{\bar{p}}_e \end{Bmatrix} \right\} = \begin{Bmatrix} 0 \\ 0 \end{Bmatrix}, \tag{66}$$

from where we eliminate the nodal stresses on element level according to $\hat{s}_e = f_e^{-1} l_e \hat{u}_e$ and resulting in

$$\underbrace{l_e^T f_e^{-1} l_e}_{\overset{*}{k}_e} \hat{u}_e = \hat{\bar{p}}_e. \tag{67}$$

Then the assembling of the elements to the global system is purely kinematic, as in primal FEM, using the Boolean matrices a_e and the global reduced displacement vector $\hat{U}$ as

$$\hat{u}_e = a_e \hat{U}, \tag{68}$$

yielding

$$\underbrace{\sum_e (a_e^T \overset{*}{k}_e a_e)}_{\overset{*}{K} = \overset{*}{K}^T;\ \det \overset{*}{K} \neq 0} \hat{U} = \underbrace{\sum_e a_e \hat{\bar{p}}_e}_{\hat{P}}, \tag{69}$$

from which the unknown global nodal displacement vector $\hat{U}$ is calculated by solving the linear algebraic equation system as

$$\hat{U} = \overset{*}{K}{}^{-1} \hat{P}. \tag{70}$$

It has to be noted that in order to prove the existence and uniqueness of the solution $(\sigma, u) \in \mathcal{T} \subset \mathcal{V}$, apart from the ellipticity of the bilinear form a, the numerical stability condition of the saddle point problem has to be proven; this is the so-called BB-inf-sup-condition (BB: Brezzi and Babuška), [64], [63], Brezzi, Fortin (1991), [70], for the bilinear form $b_e(\tau_h, u_h)$, (60a,b) and (61a,b), i.e. the Lagrangian coupling matrix for the dual mixed stress and displacement ansatz. This condition reads

$$\inf_{v \in \mathcal{V}} \sup_{\tau \in \mathcal{T}} \frac{b(\tau, v)}{\|\tau\|_\mathcal{T} \|v\|_\mathcal{V}} \geq \beta, \tag{71}$$

with a positive constant β according to Korn's second inequality.

The Hellinger-Reissner mixed variational functional reads

$$\mathcal{F}_{HR,h}(\tau_h, v_h) = \bigcup_e \{\frac{1}{2}a_e(v,v) - F(v) - \int_{\Omega_e} \lambda^{\mathrm{T}}[\epsilon_{geom}(v) - \mathbb{C}^{-1}\tau]\,\mathrm{d}\Omega\}, \tag{72}$$

with

$$\lambda = \tau \tag{73}$$

resulting from the first variation of this functional.

Notable early contributions to mixed plate and shell elements were given by Prager (1968), [66], Herrmann (1967), [71], Prato (1969), [72], and Bathe, Dvorkin (1985), [73].

Some famous mixed elements are the Crouzeix-Raviart element, Crouzeix, Raviart (1973), [74], the Raviart-Thomas element, Raviart, Thomas (1977), [75], the PEERS element by Stenberg (1984), [76], Arnold, Brezzi, Douglas (1984), [77], Arnold, Douglas, Gupta (1984), [78], and the class of the BDM-elements, Brezzi, Douglas, Marini (1985), [79], Brezzi, Douglas, Fortin, Marini (1987), [80]. Brezzi and Fortin wrote the standard textbook on mixed elements in 1991, [70], see also [69].

4.4 Numerical Locking Phenomena in Finite Elements, Especially of Thin-Walled Elastic Structures

The Hellinger-Reissner-type mixed finite element formulation is of particular importance for rigorous numerical stability proofs of finite element discretizations, especially for low order trial and test functions. This gives rise to locking phenomena and/or spurious strains in case of shear-elastic beam-, plate- and shell-theories of Timoshenko and Reissner-Mindlin type with approximated constant transverse shear strains over the thickness as well as spurious strains in case of nearly incompressible elastic deformations. In the first case the approximated differential equations of 4th order in 2D domains for thin plates and shells with smooth solutions are reduced to 2nd order PDEs with independent variables for displacements and rotations of the cross-section normals. Discretizing both with linear shape functions leads to shear-locking and thus ill-conditioning with wrong rank of the element stiffness matrix, indicating stiff differential equations. Shear-locking also appears, if thin-walled structures are modeled with the 2nd order Lamé PDEs in 3D domains with low order discretizations, i.e. independent from modeling by PDEs within approximated 2D or 3D spatial domains.

An extensive literature is available for overcoming locking phenomena, e.g. by Zienkiewicz and Taylor (2000), [24], and Bischoff et al. (2004), [81], summarizing and evaluating the following stabilization techniques:

- Reduced integration of transverse shear terms for avoiding artificial zero-energy modes: Zienkiewicz, Taylor and Too (1973), [82], Hughes (1987), [26], Belytschko, Liu and Moran (2000), [83].
- Assumed strain methods with adequate interpolations of transverse shear strains: MacNeal (1978), [84], Hughes and Tezduyar (1981), [85], extended to the MITC elements by Bathe and Dvorkin (1985, 1986), [73], [86], Bathe, Brezzi and Cho (1989), [87], Bathe, Brezzi, Fortin (1989), [88].
- B-bar techniques with changes of the kinematic relations, i.e. the B-operator matrix of partial derivatives by Hughes and Liu (1981), [89].
- Enhanced assumed strain (EAS) method by Simo and Rifai (1990), [90], with the variant of linked interpolations by Tessler and Huges (1985), [91], see also Taylor and Auricchio (1993), [92], and Bischoff and Taylor (2001), [93].

Of course, complicated locking phenomena also appear in case of finite rotations and possibly additional finite deformations of membrane or/and bending-dominated shells, discretized with low order finite elements. For stabilization of those elements Armero (2000), [94], and Reese (2003), [95], made substantial contributions on an engineering basis. A sound mathematical research is still missing.

Remark

Once more it should be pointed out that in case of non-robust BVPs a global interpolation constant in the explicit error estimator does not exist, and only a mixed variational formulation with the compulsory global infsup condition, equ. (71) has to be fulfilled.

Nevertheless, the published sophisticated engineering techniques cited above are of great importance for engineering applications, and they allow to get mechanical and numerical insight into the complicated interrelationships.

5 The Three-Field Hu-Washizu Variational Functional

5.1 *The Hu-Washizu Functional for Linear Elastic BVPs*

As we have seen, however, the mixed approach presented does not yield a continuous stress field in case of low order shape functions. Furthermore, with the Hellinger-Reissner type mixed method we still encounter difficulties in nearly incompressible elastic materials. To cope with these problems, another pairing of test and solution spaces is required, namely $\mathcal{H}(\mathrm{div}, \Omega)$ and $\mathcal{L}_2(\Omega)$ instead of $\mathcal{L}_2(\Omega)$ and $\mathcal{V}$. Therefore, a three-field variational formulation was proposed by Hu and Washizu in 1955, Washizu (1968), [96].

Rather than two independent unknown fields $\boldsymbol{\sigma} \in \mathcal{T}$ and $\boldsymbol{u} \in \mathcal{V}$, as used in the Hellinger-Reissner functional, now an additional unknown strain field $\boldsymbol{\epsilon} \in \mathcal{E}$ is introduced, which leads to the Hu-Washizu functional

$$\mathcal{F}_{HW} : \mathcal{E} \times \mathcal{V} \times \mathcal{T} \to \mathbb{R}, \tag{74}$$

with $\mathcal{E} = \mathcal{T} = [\mathcal{L}_2(\Omega)]^{3\times 3}$ in the simplest case, whereas furthermore $\mathcal{V} = [\mathcal{H}^1(\Omega)]^{3\times 3}$ must hold. The field equations follow from three variational conditions for the existence of a saddle point problem, namely $D_\epsilon \mathcal{F}_{HW}(\boldsymbol{\epsilon},\boldsymbol{u},\boldsymbol{\sigma}) \cdot \boldsymbol{\eta} = 0$, $D_{\boldsymbol{u}} \mathcal{F}_{HW}(\boldsymbol{\epsilon},\boldsymbol{u},\boldsymbol{\sigma}) \cdot \boldsymbol{v} = 0$ and $D_{\boldsymbol{\sigma}} \mathcal{F}_{HW}(\boldsymbol{\epsilon},\boldsymbol{u},\boldsymbol{\sigma}) \cdot \boldsymbol{\tau} = 0$, or alternatively $\delta_\epsilon \mathcal{F}_{HW} = 0$, $\delta_{\boldsymbol{u}} \mathcal{F}_{HW} = 0$ and $\delta_{\boldsymbol{\sigma}} \mathcal{F}_{HW} = 0$.

Thus, in the associated mixed variational problem, we seek a solution $\boldsymbol{\epsilon}, \boldsymbol{u}, \boldsymbol{\sigma}) \in \mathcal{E} \times \mathcal{V} \times \mathcal{T}$, satisfying in the continuous version the weak constitutive equations

$$\int_\Omega (\mathbb{C} : \boldsymbol{\epsilon} - \boldsymbol{\sigma}) : \boldsymbol{\eta} \, d\Omega = 0 \quad \forall \boldsymbol{\eta} \in \mathcal{E}, \tag{75a}$$

the equilibrium conditions

$$\int_\Omega \boldsymbol{\sigma} : (\mathrm{grad}_{sym} \boldsymbol{v}) \, d\Omega = \int_\Omega \boldsymbol{f} \cdot \boldsymbol{v} \, d\Omega + \int_{\Gamma_N} \bar{\boldsymbol{t}} \cdot \boldsymbol{v} \, d\Gamma \quad \forall \boldsymbol{v} \in \mathcal{V} \tag{75b}$$

and the kinematic compatibility conditions

$$\int_\Omega [(\mathrm{grad}_{sym} \boldsymbol{u}) - \boldsymbol{\epsilon}] : \boldsymbol{\tau} \, d\Omega = 0 \quad \forall \boldsymbol{\tau} \in \mathcal{T}. \tag{75c}$$

The solution of the mixed problem is a saddle point

$$\mathcal{F}_{HW}(\boldsymbol{\epsilon},\boldsymbol{u},\boldsymbol{\sigma}) = \sup_{\boldsymbol{\eta} \in \mathcal{E}} \sup_{\boldsymbol{v} \in \mathcal{V}} \inf_{\boldsymbol{\tau} \in \mathcal{T}} \mathcal{F}_{HW}(\boldsymbol{\eta},\boldsymbol{v},\boldsymbol{\tau}). \tag{76}$$

The introduction of finite element spaces and the calculation of the bilinear forms follow from the methods outlined above for the Hellinger-Reissner mixed method.

The discrete variational conditions of the Hu-Washizu mixed finite element method then result in

$$\bigcup_e \{a_e(\boldsymbol{\eta}_h, \boldsymbol{\epsilon}_h) + b_e(\boldsymbol{\sigma}_h, \boldsymbol{\eta}_h) = 0 \quad \forall \boldsymbol{\eta}_h \in \mathcal{E}_h\} \tag{77a}$$

$$\bigcup_e \{c_e(\boldsymbol{\sigma}_h, \boldsymbol{v}_h) = F_e(\boldsymbol{v}_h) \quad \forall \boldsymbol{v}_h \in \mathcal{V}_h\} \tag{77b}$$

$$\bigcup_e \{b_e(\boldsymbol{\epsilon}_h, \boldsymbol{\tau}_h) + c_e(\boldsymbol{\tau}_h, \boldsymbol{u}_h) = 0 \quad \forall \boldsymbol{\tau}_h \in \mathcal{T}_h\}. \tag{77c}$$

The associated inf-sup condition reads

$$\inf_{\tau \in \mathcal{T}} \left\{ \sup_{\eta \in \mathcal{E}} \frac{b(\tau, \eta)}{\|\tau\|_\mathcal{T} \|\eta\|_\mathcal{E}} + \sup_{v \in \mathcal{V}} \frac{c(\tau, v)}{\|\tau\|_\mathcal{T} \|v\|_\mathcal{V}} \right\} \geq \beta, \tag{78}$$

with a positive constant β.

5.2 Finite Element Discretization

Next, shape functions for the displacement, stress and strain fields have to be chosen under the important restriction that the global inf-sup condition is fulfilled. Similar to the global equation system for the Hellinger-Reissner principle, (66), we arrive at the global algebraic equation system for the Hu-Washizu functional

$$\bigcup_e \left\{ \begin{bmatrix} f_e & 0 & l_{\epsilon,e} \\ 0 & 0 & l_{u,e} \\ l_{\epsilon,e}^T & l_{u,e}^T & 0 \end{bmatrix} \begin{Bmatrix} \hat{s}_e \\ \hat{\epsilon}_e \\ \hat{u}_e \end{Bmatrix} \right\} = \begin{Bmatrix} 0 \\ 0 \\ \hat{p}_e \end{Bmatrix}. \tag{79}$$

Since the stress and strain shape functions are in $\mathcal{L}_2$ spaces, we can eliminate f_e on element level, thus getting the ficticious element stiffness matrix $k_e^* = l_{u,e}^T f_e^{-1} l_{u,e}$ and finally arrive at the system

$$\bigcup_e \left\{ \begin{bmatrix} k_e^* & l_{u,e} \\ l_{u,e}^T & 0_e \end{bmatrix} \begin{Bmatrix} \hat{\sigma}_e \\ \hat{u}_e \end{Bmatrix} \right\} = \begin{Bmatrix} 0_e \\ \hat{\bar{p}}_e \end{Bmatrix}. \tag{80}$$

Alternatively, the elimination of $\hat{\epsilon}_e$ is possible, depending on the ansatz functions.

Remark

A further reduction is not possible because k_e^* is non-regular in general. Stolarski and Belytschko (1987), [97], have shown that for certain choices of pairings, the finite element solution (σ_h, u_h) of the Hu-Washizu formulation equals the Hellinger-Reissner formulation with the same test spaces $\mathcal{T}_h$ and $\mathcal{V}_h$.

Finally, we remark that other mixed functionals and associated finite element methods, which are balanced on element level and also provide global numerical stability and well-posedness, have been developed in the 1980s and 90s, especially the PEERS-element (plane elasticity with reduced symmetry), Arnold, Brezzi, Douglas (1984), [77], and Stenberg (1984), [76], and especially the class of BDM-elements by Brezzi, Douglas and Marini (1985), [79].

The Hu-Washizu-based FEM was applied to special material properties in order to circumvent numerical stability problems, but it has to be pointed out that special pairings of test and solution spaces are required for fulfilling the global numerical stability conditions.

6 Development of General Purpose Finite Element Program Systems

Early finite element programs for static and dynamic analysis with primal FEM (with some exceptions) were developed in the 1960s and 1970s mostly using FØRTRAN program language and some were written in ALGOL. Most of them were operational in 1971. A selection of those is, see Buck, Scharpf, Schrem, Stein (1973), [98]:

1. AMSA 9 and AMSA 20, FIAT, Divisione Aviazioni – SCV, Turin, Italy
2. ASAS, Atkins Research and Development, Woodcote Grove, England
3. ASKA, Institut für Statik und Dynamik der Luft- und Raumfahrtkonstruktionen, Universität Stuttgart, Germany and IKOSS - Software Service Stuttgart
4. BERSAFE, Central Electricity Generating Board, Berkeley Nuclear Laboratories, Berkeley, Gloucestershire, England
5. EASE (FIDES), FIDES Rechenzentrum, Dr. D. Pfaffinger, Zürich, Switzerland
6. MARC 2, Marcal, Div. of Engg., Brown University, Providence, RI, USA and Control Data GmbH, Stuttgart, Germany
7. NASTRAN, NASTRAN Systems, Langley Research Center, Hampton, VA, USA
8. PRAKSI-RIB, Rechen-Institut für EDV im Bauwesen, Stuttgart, Germany
9. SAP, Prof. E. L. Wilson, Dept. of Civil Engg., University of California at Berkeley, CA, USA
10. SESAM 69, A/S Computas, Økernvein 145, Oslo 5, Norway

Internationally available and competitive general purpose finite element program systems for linear and non-linear thermomechanical (and today also for multi-physical) mathematical models were developed since the 1980s. They are mostly written in FØRTRAN 77 and FØRTRAN 90; those are, e.g.: NASTRAN, ANSYS, Abaqus and ADINA.

Different from these commercial program systems the general FE program FEAP by Taylor, University of California at Berkeley, has a command language and is written in FØRTRAN 77. The complete source code is available for the users mostly working in academia.

Remark

It has to be emphasized that various object-oriented finite element programs with distributed vector and matrix data structures have been developed especially in university institutes since the 1990s, usually with the C++ program language. Within this concept, strict seperation of topology, geometry, mathematical modeling, numerical methods, equation solvers, data evaluation and visualization was realized, e.g. in Niekamp and Stein, [33]. This provides many fundamental benefits, but is not realized so far in the above mentioned commercial program systems, causing crucial disadvantages for the users; some examples are:

(a) optimization problems with topological, geometrical and material design parameters are hard to realize because the geometrical data should be based on independent topological and parametric representations as separated logical objects. The same holds for the separation of mathematical modeling, numerical method and numerical calculus including equation solvers and extraction of wanted result datas.
(b) mesh adaptivity with residual a posteriori error estimators requires the access to neighboured elements in order to calculate the difference of interface tractions, but non of the mentioned program systems admits this. Only access to element data within element domains is provided. On the other hand, gradient smoothing of stresses and their recovery as well as element equilibration techniques are applied today.

Therefore, it is due time to develop new object-oriented program systems for the next technological generation with virtual product development and multi-coupled problems. These new developments are necessary to keep compuational methods economical and payable.

7 Further Important Milestones of FEM

So far mainly primal finite element methods for linear elastic static BVPs were presented here, but beginning in the 1920s with papers by Hellinger, Reissner, Prager and Washizu generalized variational principles were applied especially for parameter-dependent (i.e. non-robust) problems, leading to mixed finite element methods by adding relaxed field and/or interface conditions, multiplied with Lagrangian parameters, to the related variational principle. These methods yield saddle point problems and need to fulfill global inf-sup stability conditions for the nodal values of displacements and stresses.

Further generalizations of primal FEM for elastic static problems concern elastic dynamic, elastoplastic and time-dependent visco-elastic and visco-plastic deformations.

Since the 1990s multiscale FEMs for multiphysical problems were developed, and concerning fracture mechanics, the so-called XFEM, was first published by Belytschko and Black (1999), [99]. Herein, the FE mesh is chosen theoretically independent from progressing cracks, and meshless FE discretizations are applied for complex problems. Furthermore, particle methods, which are equivalent to element-free Gakerkin methods are developed very fast and get theoretical foundation.

Table 1 Further important milestones of generalized FEMs

Discretization method	Main features	Ref.
Parameter-dependent BVPs with locking or spurious modes (numerical instabilities) for elastic problems, since the 1980s	• global inf sup condition has to be fulfilled • mixed FEs required in principle • engineering methods: reduced integr. enhanced assumed strains	[94] [95] [100]
FEM for incremental kinematically nonlinear elastic problems, e.g. structural instabilities of beams, plates and shells	• geometrical stiffness matrix • incremental methods, Lagrangian, Eulerian, ALE • consistent numerical tangent matrix • step length control	[101] [102]
FEM for incremental inelastic and time-dependent visco-elastic-plastic deformations	• predictor-corrector methods with projections to actual yield surface; continuous and discontinuous time discretization	[103] [104] [105]
FEM for elasto-dynamic problems	• linear dynamics in time or in phase space • FEs in time and space, numerical integration in time, problem: stability, conservation of energy and momentum	[106] [107] [108] [109]
Computational contact mechanics	• Classical penalty, augmented lagrangian methods and mortar methods on macroscale • Micromechanical and thermal effects of static and dynamic contact	[110] [111] [112]
FE^2 for multiscale FEM	• homogenization from micro and meso to macro-scale via Hill's equation and projection methods	[113] [114] [115]
XFEM, esp. for fracture mechanics	• singular functions from analytical solution at crack tip as enrichment functions	[99] [116] [115] [117] [118]
Particle Methods	• equivalent to element-free Galerkin method; problems: stability analysis • use of meshless discretizations	[119] [120] [121] [122] [123]
Isogeometric analysis	• FEM with real geometry, integration of computer aided design and finite element analysis, using NURBS, T-splines, ...	[124] [125]
Meshless/Meshfree Methods	• Selecting an arbitrary grid of points with a background mesh for integration	[126] [127]

Level set methods and isogeometric analysis, especially by Hughes et al., were developed very fast in the last decade for applications with high continuity requirements for structural surfaces, using NURBS and T-splines. The same discretization of undeformed and deformed bodies, shells, etc., requires at least C^1-continuity at element interfaces and can also realize higher order continuity.

Furthermore, contact problems are an important class of problems with a long tradition. Today, the micro-properties of contact zones are crucial for modeling and computation, using multi-physical and multi-scale concepts.

References

1. Leibniz, G.: Communicatio suae pariter, duarumque alienarum ad adendum sibi primum a Dn. Jo. Bernoullio, deinde a Dn. Marchione Hospitalio communicatarum solutionum problematis curvae celerrimi descensus a Dn. Jo. Bernoullio geometris publice propositi, una cum solutione sua problematis alterius ab eodem postea propositi. Acta Eruditorum, 201–206 (May 1697)
2. Stein, E.: The origins of mechanical conservation principles and variational calculus in the the 17th century. GAMM-Mitt 2, 145–163 (2011)
3. Stein, E., Wichmann, K.: New insight into optimization an variational problems in the 17th century. Engineering Computations 20, 699–724 (2003)
4. Schellbach, K.H.: Probleme der Variationsrechnung. Crelle's Journal für die reine und angewandte Mathematik 41, 293–363 + 1 table (1851)
5. Strutt, J.W., Lord Rayleigh, F.R.S.: The theory of sound, vol. I. Macmillan and Co., London (1877)
6. Ritz, W.: Über eine neue Methode zur Lösung gewisser Probleme der mathematischen Physik. Journal für die reine und angewandte Mathematik 135, 1–61 (1909)
7. Galerkin, B.G.: Beams and plates, series for some problems of elastic equilibrium of beams and plates (article in russian language). Wjestnik Ingenerow 10, 897–908 (1915)
8. Courant, R.: Variational methods for the solution of problems of equilibrium and vibrations. Bull. Amer. Math. Soc. 49, 1–23 (1943)
9. Langefors, B.: Analysis of elastic structures by matrix transformations. J. Aeron. Sci. 19, 451–458 (1952)
10. Argyris, J.H.: Energy theorems and structural analysis. Aircraft Engineering 26 and 27, 347–356, 383–387, 394 and 42–58, 80–94, 125–134, 154–158 (1954 and 1955)
11. Livesley, R.: Matrix methods of structural analysis. Pergamon Press (1964)
12. Zurmühl, R.: Ein Matrizenverfahren zur Behandlung von Biegeschwingungen. Ingenieur-Archiv 32, 201–213 (1963)
13. Fenves, S.: Structural analysis by networks, matrices and computers. J. Struct. Div. ASCE 92, 199–221 (1966)
14. Zuse, K.: Entwicklungslinien einer Rechengeräteentwicklung von der Mechanik zur Elektronik (special print). Friedrich Vieweg & Sohn Braunschweig (1962)
15. Trefftz, E.: Ein Gegenstück zum Ritzschen Verfahren. In: Verhandlungen des 2. Internationalen Kongresses für technische Mechanik, Zürich, pp. 131–137 (1926)

16. Jirousek, J., Guex, L.: The hybrid-Trefftz finite element model and its application to plate bending. Int. J. Num. Meth. Engg. 23, 651–693 (1986)
17. Peters, K., Stein, E., Wagner, W.: A new boundary type finite element for 2D- and 3D-elastic solids. Int. J. Num. Meth. Engg. 37, 1009–1025 (1994)
18. Hrennikoff, A.P.: Plane stress and bending of plates by method of articulated framework. Ph.D. thesis, Dpt. of Civil and Sanitary Engg., M.I.T., Boston, USA (May 1940)
19. Turner, M.J., Clough, R.W., Martin, H.C., Topp, L.J.: Stiffness and deflection analysis of complex strucutres. Journal of the Aeronautical Sciences 23, 805–823 (1956)
20. Melosh, R.J.: Basis for derivation of matrices for the direct stiffness method. AIAA J. 1, 1631–1637 (1963)
21. Wilson, E., Farhoomand, I., Bathe, K.: Nonlinear dynamic analysis of complex structures. Earth-quake Engineering and Structural Dynamics 1, 241–252 (1973)
22. Argyris, J., Scharpf, D.: Some general consideration on the natural mode technique. Aeron. J. of the Royal Aeron. Soc. 73, 218–226, 361–368 (1969)
23. Zienkiewicz, O., Cheung, Y.: The finite element method, vol. 1. McGraw-Hill (1967)
24. Zienkiewicz, O., Taylor, R.: The finite element method, 5th edn., vol. 1-3. Butterworth-Heinemann (2000)
25. Bathe, K.-J., Wilson, E.: Numerical Methods in finite element analysis. Prentice-Hall, Inc. (1976)
26. Hughes, T.J.R.: The finite element method: Linear static and dynamic finite element analysis. Prentice-Hall Internat. (1987)
27. Szabó, B., Babuška, I.: Finite Element Analysis. John Wiley & Sons, New York (1991)
28. Delaunay, B.N.: Sur la sphère vide. Bulletin of Academy of Sciences of the USSR, 793–800 (1934)
29. George, P.L.: Automatic mesh generation. Applications to finite element methods. Wiley (1991)
30. Löhner, R., Parikh, P.: Three-dimensional grid generation by the advancing front method. Int. J. Numer. Methods Fluids 8, 1135–1149 (1988)
31. Shephard, M., Georges, M.: Automatic three-dimensional mesh generation by the finite octree technique. Tech. Rep. 1, SCOREC Report (1991)
32. Rivara, M.: Mesh refinement processes based on the generalized bisection of simplices. SIAM Journal on Numerical Analysis 21, 604–613 (1984)
33. Niekamp, R., Stein, E.: An object-oriented approach for parallel two- and three-dimensional adaptive finite element computations. Computers & Structures 80, 317–328 (2002)
34. Melenk, J.: The Partition of Unity Finite Element Method: Basic Theory and Applications. TICAM report, Texas Institute for Computational and Applied Mathematics, University of Texas at Austin (1996)
35. Apel, T.: Interpolation in h-version finite element spaces. In: Stein et al. [44], vol. 1, ch. 3, pp. 55–70 (2004)
36. Brenner, S.C., Carstensen, C.: Finite element methods. In: Stein, E., de Borst, R., Hughes, T.J.R. (eds.) Encyclopedia of Computational Mechanics. Fundamentals, vol. 1, ch. 4, pp. 73–118. Wiley (2004)
37. Szabó, B., Düster, A., Rank, E.: The p-version of the finite element method. In: Stein et al. [44], vol. 1, ch. 5, pp. 119–139 (2004)

38. Borouchaki, H., George, P.L., Hecht, F., Laug, P., Saltel, E.: Delaunay mesh generation governed by metric specifications. Part I. algorithms. Finite Elem. Anal. Des. 25, 61–83 (1997)
39. Irons, B.M., Zienkiewicz, O.C.: The isoparametric finite element system – a new concept in finite element analysis. In: Proc. Conf.: Recent Advances in Stress Analysis, Royal Aeronautical Society (1968)
40. Gauß, C.F.: Werke, vol. 1-6. Dieterich, Göttingen (1863-1874)
41. Buck, K.E., Scharpf, D.W.: Einführung in die Matrizen-Verschiebungsmethode. In: Buck, K.E., Scharpf, D.W., Stein, E., Wunderlich, W. (eds.) Finite Elemente in der Statik, pp. 1–70. Wilhelm Ernst & Sohn, Berlin (1973)
42. Stein, E., Wunderlich, W.: Finite-Element-Methoden als direkte Variationsverfahren in der Elastostatik. In: Buck, K.E., Scharpf, D.W., Stein, E., Wunderlich, W. (eds.) Finite Elemente in der Statik, pp. 71–125. Wilhelm Ernst & Sohn, Berlin (1973)
43. Bathe, K.J.: Finite element procedures, 1st edn. Prentice-Hall (1982) (2nd ed. Cambridge 2006)
44. Stein, E., de Borst, R., Hughes, T. (eds.): Encyclopedia of computational mechanics (ECM), vol. 1, Fundamentals, vol. 2, Solids and Structures, vol. 3, Fluids. John Wiley & Sons, Chichester (2004)
45. Stein, E., Rüter, M.: Finite element methods for elasticity with error-controlled discretization and model adaptivity. In: Stein et al. [44], vol. 2, pp. 5–58 (2004)
46. Braess, D.: Finite elements, 2nd edn. Cambridge University Press (2001) (1st ed. in German language, 1991)
47. Menabrea, F.L.: Sul principio di elasticità, delucidazioni di L.F.M. (1870)
48. Betti, E.: Teorema generale intorno alle deformazioni che fanno equilibrio a forze che agiscono alla superficie. Il Nuovo Cimento, Series 2(7-8), 87–97 (1872)
49. Castigliano, C.A.: Théorie de l'équilibre des systèmes élastiques et ses applications. Nero, Turin (1879)
50. Ostenfeld, A.S.: Teknisk Statik I. 3rd edn. (1920)
51. Klöppel, K., Reuschling, D.: Zur Anwendung der Theorie der Graphen bei der Matrizenformulierung statischer Probleme. Der Stahlbau 35, 236–245 (1966)
52. Möller, K.-H., Wagemann, C.-H.: Die Formulierungen der Einheitsverformungs- und der Einheitsbelastungszustände in Matrizenschreibweise mit Hilfe der Graphen. Der Stahlbau 35, 257–269 (1966)
53. de Veubeke, B.F. (ed.): Matrix methods of structural analysis. Pergamon Press, Oxford (1964)
54. Robinson, J.: Structural Matrix Analysis for the Engineer. John Wiley & Sons, New York (1966)
55. Pian, T.H.-H.: Derivation of element stiffness matrices by assumed stress distributions. AIAA J. 2, 1533–1536 (1964)
56. Pian, T.H.-H., Tong, P.: Basis of finite element methods for solid continua. Int. J. Num. Meth. Engng. 1, 3–28 (1964)
57. Hellinger, E.: Die allgemeinen Ansätze der Mechanik der Kontinua. In: Klein, F., Müller, C. (eds.) Enzyklopädie der Mathematischen Wissenschaften 4 (Teil 4), pp. 601–694. Teubner, Leipzig (1914)
58. Prange, G.: Die Variations- und Minimalprinzipe der Statik der Baukonstruktion. Habilitationsschrift, Technische Hochschule Hannover (1916)

59. Reissner, E.: On a variational theorem in elasticity. J. Math. Phys. 29, 90–95 (1950)
60. Legendre, A.-M.: Exercises du calcul intégral, Paris, vol. 1-3 (1811/1817)
61. Prager, W., Synge, J.: Approximations in eleasticity based on the concept of function spaces. Quart. Appl. Math. 5, 241–269 (1949)
62. Synge, J.: The hypercircle in mathematical physics. Cambridge University Press (1957)
63. Babuška, I.: The finite element method with lagrangian multipliers. Numer. Math. 20, 179–192 (1973)
64. Brezzi, F.: On the existence, uniqueness and approximation of saddle-point problems arising from lagrangian multipliers. Rev. Fr. Automat. Inf. Rech. Opérat. Sér. Rouge 8, 129–151 (1974)
65. Zienkiewicz, O.C., Qu, S., Taylor, R.L., Nakazawa, S.: The patch test for mixed formulations. Int. J. Num. Meth. Eng. 23, 1873–1883 (1986)
66. Prager, W.: Variational principles for elastic plates with relaxed continuity requirements. Int. J. Solids & Structures 4, 837–844 (1968)
67. Oden, J., Carey, G.: Finite elements. Mathematical aspects, vol. IV. Prentice Hall, Inc. (1983)
68. Babuška, I.: The finite element method with lagrangian multipliers. Numerische Mathematik 20, 179–192 (1973)
69. Auricchio, F., Brezzi, F., Lovadina, C.: Mixed finite element methods. In: Stein et al. [44], vol. 1, ch. 9, pp. 237–278 (2004)
70. Brezzi, F., Fortin, M.: Mixed and hybrid finite element methods. Springer, New York (1991)
71. Herrmann, L.R.: Finite-element bending analysis for plates. J. Engg. Mech. EM 5 93, 13–26 (1967)
72. Prato, C.A.: Shell finite element method via Reissner's principle. Int. J. Solids & Structures 5, 1119–1133 (1969)
73. Bathe, K.-J., Dvorkin, E.: A four-node plate bending element based on mindlin reissner plate theory and a mixed interpolation. Int. J. Num. Meth. Eng. 21, 367–383 (1985)
74. Crouzeix, M., Raviart, P.A.: Conforming and non-conforming finite element methods for solving the stationary Stokes equations. RAIRO Anal. Numér. R3, 33–76 (1973)
75. Raviart, P., Thomas, J.: A mixed finite element method for second order elliptic problems. In: Galigani, I., Magenes, E. (eds.) Mathematical Aspects of the Finite Element Method. Lecture Notes in Mathematics, vol. 606, pp. 292–315. Springer, New York (1977)
76. Stenberg, R.: Analysis of mixed finite element methods for the Stokes problem: A unified approach. Math. Comp. 42, 9–23 (1984)
77. Arnold, D.N., Brezzi, F., Douglas Jr., J.: PEERS: A new mixed finite element for plane elasticity. Japan. J. Appl. Math. 1, 347–367 (1984)
78. Arnold, D.N., Douglas Jr., J., Gupta, C.P.: A family of higher order finite element methods for plane elasticity. Numer. Math. 45, 1–22 (1984)
79. Brezzi, F., Douglas, J.J., Marini, L.D.: Two families of mixed finite elements for second-order elliptic problems. Numer. Math. 47, 217–235 (1985)
80. Brezzi, F., Douglas, J.J., Fortin, M., Marini, L.D.: Efficient rectangular mixed finite elements in two and three space variables. RAIRO M_2AN 21, 581–604 (1987)

81. Bischoff, M., Wall, W., Bletzinger, K.-U., Ramm, E.: Models and finite elements for thin-walled structures. In: Stein et al. [44], vol. 2-3, pp. 59–137 (2004)
82. Zienkiewicz, O.C., Taylor, R.L., Too, J.M.: Reduced integration technique in general analysis of plates and shells. Int. J. Num. Meth. Eng. 3, 275–290 (1973)
83. Belytschko, T., Liu, W., Moran, B.: Nonlinear finite elements for continua and structures. Wiley (2000)
84. MacNeal, R.: A simple quadrilateral shell element. Computers & Structures 183, 175–183 (1978)
85. Hughes, T.J.R., Tezduyar, T.: Finite elements based upon Mindlin plate theory with particular reference to the 4-node isoparametric element. J. Appl. Mech. 48, 587–596 (1981)
86. Bathe, K.-J., Dvorkin, E.: A formulation of general shell elements – the use of mixed interpolation of tensorial components. Int. J. Num. Meth. Eng. 22, 697–722 (1986)
87. Bathe, K.-J., Brezzi, F., Cho, S.: The MITC7 and MITC9 plate elements. Computers & Structures 32, 797–814 (1989)
88. Bathe, K.-J., Brezzi, F., Fortin, M.: Mixed-interpolated elements for Reissner-Mindlin plates. Int. J. Num. Meth. Eng. 28, 1787–1801 (1989)
89. Hughes, T.J.R., Liu, W.K.: Nonlinear finite element analysis of shells, part i: three-dimensional shells. Comput. Methods Appl. Mech. Engng. 26, 331–361 (1981)
90. Simo, J.C., Rifai, M.S.: A class of mixed assumed strain methods and the method of incompatible modes. Int. J. Num. Meth. Eng. 29, 1595–1638 (1990)
91. Tessler, A., Hughes, T.: A three-node mindlin plate element with improved transverse shear. Comput. Methods Appl. Mech. Engng. 50, 71–101 (1985)
92. Taylor, R., Auricchio, F.: Linked interpolation for Reissner-Mindlin plate elements: part ii. a simple triangle. Int. J. Num. Meth. Eng. 36, 3057–3066 (1993)
93. Bischoff, M., Taylor, R.: A three-dimensional shell element with an exact thin limit. In: Waszczyszyn, Z., Parmin, J. (eds.) Proceedings of the 2nd European Conference on Computational Mechanics. Fundacja Zdrowia Publicznego, Cracow (2001)
94. Armero, F.: On the locking and stability of finite elements in finite deformation plane strain problems. Computers & Structures 75, 261–290 (2000)
95. Reese, S.: On a consistent hourglass stabilization technique to treat large deformations and thermomechanical coupling in plane strain problems. Int. J. Num. Meth. Eng. 57, 1095–1127 (2003)
96. Washizu, K.: Variational Methods in Elasticity and Plasticity. Pergamon Press (1968)
97. Stolarski, H., Belytschko, T.: Limitation principles for mixed finite elements based on the Hu-Washizu variational formulation. Comput. Methods Appl. Mech. Engng. 60, 195–216 (1987)
98. Buck, K.E., Scharpf, D.W., Schrem, E., Stein, E.: Einige allgemeine Programmsysteme für finite Elemente. In: Buck, K.E., Scharpf, D.W., Stein, E., Wunderlich, W. (eds.) Finite Elemente in der Statik, pp. 399–454. Wilhelm Ernst & Sohn, Berlin (1973)
99. Belytschko, T., Black, T.: Elastic crack growth in finite elements with minimal remeshing. Int. J. Num. Meth. Eng. 45, 601–620 (1999)

100. Reese, S., Tini, V., Kiliclar, Y., Frischkorn, J., Schwarze, M.: Stability of mixed finite element formulations – a new approach, ch. 7, pp. 51–60. Springer (2011)
101. Wriggers, P.: Nonlinear Finite Element Methods. Springer, Berlin (2008)
102. Ibrahimbegovic, A.: Nonlinear solid mechanics – theoretical formulations and finite element solution methods. Springer (2009)
103. Stein, E. (ed.): Error-controlled adaptive finite elements in solid mechanics. Wiley (2003)
104. Armero, F.: Elastoplastic and viscoplastic deformations in solids and structures. In: Stein et al. [44], vol. 2, pp. 227–264 (2004)
105. Miehe, C., Schotte, J.: Crystal plasticity and evolution of polycrystalline microstructure. In: Stein et al. [44], vol. 2, ch. 8, pp. 267–290 (2004)
106. Hulbert, G.M.: Computational Structural Dynamics. In: Stein et al. [44], vol. 2, ch. 5, pp. 169–194 (2004)
107. Chen, S., Hansen, J.M., Tortorelli, D.A.: Unconditional energy stable implicit time integration: application to multibody systems analysis and design. Int. J. Num. Meth. Eng. 48, 791–822 (2000)
108. Hughes, T.J.R., Hulbert, G.M.: Space-time finite element methods for elastodynamics: formulation and error estimates. Comput. Methods Appl. Mech. Engng. 66, 339–363 (1988)
109. Simo, J.C., Tarnow, N.: The discrete energy momentum method conserving algorithms for nonlinear elastodynamics. ZAMP 43, 757–793 (1992)
110. Zavarise, G., Wriggers, P. (eds.): Trends in Computational Contact Mechanics. LNACM, vol. 58. Springer, Berlin (2011)
111. Wriggers, P.: Computational Contact Mechanics, 2nd edn. Springer, Heidelberg (2006)
112. Wriggers, P., Zavarise, G.: Computational Contact Mechanics. In: Stein et al. [44], vol. 2, ch. 6, pp. 195–226 (2004)
113. Zohdi, T.I., Wriggers, P.: Introduction to Computational Micromechanics. LNACM, vol. 20. Springer, Berlin (2005)
114. Zohdi, T.I.: Homogenization Methods and Multiscale Modeling. In: Stein et al. [44], vol. 2, ch. 12, pp. 407–430 (2004)
115. Loehnert, S., Belytschko, T.: A multiscale projection method for macro/microcrack simulations. Int. J. Num. Meth. Eng. 71, 1466–1482 (2007)
116. Fries, T.-P.: A corrected xfem approximation without problems in blending elements. Int. J. Num. Meth. Eng. 75, 503–532 (2008)
117. Loehnert, S., Mueller-Hoeppe, D.: 3D multiscale projection method for micro-/Macrocrack interaction simulations. In: Mueller-Hoeppe, D., Loehnert, S., Reese, S. (eds.) Recent Developments and Innovative Applications, vol. 59, pp. 223–230. Springer, Heidelberg (2011)
118. Rüter, M., Stein, E.: Goal-oriented residual error estimates for XFEM approximations in LEFM. In: Mueller-Hoeppe, D., Loehnert, S., Reese, S. (eds.) Recent Developments and Innovative Applications, vol. 59, pp. 231–238. Springer, Heidelberg (2011)
119. Aharonov, E., Rothman, D.: Non-newtonian flow (through porous media): a lattice-boltzmann method. Geop. 20, 679–682 (1993)
120. Chen, S., Doolen, G.: Lattice boltzmann method for fluid flows. Annu. Rev. Fluid Mech. 30, 329–364 (1998)
121. Han, K., Feng, Y.T., Owen, D.R.J.: Numerical simulation of irregular particle transport in turbulent flows using coupled lbm dem. Comput. Model. Eng. Sci. 18(2), 87–100 (2007)

122. Cundall, P., Strack, O.D.L.: Discrete numerical model for granular assemblies. Geotechnique 29, 47–65 (1979)
123. Hu, H.H., Joseph, D.D., Crochet, M.J.: Direct simulation of fluid particle motions. Theor. Comp. Fluid Dyn. 3, 285–306 (1992)
124. Hughes, T.J.R., Cottrell, J.A., Bazilevs, Y.: Isogeometric Analysis: Toward Integration of CAD and FEA. Wiley, Chichester (2009)
125. Hughes, T.J.R., Cottrell, J.A., Bazilevs, Y.: Isogeometric analysis: Cad, finite elements, nurbs, exact geometry and mesh refinement. Comput. Methods Appl. Mech. Engng. 194, 4135–4195 (2005)
126. Osher, S., Fedkiw, R.: Level Set Methods and Dynamic Implicit Surfaces. Applied Mathematical Sciences, vol. 153. Springer (2003)
127. Hu, H.Y., Chen, J.S., Hu, W.: Reproducing kernel enhanced local radial collocation method. In: Ferreira, A., Kansa, E., Fasshauer, G., Leitão, V. (eds.) Meshless Method, pp. 175–188. Springer (2008)

History of the Finite Element Method – Mathematics Meets Mechanics – Part II: Mathematical Foundation of Primal FEM for Elastic Deformations, Error Analysis and Adaptivity

Erwin Stein

Abstract. This chapter treats the history of mathematical foundation of primal FEM, especially a posteriori error estimates and adaptivity, based on functional analysis in Sobolev spaces. This is of equal importance as the creation of multifarious computational methods and techniques in engineering and computer sciences. BVPs for linear elliptic PDEs, mainly the Lamè equations for linear static elasticity are treated.

Bounded residual explicit and various implicit error estimators of primal FEM were mainly developed by Babuška and Rheinboldt (1978), Bank and Weiser (1985), Babuška and Miller (1987) and Aubin (1967) and Nietsche (1977).

Mechanically motivated explicit and implicit error estimators were created by Zienkiewicz and Zhu (1987), using gradient smoothing of the C^0-continuous displacements and stress recovery for which convergence and upper bound property were proven by Carstensen and Funken (2001).

A variant of implicit a posteriori error estimators is the error of consitutive equations by Ladevèze et al. (1998). Equilibrated test stresses on element and patch levels are required, Ladevèze, Pelle (2005). Gradient-free formulations, e.g. by Cottereau, Díez and Huerta (2009), are also competitive. Generalizations of a priori and a posteriori error estimates, using the three-functional-theorem by Prager and Synge (1947), are very useful.

Goal-oriented error estimators for quantities of interest (as linear or nonlinear functionals, defined of closed finite supports) are of practical importance, Eriksson et al. (1995), Rannacher and Suttmeier (1997), Cirac and Ramm (1998), Ohnimus et al. (2001), Stein and Rüter (2004) and others. Textbooks

---

Erwin Stein
Institute of Mechanics and Computational Mechanics,
Leibniz Universität Hannover, Appelstr. 9A, 30167 Hannover, Germany
e-mail: stein@ibnm.uni-hannover.de

by Verfürth (1996, 1999, 2013), Ainsworth and Oden (2000), Babuška and Strouboulis (2001), are available.

Verification with prescribed error tolerances is realized with the above cited bounded error estimators and related discretization adaptivity, provided that the solution exists in the used test space.

Moreover, *model validation* requires model adaptivity of the adequate physical and mathematical modeling which additionally needs experimental verification, requiring a posteriori model error estimators combined with discretization error estimators.

Model reductions, e.g. for reinforced laminates, were treated by Oden (2002), and *model expansions*, e.g. for 3D boundary layers of 2D plate and shell theories by Stein and Ohnimus (1997), and Stein, Rüter and Ohnimus (2011).

1 Foundation of the Mathematical Theory of Primal FEM and Error Analysis in the 20th Century

The mathematical foundation of the finite element method for elliptic partial differential equations of 2nd order began in the 1960s and 1970s with the eminent mathematicians Bramble, Babuška, Rheinboldt, Strang, Fix, Aubin, Nitsche, Brezzi, Ciarlet, Johnson, Oden and many others, based on modern functional analysis by Hilbert, Korn, Poincaré, and Sobolev.

The algorithmic simplicity, robustness and generality (especially for applications) of primal FEM compared with the finite difference method (FDM) – especially the choice of test and trial functions and their algorithmic treatment within element domains only and the simple formulation of BVPs – is accompanied by the severe problem that for 2nd order PDEs the C^0-continuous test and trial functions at element interfaces provide jumps of derivatives, i.e. of strains, stresses, and tractions, and therefore require Sobolev function spaces for the mathematical analysis. This means that the mathematical foundation is by far not trivial. The analysis and error analysis began in the 1960s with first highlights in the 1970s and the 1980s. The strong mathematical research of FEM and its various generalizations is going on strongly, also including singularities (as in fracture mechanics) as well as kinematic and physical nonlinearities for describing mechanical instabilities and complex material deformations on macro- and microscales.

Major monographs on the mathematical foundation of FEM can be found from Strang and Fix (1973), [91], Oden and Carey (1983), [66], Brenner and Scott (1994), [30], Verfürth (1996), [92], Eriksson et al. (1996), [42], and in extended form with new subjects Verfürth (2013), [93], Braess (1992), [25], Neittaanmäki and Repin (2004), [63], Babuška, Whiteman, Strouboulis (2011), [16], as well as volume 1 of the Encyclopedia of Computational Mechanics, Stein, de Borst, Hughes (eds.) (2004), [84], especially Apel (2004), [3], Brenner, Carstensen (2004), [29], Auricchio, Brezzi Lovadino (2004), [6].

Of importance are also the five proceedings from the ADMOS conferences (Adaptive Modeling and Simulation) from 2003 to 2011, published by CIMNE, Barcelona, [38].

1.1 Dirichlet Principle of Minimum of Total Potential Energy

The following outline of the development of convergence theory and error analysis holds for linear V-elliptic partial differential equations, usually of 2nd order, i.e. with self-adjoint operators. An important problem of this type is linear static theory of elastically deformable bodies where the displacements $\boldsymbol{u}(\boldsymbol{x}) \in \Omega \subset \mathbb{R}^3$ are C^1 manifolds.

The variational formulation of the 3D-Lamé theory of linear elasticity was given by Dirichlet (1805-1859) in 1835, [39]. For mixed boundary conditions, the symmetric variational form with a test function $\boldsymbol{v}(\boldsymbol{x}) \in \mathcal{H}^1(\Omega)$ reads in tensor notation – since the 1960s – e.g. in Strang and Fix (1973), [91],

$$a(\boldsymbol{v}, \boldsymbol{v}) = \int_\Omega \boldsymbol{\epsilon}(\boldsymbol{v}) : \mathbb{C} : \boldsymbol{\epsilon}(\boldsymbol{v}) \, \mathrm{d}V, \tag{1}$$

with the admissible test space $\mathcal{V} = \{\boldsymbol{v} \in [\mathcal{H}^1(\Omega)]^3; \boldsymbol{v} = \boldsymbol{0} \text{ at } \Gamma_D\}$; $\boldsymbol{\epsilon}(\boldsymbol{v})$ is the linear symmetric strain tensor and $\mathbb{C}$ the symmetric and positive definite 4th order elasticity tensor.

The linear forms for the given loads are

$$F(\boldsymbol{v}) = \int_\Omega \boldsymbol{f} \cdot \boldsymbol{v} \, \mathrm{d}V; \quad g(\boldsymbol{v}) = \int_{\Gamma_N} \bar{\boldsymbol{t}} \cdot \boldsymbol{v} \, \mathrm{d}S, \tag{2}$$

with the body force $\boldsymbol{f}(\boldsymbol{x}) \in \mathcal{L}_2(\Omega)$ and the conservative given boundary tractions $\bar{\boldsymbol{t}}(\boldsymbol{v}) \in \mathcal{L}_2(\Gamma_N)$, i.e. at Neumann boundaries.

The variational functional, the total potential energy of the system,

$$J(\boldsymbol{v}) := \underbrace{1/2 a(\boldsymbol{v}, \boldsymbol{v})}_{\Pi_{int}} \underbrace{-[F(\boldsymbol{v}) + g(\boldsymbol{v})]}_{\Pi_{ext}} \to \min_{\boldsymbol{v} \in \mathcal{V}}, \tag{3}$$

can be constructed from the bilinear and linear forms, eqs. (1) and (2), and yields a minimum for analytical solutions $\boldsymbol{u}(\boldsymbol{x}) \in \mathcal{H}^2(\Omega)$ for well-posed mixed BVPs in stable static equilibrium in case of square-integrable loads in the domain and at Neumann boundaries and Lipschitz-continuous boundaries.

The necessary stationarity condition requires the vanishing of the 1st variation of the Dirichlet functional, as

$$\delta J(\boldsymbol{v}) = 0 \Rightarrow \boldsymbol{u}, \tag{4}$$

and the necessary condition for the minimal property is that the 2nd variation is larger than zero at the solution point, as

$$\delta^2 J(v) > 0 \text{ for } v = u. \tag{5}$$

1.2 Theorems of Existence and Uniqueness of Variational Solutions, The Lax-Milgram Lemma

Theorem: A unique minimal solution $u(x)$ exists for the variational problem eqs. (1) to (5). It is proven by the Lagrangian embedding method as follows, Eriksson et al. (1996), [42], Braess (2001), [25].

Assuming a neighbored solution $\tilde{u} = u + tv$, $t \in \mathbb{R}$, yields the minimal functional

$$J(u + tv) = 1/2 \, a(u + tv, u + tv) - F(u + tv) - g(u + tv) \tag{6}$$
$$= J(u) + t[a(u, v) - F(v) - g(v) + 1/2 \, t^2 a(v, v)] \tag{7}$$

and for $t = 1$

$$J(u + v) = J(u) + 1/2 \, a(v, v) \, \forall \, v \in \mathcal{V} \tag{8}$$

$$J(u + v) > J(u); \quad J(u) = \min_{v \in \mathcal{V}} J(v). \tag{9}$$

The 2nd variation of J yields the quadratic form

$$\delta^2 J(v) = \int_\Omega \delta\epsilon(v) : \mathbb{C} : \delta\epsilon(v) \, \mathrm{d}\Omega > 0 \text{ for } \mathbb{C} \text{ positive definite}, \tag{10}$$

which proves the minimal property of J.

Under the presumptions of coercivity of the internal energy, i.e. the bilinear form of the test function v, i.e. $a(v, v) \geq \alpha \|v\|_\mathcal{V}^2$, $a(v, v) \to \infty$ as $|v| \to \infty$, fig. 1 (a), and furthermore of the existence of a continuous linear functional $F(v)$, $F(v) \leq C\|v\|_\mathcal{V} < \infty$ with a linear growth of $F(v)$ with $\|f\|$, fig. 1 (b), postulating that the displacements are bounded by the load, i.e. that equilibrium is stable, the following Lax-Milgram theorem, Lax (republished 2005), [60], holds which was completely proven by Babuška.

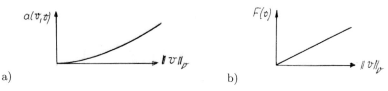

Fig. 1 a) Coercivity of strain energy $a(v, v)$; b) continuous linear functional $F(v)$

Theorem: For the bilinear form $a(\bm{v}, \bm{v}) : \mathcal{H}^1 \times \mathcal{H}^1 \to \mathbb{R}$ and $F(\bm{v}) \in \mathcal{H}_0 \to \mathbb{R}$, the variational problem

$$J(\bm{u}) = \min_{\bm{v}} J(\bm{v}); \quad J(\bm{v}) := 1/2 a(\bm{v}, \bm{v}) - F(\bm{v}) \tag{11}$$

has a unique solution in $\mathcal{V}$.

The proof uses the Poincaré-Friedrichs-inequality, Sect. 1.4, and yields:

1. existence:

$$a(\bm{v}, \bm{v}) \geq 1/2\alpha \|\bm{v}\|^2 - \|\bm{f}\| \|\bm{v}\| \tag{12}$$

$$\alpha \|\Delta \bm{v}\|^2 \leq a(\Delta \bm{v}, \Delta \bm{v}), \text{ for } \Delta \bm{v} := \bm{v}_n - \bm{v}_m, n > m \tag{13}$$

and
2. uniqueness: The solution $\bm{u}$ is unique, because each minimal sequence $a(\bm{v}, \bm{v}) \geq \alpha \|\bm{v}\|^2$ is a Cauchy sequence.

1.3 Test and Trial Spaces for Variational Calculus

1.3.1 Hilbert and Sobolev Spaces

By the scalar (or inner) product

$$(u, v)_0 := (u, v)_{\mathcal{L}_2(\Omega)} = \int_\Omega u(\bm{x}) \, v(\bm{x}) \, \mathrm{d}\bm{x} \tag{14}$$

the space $\mathcal{L}_2(\Omega)$ of quadratically integrable functions in Ω is completed to a Hilbert space (complete vector space with an inner product) with the norm

$$\|u\|_0 = \sqrt{(u, u)_0}. \tag{15}$$

This important function space was introduced by Hilbert (1928), [49].

Primal FEM with C^0 continuous trial and test functions for the displacements at all element boundaries has the crucial problem of jumps of derivatives at element interfaces. Thus, the strains, stresses and boundary tractions are discontinuous. By this, Hilbert spaces are not sufficient for finite element analysis, and this was the reason why the mathematics of FEM was only developed since the 1970s, based on Sobolev spaces $\mathcal{H}^m(\Omega)$, discovered by Sobolev in his famous work from 1963, [80].

A Sobolev space $\mathcal{H}^m(\Omega)$ is the set of all functions $\bm{u}(\bm{x}) \in \mathcal{L}_2(\Omega)$ which have weak derivatives $\partial^\alpha u \, \forall \, |\alpha| \leq m$.

In $\mathcal{H}^m(\Omega)$ a scalar product

$$(u, v)_m := \sum_{|\alpha| \leq m} (\partial^\alpha u, \partial^\alpha v)_0 \tag{16}$$

is defined with the complete norm

$$\|u\|_m := \sqrt{(u,u)_m} = \sqrt{\sum_{|\alpha|\leq m} \|\partial^\alpha u\|^2_{\mathcal{L}_2(\Omega)}}; \|u\|_m > 0 \text{ for } \boldsymbol{u} \neq \boldsymbol{0} \quad (17)$$

and the seminorm (the internal energy with allowed rigid body modes in case of the Lamé PDEs)

$$|u|_m := \sqrt{\sum_{|\alpha|=m} \|\partial^\alpha u\|^2_{\mathcal{L}_2(\Omega)}}; |u|_m \geq 0 \text{ for } \boldsymbol{u} \neq \boldsymbol{0}. \quad (18)$$

Inclusions of the Sobolev spaces are:

$$\begin{matrix} \mathcal{L}_2(\Omega) = \mathcal{H}^0(\Omega) \supset \mathcal{H}^1(\Omega) \supset \mathcal{H}^2(\Omega) \supset \ldots \\ \| \quad\quad \cup \quad\quad \cup \\ \mathcal{H}_0^0(\Omega) \supset \mathcal{H}_0^1(\Omega) \supset \mathcal{H}_0^2(\Omega) \supset \ldots \end{matrix}, \quad (19)$$

which are useful for transformations of grad, div and rot operators.

1.4 The Poincaré-Friedrichs Inequality

For estimating the interpolation error of FE discretizations the inequality by Poincaré (1885), [69], and Friedrichs (1947), [45], is of great importance.

Suppose that the domain Ω is contained in an n-dimensional cube. Then the theorem holds

$$\|v\|_0 \leq s\,|v|_1 \;\forall\, v \in \mathcal{H}_0^1(\Omega), \quad (20)$$

with s the edge length of the cube and

$$v_\Omega = \frac{1}{|\Omega|} \int_\Omega v(\boldsymbol{x})\,\mathrm{d}\Omega \quad (21)$$

the average value of v over Ω, with $|\Omega|$ standing for the Lebesgue measure of the domain Ω. The proof only requires zero boundary conditions on Γ_D; $\Gamma = \Gamma_D \cup \Gamma_N = \partial\Omega$; $\Gamma_D \cap \Gamma_N = \emptyset$.

This theorem states that the $\mathcal{L}_2$ norm of the test function can be estimated from above by the seminorm in $\mathcal{H}_0^1$, i.e. for test functions which are C^0 continuous at interfaces and zero at Dirichlet boundaries.

This theorem was proven by Poincaré (1885) and later in an alternate form by Friedrichs (1947). It is important for proving upper bound properties of finite element approximations for growing numbers of degrees of freedom (DOFs).

1.5 The Céa-Lemma – Optimality of the Galerkin Method

The Céa-lemma states, Céa (1964), [33], and e.g. Braess, [25]:
The Galerkin approximation of the variational $\mathcal{V}$-elliptic form, eqs. (1) and (2),

$$a(\boldsymbol{u}, \boldsymbol{v}) = F(\boldsymbol{v}) + g(\boldsymbol{v}), \tag{22}$$

with $\boldsymbol{v} \in \mathcal{V}$, namely

$$a(\boldsymbol{u}_h, \boldsymbol{v}_h) = F(\boldsymbol{v}_h) + g(\boldsymbol{v}_h) \text{ with } \boldsymbol{v}_h \in \mathcal{V}_h \subset \mathcal{V}, \tag{23}$$

yielding

$$\boldsymbol{u}_h \to \boldsymbol{u} \text{ for } N \to \infty, \tag{24}$$

is quasi-optimal, i.e. better than any other approximation.
The Céa-lemma reads

$$\|\boldsymbol{u} - \boldsymbol{u}_h\|_m \leq \frac{C}{\alpha} \inf_{\boldsymbol{v}_h \in \mathcal{V}_h} \|\boldsymbol{u} - \boldsymbol{v}_h\|_m, \tag{25}$$

with $M = \frac{C}{\alpha}$ the ellipticity constant and the projection theorem

$$\|\boldsymbol{u} - \Pi_h \boldsymbol{u}\|_m = \inf_{\boldsymbol{v}_h \in \mathcal{V}} \|\boldsymbol{u} - \boldsymbol{v}_h\|_m, \tag{26}$$

with Π_h the orthogonal projection operator.
The simple proof for this important theorem starts with the subtraction of the variational forms for the true and the approximated solution and thus from the error $\boldsymbol{e} := \boldsymbol{u} - \boldsymbol{u}_h$, yielding the Galerkin orthogonality

$$a(\underbrace{\boldsymbol{u} - \boldsymbol{u}_h}_{=: \boldsymbol{e}}, \boldsymbol{v}) = 0 \ \forall \ \boldsymbol{v} \in \mathcal{V}_h; \ a(\boldsymbol{e}, \boldsymbol{v}) = R \neq 0 \ \forall \ \boldsymbol{v} \in \mathcal{V}, \tag{27}$$

with the evident result that the residuum of the weak form of equilibrium is unequal zero for the approximated solution. Postulating the coercivity of the internal energy, i.e. $a(\boldsymbol{v}, \boldsymbol{v}) \to \infty$ for $\boldsymbol{v} \to \infty$, introducing a further test space $\boldsymbol{w} := \boldsymbol{v}_h - \boldsymbol{u}_h \in \mathcal{V}_h$ with $a(\boldsymbol{u} - \boldsymbol{u}_h, \boldsymbol{v} - \boldsymbol{v}_h) = 0$, the optimality of the Galerkin method is got as follows

$$\alpha \|\boldsymbol{u} - \boldsymbol{u}_h\|_m^2 \leq a(\boldsymbol{u} - \boldsymbol{u}_h, \boldsymbol{u} - \boldsymbol{u}_h) = a(\boldsymbol{u} - \boldsymbol{u}_h, \boldsymbol{u} - \boldsymbol{v}_h) + a(\boldsymbol{u} - \boldsymbol{u}_h, \boldsymbol{v}_h - \boldsymbol{v}),$$

$$\alpha \|\boldsymbol{u} - \boldsymbol{u}_h\|_m^2 \leq C \|\boldsymbol{u} - \boldsymbol{u}_h\|_m \|\boldsymbol{u} - \boldsymbol{v}_h\|_m. \tag{28}$$

2 Error Analysis and Adaptivity

2.1 General Remarks

A priori and a posteriori error analysis of FEM are the basis for proper error estimators, at least with upper bounds and desirably also with strict lower bounds, and thus for adaptive mesh refinements with good effectivity indices, desirably between 1.2 and 1.4. Analysis and error analysis of FEM, especially both explicit and implicit residual a posteriori error estimators for self-adjoint elliptic BVPs were mainly derived by Babuška and Rheinboldt since the 1970s, and a bounded explicit estimator in the energy norm. $|||e||| = Ch^p$, was published by Babuška and Miller (1987), [9], requiring Sobolev-spaces for the test and trial functions (usually Lagrange or Legendre polynomials) because they have to be only C^0-continuous at element interfaces, and thus the derivatives of displacements (the strains, related stresses and tractions) have jumps at interfaces.

Especially nowadays, the further development of FEM and its various complicated generalizations, such as Generalized FEM (GFEM), Duarte, Babuška, Oden (2000), [40], requiring the partition of unity condition by Babuška, Melenk (1997), [7], and other variants of FEM, especially the Extended FEM (XFEM) by Belytschko and Black (1999), [20], as well as the Singular Function Method (SFM) by Fix, Gulati, Wakoff (1973), [44]. They require both mathematical and mechanical access for guaranteed consistency, stability and optimal convergence orders of well posed problems. Especially error analysis and quasi-optimal effectivity of adaptive finite element meshes and – more general – of the ansatz techniques require sound mathematical analysis which usually cannot be replaced by engineering intuition.

2.2 A Priori Interpolation Error Estimation – The Bramble-Hilbert Lemma

This concerns the error estimation for interpolation by polynomials. Let $\Omega \subset \mathbb{R}^2$ be a 2D domain with a Lipschitz continuous boundary which satisfies a cone condition. In addition let $t \geq 2$, and suppose that $z_1, z_2, \ldots, z_s$ are in total $s := t(t+1)/2$ prescribed points in $\bar{\Omega}$ such that the interpolation operator $I; H^t \to P_{t-1}$ is well defined for polynomials of degree $\leq t - 1$. The Bramble-Hilbert Lemma (1970), [28], states that there exists a constant $c = c(l, z_1, \ldots, z_s)$, such that the half norm is smaller than or equal to the norm of the interpolation error, as

$$\|u - Iu\|_t \leq c|u|_t \quad \forall u \in H^1(\Omega). \tag{29}$$

2.3 A Priori Error Estimates in the Energy Norm

The following theorem holds for a given quasi-uniform triangulation $\mathcal{T}$ of Ω; then the FE approximation $u_h \in \mathcal{V}_h$ with linear shape functions yields the a priori error estimate

$$\|e\|_1 = \|u - u_h\|_1 \leq Ch \|u\|_2 \leq Ch \|f\|_0, \tag{30}$$

$$\||e\|| = \int_\Omega \epsilon(e) : \mathbb{C} : \epsilon(e) \, d\Omega, \text{ the energy norm.} \tag{31}$$

An important result by Babuška and Reinboldt (1978), [10], for the explicit residual error estimator in the energy norm is the convergence order h^p for the characteristic element length h and the polynomial order p of the shape functions, saying that the stresses (essentially the first derivatives of the displacements) converge with the power p; C is the global unknown interpolation constant in case of well-posed problems. Thus it holds

$$\||e\|| = C\mathcal{O}(h^p). \tag{32}$$

According to Aubin (1967), [5], and Nitsche (1977), [65], the approximated displacements converge with power $p+1$ in the $\mathcal{L}_2$-norm

$$\|e\|_{\mathcal{L}_2} = C\mathcal{O}(h^{p+1}), \tag{33}$$

With $v_h(x) \in [\mathcal{H}^1(\Omega)]^3$, $f(x) \in [\mathcal{L}_2(\Omega)]^3$ and sufficient regularity of the b.c.'s, the approximated displacements u_h converge to $u(x) \in [\mathcal{H}^2(\Omega)]^3$ for BVPs of the Lamé equations.

The a priori residual error estimator is given by

$$\|e\|_{H^1(\Omega)} \leq C \inf_{v_h \in \mathcal{V}} \|u - v_h\|_{\mathcal{H}^1(\Omega)}; \quad v_h = \pi_h u, \tag{34}$$

and further in the energy norm

$$\||e\|| \leq \inf_{\mathcal{V}_h} \||u - u_h\||; \quad \||e\|| \leq Ch^p |u|_{\mathcal{H}^{p+1}(\Omega)}. \tag{35}$$

2.4 Residual (Energy-Based) Explicit (Relative) and Implicite (Absolute) a Posteriori Error Estimators for Mixed BVPs of the Lamé Equations by Babuška and Miller (1987)

The weak form for primal FEM of mixed BVPs reads

$$a(u_h, v_h) = F(v_h) + g(v_h) \; \forall \; v_h \in \mathcal{V}_h. \tag{36}$$

Introducing the discretization error $e := u - u_h$, the residuum of the weak form is

$$a(e, v) - a(e, v - \underbrace{w_h}_{I_h v}) := R(v - w_h) \ \forall \ v \in \mathcal{V}, \ \forall \ w_h \in \mathcal{V}_h \quad (37)$$

and yields the inequality, [9],

$$a(e, v) \leq \sum_{e=1}^{n_e} R_e(u_h) \cdot v \ dV + \sum_{r=1}^{n_r} J_r(u_h) \cdot v \ dA , \quad (38)$$

with the residual of equilibrium R_e in Ω_e and the jumps of tractions $J_r = [\![t(u_h)]\!]_{\Gamma_r}$.

The generalization for error analysis of stress intensity factors for crack propagation problems is also related to Babuška, Miller (1984), [8].

2.4.1 Explicit Error Estimators with Strict Upper and Lower Bounds

The a priori upper bound estimate in the energy norm reads, eq. (35),

$$|||e||| \leq C \, \mathcal{O}(h^p), \quad C = C_{int} \cdot C_{stab}, \quad (39)$$

where the global interpolation constant C is the product of a proper interpolation and a stability constant, depending on the differential operator and the BVP. It is unknown for explicit residual a posteriori error estimators and can become large, depending on the ellipticity property and the well-posedness of the BVP. For example, it may be large for strong elastic anisotropy, Babuška, Suri (1992), [14], Babuška, Strouboulis (2001), [11].

In double-logarithmic coordinates this a priori estimator yields a straight line with the slope $\tan \alpha(p) = \frac{p}{1}$, i.e. 1 for $p = 1$; the stresses converge linearly, fig. 2.

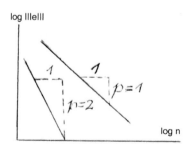

Fig. 2 Linear and quadratic convergence of the stresses for $p = 1$ and $p = 2$

With the lower bound interpolation estimate according to the Poincaré-Friedrichs-inequality,

$$\|u - Iu\|_{\mathcal{H}_1(\Omega)} \leq \||u - Iu|\|; \quad Iu = u_h, \tag{40}$$

the energy norm of the exact error $e = u - u_h \in \mathcal{V}$ is bounded from above and from below by the important inequality of Babuška, Rheinboldt (1978), [10],

$$C_l\||\Pi_h e|\| \leq \||e|\| \leq C_u\||\Pi_h e|\|; \quad \Pi_h e = e_h, \tag{41}$$

where $\Pi_h e$ is the computable projection of $e \in \mathcal{V}$ onto the ansatz space $\mathcal{V}_h$.

The residual error estimator in the energy norm results in

$$\||e|\| = \sum_{e=1}^{n_e} [\int_{\Omega_e} (div\,\boldsymbol{\sigma}(u_h) + f) \cdot v\,d\Omega + \int_{\partial\Omega_e} [\![t(u_h)]\!] \cdot v dS]. \tag{42}$$

The effectivity index quantifies the quality of the estimated error with respect to the real error as, Babuška et al. (1980), [15], fig. 3,

$$\theta = \frac{\eta}{\||e|\|}; \quad \eta = (\sum_{i=1}^{n} e_i^2)^{1/2}. \tag{43}$$

Fig. 3 Ivo M. Babuška (*1926), who contributed seminal theoretical and numerical results on a priori and a posteriori error estimation of FEM and to the whole methodology

In case of explicit residual error estimators the calculation of θ does not make sense because the global constant C is unknown, and the choice $C = 1$ can strongly overestimate the error and yield large effectivity indices, which – for upper bound estimators – should not be larger than about 1.5 for efficient adaptivity.

The Babuška-Miller explicit residual a posteriori error estimator controls the approximated equilibrium conditions in each element, at all element interfaces and at Neumann boundaries.

It should be mentioned that recently, explicitly computable constants C have been derived which hold for specific elements and p-orders for

Lamé-BVPs, Gerasimov, Stein, Wriggers (2013), [47]; the derivation uses the Poincaré inequality, the 2nd Korn inequality and the Lax-Milgram theorem. Effectivity indices of $1.2 \leq \theta \leq 1.4$ were achieved for various problems with singularities.

2.4.2 Implicit (Residual) a Posteriori Error Estimates

These constant-free (absolute) estimators, developed since the 1980s, are calculated on element patches or on single elements. They need the solution of local equation systems. Upper and lower bounds can be guaranteed using Cauchy-Schwarz inequality; other intuitive techniques for error indication may not yield guaranteed upper bounds.

Besides the requested strict upper bound property, i.e. $\eta \leq |||e|||$, small effectivity indices $\theta = \eta/|||e|||$, $1 \leq \theta \leq 2$, are requested in order to get quasi-optimal adaptive FE meshes with optimal convergence rates. But it also has to be emphasized that estimators with strict upper bounds may overestimate the a posteriori error estimator considerably, due to the Cauchy-Schwarz inequality, which especially holds in case of estimators for quantities of interest, where the vector of errors of the primal solution can enclose an angle up to 90° with the error-vector of the dual solution.

Numerous implicit estimators have been developed, and they are subjects of current research.

Element-Wise Error Estimation with Equilibrated Boundary Tractions via Local Neumann Problems by Bank and Weiser

This type of a posteriori error estimation at element boundaries was first given by Bank and Weiser (1985), [17], because u_h is only zero-continuous at $\partial \Omega_e$, the boundary tractions $t_h = \sigma(u_h) \cdot n$ at $d\Omega_e$ are discontinuous at $d\Omega_e$. Approximately equilibrated tractions $\tilde{t}_e$ are calculated by local Neumann problems. From the weak residuum of equilibrium for discretized linear-elastic problems

$$R(u_h, v_h) = F(v_h) - a(u_h, v_h) \stackrel{!}{=} 0 \quad \forall \, v_h \in \mathcal{V}_h \subset \mathcal{V}, \tag{44}$$

with the discrete test space

$$\mathcal{V}_{h,e} = \{v_h \in H^1(\Omega_e), [\![v_h]\!] = 0 \text{ on } \Gamma_e, \ v_e = 0 \text{ on } \Gamma_{e,D}\}, \tag{45}$$

and for the total test space $u \in \mathcal{V}$ it holds

$$R(u, v_h) = F(v_h) - a(u, v_h), \tag{46}$$

yielding the residuum of equilibrium with the exact error representation

$$R(u, v_h) = a(\underbrace{u - u_h}_{e}, v_h) = 0; \quad e := u - u_h, \tag{47}$$

i.e. the Galerkin orthogonality and the construction principle of FEM. The bilinear form of the discretization error for each element Ω_e reads

$$a_e(e_e, v_e) = \int_{\Omega_e} f_e \cdot v_e d\Omega + \int_{\Gamma_e} \underbrace{t_e(\sigma(u_h))}_{\sigma_e(u_h)\cdot n} \cdot v_e dS + \int_{\Gamma_{e,N}} t_{e,N} \cdot v_e dS$$
$$- \int_{\Omega_e} \sigma(u_h)|_{\bar{\Omega}_e} : \epsilon(v_e) d\Omega \quad \forall v_e \in V_e \tag{48}$$

Remarks:

- This is a pure Neumann problem, because it is gained by cutting a single element $\bar{\Omega}_e = \Omega_e \cup \Gamma_e$ out of the entire body $\bar{\Omega}$.
- Although the problem is purely local, the data comprise information from the entire body $\bar{\Omega}$, as t_e on $\partial \Omega_e$ are the exact tractions.
- As a consequence the exact tractions $t(\partial \Omega_e)$ can only be approximately determined, denoted by $\tilde{t}(\partial \Omega_e)$ and referred to as *equilibrated tractions*.

The idea of deriving an a posteriori error estimator is to substitute the exact tractions t_e by the approximated C^0 continuous $\tilde{t}_e$, permitting to calculate the linearized error ψ_e of the true error e_e of an interior element

$$a_e(\psi_e, v_e) = \int_{\Omega_e} f_e \cdot v_e d\Omega + \int_{\partial \Omega_e} \tilde{t}_e \cdot v_e dS - \int_{\Omega_e} \sigma(u_{h,e}) : \epsilon(v_e) d\Omega. \tag{49}$$

Summing up the righthand sides of all elements yields the global error representation and the equilibrated residual $\tilde{R}$

$$a(\psi, v) = \sum_{n_e} a_e(\psi_e, v_e) = \sum_{n_e} \tilde{R}_{h,e}(\psi_e, v_e). \tag{50}$$

The weak form of equilibrium for the equilibrated tractions is

$$\sum_{\Gamma_{N,e}} \int_{\Gamma_{N,e}} \bar{t}_e \cdot v_e \, dS = \sum_{n_e} \int_{\Gamma_e} \tilde{t}_e \cdot v_e \, dS \quad \forall v_e \in V. \tag{51}$$

The computation of the $\tilde{t}_e$ is not unique; several approximation methods are known and discussed by Babuška, Strouboulis (2001), [11], Stein, Ohnimus (1999), [87].

The bilinear form of the linearized error is represented by the approximated linearized errors ψ_e,

$$a(e, v) \approx \sum_{n_e} a_e(\psi_e, v_e|_{\bar{\Omega}_e = \Omega_e \cup \Gamma_e}), \tag{52}$$

and due to Ainsworth, Oden (2000), [2], applying the Cauchy-Schwarz inequality twice, we get the implicit a posteriori error estimator in the energy norm with the guaranteed upper bound as

$$|||e||| \leq \{\sum_{n_e}|||\psi_e|||^2\}^{1/2}. \tag{53}$$

Implicit Residual Error Estimators Based on Equilibrated Interface Tractions via Local Neumann Problems by Stein and Ohnimus Using Orthonormal Higher Test Functions

This estimator is based on the calculation of improved interface tractions for 2nd order elliptic PDEs, especially for plates in bending and transverse shear, Bufler, Stein (1970), [31], Stein, Ahmad (1977), [83].

The discrete global variational problem with the residuum of equilibrium reads

$$a(e, v) = l(v) - a(u_h, v) := R(v) \quad \forall v \in V, v_h, u_h \in V_h \subset V, \tag{54}$$

yielding the representation of the error in the energy norm

$$|||e||| = \sup_{v \in V(\Omega)} \frac{R(v)}{|||v|||}. \tag{55}$$

The residuum $R(v) := \sum_{n_e} R_e(v)$ is replaced by the improved one $\tilde{R}(\tilde{v}) := \sum_{n_e} \tilde{R}_e(\tilde{v}_e)$ via C^0-continuous equilibrated boundary tractions $\tilde{t}$ at all interfaces Γ_e, i.e. $\tilde{t}(x)_{\Gamma_i} + \tilde{t}(x)_{\Gamma_j} = 0, i \neq j$, and $\tilde{t}_{\Gamma_k} = \bar{t}_{\Gamma_k}$ must hold at Neumann element boundaries.

A simple approximation for the $\tilde{t}$ is averaging of the discontinuous tractions $t_{h,e}(u_h)$ at element interfaces. More general, also valid for anisotropic problems and dimensional (model) adaptivity, the improved tractions are calculated by overdetermined local algebraic equation systems on element levels.

The current test space V_h does not permit to determine improved interface tractions and constant-free a posteriori error estimators. Therefore, an enhanced test space $V_{h+} \subset V \ominus V_h$, $\tilde{v} = v_h + v_{h+} \in \tilde{V}_e = V_{h,e} \cup V_{h+,e}$ with the computable discretization error $\tilde{e} = e_h + e_{h+}, \tilde{e} \in \tilde{V}_e$ is introduced for calculating equilibrated interface tractions $\tilde{t} = t_h + t_{h+}$ with $\tilde{v}_{\Gamma_e} \subset H^{1/2}(\Gamma_e)^2$ and $\tilde{t}_{\Gamma_e} \subset H^{-1/2}(\Gamma_e)^2$ for 2D problems.

The solvability presumes that the righthand side of the local variational problem becomes zero if the test functions v are in the kernel Z of rigid body motions, given by

$$Z(\Omega_e) := \{Z \in V(\Omega_e) \text{ for } a(z, v_e)_{\Omega_e} = 0 \; \forall v \in V(\Omega_e)\}, \tag{56}$$

for which $\tilde{R}_e(\tilde{v}) = 0 \; \forall \tilde{v} \in Z(\Omega_e)$ holds.

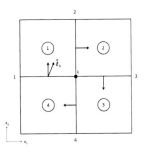

Fig. 4 Patch $\mathcal{P}_k$ for element node k with normal vectors at the element sides and C^0-constinuous tractions acting at element sides

The equation for determining improved (linearized) displacement errors $\boldsymbol{\psi}_e$, according to $\tilde{e}_e = \boldsymbol{\psi}_e + \text{n.l.t.}$, of error analysis for an element e reads

$$a_e(\boldsymbol{\psi}_e, \boldsymbol{v}_{h+}) = l(\boldsymbol{v}_{h+}) - a_e(\boldsymbol{u}_h, \boldsymbol{v}_{h+}) = \tilde{R}_e(\boldsymbol{v}_{h+}) \text{ in } \Omega = \bigcup_{n_e} \Omega_e, \qquad (57)$$

and the global improved residuum of equilibrium results in

$$\tilde{R}(\boldsymbol{v}_{h+}) = \sum_{n_e} [\tilde{R}_e(\boldsymbol{v}_{h+}) + \int_{\Gamma_e} \tilde{\boldsymbol{t}}(\boldsymbol{v}_{h+}) \cdot \boldsymbol{v}_{h+} \mathrm{d}\boldsymbol{x}] = 0. \qquad (58)$$

Solving the elementwise variational problems

$$a_e(\boldsymbol{\psi}_e, \boldsymbol{v}_{h+}) = \tilde{R}_e(\boldsymbol{v}_{h+}) \text{ in } \Omega_e \; \forall \boldsymbol{v}_{h+} \in \boldsymbol{v}_{h+}(\mathcal{V}_{h+} \in \mathcal{V}), \qquad (59)$$

and summing them up yields the global constant-free error estimator with the upper bound property
$|||\boldsymbol{\psi}||| = (\sum_{n_e} |||\boldsymbol{\psi}_e|||^2)^{1/2} \leq |||e|||$.

The requested equilibrated interface tractions $\tilde{\boldsymbol{t}}_h \in H^{-1/2}(\Gamma_e)$ are defined on inner sides of element patches around an element edge node k, for 2D problems, fig. 4, [86], [82], [85], and approximated by Lagrangean trial functions N_t as

$$\tilde{\boldsymbol{t}}_h = \boldsymbol{N}_t \hat{\tilde{\boldsymbol{t}}} = \begin{bmatrix} N_t(r,s) & 0 \\ 0 & N_t(r,s) \end{bmatrix} \begin{Bmatrix} \hat{\tilde{t}}_x \\ \hat{\tilde{t}}_y \end{Bmatrix} ; \; N_t(r,s) = N_H(r) \otimes N_H(s). \qquad (60)$$

Using the same polynomial basis for the extended test displacements $\boldsymbol{v}_h^+$, but with contravariant shape functions, yields diagonal matrices for the determination of the improved tractions, according to the equilibrium condition

$$\int_{\partial\Omega_e} \tilde{\boldsymbol{t}}_h^T \boldsymbol{v}_h \mathrm{d}\boldsymbol{s} = (\boldsymbol{u}_h, \boldsymbol{v}_h)_{\Omega_e} - \int_{\Omega_e} \boldsymbol{b}^T \boldsymbol{v}_h \mathrm{d}\boldsymbol{x} \quad \forall \boldsymbol{v}_h \in \boldsymbol{V}_h(\Omega_e), \qquad (61)$$

and with the energy-consistent fictitious nodal load vectors $\hat{\boldsymbol{p}}_{he}$ as well as the related nodal test displacement vector $\hat{\boldsymbol{v}}_h$, one gets

$$a(\boldsymbol{u}_h, \boldsymbol{v}_h)_{\Omega_e} - \int_{\Omega_e} \boldsymbol{b}^T \boldsymbol{v}_h \, \mathrm{d}\boldsymbol{x} := \hat{\boldsymbol{p}}_{he}^T \hat{\boldsymbol{v}}_h, \qquad (62)$$

yielding

$$\hat{\boldsymbol{p}}_{he}(\boldsymbol{u}_h) = \overbrace{\int_{\Omega_e} \boldsymbol{N}_v^T \bar{\boldsymbol{N}}_t \mathrm{d}s}^{I} \; \hat{\boldsymbol{t}}_h. \qquad (63)$$

The transformation of the parametric to the real element geometry is got by the Jacobi matrix of coordinate derivatives.

The technique with orthonormal co- and contravariant test functions is shortly explained in the sequel:

[Isoparametric elements with Lagrange interpolation polynomials $N_i(\xi)$ and their tensor products for 2D and 3D problems, as used for FEM, can be hierarchically extended by orthonormal co- and contravariant Legendre polynomials $L_k(\xi)$ as $\int_{-1}^{1} L_i(\xi)\bar{L}^j(\xi) \, \mathrm{d}\xi = \delta_i^j$, defined in the parameter space of isoparametric elements.

Likewise also enhanced orthonormal test functions $H_k(\xi)$ with halvened element lengths were derived for h-adaptivity via piecewise linear combinations of Legendre polynomials for $h^+ = h/2, h/4, h/8$, as $\int_{-1}^{1} H_i(\xi)\bar{H}^j(\xi) \, \mathrm{d}\xi = \delta_i^j$, with the condition that they are as near as possible to each other, realized by $\sum_k (H_k(\xi) - L_k(\xi))^2 \to \min$.

These extended hierarchical functions are presented in figs. 5 and 6, [86], [87], [82], [85].]

The equilibrated tractions are determined from the equilibrium conditions at element interfaces including the interior load of an element as

$$\tilde{R}_e(\boldsymbol{v}_{\mathcal{Z}_e}) = 0 \; \forall \boldsymbol{v}_{\mathcal{Z}_e} \in \mathcal{Z}_e = \{\boldsymbol{v}_e \in \mathcal{V}_e; \; a_e(\boldsymbol{w}_e, \boldsymbol{v}_e) \equiv 0, \; \forall \boldsymbol{w}_e \in \mathcal{V}_e\}, \qquad (64)$$

with $\mathcal{Z}_e$ the space of admitted rigid body modes.

The resulting equation matrix for nodal values of $\tilde{\boldsymbol{t}}_e$ has one zero eigenvalue for bilinear isoparametric 2D- and five zero eigenvalues for trilinear 3D hexahedral elements. The approximation by the least squares method

$$\int_{\Gamma_e} [\tilde{\boldsymbol{t}}_e - \boldsymbol{\sigma}(\boldsymbol{u}_{h,e}) \cdot \boldsymbol{n}]^2 \to \min_{\tilde{\boldsymbol{t}}_e} \qquad (65)$$

Linear actual test functions

$N_0^1 = \frac{1}{2}(1-\xi)$

$N_1^1 = \frac{1}{2}(1+\xi),$

hierarchical p-extension:

$L_2 = (1-\xi^2)$

$L_3 = \xi(1-\xi^2)$

$L_4 = 5\xi^4 - 6\xi^2 + 1,$

hierarchical h-extension:

$H_2^1 = (1+\xi)$ for $\xi \leq 0$

$H_2^1 = (1-\xi)$ for $\xi \geq 0$

$H_3^1 = (2+2\xi)$ for $-1 \leq \xi \leq -0.5$

$H_3^1 = (-2\xi)$ for $-0.5 \leq \xi \leq 0.5$

$H_3^1 = (2\xi - 2)$ for $0.5 \leq \xi \leq 1$

Quadratic actual test functions

$N_0^2 = \frac{1}{2}(\xi^2 - \xi)$

$N_1^2 = (1-\xi^2)$

$N_2^2 = \frac{1}{2}(\xi^2 + \xi),$

$\int_{-1}^{+1} N_L^{p_i}(\xi) \bar{N}_L^{p_j}(\xi) \, d\xi = \delta_i^j, \; j \geq i$

$\bar{N}_L^{p_j}(\xi) = \det J_e N_L^{p_j}(\xi)$

$L_3 = \xi(1-\xi^2)$

$L_4 = 5\xi^4 - 6\xi^2 + 1,$

$\int_{-1}^{+1} H_n(\xi) \cdot \bar{P}_{n\pm k}(\xi) d\xi = 0$ for $k \in \{1, 3, 4, ...\}$

$\bar{P}_{n\pm k}(\xi) = \det J_e P_{n\pm k}(\xi)$

$H_3^2 = (-4\xi - 4\xi^2)$ for $\xi \leq 0$

$H_3^2 = (-4\xi + 4\xi^2)$ for $\xi \geq 0$

$H_4^2 = (1 + 4\xi + 3\xi^2)$ for $\xi \leq 0$

$H_4^2 = (1 - 4\xi + 3\xi^2)$ for $\xi \geq 0$.

Fig. 5 Co- and contravariant hierarchical orthonormal p- and h-extensions of Lagrangian polynomials. $N_L^{p_j}(\xi)$ are covariant, and $\bar{N}_L^{p_i}(\xi)$ are related contravariant test functions for p-extension. $H_n(\xi)$ are corresponding h-extensions. $P_{n\pm k}(\xi)$ and $\bar{P}_{n\pm k}(\xi)$ are co- and contravariant Legendre polynomials. J_e is the Jacobi determinant of the isoparametric element.

with the stationarity condition, yields the symmetric and positive definite Gauß transformation $\boldsymbol{A}^T \boldsymbol{A} = \boldsymbol{M}$ of a row-regular rectangular matrix $\boldsymbol{A}$ for determining the improved nodal values of the equilibrated tractions $\tilde{\boldsymbol{t}}_e$.

Several other approximations are possible to determine the equilibrated tractions, e.g. by solving local Laplace problems, [11].

The determination of equilibrated interface tractions – instead of averaging the left and right tractions at an interface, e.g. by Ladevèze, Leguillon (1983), [55] – is of importance for anisotropic problems, e.g. fiber-reinforced epoxid laminates, [86], [88].

For the equilibrated C^0 interface tractions $\tilde{\boldsymbol{t}}|_{\Gamma_e}$ the orthonormal polynomials, fig. 5, are used, in matrix notation $\tilde{\boldsymbol{t}} = \boldsymbol{N}_t(\xi)\hat{\tilde{\boldsymbol{t}}}$, and the least squares condition for the overdetermined local problem reads

$$(\tilde{\boldsymbol{t}} - \boldsymbol{\sigma}(\boldsymbol{u}_h) \cdot \boldsymbol{n})^2_{\bar{\Omega}_e} \to \min_{\hat{\tilde{\boldsymbol{t}}}}, \quad \bar{\Omega}_e = \Omega_e \cup \Gamma_e. \tag{66}$$

The derivative with respect to the nodal traction components, $\partial(\cdot)/\partial\hat{\tilde{\boldsymbol{t}}} = 0$, yields in matrix notation an equation system with a symmetric, positive definite matrix $\boldsymbol{N}_t^T \boldsymbol{N}_t$ from which $\hat{\tilde{\boldsymbol{t}}}$ follows as

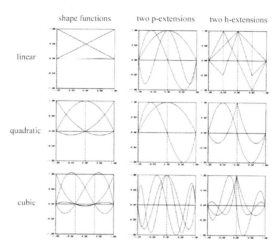

Fig. 6 Graphs of linear, quadratic and cubic test functions, each with two p- and h-extensions

$$N_t^T N_t \hat{\tilde{t}} - N_t^T \bar{t} = 0 \rightsquigarrow \hat{\tilde{t}} = (N_t^T N_t)^{-1} N_t^T \bar{t}. \qquad (67)$$

Implicit Estimators, Especially of Constitutive Equation Errors by Ladevèze et al.

The Cachan-school in France, inspired and directed by Pierre Ladevèze, contributed since the 1980s to error estimation with upper bound properties and related adaptive FEM in a broad field of applications in solid and structural engineering. Implicit error estimates for primal FEM using equilibrated element residuals of equilibrium and averaged boundary tractions $\tilde{t}_h$ (derived from the precalculated displacements u_h) followed by global equilibration was published by Ladevèze, Leguillon (1983), [55], by Ladvèze, Maunder (1996), [56], and as a textbook by Ladvèze, Pelle (2005), [58].

Main efforts were related to error estimators for primal FEM, using the *constitutive relation error* of the a priori not fulfilled constitutive equations. For linear elastic material and linear kinematics, there is no mathematical difference with respect to the above treated implicit residual error estimators, but for nonlinear problems the analysis is different; the a posteriori estimator yields good effectivity indices, Ladevèze, Moës (1998), [57] (for time-dependend problems), and Ladevèze (2001), [52].

The constitutive relation error can be motivated and derived from the two-functional Prager-Synge hypercircle theorem, Prager, Synge (1949), [70], used in the book of Ladevèze, Pelle (2005), [58], and in several articles by Ladevèze, Moës, e.g. (1998), [57], also extended to time-dependent problems. Error bounds without generic constants are achieved by comparing the primal variational problem with the mixed dual formulation. This method was further developed by Neittaanmäki and Repin (2004), [63]. A related a posteriori error estimate – as used by Ladevèze, [58] – was mathematically proven by Braess and Schöberl (2008), [27]. The mathematical treatment can also be found in Verfürth (2013), [93], and in the 5th edition of Braess (2013), [26].

Additionally, Ladevèze et al. extended implicit constitutive relation error estimates to those for quantities of interest, Ladevèze (2008), [53], (2011), [54], see also Stein et al. (2011), [89]. Therein the product of the lengths of the primal and the dual error estimators – due to the Cauchy-Schwarz inequality – appears, which can lead to significant overestimation, because these vectors may enclose an angle of up to 90°.

Implicit Residual Error Estimators Based on Local Dirichlet Problems

This type of error estimators was derived by Babuška & Reinboldt (1978). The basic idea again follows the error representation formula eq. 47 and considers node-oriented element-patches for each node $X_n \in \bar{\Omega}$ with node-based test functions $\boldsymbol{v}_h = \boldsymbol{N}_v \hat{\boldsymbol{v}}_e$, yielding the matrix of shape functions $\boldsymbol{N}_v(\boldsymbol{x})$ and the vector $\hat{\boldsymbol{v}}_e$ of Lagrangian nodal test displacements of the patch. Then the error can be represented as

$$\sum_{n_n} a(\boldsymbol{e}, \boldsymbol{v}_n) = \sum_{n_n} R(\boldsymbol{v}_n) \quad \forall \boldsymbol{v}_n \in \mathcal{V}_p = [\mathcal{H}_0^1(\bar{\Omega}_p)]^3. \tag{68}$$

The corresponding Dirichlet problem reads: Find the linearized error $\psi_p \in \mathcal{V}_p$ by the problem

$$a_p(\boldsymbol{\psi}_p, \boldsymbol{v}_p) = \int_{\Omega_p} \boldsymbol{f} \cdot \boldsymbol{v}_p d\Omega + \int_{\partial \Omega_p \cap \Gamma_n} \bar{\boldsymbol{t}} \cdot \boldsymbol{v}_p dS - \int_{\Omega_p} \boldsymbol{\sigma}(\boldsymbol{u}_h)|_{\bar{\Omega}_p} : \boldsymbol{\epsilon}(\boldsymbol{v}_p) d\Omega \quad \forall \boldsymbol{v}_p \in \mathcal{V}_p, \tag{69}$$

and the global bilinear form of the error reads

$$a(\boldsymbol{e}, \boldsymbol{v}) = \sum_{n_n} a_p(\boldsymbol{\psi}_p, \boldsymbol{v}_n). \tag{70}$$

Using the Galerkin orthogonality, the Cauchy-Schwarz inequality and an interpolation estimate one arrives at the guaranteed upper bound estimator with the generic constant C_1

$$|||\boldsymbol{e}||| \leq C_1 (\sum n_n a_p(\boldsymbol{\psi}_p, \boldsymbol{\psi}_p))^{\frac{1}{2}}. \tag{71}$$

One also can derive a lower bound estimator, applying again the Cauchy-Schwarz inequality and a partitioning of the nodes X_n as

$$C_2 (\sum_{n_n} a(\boldsymbol{\psi}_p, \boldsymbol{\psi}_p)) \leq |||\boldsymbol{e}|||, \tag{72}$$

with the global multiplicative constant C_2 for well-posed problems.

2.4.3 Gradient Smoothing and Stress Recovery "ZZ" Explicit and Implicit a Posteriori Error Estimators by Zienkiewicz and Zhu (1987)

Different from the explicit residual error estimator by Babuška and Miller (1987), [9], (38) and (39), Zienkiewicz and Zhu (1987), [94], published an a posteriori error estimator based on gradient-smoothing of the displacements and recalculation of recovered C^0-continuous stresses at element interfaces for which the shape functions are chosen equal to those for the primal C^0-continuous displacement ansatz functions, fig. 7.

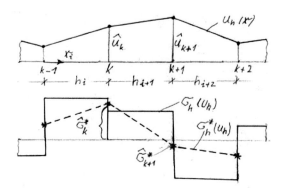

Fig. 7 1D linear C^0-continuous displacement ansatz (top), and discretized stresses $\sigma_h(u_h)$ as well as recovered C^0-continuous stresses $\sigma_h^*(u_h)$ (bottom)

Explicit Smoothing and Recovery Estimator

Primal FEM with (e.g. linear) discretized displacements $u_h(x) = N(x)\hat{u}_e$ in Ω_e, $\Omega = \cup_e \Omega_e$, and discretized strains $\epsilon_h(u_h) = DN\hat{u}_e = B\hat{u}_e$ yields the discretized stresses $\sigma_h(u_h(x)) = \mathbb{C}B(x)\hat{u}_e$, according to the symmetric and positive definite elasticity tensor $\mathbb{C}$.

The idea of gradient smoothing is to introduce so-called *recovered* stresses

$$\sigma_h^*(x) := N(x)\,\hat{\sigma}_e^* \; \forall \; \Omega_e \tag{73}$$

with same shape functions $N(x)$ (e.g. linear) as for the displacement ansatz $u_h(x)$ and with C^0-continuity at element interfaces, fig. 7. Of course, it has to be proven that the recovered stresses σ_h^* are more accurate than the discretized stresses σ_h. Then the least squares conditions for all elements read

$$\bigcup_{e=1}^{N_{el}} \int_{\Omega_e} \underbrace{[\sigma_h^*(x) - \sigma_h(u_h(x))]^2}_{e_\sigma^*(\Omega_e)} \, d\Omega_e \to \min_{\hat{\sigma}_k^*} \hookrightarrow \hat{\sigma}_k^*, \tag{74}$$

with the a posteriori gradient smoothing stress error

$$e_\sigma^*(x) := \sigma_h^*(x) - \sigma_h(u_h(x)) \text{ in } \Omega_e \in \Omega. \tag{75}$$

The stationarity condition yields a symmetric system of linear equations for the nodal values $\hat{\sigma}_k^*$ of nodes k. Approximated explicit solutions are gained by regarding only the diagonal matrix or by nodal averaging.

The global error estimator in the complementary energy norm then reads

$$|||e_\sigma^*|||_\Omega = [\sum_{e=1}^{N_{el}} (\int_{\Omega_e} e_\sigma^* : \mathbb{C}^{-1} : e_\sigma^* \, d\Omega)]^{1/2}. \tag{76}$$

Numerous discussions amongst mathematicians and mechanicians took place in the 1980s and 1990s whether the ZZ estimator was mathematically sound, yielding upper bounds for well-posed BVPs for sufficient regularity properties and to which extend the recovered stresses are more accurate than the discretized ones. On the other hand, numerous comparative convergence studies showed the superiority or at least equal convergence behaviour compared with explicit residual error estimators. An upper bound was proven by Rodríguez (1994), [76], as

$$\|\sigma(u) - \sigma^*(u_h)\|_{L_2(\Omega)} \leq C \|\sigma(u) - \sigma(u_h)\|_{L_2(\Omega)}, \tag{77}$$

where C is a global interpolation constant. Upper and lower bounds were proven by Carstensen and Funken (2001), [32], as

$$\frac{1}{1+C} \underbrace{\|\sigma^*(u_h) - \sigma(u_h)\|_{L_2(\Omega)}}_{e^*(\sigma^*)} \leq \underbrace{\|\sigma(u) - \sigma(u_h)\|_{L_2(\Omega)}}_{e(\sigma(u_h))} \leq \frac{1}{1-C} \underbrace{\|\sigma^*(u_h) - \sigma(u_h)\|_{L_2(\Omega)}}_{e^*(\sigma^*)}. \tag{78}$$

Implicit "ZZ" SPR Smoothing Estimator

Zienkiewicz and Zhu published in 1992 an important improvement of their explicit estimator by the so-called *SPR* (superconvergent patch recovery) which is performed with postprocessing on element patches $\bar{\Omega}_P$ around the considered element nodes (instead of nodal averaging of stress components), Zienkiewicz (1992), [95], fig. 8.

SPR is based on the fact that – in case of sufficient regularity – the discretized stresses $\sigma(u_h)|_{\Omega_e}$ are more accurate at Gauss points (GPs) of elements Ω_e than at element interfaces $\partial \Omega_e$ and at nodal points of the elements. Choosing these GPs as optimal sampling points $\xi_n(\bar{\Omega}_P)$, superconvergence can be achieved for the improved nodal stress components $\hat{\sigma}_I^*(X_k)|_{\bar{\Omega}_P}$.

The vector of the searched nodal stress components at patch node k is presented by bilinear monomials $P(x,y)$, e.g. for 2D problems using $\mathbb{Q}_1$-quadrilateral elements, as

$$\hat{\sigma}_I^*(x_k)|_{\bar{\Omega}_P} := P^T(x_k)\hat{a}_I; \quad P = [1 \ x \ y \ xy]^T \text{ for } \Omega_e \subset \mathbb{R}^2. \tag{79}$$

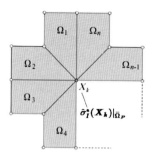

Fig. 8 SPR for element patch $\bar{\Omega}_P$ surrounding node k

The unknown coefficients $\hat{a}_I$ are determinded by the condition

$$\sum_{n_{GP|\bar{\Omega}_P}} [\sigma_I(\boldsymbol{u}_h(\boldsymbol{\xi}_n)) - \boldsymbol{P}^T(\boldsymbol{\xi}_n)\hat{\boldsymbol{a}}_I] = 0, \tag{80}$$

with the row-regular rectangular matrix $\boldsymbol{P}$. In case of more sampling points than coefficients of the vector $\hat{\boldsymbol{a}}_{I,k}$ the least squares method

$$\sum_{n_{GP|\bar{\Omega}_P}} \boldsymbol{P}(\boldsymbol{\xi}_n)\boldsymbol{P}^T(\boldsymbol{\xi}_n)\hat{\boldsymbol{a}}_I = \sum_{n_{GP|\bar{\Omega}_P}} \boldsymbol{P}(\boldsymbol{\xi}_n)\sigma_I(\boldsymbol{u}_h(\boldsymbol{\xi}_n)) \tag{81}$$

yields $\hat{\boldsymbol{a}}_I$ from a small symmetric, positive definite system of linear algebraic equations for each patch. Boundary nodes $\boldsymbol{x}_n(\partial\bar{\Omega})$ need a special treatment.

Superconvergence could be shown for many applications by applying quasi-optimal mesh refinements for the calculated a posteriori error estimators.

Important Remark

From the state of the art of the available general purpose finite element programs there is a big advantage of the explicit ZZ estimator compared with implicit residual estimators, because the smoothing estimator $|||e_{ZZ}^*|||$ only needs data from the considered elements themselves but not from the neighboring elements – whereas the residual estimator $|||e_{Res}|||$ requires the jumps of the tractions to neighboring elements in the residuum of equilibrium. However, so far the access to element neighborhoods is not available in all industrial finite element programs, and therefore, the ZZ estimators are predominant in practice. They are available in most program systems, combined with either local mesh refinements – mostly for triangular and tetrahedral elements – or for global mesh refinements according to the distribution of scaled error indicators of the system, e.g. in Abaqus.

Also gradient-free implicit error estimates with equilibrated test stresses in elements/patches, e.g. by Diez et al. (2010), [37], do not require data from neighboring elements/patches.

3 Error Estimators for Quantities of Interest Based on Primal and Dual Local Solutions

In this section, goal-oriented error estimators are treated which play a big role in engineering. Thus, e.g. for a crack propagation problem, one is not interested in the convergence and the tolerance of the error in the energy norm but in related functionals on local supports, e.g. the Rice integral for crack propagation in linear elasticity. Further, averaged displacements and stresses, defined on small supports, are quantities of interest.

Duality techniques were developed by Johnson et al. (1993), [51], Ericksson et al. (1995), [43], and extended by Becker and Rannacher (1996, 1998), [18], [19], as well as Rannacher, Suttmeier (2003), [71], as the dual weighted residual method.

3.1 Primal and Dual Elliptic Variational Problems

A goal-oriented error measure is given by the Gâteaux differentiable generally nonlinear or linear functional $E : \mathcal{V} \times \mathcal{V} \to \mathbb{R}$, satisfying the condition $E(v, v) = 0$, $\forall\, v \in \mathcal{V}$, introduced by Larsson et al. (2002), [59]. With the quantity of interest $Q : \mathcal{V} \to \mathbb{R}$, the error measure E is equal to the error of the quantity of interest, as

$$E(u, u_h) = Q(u - u_h) = Q(e_u) = Q(u) - Q(u_h). \tag{82}$$

If E is a linear functional, the quantity of the error and the error of the quantity coincide. Due to $E(u_h, u_h) = 0$, it holds

$$E(u, u_h) = E(u, u_h) - E(u_h, u_h), \tag{83}$$

from where some mathematical properties for nonlinear functionals E can be deduced.

3.2 Duality Technique

For estimating E, an auxiliary "dual problem" next to the "primal problem" has to be solved. It can be understood mechanically by the Betti-Maxwell reciprocity theorem, because we need a further solution with a loading (right-hand side) which is reciprocal to the considered averaged quantity (functional) of interest. Mathematically, we consider for simplicity a linear elliptic partial differential operator $A : \mathcal{V} \to \mathcal{V}'$, defined by

$$\langle A(u), v \rangle_{\mathcal{V}' \times \mathcal{V}} = a(u, u_h; w, v)\ \forall w, v \in \mathcal{V}. \tag{84}$$

The corresponding dual operator $A' : \mathcal{V} \to \mathcal{V}'$ is defined by

$$\langle A(\boldsymbol{w}), \boldsymbol{v}\rangle_{\mathcal{V}' \to \mathcal{V}} = \langle \boldsymbol{w}, A'(\boldsymbol{v})\rangle_{\mathcal{V} \to \mathcal{V}'} \; \forall \boldsymbol{w}, \boldsymbol{v} \in \mathcal{V}, \tag{85}$$

which, in turn, implies the dual bilinear form $\overset{*}{a} : \mathcal{V} \times \mathcal{V} \to \mathbb{R}$, defined by

$$\overset{*}{a}(\boldsymbol{u}, \boldsymbol{u}_h; \boldsymbol{w}, \boldsymbol{v}) = \langle A'(\boldsymbol{w}), \boldsymbol{v}\rangle_{\mathcal{V}' \times \mathcal{V}}. \tag{86}$$

Therefore it holds

$$\overset{*}{a}(\boldsymbol{u}, \boldsymbol{u}_h; \boldsymbol{w}, \boldsymbol{v}) = \langle A'(\boldsymbol{w}), \boldsymbol{v}\rangle_{\mathcal{V}' \times \mathcal{V}} = a(\boldsymbol{u}, \boldsymbol{u}_h; \boldsymbol{v}, \boldsymbol{w}), \tag{87}$$

stating that if $A = A'$, then a is symmetric and $a = \overset{*}{a}$ holds.

It can be shown that for a non-linear elliptic operator A_{nonlin}, for which A_s is the tangential, self-adjoint operator (according to the Gâteaux derivative), the dual operator A' is always linear such that for nonlinear BVPs the dual solution is linear, Larsson (2002), [59] and Rüter (2003), [77].

To get E one has to find the discretized dual solution $\overset{*}{\boldsymbol{u}}_h \in \mathcal{V}_h \subset \mathcal{V}$ for the dual problem

$$\overset{*}{a}(\boldsymbol{u}, \boldsymbol{u}_h; \overset{*}{\boldsymbol{u}}_h, \boldsymbol{v}) = E(\boldsymbol{u}, \boldsymbol{u}_h; \boldsymbol{v}) \; \forall \boldsymbol{u}_h, \overset{*}{\boldsymbol{u}}_h \subset \mathcal{V}_h. \tag{88}$$

Substituting $\overset{*}{\boldsymbol{u}} \in \mathcal{V}$ in (88) with the finite element approximation $\overset{*}{\boldsymbol{u}}_h \in \mathcal{V}_h$ yields the weak form of the residual $R_{\overset{*}{\boldsymbol{u}}_h} : \mathcal{V} \to \mathbb{R}$ of the dual problem

$$R_{\overset{*}{\boldsymbol{u}}_h}(\boldsymbol{v}) = E(\boldsymbol{u}, \boldsymbol{u}_h; \boldsymbol{v}) - \overset{*}{a}(\boldsymbol{u}, \boldsymbol{u}_h; \overset{*}{\boldsymbol{u}}_h, \boldsymbol{v}). \tag{89}$$

With the associated Galerkin orthogonality of the exact linearization E_S

$$R_{\overset{*}{\boldsymbol{u}}_h}(\boldsymbol{v}_h) = E_S(\boldsymbol{u}, \boldsymbol{u}_h; \boldsymbol{v}_h) - \overset{*}{a}_S(\boldsymbol{u}, \boldsymbol{u}_h; \overset{*}{\boldsymbol{u}}_h, \boldsymbol{v}_h) = 0 \tag{90}$$

and the dual discretization error $\boldsymbol{e}_{\overset{*}{\boldsymbol{u}}} = \overset{*}{\boldsymbol{u}} - \overset{*}{\boldsymbol{u}}_h$, it holds

$$\overset{*}{a}_S(\boldsymbol{u}, \boldsymbol{u}_h; \boldsymbol{e}_{\overset{*}{\boldsymbol{u}}}, \boldsymbol{v}) = R_{\overset{*}{\boldsymbol{u}}(v)}. \tag{91}$$

The approximated dual variational problems

$$\overset{*}{a}_T(\overset{*}{\boldsymbol{u}}, \boldsymbol{v}) = E_T \text{ and } \overset{*}{a}_T(\overset{*}{\boldsymbol{u}}_h, \boldsymbol{v}_h) = E_T(\boldsymbol{v}_h) \tag{92}$$

result in the dual error, represented as

$$\overset{*}{a}_T(\boldsymbol{e}_{\overset{*}{\boldsymbol{u}}}, \boldsymbol{v}) = \tilde{R}_{\overset{*}{\boldsymbol{u}}}(\boldsymbol{v}) \; \forall \; \boldsymbol{v} \in \mathcal{V}. \tag{93}$$

From this representation several techniques for implicit error estimation can be applied, Sect. 2, see [77], [88] where equilibration of the residuals via equilibrated interface tractions is used.

Pioneering contributions came from Cirak, Ramm (1998, 2000), [35] for elastic deformations and [36] for elastoplastic deformations, as well as essential mathematical contributions from Becker, Rannacher (1996), [18], and Rannacher, Suttmeier (2003), [72], for elastoplastic deformations of Hencky type, in Stein (ed.) (2003), [81].

Error estimates for quantities of interest can also be successfully used for coupled validation and verification, treated in Sect. 4, see [77] and Stein, Rüter, Ohnimus (2011), [89].

Error measures for quantities of interest were also substantially influenced by the so-called pollution error in the h-version of FEM and the local quality of a posteriori error estimators by Babuška et al. (1994), [12], and Babuška et al. (1995), [13].

4 Combined Discretization and Model Adaptivity

4.1 Scope and Methods

Mathematical models for engineering problems are as simple as possible and as complicated as necessary. It is the art of engineers to select those models for specific problems with the consequence that such a mathematical model has to be enhanced in the loading history or for new effects. Such an expansive model adaptivity – aiming at the validation of a mathematical model – is reasonable for 3D boundary layers of 2D plate and shell theories and the transition from elastic to elastic-plastic material behavior, Roache (1998), [75], AIAA guide (1998) [1].

In contrast to this, model order reductions are also used in the wider frame of multiscale modeling, e.g. homogenized anisotropic modeling of fiber-reinforced laminated thin-walled epoxid structures, investigated by Oden, Prudhomme (2002), [67].

It is obvious that from overall safety and reliability reasons, model adaptivity for describing physical reality as accurate as necessary is of equal importance as discretization adaptivity – realizing verification – which both have to be coupled in order to get model and discretization error tolerances of equal magnitude. These aspects are of special importance for virtual production techniques where numerical results have to replace physical test results. The rules for this new field of study were also published in the ASME V&V code 10-2006, Schwer (2007), [78].

A scheme of these coupled processes is shown in fig. 9.

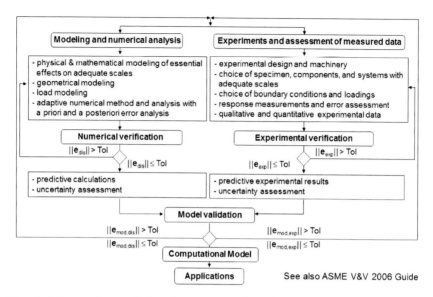

Fig. 9 Scheme of numerical and experimental verification coupled with validation

Verification Coupled with Validation

Verification of FE calculations means convergent numerical approximations via bounded a posteriori discretization error estimators and related mesh adaptivity up to a given tolerance, presumed that the analytical solution exists in the introduced test space.

Moreover, **validation** intends to calculate model error estimates for stationary or process-depending physical-mathematical models. This requires experimental feedback, such that numerical and experimental verifications are necessary to achieve a prescribed tolerance of the coupled model and discretization error tolerances.

4.2 Primal Variational Problems with Model Expansion

We define a simplified "coarse" model, e.g. a 2D shell theory, by the index m and the adapted expanded "fine" model of a 3D continuum by $m+1$. Then, the primal variational formulation for the coarse solution $\boldsymbol{u}_m \in \mathcal{V}_m$ of a linear elastic problem reads

$$a_m(\boldsymbol{u}_m, \boldsymbol{v}_m) = F_m(\boldsymbol{v}_m) \; \forall \boldsymbol{v}_m \in \mathcal{V}_m. \tag{94}$$

Likewise, the variational form for the fine 3D model reads

$$a_{m+1}(\boldsymbol{u}_{m+1}, \boldsymbol{v}_{m+1}) = F_{m+1}(\boldsymbol{v}_{m+1}) \; \forall \boldsymbol{v}_{m+1} \in \mathcal{V}_{m+1}. \tag{95}$$

The discretized weak forms are

$$a_{m,h}(\boldsymbol{u}_{m,h}, \boldsymbol{v}_{m,h}) = F_{m,h}(\boldsymbol{v}_{m,h}) \;\forall \boldsymbol{v}_{m,h} \in \mathcal{V}_{m,h} \in \mathcal{V}_m, \tag{96}$$

$$a_{m+1,h}(\boldsymbol{u}_{m+1,h}, \boldsymbol{v}_{m+1,h}) = F_{m+1,h}(\boldsymbol{v}_{m+1,h}) \;\forall \boldsymbol{v}_{m+1,h} \in \mathcal{V}_{m+1,h} \in \mathcal{V}_{m+}. \tag{97}$$

The discretization errors are defined as

$$\boldsymbol{e}_{m,dis} = \boldsymbol{u}_m - \boldsymbol{u}_{m,h} \in \mathcal{V}_m;\; \boldsymbol{e}_{m+1,dis} = \boldsymbol{u}_{m+1} - \boldsymbol{u}_{m+1,h} \in \mathcal{V}_{m+1} \tag{98}$$

The combined modeling and discretization error $\boldsymbol{e}_{m+1,dis}$ has to be presented in the test space of the fine model and therefore needs a prolongation operator $\boldsymbol{P}$, mapping the test space $\mathcal{V}_m$ into $\mathcal{V}_{m+1}$, as well as the related operator for the discretized problem $\boldsymbol{P}_h$, mapping $\mathcal{V}_{m,h}$ into $\mathcal{V}_{m+1,h}$, yielding

$$\boldsymbol{e}_{m+1,dis} = \boldsymbol{P}\boldsymbol{u}_m - \boldsymbol{P}_h \boldsymbol{u}_{m,h} = u_{m+1}^{(m)} - u_{m+1,h}^{(m)},\; \boldsymbol{P} := \boldsymbol{P}_{m+1,m}. \tag{99}$$

Then we get the combined error without the unknown operator $\boldsymbol{P}$ for the analytical solutions

$$\boldsymbol{e}_{mod,dis} = \boldsymbol{u}_{m+1} - \boldsymbol{P}\boldsymbol{u}_m + \boldsymbol{P}\boldsymbol{u}_m - \boldsymbol{P}_h\boldsymbol{u}_{m,h} = \boldsymbol{u}_{m+1} - \boldsymbol{P}_h\boldsymbol{u}_{m,h}. \tag{100}$$

The Galerkin orthogonality for the discretization error of model m is

$$a_m(\boldsymbol{u}_m - \boldsymbol{u}_{m,h}, \boldsymbol{v}_{m,h}) = a_m(\boldsymbol{e}_{m,dis}, \boldsymbol{v}_{m,h}) = 0. \tag{101}$$

Now the key problem is to achieve such an orthogonality condition as (101) that is satisfied by the prolongation operator $\boldsymbol{P}$.

Substracting $a_{m+1}(\boldsymbol{P}\boldsymbol{u}_m, \boldsymbol{P}\boldsymbol{v}_m) = F_{m+1}(\boldsymbol{P}\boldsymbol{v}_m)$ from $a_{m+1}(\boldsymbol{u}_{m+1}, \boldsymbol{v}_{m+1}) = F_{m+1}(\boldsymbol{v}_{m+1})$, with $\boldsymbol{v}_{m+1} \stackrel{!}{=} \boldsymbol{P}\boldsymbol{v}_m$, yields

$$a_{m+1}(\boldsymbol{u}_{m+1} - \boldsymbol{P}\boldsymbol{u}_m, \boldsymbol{P}\boldsymbol{v}_m) = a_{m+1}(\boldsymbol{e}_{mod}, \boldsymbol{P}\boldsymbol{v}_m) \stackrel{!}{=} 0. \tag{102}$$

which means that the model error has to be orthogonal to the prolongated test space $\mathcal{V}_m$ with respect to the bilinear form a_{m+1}. Thus, discrete elementwise orthonormal prolongating operators $\boldsymbol{P}_{h,e}$ have to be constructed which fulfill this enhanced orthogonality condition, treated in Stein, Rüter, Ohnimus (2011), [89], Stein, Rüter (2004), [88].

Then the global combined error, eq. (100), can be represented by

$$|||\boldsymbol{e}_{mod,dis}||| \leq \{\sum_{n_e} |||\boldsymbol{e}_{mod,e}|||_{m+1}^2\}^{1/2} + \{\sum_{n_e} |||\boldsymbol{e}_{mod,e}|||_m^2\}^{1/2}$$
$$\leq |||\boldsymbol{e}_{mod}|||_{m+1} + |||\boldsymbol{e}_{m,dis}|||_m. \tag{103}$$

The a posteriori error estimation is again organized with locally equilibrated residua via improved interface tractions as in Sect. 2.4.2. This results for the approximated model error $\psi_{mod,e}$ in the error representation for the equilibrated residual $\tilde{R}_{m+1,e}$

$$a_{m+1,e}(\psi_{mod,e}, v_{m+1,e}) = \tilde{R}_{m+1,e}(v_{m+1,e}) \tag{104}$$

with

$$\tilde{R}_{m+1,e} = \int_{\Omega_{m+1,e}} \tilde{F}^T_{m+1,e} v_{m+1,e}\,dV + \int_{\Gamma_{m+1,e}} \tilde{t}^T_{m+1,e} v_{m+1,e}\,dA \tag{105}$$

in which $F_{m+1,e}$ is replaced by the equilibrated linear functional $\tilde{F}_{m+1,e}$.
The global bilinear form of the error then reads

$$a_{m+1}(e_{mod}, v_{m+1}) \leq \sum_{n_e} a_{m+1,e}(\psi_{mod,e}, v_{m+1,e}). \tag{106}$$

Applying the Cauchy-Schwarz inequality yields the upper bound estimate

$$|||e_{mod}|||_{m+1} \leq \{\sum_{n_e} a_{m+1,e}(\psi_{mod,e}, \psi_{mod,e})\}^{1/2} \tag{107}$$

5 Techniques for Adaptive Finite Element Discretizations

5.1 Triangular Meshes for 2D Problems

Fig. 10 shows refinements of triangular elements.

One can easily prove that the condition number of the stiffness matrices of refined element patches depend on the ratio of side lengths and the angles between adjacent element sides. Therefore, bounds for these quantities have to be applied for local mesh refinements, Rivara (1984), [74], Borouchaki (1997), [24]. Also the book by Edelbrunner (2001), [41], over the good theoretical and practical introduction to this topic.

For global remeshing of triangular elements according to the distribution of scaled error indicators we especially refer to Delaunay-type triangulation algorithms, [24].

5.2 Quadrilateral Meshes for 2D Problems

Fig. 11 shows refinements of quadrilateral elements, e.g. $\mathbb{Q}_1$-elements.

Two techniques are indicated: first, avoiding irregular nodes; irregular elements are introduced and resolved in the next refinement step, fig. 11 a, b,

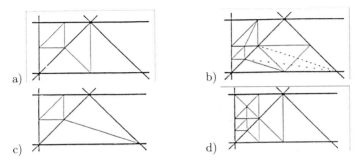

Fig. 10 a) Standard refinements of triangles without irregular nodes; b) Resolving of elements for the next refined step in order to avoid small angles; c) Halvening of longest side (Rivara); d) No resolving of elements

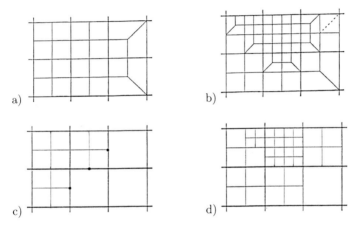

Fig. 11 Local refinement techniques for quadrilateral elements; **a)** and **b)** Transition elements for avoiding irregular nodes, irregular elements are resolved in the next refinement step; **c)** and **d)** Preserving regular elements by eliminating irregular nodes via hanging node technique

and secondly, the so-called hanging node technique where irregular nodes of regular elements are eliminated by proper interpolation of the shape functions, fig. 11 c, d.

It is obvious that the hanging node technique yields much better locality of the refinements and thus is recommendable.

In case of hexahedral elements for 3D problems also hanging node techniques on element surfaces should be applied for locality and simple algorithms, as well as regarding the fact that refinements with regular nodes yield some degenerated elements with hyperbolic surfaces. A local refinement based on preserving regular elements creates 84 new elements, Niekamp, Stein (2002), [64].

6 Table of a Posteriori Error Estimates for Primal FEM of Lamé PDEs

Table 1 shows a systematic classification of a posteriori error estimates for primal FEM. Next to residual and gradient smoothing estimators hierarchical estimators by Bank and Smith and gradient free estimators by Huerta, Diez et al. are included. The other arrangement shows explicit and implicit estimators.

Table 1 Survey of a posteriori error estimates for primal FEM

	residual (Babuška, Rheinboldt, Miller, Bank, Weiser, Ladevèze, Oden, Stein)	hierarchical (Bank, Smith)	gradient free (Huerta, Diez, et.al.)	gradient smoothing (Zienkiewicz, Zhu; Carstensen, Funken)
explicit (relative) dep. from interpolation and stability constants	• upper and lower bounds • estimation of residua of equilib. in all Ω_e and at all Γ_e and at $\Gamma_{N,e}$	• lower and upper bound approx. • approx. in a hierarchically expanded subspace	• with equilibrated test stresses on element level • no data required at Γ_e	• upper bound property • element-wise C^0-cont. stress approximation • data only from Ω_e, not from Γ_e
implicit (absolut), with enhanced test space on el. or patch level	• equilib. of interface tractions, solv. local Neumann problems • const.-free error bounds		• lower bound property • enhanced test space required	• SPR with higher converg. property in Gauß integr.-points compared to nodal points • using higher polynomials for stresses on patches

7 Finite Element Approximations and Error Estimators for Nonlinear and Elastoplastic Deformations, A Brief Survey

Extensions of error estimation to nonlinear elastic problems including structural instabilities are usually treated by incremental (tangential) algorithms, based on Gâteaux derivatives of the energy functionals, Stein, Rüter (2004), [88], Bischoff et al. (2004), [23], and Rheinboldt (2004), [73].

Error estimation for elasto-plastic deformations, based on convex yield surfaces and Hill's flow rule, are also treated incrementally, effectively by elastic predicitons and orthogonal plastic projections to the actual yield surface, Ortiz, Simo (1986), [68], Simo, Hughes (1998), [79], Stein (ed.) (2003), [81], Armero (2004), [4], and Ibrahimbegovic (2009), [50]. Adaptivity in space and time (for the load history) was treated by Gallimard et al. (1996), [46], and Stein, Schmidt (2003), [90]. A combined variational time-space integration was proposed by Lew et al. (2004), [61].

8 Meshfree and Reproducing Kernel Particle Methods, A Short Survey

So-called meshfree (elementfree) discretization methods with reproducing kernel approximations were first developed at the Northwestern University, Illinois, USA, by Liu, Jun and Zhang (1995), [62], Chen, Pan, Wu and Liu (1996), [34], Han and Meng (2001), [48], Belytschko and Huerta (2004), [22], Belytschko and Chen (2007), [21], and by many others.

Numerous extensions and applications to problems in solid and fluid mechanics were published in recent years.

The basic idea is to introduce for each chosen point (particle) in the open cover $C = \{\omega_i\}_{i=1}^{NP}$, $\Omega \subset \cup_{i=1}^{NP} \omega_i$, approximations with reproducing kernel functions $\psi_i(x)$, fulfilling the parity of unity condition $\sum_{i=1}^{NP} \psi_i(x) = 1$, $\forall x \in \bar{\Omega}$. The order p of these functions yields C^p-continuity for displacement fields. Dirichlet boundaries D have to be approximated.

A crucial problem is the integration of the approximated bilinear forms with sufficient accuracy. Gaussian quadrature and collocation methods are applied. The integration for each point i requires the determination of all neighboring points within the circle (for 2D) or the sphere (for 3D) of influence of the RK shape functions for this point. The mathematical theory was developed by Han and Meng (2001), [48].

Applications for numerous challenging linear and nonlinear problems have been published, including fracture and contact problems for small and large elastoplastic deformations.

Remarkably, C^p-continuity for contact problems can be achieved which is important for fast convergence. Locking and spurious mode phenomena in FEM can be avoided without special regularizations.

So far, only error indicators for filtering methods were developed; upper bound error estimators have not yet been published.

One can expect that this methodology will be developed strongly in the future, also in conjunction with particle methods. The application of adaptive discretizations with additional support points is simpler for the user than adaptive mesh refinements of finite elements, but the numerical algorithms and the mathematical analysis are complicated and need further research.

References

1. AIAA guide for the verification and validation of computational fluid dynamics simulations. AIAA Report G-077-1988, AIAA (1998)
2. Ainsworth, M., Oden, J.T.: A posteriori error estimators in finite element analysis. John Wiley & Sons, Chichester (2000)
3. Apel, T.: Interpolation in h-version finite element spaces. In: Stein et al. [84], vol. 1, ch. 3, pp. 55–70 (2004)
4. Armero, F.: Elastoplastic and viscoplastic deformations in solids and structures. In: Stein et al. [84], vol. 2, ch. 7, pp. 227–266 (2004)
5. Aubin, J.P.: Behaviour of the error of the approximate solution of boundary value problems for linear elliptic operators by Galerkin's and finite difference methods. Ann. Scuola Norm. Sup. Pisa 21, 599–637 (1967)
6. Auricchio, F., Brezzi, F., Lovadina, C.: Mixed finite element methods. In: Stein, E., de Borst, R., Hughes, T.J.R. (eds.) Encyclopedia of Computational Mechanics. Fundamentals, ch. 9, pp. 237–278. Wiley (2004)
7. Babuška, I., Melenk, J.M.: The partition of unity method. Int. J. Num. Meth. Eng. 40, 727–758 (1997)
8. Babuška, I., Miller, A.: The post-processing approach in the finite element method – part 2: the calculation of stress intensity factors. Int. J. Num. Meth. Eng. 20, 1111–1129 (1984)
9. Babuška, I., Miller, A.: A feedback finite element method with a posteriori error estimation: Part i. the finite element method and some basic properties of the a posteriori error estimator. Comput. Methods Appl. Mech. Engng. 61, 1–40 (1987)
10. Babuška, I., Rheinboldt, W.C.: A-posteriori error estimates for the finite element method. Int. J. Num. Meth. Engng. 12, 1597–1615 (1978)
11. Babuška, I., Strouboulis, T.: The finite element method and its reliability. Oxford University Press, New York (2001)
12. Babuška, I., Strouboulis, T., Mathur, A., Upadhyay, C.S.: Pollution-error in the h-version of finite element method and the local quality of a-posteriori error estimators. Finite Elem. Anal. Des. 17, 273–321 (1994)
13. Babuška, I., Strouboulis, T., Upadhyay, C.S., Gangaraj, S.K.: A posteriori estimation and adaptive control of the pollution error in the h-version of the finite element method. Int. J. Num. Meth. Eng. 38, 4207–4235 (1995)
14. Babuška, I., Suri, M.: On locking and robustness in the finite element method. SIAM J. Numer. Anal. 29, 1261–1293 (1992)
15. Babuška, I., Szabó, B., Katz, I.N.: The p-version of the finite element method. SIAM J. Numer. Anal. 18, 515–545 (1981)
16. Babuška, I., Whiteman, J.R., Strouboulis, T.: Finite elements. An introduction to the method and error estimation. Oxford University Press (2011)
17. Bank, R.E., Weiser, A.: Some a posteriori error estimators for elliptic partial differential equations. Math. Comp. 44, 283–301 (1985)
18. Becker, R., Rannacher, R.: A feed-back approach to error control in finite element methods: Basic analysis and examples. East-West J. Numer. Math. 4, 237–264 (1996)
19. Becker, R., Rannacher, R.: Weighted a posteriori error control in FE methods. In: Bock, H.G., et al. (eds.) Proc. ENUMATH 1997, pp. 621–637. World Scient. Publ., Singapore (1998)

20. Belytschko, T., Black, T.: Elastic growth in finite elements with minimal remeshing. Int. J. Num. Meth. Eng. 45, 601–620 (1999)
21. Belytschko, T., Chen, J.S.: Meshfree and particle methods. John Wiley (2007)
22. Belytschko, T., Huerta, A., Fernandez-Méndez, S., Rabczuk, T.: Meshless methods. In: Stein et al. [84], vol. 1, ch. 10 (2004)
23. Bischoff, M., Wall, W.A., Bletzinger, K.-U., Ramm, E.: Models and finite elements for thin-walled structures. In: Stein et al. [84], vol. 2, ch. 3, pp. 59–138 (2004)
24. Borouchaki, H., George, P.L., Hecht, F., Laug, P., Saltel, E.: Delaunay mesh generation governed by metric specifications. Part I. algorithms. Finite Elem. Anal. Des. 25, 61–83 (1997)
25. Braess, D.: Finite elements, 2nd edn. Cambridge University Press (2001), 1st ed. (1997)
26. Braess, D.: Finite Elemente, 5th edn. Springer Spektrum (2013), 1st edn. (1992)
27. Braess, D., Schöberl, J.: Equilibrated residual error estimator for hedge elements. Math. Comp. 77, 651–672 (2008)
28. Bramble, J.H., Hilbert, A.H.: Estimation of linear functionals on Sobolev spaces with applications to fourier transforms and spline interpolation. SIAM J. Numer. Anal. 7, 113–124 (1970)
29. Brenner, S.C., Carstensen, C.: Finite Element Methods. In: Stein et al. [84], vol. 1, ch. 4, pp. 73–118 (2004)
30. Brenner, S.C., Scott, L.R.: The mathematical theory of finite element methods. Springer, 1994 (2002)
31. Bufler, H., Stein, E.: Zur Plattenberechnung mittels finiter Elemente. Ingenieur-Archiv 39, 248–260 (1970)
32. Carstensen, C., Funken, S.A.: Averaging technique for FE-a posteriori error control in elasticity. Comput. Methods Appl. Mech. Engng. 190, 2483–2498, 4663–4675; 191, 861–877 (2001)
33. Céa, J.: Approximation variationnelle des problèmes aux limites (phd thesis). Annales de l'institut Fourier 14(2), 345–444 (1964)
34. Chen, J.S., Pan, C., Wu, C.T., Liu, W.K.: Comp. Meth. in Appl. Mech. and Eng. 139, 195–227 (1996)
35. Cirak, F., Ramm, E.: A posteriori error estimation and adaptivity for linear elasticity using the reciprocal theorem. Comput. Methods Appl. Mech. Engng. 156, 351–362 (1998)
36. Cirak, F., Ramm, E.: A posteriori error estimation and adaptivity for elastoplasticity using the reciprocal theorem. Int. J. Num. Meth. Eng. 47, 379–394 (2000)
37. Dìez, P., Parés, N., Huerta, A.: Error estimation and quality control. In: Encyclopedia of Aerospace Engineering, vol. 3, ch. 144, pp. 1725–1734. Wiley (2010)
38. Dìez, P., Wiberg, N.-E., Bouillard, P., Moitinho de Almeida, J.P., Tiago, C., Parés, N. (eds.): Adaptive Modeling and Simulation. CIMNE, Barcelona (2003, 2005, 2007, 2009, 2011, 2013)
39. Dirichlet, P.G.L.: Gustav Lejeune Dirichlet's Werke. Collection of the University of Michigan (1889)
40. Duarte, C.A., Babuška, I., Oden, J.T.: Generalized finite element methods for three-dimensional structural mechanics problems. Computers & Structures 77, 215–232 (2000)
41. Edelsbrunner, H.: Geometry and topology for mesh generation. Cambridge University Press, U.K (2001)

42. Eriksson, K., Estep, D., Hansbo, P., Johnson, C.: Computational differential equations. Cambridge University Press, USA (1996)
43. Eriksson, K., Estep, D., Hansbo, P., Johnson, J.: Introduction to adaptive methods for differential equations. Acta Numerica, 105–158 (1995)
44. Fix, G.J., Gulati, S., Wakoff, G.I.: On the use of singular functions with finite element approximations. Journal of Computational Physics 13, 209–228 (1973)
45. Friedrichs, K.O.: On the boundary value problems of the theory of elasticity and Korn's inequality. Ann. Math. 48, 441–471 (1947)
46. Gallimard, L., Ladevèze, P., Pelle, J.-P.: Error estimation and adaptivity in elastoplasticity. Int. J. Num. Meth. Eng. 39, 189–217 (1996)
47. Gerasimov, T., Stein, E., Wriggers, P.: New simple, cheap and efficient constant-free explicit error estimator for adaptive FEM analysis in linear elasticity and fracture. Int. J. Num. Meth. Eng. (submitted 2013)
48. Han, W., Meng, X.: Comp. Meth. in Appl. Mech. and Eng. 190, 6157–6181 (2001)
49. Hilbert, D.: Die Grundlagen der Mathematik. Abhandlungen aus dem mathematischen Seminar der Hamburgischen Universität, Band VI (1928)
50. Ibrahimbegovic, A.: Nonlinear solid mechanics – theoretical formulations and finite element solution methods. Springer (2009)
51. Johnson, C.: A new paradigm for adaptive finite element methods. In: Whiteman, J. (ed.) Proc. MAFLEAP 1993. John Wiley (1993)
52. Ladevèze, P.: Constitutive relation error estimations for finite element analyses considering (visco) plasticity and damage. Int. J. Num. Meth. Eng. 52, 527–542 (2001)
53. Ladevèze, P.: Strict upper error bounds on computed outputs of interest in computational structural mechanics. Comp. Mech. 42, 271–286 (2008)
54. Ladevèze, P.: Model verification through guaranteed upper bounds: state of the art and challenges. In: Aubry, D., Díez, P., Tie, B., Parès, N. (eds.) Adaptive Modeling and Simulation 2011, pp. 20–29. CIMNE, Barcelona (2011)
55. Ladevèze, P., Leguillon, D.: Error estimate procedure in the finite element method and applications. SIAM J. Numer. Anal. 20, 485–509 (1983); Translated from the French edition by Hermes-Lavoisier Science Publishers, Paris (2001)
56. Ladevèze, P., Maunder, E.A.W.: A general method for recovering equilibrating element tractions. Comp. Meth. in Appl. Mech. and Eng. 137, 111–151 (1996)
57. Ladevèze, P., Moës, N.: A new a posteriori error estimation for nonlinear time-dependent finite element analysis. Comp. Meth. in Appl. Mech. and Eng. 157, 45–68 (1998)
58. Ladevèze, P., Pelle, J.-P.: Mastering calculations in linear and nonlinear mechanics.Springer Science+Business Media, Inc. (2005); Translated from the French edition by Hermes-Lavoisier Science Publishers, Paris (2001)
59. Larsson, F., Hansbo, P., Runesson, K.: On the computation of goal-oriented a posteriori error measures in nonlinear elasticity. Int. J. Num. Meth. Eng. 55, 379–394 (2002)
60. Lax, P.D.: Selected papers, vol. I, II. Springer, Berlin (2005)
61. Lew, A., Marsden, J.E., Ortiz, M., West, M.: Variational time integrators. Int. J. Num. Meth. Eng. 60, 153–212 (2004)
62. Liu, W.K., Jun, S., Zhang, Y.F.: Int. J. Num. Meth. Eng. 20, 1081–1106 (1995)
63. Neittaanmäki, P., Repin, S.: Reliable methods for computer simulation, Error control and a posteriori estimates. Elsevier, Amsterdam (2004)

64. Niekamp, R., Stein, E.: An object-oriented approach for parallel two- and three-dimensional adaptive finite element computations. Computers & Structures 80, 317–328 (2002)
65. Nitsche, J.A.: l_∞-convergence of finite element approximations. In: Mathematical Aspects of Finite Element Methods. Lecture Notes in Mathematics, vol. 606, pp. 261–274. Springer, New York (1977)
66. Oden, J.T., Carey, G.F.: Finite elements. Mathematical aspects, vol. IV. Prentice Hall, Inc. (1983)
67. Oden, J.T., Prudhomme, S.: Estimation of modeling error in computational mechanics. J. Comput. Phys. 182, 496–515 (2002)
68. Ortiz, M., Simo, J.C.: Analysis of a new class of integration algorithms for elastoplastic constitutive equations. Int. J. Num. Meth. Eng. 21, 353–366 (1986)
69. Poincaré, H.: Cours professé à la Faculté des Sciences de Paris - mécanique physique. L'Association amicale des élèves et anciens élèves de la Faculté des sciences - Cours de Physique Mathématique (1885)
70. Prager, W., Synge, J.L.: Approximations in eleasticity based on the concept of function spaces. Quart. Appl. Math. 5, 241–269 (1949)
71. Rannacher, R.: Duality techniques for error estimation and mesh adaptation in finite element methods. In: Stein [82], ch. 1, pp. 1–58 (2005)
72. Rannacher, R., Suttmeier, F.-T.: Error estimation and adaptive mesh design for FE models in elasto-plasticity theory. In: Stein [81], ch. 2, pp. 5–52 (2003)
73. Rheinboldt, W.C.: Nonlinear systems and bifurcations. In: Stein et al. [84], vol. 1, ch. 23, pp. 649–674 (2004)
74. Rivara, M.C.: Mesh refinement processes based on the generalized bisection of simplices. SIAM Journal on Numerical Analysis 21, 604–613 (1984)
75. Roache, P.J.: Verification and Validation in Computational Science and Engineering. Hermosa Publishers (1998)
76. Rodríguez, R.: Some remarks on Zienkiewicz-Zhu estimator. Numer. Methods Partial Diff. Equations 10, 625–635 (1994)
77. Rüter, M.: Error-controlled adaptive finite element methods in large strain hyperelasticity and fracture mechanics. Institute report F03/1 Institut für Baumechanik und Numerische Mechanik, Leibniz Universität Hannover (2003)
78. Schwer, L.E.: An overview of the PTC 60/V&V 10: guide for verification and validation in computational solid mechanics. Engineering with Computers 23(4), 245–252 (2007)
79. Simo, J.C., Hughes, T.J.R.: Computational Inelasticity. Springer, New York (1998)
80. Sobolev, S.L.: Applications of Functional Analysis in Mathematical Physics. Mathematical Monographs, vol. 7. AMS, Providence (1963)
81. Stein, E. (ed.): Error-controlled adaptive finite elements in solid mechanics. Wiley (2003)
82. Stein, E. (ed.): Adaptive finite elements in linear and nonlinear solid and structural mechanics. CISM courses and lectures (Udine), vol. 416. Springer, Wien (2005)
83. Stein, E., Ahmad, R.: An equilibrium method for stress calculation using finite element methods in solid and structural mechanics. Comp. Meth. in Appl. Mech. and Eng. 10, 175–198 (1977)
84. Stein, E., de Borst, R., Hughes, T.J.R. (eds.): Encyclopedia of Computational Mechanics, vol. 1: Fundamentals, vol. 2: Solids and Structures, vol. 3: Fluids. John Wiley & Sons, Chichester (2004) (2nd edition in Internet 2007)

85. Stein, E., Niekamp, R., Ohnimus, S., Schmidt, M.: Hierarchical Model and Solution Adaptivity of thin-walled Structures by the Finite-Element-Method. In: Stein [82], ch. 2, pp. 59–147 (2005)
86. Stein, E., Ohnimus, S.: Coupled model- and solution adaptivity in the finite element method. Comp. Meth. in Appl. Mech. and Eng. 150, 327–350 (1997)
87. Stein, E., Ohnimus, S.: Anisotropic discretization- and model- error estimation in solids mechanics by local neumann problems. Comp. Meth. in Appl. Mech. and Eng. 176, 363–385 (1999)
88. Stein, E., Rüter, M.: Finite Element Methods for elasticity with error-controlled discretization and model adaptivity. In: Stein et al. [84], vol. 2, ch. 2 (2004)
89. Stein, E., Rüter, M., Ohnimus, S.: Implicit upper bound error estimates for combined expansive model and discretization adaptivity. Comput. Methods Appl. Mech. Engng. 200, 2626–2638 (2011)
90. Stein, E., Schmidt, M.: Adaptive FEM for elasto-plastic deformations. In: Stein [81], ch. 3, pp. 53–107 (2003)
91. Strang, G., Fix, G.J.: An Analysis of the Finite Element Method. Prentice Hall, Inc. (1973); Reprinted by Wellesly-Cambridge Press (1988)
92. Verfürth, R.: A review of a posteriori error estimation and adaptive mesh refinement technis. Wiley-Teubner, Chichester (1996)
93. Verfürth, R.: A Posteriori Error Estimation Techniques for Finite Element Methods. Oxford University Press (2013)
94. Zienkiewicz, O.C., Zhu, J.Z.: A simple error estimator and adaptive procedure for practical engineering analysis. Int. J. Num. Meth. Eng. 24, 337–357 (1987)
95. Zienkiewicz, O.C., Zhu, J.Z.: The superconvergent patch recovery (SPR) and adaptive finite element refinements. Comput. Methods Appl. Mech. Engng. 101, 207–224 (1992)

Author Index

Bertram, Albrecht 119
Bremer, Hartmut 45, 73
Bruhns, Otto T. 133

Döge, Torsten 249

Ehlers, Wolfgang 211

Gaul, Lothar 385
Gebbeken, Norbert 249
Gerstner, Maximilian 61
Gray, J.A.T. 153
Gross, Dietmar 195

Krause, Egon 317

Lehmann, Eike 267

Mahnken, Rolf 229
Mahrenholtz, Oskar 299

Ponter, Alan 169
Popov, Valentin L. 153

Ramm, Ekkehard 23

Schiehlen, Werner 101
Schmitt, Patrick R. 61
Selvadurai, A. Patrick S. 343
Sockel, Helmut 355
Stein, Erwin 3, 399, 443
Steinmann, Paul 61

Trierenberg, Andor 81

Wagner, Jörg F. 81
Weichert, Dieter 169

Subject Index

A

a posteriori error estimator 449ff
a posteriori error estimators by gradient smoothing and stress recovery 443ff
a priori error estimates 451
Abbé Marie 55
Ackermann, M. 112
action functional 68, 71
activation dynamics 111
adaptive quadrilateral elements 470, 471
adaptive triangular meshes 470
adaptivity 399ff.
adjustable blades 364
Adkins, J.E. 221
admissible functions 394
advanced modeling 238
aerodynamic drag 375
aesthetic cost function 114
Ainsworth, M., Oden, J.T. 450, 462
airy differential equation 280
Akakia, Diatribe du Docteur 30
Akakia, Histoire du Docteur 31
Al Ğazarī, Ibn-Razzaz (al Dschazarī) 301
Algarotti, F. 46
algebraic structure of mixed FEM 425
analytical mechanics 45, 48, 54, 61, 62, 68, 70, 71, 73–76
analytical procedures 77
Anderson, F.C. 109
angular-momentum balances 220
Anschütz-Kaempfe, H. 95
any body 107
aqueduct 300
Arago, F.J.D. 82, 93, 98
Archimede screw 358

Archimedes 56, 299
Argyris, J.H. 399
Aristotle 47, 251
Aronovich, G.D. 177
artificial horizon 82
Arzt, V. 74
assumed strain methods 430
atomic force microscope 156, 164
Aubin, T. 399
axioms of mechanics 62
axisymmetric problems in elasticity 347

B

Babuška, I. 399, 406, 422, 429
Babuška, I., Rheinboldt, W. 444, 450, 453
balance equations 212, 218, 220
banded matrices 397
Bank, R.E., Weiser, A. 454
Barnes, F.K. 272
barometer 304
Barthez, P.J. 103
barycentre 106, 107
basic principles 223
Bathe, K.-J. 399
b-bar techniques 430
Beletsky, V.V. 109
Beltrami, E. 200
Beltrami-Michell compatibility equations 348
Belytschko, T. 450
Benzenberg, J.F. 89–91, 94
Bernoulli equation 309, 319
Bernoulli, Daniel 253, 308–310, 357, 389
Bernoulli, Jacob 197, 231, 233, 407
Bernoulli, Johann I. 47–48, 61, 63, 253, 307–309, 407

Bernoulli's solution (Jacob Bernoulli) 5, 7, 12
Bernoulli's solution (Johann Bernoulli) 5, 14
Bernoulli-Euler beam theory 272
Bertram, A. 127
Betti-Maxwell reciprocity theorem 465
Betz, A. 375
Biles, J. 280
biomechanics 224
Biot, M.A. 216f
black magic 56–57
blade-wheel propeller 370
Bleich, H.H. 172
Blücher (vessel name) 292
Bohnenberger, J.G.F. 84, 81–91, 93
Boltzmann, S. 73
Booker, J.R. 181
Borelli, G.A. 102
Borouchaki, H. 470
Bouguer, P. 273
boundary layer blowing 379
boundary lubrication 166
Boussinesq, J. 343
Bowen, R.M. 212, 222
Boyle, R. 251
Boyle-Mariotte law 198
brachistrochrone problem (BCP) 8, 11, 48
Braess, D. 444
Bramble-Hilbert lemma 446
Brezzi, F. 399
Bridgman, P.W. 124
Brno College of Technology 362
Brush, S.G. 253
Bryan, G.H. 282
Buoyancy forces 211, 212, 217
Burgers, J.M. 237
Buzengeiger, J.W.G. 83, 87–89, 91, 93, 95, 98

C

Caldwell, J. 292
canonical momentum 70
Cardan angles 92
Carnot, N.L.S. 259
Cartan, É.J. 71
Cartesian grid 336
catenary curve 7

Cauchy continuum 221
Cauchy, A.-L. 199–200, 232, 233, 315
Cauchy-Schwarz inequality 454, 456, 461
Cavendish, H. 230–231
cavitation 367
Céa-lemma 449
central equation 56, 73, 76, 77, 79
chaos 74–75
chemical potential 222
chemical reaction 215
Chen, J.S. 473
Chladni, F.F. 53
Cirak, F. Ramm, R. 443, 467
circulation 330
Civetta (vessel name) 360
Clapeyron, É. 232
classical elasticity 343
classical mechanics 73--75, 79
classification societies 267
Clausius, R.J.E 216, 220, 260
Clausius-Duhem inequality 129
Clay Mathematics Institute 317
closed system 215, 220
Clough, R.W. 399, 405, 408
cohesive zone model 195, 208
Coleman, B.D. 123, 221, 223
collision tests 289
common cycloid, properties 6, 7
component-mode synthesis 397
computational grid 317
concentrated force problem 349
concentration factor 343
condition of consistency 145
configuration manifold 70
conservation of angular momentum 260
conservation of energy 255
conservation of mass 311
conservation of momentum 254
constitutive equation 214, 221
constitutive equation errors 460
constraint force 62–66
constraints 1, 5
contact force 112
continuity equation 308, 309
contraction dynamics 112
contributions to aeronautics 381
Coriolis, G.G. de 55
Cosserat, E. and F. 221

Subject Index 483

cotangent bundle 70
Cotta, J.F. 86, 89
Cotterell, B. 196
Couette flow 324
Coulomb, C.-A. 74, 216, 199
coupled problems 223, 225
Courant, R. 396
Courant-Friedrichs-Lewy condition 334
crack extension force 204
crack tip-field 201
cracks 170, 179
Cummeron, C. 359
cycloidal propeller 370

D

d'Alembert, J.-B. le Rond 38, 45, 50–52, 66, 307
d'Alembert's principle 38, 45, 64, 66
d'Alembert's paradoxon 318
Darcy, H. 214, 235
Darcy's law 214
Darwin, C. 235
de Boer, R. 212, 217
de Boyer, C. 46, 47
de Saint-Venant, B. 30, 137, 138
de Sapio, V. 110
de Veubeke, B.F. 180
deformation law 139
deformation theory 133, 144–146
Delauney triangulation 451
Delesse, A.E.O.J. 213, 214, 215
Descartes, R. 304, 308, 309
Desoyer, K. 218
determinism 121, 223
diffusion 214, 221
diffusion of load 343
Dirac, P. 71
direct methods 182
direct numerical simulation 335
directional derivatives 77
directional gyro 95
Dirichlet principle of minimum potential energy 445
discretization 333
dissipation 223
drag prediction workshops 337
duality technique 465
Duhem, P.M.M. 216, 221
dynamic viscosity 314, 315

E

Eckart, C. 124
Ecole Polytechnique 82, 91, 93
effective breadth 280
effectivity index 453
eigenfunction 395
eigenvalue 394
eigenvalue problem 394
elastic instability 156
element and global stiffness matrices 412ff
elementary particle 72
endochronic 124
energetic cost function 112
energy approach 202, 204
energy conservation 52, 55
energy flux 222
energy norm 450–457
energy release rate 204
Eneström, G. 253
engine braking 166
enhanced assumed strain method 430
entropy 260
entropy inequality 220, 221
environment 211, 223
equations of motion 51, 54, 77, 79, 101, 106
equations of state 259
equilibrated boundary tractions 454ff
equilibration of interface tractions 460
equinoxes 47
equipresence 223
Eriksson, K. 443, 444, 446
Eringen, A.C. 221
error estimators by local Dirichlet problems 443
error estimators for elastoplastic deformations 465
error estimators for nonlinear elasticproblems 472
error measures for quantities of interest 467
Escher Wyss (company) 365
Esso Norway (vessel name) 289
Esso Scotia (vessel name) 288
Euler equations 255, 317
Euler, L. 26, 45, 47–48, 50, 54, 56, 61, 74, 76 239, 253, 307–309, 311, 385
Euler-Lagrange equations 62, 69, 76

evolution equation   145
evolution strategy   239, 240–242
experiments   230-233, 243
extended p- and h-test spaces   444
extremum problem   390

F

fading memory   123
Fairbairn, W.   267
FEM for elasto-dynamic problems   435
FEM for non-linear elastic problems   406
Fermat's principle of light path in minimal time   6
Fessel gyroscope   95
Feynman, R.   71
Fick, A.E.   214, 235
Fick's laws   214
Fillunger, P.   217, 222
finite deformation   146–147
finite element method   212, 244, 281, 336, 394
finite element solution of the BCP   20ff.
finite elements   385
Fischer, O.   106, 107
flexible multibody systems   397
Fontana   358
force laws   51, 55
Foucault, J.B.L.   81, 82, 93–95
founder crisis   73
fracture concept   201–204
fracture stress   202, 206, 207
fracture surface energy   203
frames (w.r.t. shakedown)   174
Francis turbine   355
Frederick II, King of Prussia   25, 46
free-energy transport   221
free-jet wind tunnel   355
French revolution   47, 54
Frenkel-Kontorova model   155
friction
   Coulomb's law   164
   dependence on temperature   153
   dependence on velocity   153
   dry   153
   kinetic   155
   static   155
Friedrich Krupp AG   375
Fröhlich, O.K.   217, 218, 343

fundamental balance law   220
further important milestones   434–435

G

gait disorder   101
Galerkin approximation   449
Galerkin orthogonality   403, 449, 455, 461
Galerkin, B.G.   142, 143, 399
Galilei's approximated solution of the BCP   8
Galilei, G.   62, 235, 250, 197–198
Gauss, C.F.   35, 83, 86, 239
Gay-Lussac, J.L.   255
Geiringer, H.   141, 145, 149
general formula (Lagrangian principle)   50, 54
general purpose finite element problems   433
generalized coordinates   62, 66, 67, 108
generalized force   66, 77, 204
generalized momentum, conjugate momentum   70
generalized velocities   62, 66, 69
geodesy   81, 86, 95, 97
geomechanics   216, 343
geometrical optics   68
Germanischer Lloyd   276
Gibbs, J.W.   215, 216
Gleick, J.   74
Goethe, von, J.W.   46
Gokhfeld, D.A.   177
Göttingen, University of   85, 95
Göttingen-type tunnel   375
Grantham, J.   272
Great Eastern (vessel name)   270
Green, A.E.   121, 124, 221
Griffith, A.A   195, 202
ground reaction forces   114
Grüning, B.M.   171, 172
Guericke, von, O.   251, 305
Guest, J.J.   200, 205
Gutehoffnungshütte   372
gyro with cardanic suspension   81, 94–97
gyroscope   81, 94, 95

H

Habsburg empire   355, 356
Hagen, G.H.L.   313

Subject Index

Hagen-Poiseuille flow 323
Hamel, G. 51, 57, 73–79
Hamel-coeffcients 77
Hamilton, Sir, W.R. 34, 61, 68
Hamilton's equations 69, 70
Hamilton's function, Hamiltonian 62, 68–69
Han, W. 473
hand-fitting 229, 234, 236–238
hanging node technique 471
Hassler, F.R. 85
heat capacity ratio 255
heat flux 222
Heinrich, G. 217, 222
Heisenberg, W. 71
Helgoland (ship) 374
Hellinger-Pranger-Reissner two-field functional 421
Hellinger-Reissner dual mixed FEM 426
Helmholtz, von, H.L.F. 52, 216
Helmholtz's vorticity theorems 317
Hencky, H. 33–134, 139, 141–142, 146, 200
Hermann, J. 61, 63
Heron of Alexandria 300
Hertz, H. 74, 76
Heun, K. 51, 73–77, 79
Hiemenz flow 324
high performance computing 333
historical fairness 77
history functional 119
HMS Cobra (vessel name) 267
HMS Minotaur (vessel name) 273
HMS Wolf (vessel name) 267
homogenization 211, 221, 223
Hooke, R. 199, 234, 251
Hooke's law 231, 234–235
Hopkinson, B. 195, 204, 205
Horn, F. 276
Horner, J.C. 88, 89
Huber, T.M. 139, 177, 200
Hughes, T.J.R. 400
Hugoniot, P.H. 262
hull bending moment 267
human walking 101, 102, 109
Humboldt, von, A. 91
Hu-Washizu functional 430–432
hybrid stress method 416, 424
Hydraulica (by Johann Bernoulli) 308, 309
hydraulics 312

hydrostatic paradox 303
hypergeometric function 349

I

imperial airforce 379
implicit a posteriori error estimators 399ff
implicit SPR smoothing estimator 463, 464
impressed force 49, 50
incompressible elasticity 344
incremental FEM for inelastic and timedependent deformations 435
inertial sensor, inertial platform 81, 95
infinitesimal calculus 308, 309
infsup conditions for two- and three-field functionals 428
Inglis, C.E. 195, 202, 206
Ingram, J.D. 221
Institute of Advanced Studies 332
Institution of Naval Architects 270
instrument maker 83, 87, 93
integrability 77
interaction terms 223
interior pressure 311
intermediate variables 77
internal variable 119
interpolation polynomials 409
inverse problem for central forces 61
Irwin, G.R. 195, 203–204, 206
isogeometric analysis 400, 435, 436
isotropic hardening 145

J

Jacobi, C.G.J. 55
Jacobians 79
Jaumann derivative 215
Jaumann, G. 147, 215, 219
Journal für die reine und angewandte Mathematik 23, 36
Jugendstil 73

K

Kane, T. 56
Kaplan turbine 355
Kaplan, V. 355, 361
Kármán equations 282
Kármán, von, T. 218, 375, 392
Kästner, A.G. 53
Katharina II (the Great) of Russia 47

Katzmayr, R.   355, 380
Keith, J.   46, 47
Kelly, P.D.   220
Kelvin, Lord, W.T.   74, 237, 259
kinematic hardening   145
kinematic tree   111
kinematic viscosity   322
kinematical gait analysis   102
kinetic energy   26, 29, 34, 56, 67, 77, 390
Kirchhoff, G.R.   232
Kirchhoff's uniqueness theorem   349
Klawitter, G.D.   269
Klein, F.C.   51, 71, 76, 201, 262
Knoller airplanes   355
Knoller, R.   355, 374
Knoller-Betz effect   374
Koiter, W.T.   174, 175, 178
König, J.A.   178, 179
König, J.S.   28
Kort nozzle   368
Kramers, H.A.   163
Kryloff, A.   275

L

Ladevèze, P., Leguillon, D.   459
Lagrange multiplier   64, 65, 112
Lagrange, J.-L.   45, 47–50, 57, 61, 63, 64, 74–79, 255
Lagrange's equations   62, 64, 65–66, 70
Lagrange's principle   45, 51
Lagrangian function, Lagrangian   62
Lamé, G.   232
Laplace, P.-S.   74, 82, 90, 94, 98, 257
large crude oil tanker   288
large-eddy simulation   337
Larsson, F.   465
Lax-Milgram lemma   452
least-squares method   35, 38, 229, 238, 239
Lee, E.H.   124
Legendre transformation   62–69, 422, 424
Legendre, A.-M.   35
Leibniz, G.W.   28, 62, 399
Leibniz' discrete solution draft of the BCP   13, 17
Leonardo da Vinci   197, 229
Les Loix du Mouvement et du Repos (book)   25

lever   55, 56
Lévy, M.   138, 282
Lichtenberg, G.C.   53, 85
Lie, M.S.   71, 77
Lienau, O.   276
lifting-line theory   317
Lilla Edit turbine   367
limit analysis   170, 174, 176, 178, 180, 181
linear momentum   389
link system   106
Lloyd's Register   271
local action   223
local mesh refinements   408–409
local Neumann problems   454, 456
locking phenomena in finite elements   429
logarithmic strain   146
lower bound   172, 174, 176, 180
Lur'e, A.   76
Lyttleton, W.   358

M

Mach, E.   74, 251, 385
Magdeburg half spheres   306
Magnus, K.   56, 74, 79
Mahrenholtz, O.   173, 179, 183
Maier, G.   175, 176, 183
Mandel, J.   124
maneuverability of ships   372
Mansell, R.   272
Maquet, P.   102
marine screw propeller   355
Mariotte, E.   198–199, 251
material damage   180, 182
material frame indifference   223
material science   148
Matsko, J.M.   53
Maupertuis, de, P.-L.M.   47, 48, 23, 25, 69
Maupertuisiana (book)   32, 33
maximum distortion energy criterion   200
maximum normal-stress yield criterion   216
maximum stress criterion   200
Maxwell model   123
Maxwell, J.C.   200, 237
Mayer, J.R.   257
Méchanique Analytique   45, 47, 52-53
mechanism model   101, 106, 109

Meirovitch, L.   394
Melan, E.   172–174, 178, 182
Mersenne, M.   250
mesh-free, esp. reproducing particle methods   473
metabolical cost   114
method of fluxions   308
Mettrie, la, J.   46–47
Miller, A.   399
minesweepers   374
minimal coordinates   76, 54
minimal principles   49
minimal state space   127
minimal variables   50, 51
minimal velocities   56, 76
minimum or maximum (principle of least action)   34
Mises, von, R.   134, 149, 200
mixed finite elements   399ff.
mixture theory   220, 221
model reductions and model expansions   444ff
modes of vibrations   396
Moës, N.   460
Mohr, O.   200
Mohr-Coulomb criterion   200
moment of momentum   389
motion, free and constrained   36
Müller, I.   223
multi body system   77
multi-component materials   211, 220
Murhard, F.W.A.   52–56
muscle coordination   109
Müser, M.   154

N

Nádai, A.   146, 149
Naghdi, P.M.   221
nanomachines (for directed motions)   164
Napier, R.   272
Navier, C.-L.   137, 199–200, 232–234, 257, 314, 315, 317
Neumann, von, J.   332
Newton, Sir, I.   62, 74, 230–231, 252–253, 307, 314, 385
Newton's law   230–231, 234
Newton-Euler laws of mechanics   385
Newtonian fluid   314
Nitsche, J.   444
nodal points   397

Noll, W.   119, 121, 123, 219, 221
nonholonomic variables   73, 76–77
nonlinearity   65
normality rule   45
Nörrenberg, J.G.   88
Norton's law   235
nozzle wings   379
numerical methods   175, 178, 182
numerical solutions   317
Nussbaum, M.C.   102
nutation   47

O

objective time derivative   147
Oden, J.T., Prudhomme, S.   467
Ohm, G.S.   93, 235
Olszak, W.   178
Onat, E.T.   127
one-parameter models (for constitutive equs.)   235, 236
optical method   243
optimization   39, 229, 239, 241, 245
Oseen approximation   324

P

Pandy, M.G.   109
parameter identification   229–230, 233, 240–241, 245
parameter optimization   101, 102
Parson, C.   279
Pascal, B.   251
Paucton, M.   357, 358
pavement mechanics   181
Peitgen, H.-O.   74
pendulum   37, 38, 62–63, 65, 67, 70
pendulum of Foucault   81
pendulum of Kater, reversible pendulum   86
Person, C.C.   93
Pfleiderer, C.F.   84, 87, 89, 90
phase separation   223
phase space   62, 70
Philosophiae Naturalis Principia Mathematica   307, 385
Pian, T.H.-H.   420
Piola-Kirchhoff stress   223
plagiarism and forgery   23
Planck, M.   73, 216
plastic deformation   153

plastic hysteresis   156
plasticity   170, 171, 173–176, 215
Poincaré, H.   75
Poincaré-Friedrichs inequality   447, 453
Poinsot, L.   55, 56, 75
Poiseuille flow   312
Poisson, S.D.   232, 315, 82, 91
Ponter, A.   169, 176
post buckling effect   267
potential energy   67, 393
Power law   235, 236
Power, H.   251
Poynting   237
Prager, W.   134, 141, 145, 146, 148, 173, 460
Prager-Synge hypercircle theorem   422
Prandtl, L.   75, 134, 144, 149, 153, 195, 206, 207, 284
Prandtl's revolutionary boundary-layer theory   317
Prandtl-Reuss theory   133, 146
Prandtl-Tomlinson model   153, 154, 207
precession   47
precession, advancement of the equinoxes   89, 90, 93
primal and dual elliptic variational problems   465
primal FEM   399, 402
principle of least action   23–25, 34, 39, 45, 48, 54
principle of least constraint   23, 36–38
principle of minimum potential energy   5, 7
principle of rest   48
principle of stationary action   62, 68
process class   126
process functional   122
production terms   212, 220
projection equation   78
propeller test rig   375
Prussian Royal Academy of Sciences   23, 48
Pulte, H.   56

Q

quantum electrodynamics   71
quantum field theory   71
quantum mechanics   71
quasi-coordinates   54, 77

R

Rankine, W.J.M.   200, 216, 261, 272
Rankine-Hugoniot equations   261
Rannacher, R., Suttmeier, F.-T.   443, 465
Rasmussen, J.   109, 110
rate gyro   96, 97
rational mechanics   218
Rayleigh, Lord, Strutt, J.W.   391, 395
Rayleigh quotient   385
Rayleigh-Ritz method   385
Read, T.C.   275
Rechenberg, I.   239
Reckling, K.-A.   290
reduced integration   423, 430
Reed, E.J.   272
reference frame   79
Reissner, H.   280
Ressel, J.   355, 356
Reuss, A.   149
revolute joint   63
Reynolds number   323
Reynolds, O.   216, 313
Reynolds' hypothesis   327
rheological elements   236, 238
Richter, P.H.   74
Riemann invariants   261
Riemann problem   261
Riemann, B.   261
rigidity   204–205
Ritz ansatz   390
Ritz, W.   53, 390, 395
riveting technique   271
Rivlin, R.S.   121
Robbins, M.   154
Rossmanith, H.P.   196
Rüter, M.   449, 450

S

Saint-Venant, A.J.C.B.   199, 315
Saint-Venant torsion theory   396
San Francisco (vessel name)   276
Sang, E.   94
Sawczuk, A.   178
Schellbach, K.H.   399
Schellbach's discrete solutions   17
Schnadel, G.   267
Schneider, E.L.   368
Schott, C.   251
Schrödinger, E.   71

Schuler, M. 74
Schumacher, H.C. 86, 87, 88, 93
Schwarz, H.A. 77
Schwefel, H.-P. 239
sea-going ship 373
Segner, J.A. 53, 90
semi-elastic deformation 127
Servus, H. 47, 53–55
shakedown theory 169, 173–176
shape functions 394
shear criterion 200
shear lag 280
Šilhavý, M. 124
similarity variables 322
simple material 119
Sine-Gordon equation 165
slip-line theory 140
Sloan, S. 181, 182
Smith, F.P. 361
Smith, W.E. 275
Snell's law of light refraction 6
solid mechanics 169, 170, 176
solutions of the brachistochrone problem 12ff.
specific speed 363
speed of sound 250, 252–253, 256, 257, 261
Sperry, E.A. 95
spin tensor 79
Sprengel, K. 103
standard model of particle physics 71
state space 119, 125
state variable theory 127
steam driven car 374
Štefan 215
Stein, E. 179, 183
Stein, E. et al. 443, 444
Stein, E., Ohnimus, S. 443ff, 467
Stevin, S. 302
stick-slip, atomic 156
still water bending moment 275
stochastic modeling 233, 240–241, 245
Stokes, G.G. 137, 315, 258
Stokes' first problem 322
stréphoscope 95
stress analysis of soils 347
stress intensity factor 204
stress transfer in a soil 344
structure identification 233, 245
structures 169, 170, 174, 178
super lubrication 158

superlubricity 153
superslipperiness 159
surface energy 202–203
Symonds, P.S. 173, 174
Synge, J.L. 460
Szabó, B. 412
Szabó, I. 74, 262, 303

T

tangent bundle 70
tautochrony property of the cycloid 9
Taylor, G.I. 202
tearing (rupture) 204
temperature, absolute 259–260
Tenacity (material parameter) 204
Terzaghi, K. 216, 217
the great award (of the French Academy of Science) 47
the moon problem 50
theory of porous media 211
Theory of Sound (book) 385
theory of variations 62
thermal loads 177
thermodynamics 216, 219, 225
thrust (in a boat) 355
time-dependent fracture 195, 208
Timoshenko, S.P. 282
Tomlinson, G.A. 153
Torricelli, E. 5, 20, 304, 305
torsion of the ship hull 283
Toupin, R. 219, 220
Towneley, R. 251
traction-separation relationship (for fracture) 207
traité de dynamique (by d'Alembert) 52
transition of laminar into turbulent flow 313
Trefftz, E. 404
Tresca, H.É. 133, 144, 149, 200
trial functions (for finite elements) 395
Truesdell, C. 56, 57, 120, 123, 127, 212, 218–221, 253, 308, 391
Truesdell's metaphysical principles 220
Truesdell's mixture theory 220, 221
truncation error 335
Tübingen, University of 81–84, 86, 87
turbulent flow 317
Turner, M.J. 396

U

Ubaldi, G.   47
uncertainties of data   245
uniqueness of solution   343
upper error bound   174, 175, 178
upper and lower error bounds   452, 454, 463
Urbakh, M.   154

V

Valanis, K.C.   124
Vallejo, Galilei.   115
variational approach   77, 395
variational form   445, 449, 458
variational problem   62, 390
Varignon, P.   48
Verfürth, R.   444, 461
verification and validation   474
vertical tunnel   375
Vienna College of Technology   362, 363, 368
virtual work   56, 77
visco-plasticity   235, 238
viscous flow   320
Viviani, V.   251
Voith (company)   365, 368
Voith-Schneider propeller   355, 372
Voltaire (Arouet, F.-M.)   30, 46, 47
volume fractions   211, 213, 222

W

Wang, C.C.   123
warping torsion theory of Vlasov   284
Watt, I.   272
wave height   274
Weber, E.F.   104
Weber, M.   281
Weber, W.E.   104
Weibull, W.   203
Weichert, D.   169, 177, 179, 181
Weinblum, G.   276
Wieghardt, K.   201, 204–205
Wierzbicki, T.   290
Wiese, J.C.   221
Wilson, E. L.   399
wind tunnel experiments   379
Wolff, C.   53
Woltman, R.   213, 215

X

XFEM for fracture mechanics   435

Y

yield surface   139
Young's modulus   393

Z

Zach, von, F. X.   85
Zajac, F.E.   109
Zaremba, M.S.   147
Zinkiewicz, O.C.   399
Zinkiewicz, O.C., Zhu, J.Z.   463
Zuse, K.   219
Życzkowski, M.   178
π theorem   281